Achieving sustainable cultivation of mangoes

It is widely recognised that agriculture is a significant contributor to global warming and climate change. Agriculture needs to reduce its environmental impact and adapt to current climate change whilst still feeding a growing population, i.e. become more 'climate-smart'. Burleigh Dodds Science Publishing is playing its part in achieving this by bringing together key research on making the production of the world's most important crops and livestock products more sustainable.

Based on extensive research, our publications specifically target the challenge of climate-smart agriculture. In this way we are using 'smart publishing' to help achieve climate-smart agriculture.

Burleigh Dodds Science Publishing is an independent and innovative publisher delivering high quality customer-focused agricultural science content in both print and online formats for the academic and research communities. Our aim is to build a foundation of knowledge on which researchers can build to meet the challenge of climate-smart agriculture.

For more information about Burleigh Dodds Science Publishing simply call us on +44 (0) 1223 839365, email info@bdspublishing.com or alternatively please visit our website at www.bdspublishing.com.

Related titles:

Achieving sustainable cultivation of apples
Print (ISBN 978-1-78676-032-6); Online (ISBN 978-1-78676-035-7, 978-1-78676-034-0)

Achieving sustainable cultivation of bananas
Print (ISBN 978-1-78676-156-9); Online (ISBN 978-1-78676-159-0, 978-1-78676-158-3)

Chapters are available individually from our online bookshop: https://shop.bdspublishing.com

climate•SMART•publishing
IN AGRICULTURAL SCIENCE

BURLEIGH DODDS SERIES IN AGRICULTURAL SCIENCE

NUMBER 34

Achieving sustainable cultivation of mangoes

Edited by Dr Víctor Galán Saúco, Instituto Canario de Investigaciones Agrarias (ICIA), Spain and Dr Ping Lu, Charles Darwin University, Australia

burleigh dodds
SCIENCE PUBLISHING

Published by Burleigh Dodds Science Publishing Limited
82 High Street, Sawston, Cambridge CB22 3HJ, UK
www.bdspublishing.com

Burleigh Dodds Science Publishing, 1518 Walnut Street, Suite 900, Philadelphia, PA 19102-3406, USA

First published 2018 by Burleigh Dodds Science Publishing Limited

Library of Congress Control Number: 2017956488

British Library Cataloguing in Publication Data
A catalogue record for this book is available from the British Library

ISBN 978-1-78676-132-3 (print)
ISBN 978-1-78676-135-4 (online)
ISBN 978-1-78676-134-7 (online)
ISSN 2059-6936 (print)
ISSN 2059-6944 (online)

Typeset by Deanta Global Publishing Services, Chennai, India
Printed by Lightning Source

Contents

Part 3 Post-harvest management

Series list

Title	Series number
Achieving sustainable cultivation of maize - Vol 1 From improved varieties to local applications Edited by: Dr Dave Watson, CGIAR Maize Research Program Manager, CIMMYT, Mexico	001
Achieving sustainable cultivation of maize - Vol 2 Cultivation techniques, pest and disease control Edited by: Dr Dave Watson, CGIAR Maize Research Program Manager, CIMMYT, Mexico	002
Achieving sustainable cultivation of rice - Vol 1 Breeding for higher yield and quality Edited by: Prof. Takuji Sasaki, Tokyo University of Agriculture, Japan	003
Achieving sustainable cultivation of rice - Vol 2 Cultivation, pest and disease management Edited by: Prof. Takuji Sasaki, Tokyo University of Agriculture, Japan	004
Achieving sustainable cultivation of wheat - Vol 1 Breeding, quality traits, pests and diseases Edited by: Prof. Peter Langridge, The University of Adelaide, Australia	005
Achieving sustainable cultivation of wheat - Vol 2 Cultivation techniques Edited by: Prof. Peter Langridge, The University of Adelaide, Australia	006
Achieving sustainable cultivation of tomatoes Edited by: Dr Autar Mattoo, USDA-ARS, USA & Prof. Avtar Handa, Purdue University, USA	007
Achieving sustainable production of milk - Vol 1 Milk composition, genetics and breeding Edited by: Dr Nico van Belzen, International Dairy Federation (IDF), Belgium	008
Achieving sustainable production of milk - Vol 2 Safety, quality and sustainability Edited by: Dr Nico van Belzen, International Dairy Federation (IDF), Belgium	009
Achieving sustainable production of milk - Vol 3 Dairy herd management and welfare Edited by: Prof. John Webster, University of Bristol, UK	010
Ensuring safety and quality in the production of beef - Vol 1 Safety Edited by: Prof. Gary Acuff, Texas A&M University, USA & Prof.James Dickson, Iowa State University, USA	011
Ensuring safety and quality in the production of beef - Vol 2 Quality Edited by: Prof. Michael Dikeman, Kansas State University, USA	012
Achieving sustainable production of poultry meat - Vol 1 Safety, quality and sustainability Edited by: Prof. Steven C. Ricke, University of Arkansas, USA	013
Achieving sustainable production of poultry meat - Vol 2 Breeding and nutrition Edited by: Prof. Todd Applegate, University of Georgia, USA	014
Achieving sustainable production of poultry meat - Vol 3 Health and welfare Edited by: Prof. Todd Applegate, University of Georgia, USA	015

Acknowledgements

We wish to acknowledge the following for their help in reviewing particular chapters:

- Chapter 1: Dr David Kuhn, USDA-ARS, USA
- Chapter 12: Dr Reginaldo Baez, CIAD, Mexico
- Chapter 13: Professor Elhadi Yahia, Universidad Autónoma de Querétaro, Mexico; and Dr Robert Mangan, formerly USDA-ARS, USA

Introduction

Native to Southeast Asia, mangoes are now one of the most widely cultivated fruits, grown in over 100 countries across Asia, Africa and the Americas as well as Southern Europe. Mangoes are mainly consumed as fresh fruit, but also used widely in juices and in cooking, and are valued for their nutritional and nutraceutical properties. Drawing on an international range of expertise, this book focuses on ways of improving the cultivation of mango as a food crop at each step in the value chain, from breeding through to post-harvest storage. Part 1 discusses advances in understanding tree growth, flowering, pollination and fruit development, as well as developments in marker-assisted breeding. Part 2 reviews improvements in cultivation practices, including organic and greenhouse cultivation. Part 3 covers post-harvest management and quality, whilst Part 4 of the book assesses disease and pest management.

Part 1 Genetic improvement and plant physiology

Chapter 1 focuses on the development of molecular tools to improve understanding of the biology of mango and many other crops. This chapter reviews advances made in mango genetics using different molecular tools, including biochemical markers and DNA research employing restriction fragment length polymorphism (RFLP), randomly amplified polymorphic DNA (RAPD) and amplified fragment length polymorphism (AFLP). The chapter looks ahead to current and future developments in the field, including next-generation sequencing technologies and localization of genes of interest for breeding purposes. The chapter also offers suggestions for further reading on the subject.

Chapter 2, entitled 'The genetic diversity of mangoes', consists of a table listing over 100 mango cultivars selected on the basis of their local and global importance. The description of cultivars is based on over thirty years of research conducted by Fairchild Tropical Botanic Garden in Miami, Florida. The table lists cultivar name, tree size, fruit characteristics and fruiting season as well as additional information such as origin. The table is accompanied by colour photos of these *Mangifera indica* cultivars.

Chapter 3 makes clear the mango tree canopy is a carbohydrate factory, via photosynthesis, and a support for reproduction; it is also the place where vegetative growth occurs and the tree develops. The spatial and temporal proximity of vegetative and reproductive growth in the mango canopy leads to complex interactions. This chapter reviews the current state of knowledge of vegetative growth and deciphers these interactions, in order to inform more efficiently managed cultivation practices and future research. The chapter presents the architectural model of the mango tree. The basic structural entity, the growth unit, is defined and its morphology, growth and development are reviewed. The chapter then discusses the relationships between the growth unit and canopy development, along with the effects of environmental and endogenous factors on tree phenology. Finally, the interactions between vegetative growth and reproduction are described.

Chapter 4 addresses flowering, pollination and fruit development, beginning with vegetative shoot formation and then the plant's induction, initiation and floral differentiation. The number of fruits will depend upon the success of the pollination, fertilization and fruit set processes. The chapter also analyses the genes governing flower development and the processes of pollination and fertilization. It suggests future research trends in this area and recommends further reading on the subject.

Part 2 Cultivation techniques

Chapter 5 focuses on the fact that mango is grown in more than 100 countries, and is a commercially important fruit for many countries in the tropics. However, mango cultivation faces various problems which require urgent attention if sustainable production is to be achieved. This chapter reviews the factors and constraints that affect mango productivity in the tropics. The constraints related to soil, climate (including temperature, altitude and climate change), orchard management (from new mango orchards to high-density orchards), irrigation and nutrition, are discussed. In each case, strategies are presented to address these constraints and promote sustainable mango cultivation.

Chapter 6 builds on Chapter 5 by pointing out that although mango is well adapted to hot, tropical climates, it can also be grown in the subtropics with mild winters. Under such conditions, the cooler winter temperatures compared to tropical winter temperatures, improve flower induction and cause early bearing and lower annual growth rates, which help to control size and favour high-density plantings. There are also disadvantages to growing mango in the subtropics: cold spells and low temperatures can damage vulnerable young trees or areas of young growth, or can induce flowering of very young trees in the nursery, causing premature aging of inadequately managed plants. This chapter reviews the differences between mango cultivation in tropical and subtropical climates, including out-of-season production, and explores the factors necessary for successful subtropical cultivation.

As Chapter 7 points out, in countries such as Japan, Spain and Portugal, mangoes are cultivated in greenhouses. Specific techniques are used to maximize production and quality and to ensure efficient summer and winter harvesting. This chapter describes greenhouse practices of mango cultivation, including the control of flowering, care of fruit and pest control. Specific techniques include training and pruning, fertilization and irrigation, fruit thinning and bagging as well as harvesting and tree replacement.

Chapter 8 makes clear that there are many advantages to establishing high-density mango hedgerows using small trees. This method of cultivation means the terminal-shoots, inflorescences and fruits on such trees are within easy reach of farmworkers. The fruits, as well as branches and new shoots, can be specifically targeted for fungicide or pesticide application. This chapter describes the stages of a 3 m x 1 m Tommy Atkins mango orchard, from the time of its establishment to the time the trees fill their space in the orchard row and are fully bearing. It gives an account of management actions required for sustained maximal production, describing the benefits of the reduced time from planting until the trees attain optimal canopy cover and maximum fruit production. The chapter assesses the extent of small tree-growing systems to date.

Chapter 9 discusses the fact that in recent years the demand for organic mango has increased. Few technologies are currently available to support organic mango production systems in the main mango-growing regions. This chapter explains current technologies for sustainable organic mango production in the field and post-harvest processing. The chapter describes the importance of climate and soil selection, selection of cultivars and rootstocks, and soil preparation and planting. The chapter also addresses issues arising from the management of established mango orchards including weed control, irrigation, and pest and disease management. Finally, the chapter compares organic and conventional systems of mango production.

Chapter 10 highlights that the irrigation requirements of mango have not been adequately investigated, and very few studies have been conducted on regulated deficit irrigation (RDI) strategies at different phenological stages. The chapter suggests how research in the field of irrigation and fertilization can help solve the challenges faced by the mango industry and be translated into practical outcomes for farmers by making mango production more sustainable. In order to achieve this goal, based on an extensive and detailed review of the most relevant research on these topics, the chapter identifies potential areas for applied research that can significantly contribute to more sustainable mango agriculture in small, medium and large mango farms in developed and developing countries. The chapter also includes a detailed case study.

Part 3 Post-harvest management

Chapter 11 explores mango fruit quality from the perspectives of the grower, the packer, the retailer and the consumer. The chapter examines specifications for fruit at harvest maturity and at commercial maturity (eating stage), as well as technologies for monitoring relevant attributes, including machine vision estimation of canopy flowering, temperature logging for heat sum fruit maturation models, and tools for the estimation of fruit size, colour and dry matter content. The chapter discusses the use of dry matter content as an eating quality specification for guiding harvest decisions. The chapter also addresses the use of machine vision in the context of estimating fruit number and fruit size in the orchard and estimating fruit surface defects in the packhouse. Finally, the chapter discusses post-harvest tools to monitor fruit ripeness including the measurement of temperature, colour, firmness, ethylene and CO_2. The chapter includes an example decision support system that uses heat sums and dry matter levels to guide the decision to harvest.

Chapter 12 focuses on the fact that post-harvest deterioration in the quality of mangoes is largely determined by pre-harvest factors, ranging from the cultivar grown to orchard management and harvest practices. This chapter describes mango fruit anatomy and development, and the changes related to ripening in mangoes and the pre-harvest, in-harvest and post-harvest practices that can lead to deterioration or damage. The chapter also addresses measures that can be taken to reduce the risk of fruit deterioration and damage, including fruit thinning and individual fruit bagging or netting. The chapter discusses viable and cost-effective solutions to mango damage and deterioration and looks ahead to future trends in this area. Building on Chapter 12, Chapter 13 reviews current research on the preservation of fruit quality. It also looks at ways of reducing post-harvest damage and loss by employing suitable technologies and knowledge during post-harvest operations, storage management, transportation and marketing of mango fruit.

Chapter 14 highlights that mangoes can be considered a major source of bioactive compounds, notably vitamin C, phenolics (mainly gallic acid) and carotenoids. This chapter reviews the health benefits associated with the antioxidant properties of these compounds, which potentially offer protection against cardiovascular diseases, metabolic diseases and cancers. The chapter examines specific cell, animal and clinical studies that suggest mango pulp, juice and extract are effective against metabolic diseases and certain forms of cancer. The chapter considers approaches that can be used to increase bioactive compounds in mangoes either before or after harvest, and includes a case study on the use of pulsed light to increase concentrations of vitamin C, carotenoids and phenolics.

Chapter 15 focuses on the fact that mango production systems have seldom been studied using the technique of Life Cycle Assessment (LCA), which is an international standard for evaluating the environmental impacts of agri-food value chains. Important challenges are associated with the application of LCA to the environmental evaluation of fruit systems in general and mango in particular. This chapter describes the core principles of LCA methodology, the state of the art of LCA for fruits and associated key challenges. The chapter makes up-to-date recommendations for the use of LCA. The chapter then presents and discusses the first complete LCA case study for mango exported from Brazil. Finally, the chapter analyses the environmental challenges faced by mango systems across the world, highlighting the great potential of LCA to achieve more eco-friendly production and consumption of mango.

Part 4 Diseases and pests

Chapter 16 highlights that mango is affected by a great number of fruit, foliar, stem and root diseases. This chapter covers diseases that seriously impact the crop. Their significance, geographical distribution and history are outlined, and the symptoms, causal agent(s) and epidemiology of each are detailed with emphasis on their management.

Chapter 17 builds on Chapter 16 by providing an overview of Integrated Pest Management (IPM) in mango cultivation. IPM is the compatible use of various methods to control pests, which include biological, cultural and chemical control. Biological control is based on using predators, parasitoids and pathogens to reduce pest populations. Cultural control is based on management practices, for example, pruning to create an environment non-conducive to pests and to improve spray coverage. Chemical control should be used as a last resort and should be restricted to selective and less disruptive insecticides. Regular pest and beneficial insect monitoring is an integral component of IPM with interventions only applied when pest numbers reach a certain threshold. Four case studies are included to illustrate how IPM works in practice.

Part 1

Genetic improvement and plant physiology

Chapter 1

Exploiting the mango genome: molecular markers

V. Pérez and J. I. Hormaza, Instituto de Hortofruticultura Subtropical y Mediterránea La Mayora (IHSM La Mayora – CSIC – UMA), Spain

1 Introduction

Mango (*Mangifera indica* L., Anacardiaceae) is a woody perennial fruit crop with 40 chromosomes, and a total genome size of 439 Mb (Arumuganathan and Earle, 1991), about three times the size of the model plant *Arabidopsis thaliana* (L.) Heynh., and has been described as allotetraploid (Mukherjee, 1950). Mango belongs to the *Mangifera* genus that includes approximately 69 species, from tropical Asia, in two subgenuses, *Limus* and *Mangifera* (Kostermans and Bompard, 1993). Taxonomic and molecular evidence supports an origin of mango within a large geographical area that includes northwestern Myanmar, Bangladesh and Northeastern India (Bompard, 2009; Mukherjee, 1972; Mukherjee and Litz, 2009). Mango domestication of monoembryonic varieties probably originated in India where over 1,000 varieties are recognized, most of them selections from naturally occurring open-pollinated seedlings (Iyer and Degani, 1997). Mango cultivation spread outside its centre of origin and domestication throughout many tropical and subtropical regions of the world along trading routes, resulting in selections of genotypes adapted to particular edaphoclimatic conditions (Bompard, 2009; López-Valenzuela et al., 1997;

http://dx.doi.org/10.19103/AS.2017.0026.01

Mukherjee and Litz, 2009). Because of the crosses performed during the twentieth century in the frame of a breeding programme, Florida is often considered as a second centre of diversity of mango that has resulted in the development of important commercial mango varieties that are cultivated in new growing areas worldwide (Mukherjee, 1997). This long period of mango cultivation in different regions has resulted in a high number of varieties, traditionally identified with morphological markers. In order to standardize this phenotypic characterization, descriptors that include phenological and morphological traits of flowers, leaves, fruits and seeds have been developed for mango (IPGRI, 2006). Morphological characterization is necessary for adequate cultivar identification and, especially, good phenotyping (Rajwana et al., 2011), which is ultimately needed to efficiently link molecular markers with traits of interest and accelerate breeding programmes. However, it is also inaccurate due to the influence of the environment and the phenological status of the plants and the limiting number of discriminating traits (Khan et al., 2015). Thus, molecular markers provide a highly reliable complement to morphological markers, especially for cultivar fingerprinting and diversity studies.

During the last decades, fast and significant advances in the methods used to study nucleic acids in both animals and plants have taken place, resulting in the continuous development of different types of genetic markers. This information can be used to analyse the population structure of in situ and ex situ germplasm collections, and wild stands of cultivated species and crop wild relatives. The availability of reliable molecular tools can contribute to develop appropriate strategies to optimize the conservation of genetic diversity (Larranaga and Hormaza, 2016). As in most fruit tree crops, advances in developing molecular tools in mango have been slower than in annual crops although several works involving the development and application of molecular markers in mango have been performed. Most of those works have been focused on germplasm characterization in the different mango-growing regions. This is especially important in Eastern Asian countries where the highest mango diversity is found and must be preserved for future generations. Molecular markers have also been used to identify sexual/somatic origin of seedlings in polyembryonic mango cultivars and for paternity and pedigree analyses. As in other crops, the recent developments in the field of new sequencing and 'omics' (genomics, proteomics or metabolomics) technologies will make a qualitative change in mango breeding.

In this work, we provide an overview of the main advances in mango genomics. For this, molecular markers have been separated into biochemical, usually based in the separation of protein molecules, and DNA-based markers. In addition to the overview provided in this work, molecular tools similar to those described here are also increasingly being used in other fields of interest in mango cultivation and research, such as the study of the main pests and diseases that affect this crop.

2 Biochemical markers

Biochemical markers involve the separation of variants of proteins (allozymes and isozymes) into specific bands by electrophoresis (Scandalios, 1969). They are codominant and their main disadvantage is that the number of enzyme systems available is limited. Initial work with isozymes in mango showed differences between genotypes previously considered as clones (Gan et al., 1981). This approach was later used for the characterization of mango varieties (Degani et al., 1990, 1992) and to study the paternity of some commercial varieties (Degani et al., 1990). Results obtained in different works (Dag et al., 1997, 1998, 2001, 2009; Degani et al., 1997) suggest that, although outcrossing rates decrease with increasing distance from pollinating trees, no clear effects on yield are apparent. Isozyme systems were used to differentiate zygotic from somatic seedlings from different polyembryonic mango genotypes (Schnell and Knight, 1992; Degani et al., 1993) and to study loci linkage and segregation (Aron et al., 1997). In the last two decades isozymes have fallen into disuse due to the availability of an increasing number of DNA-based molecular markers, although some additional work has still been performed in recent years (Coroza-Almontero and Espino, 2010; Jintanawongse and Changtragoon, 2000).

3 DNA markers

Different types of DNA-based markers, commonly known as molecular markers, have been continuously developed since the 1980s. Depending on whether they are able to distinguish between heterozygous and homozygous genotypes they are classified into dominant (such as RAPDs and AFLPs) or codominant (such as RFLPs, microsatellites and SNPs) markers. Qualitative changes and milestones in developing new molecular markers in recent years include the discovery of PCR in the mid-1980s (Mullis et al., 1986) and the increasingly widespread use of next-generation sequencing technologies (Glenn, 2011).

3.1 Restriction fragment length polymorphism (RFLP)

Restriction fragment length polymorphism (RFLP) produces the polymorphism when a genome or part of the genome is digested with specific restriction endonucleases, resulting in a pattern of bands that can be detected after gel electrophoresis, blotting and hybridization with a labelled probe (Jeffreys, 1979). Although this approach was commonly used in different plant species during the 1980s and 1990s, its use for molecular analyses in plants was significantly reduced after the discovery of PCR.

In mango, a limited number of studies have used RFLP markers. For example, RFLPs were used in combination with AFLPs to construct a genetic linkage map from a cross between cultivars 'Alphonso' and 'Palmer' by Chunwongse et al. (2000). They have also been used to analyse interspecific relationships; thus, phylogenetic relationships in 13 *Mangifera* species were studied with RFLPs developed from cpDNA (chloroplast DNA) (Eiadthong et al., 1999a). These studies identified two main groups: a first group with *M. indica* and *M. sylvatica* Roxb. and a second group with *M. caloneura* Kurz., *M. cochinchinensis* Engl., *M. collina* Kosterm., *M. flava* Evrard., *M. foetida* Lour., *M. gedebe* Miq., *M. griffithii* Hook.f., *M. macrocarpa* Blume, *M. oblongifolia* Hook.f., *M. odorata* Griff. and *M. pentandra* Hook.f. Ravishankar et al. (2004) also used cpDNA RFLP analysis in combination with RAPDs to separate monoembryonic and polyembryonic Indian cultivars.

Minisatellites or variable number tandem repeat loci (VNTR) can be considered as a special class of RFLP loci; they are, highly polymorphic and composed of unit sequences that range from 10 to 40 bp and tandemly repeated from tens to thousands of times resulting in very large number of alleles. Minisatellites were used by Adato et al. (1995) to analyse 20 mango cultivars in Israel.

3.2 Randomly amplified polymorphic DNA (RAPD)

Randomly amplified polymorphic DNA (RAPD) uses primers of random sequence to amplify fragments of genomic DNA and polymorphisms depend upon the presence or absence of the amplification products (Williams et al., 1990). In mango, RAPD markers were first used to analyse phylogenetic relationships between the two subgenera (*Mangifera* and *Limus*) that are included in the genus *Mangifera* (Schnell and Knight, 1993). Schnell et al. (1995) used RAPD markers to fingerprint and estimate the genetic relationships among cultivars from the Florida breeding programme. In Australia, RAPDs were used to test differences between 'Kensington Pride' trees that showed distinct morphological characters (Bally et al., 1996); results showed identical or similar amplification patterns in the samples analysed, suggesting that the variations observed could reflect environmental rather than genetic differences. A work with 15 mango accessions in Mexico separated the genotypes studied into clusters according to their geographic origin, and a specific RAPD band that was associated with polyembryony was found (López-Valenzuela et al., 1997). Similar work was performed with Indian cultivars where local varieties could be grouped, according to their geographical origin, in two main groups that distinguished southern from northern, western and eastern cultivars (Ravishankar et al., 2000). Kumar et al. (2001) also analysed 50 Indian mango cultivars and found that the highest diversity in cultivars was from Southern India. Similar results were obtained with 29 Indian mango cultivars by Karihaloo et al. (2003). Ravishankar et al. (2004) also used RAPDs in combination with cpDNA RFLP analysis to separate monoembryonic and polyembryonic Indian cultivars. Analogous results based on RAPDs were also reported by Abirami et al. (2008).

Additional studies on mango diversity using RAPDs have been performed in different mango-growing areas, such as Mexico (De Souza et al., 2004), Pakistan (Ahmad et al., 2008) where mango genotypes are grouped with northern and northeastern Indian genotypes, Colombia (Díaz-Matallana et al., 2009), Indonesia (Fitmawati et al., 2010), India (Roy and Chattopadhyay, 2011), Brazil (Souza et al., 2011), Mauritius (Ramessur and Ranghoo-Sanmukhiya, 2011), or Egypt (Mansour et al., 2014). Moreover, RAPDs have also been used for paternity analyses, such as those performed in Brazil (Cordeiro et al., 2006a; Faleiro et al., 2009).

Several studies have included RAPDs in combination with other molecular marker systems for genetic diversity and paternity analyses. Thus, Srivastava et al. (2007; 2012) studied the genetic diversity and pedigree relationships of 20 mango cultivars in India using RAPD, ISSR and directed amplified minisatellite DNA (DMAD) markers, finding differences in the results depending on the marker system used. Samal et al. (2012) analysed 65 mango genotypes with morphological, RAPD and ISSR markers and found a high correlation between RAPD and ISSR markers. RAPDs have also been used to distinguish sexual from nucellar seedlings in polyembryonic cultivars (Cordeiro et al., 2006b; Martinez et al., 2012). However, the use of RAPDs for molecular analyses in plants, including mango, has decreased significantly in the last decade, due to irreproducibility and the emergence of more powerful molecular markers.

3.3 Amplified fragment length polymorphism (AFLP)

Amplified fragment length polymorphism (AFLP) combines RFLP and PCR techniques and is based on the selective amplification by PCR of DNA fragments previously digested with restriction enzymes (Vos et al., 1995). In mango, AFLPs were used to study the genetic relationship between 23 genotypes and the construction of a genetic linkage map based on the progeny of a cross between 'Keitt' and 'Tommy Atkins' (Kashkush et al., 2001). This linkage map consists of 13 linkage groups and covers 161.5 cM defined by 34 AFLP markers. Fang et al. (2003) also used 191 AFLPs to develop a genetic map of a cross between 'Keitt' and 'Tommy Atkins'. Additional genetic maps, in combination with RFLPs, have been developed from a cross between 'Alphonso' and 'Palmer' (Chunwongse et al., 2000). AFLPs have been used to study the genetic relationship of 105 mango accessions from local and foreign origin in Brazil (Santos et al., 2008), of 41 accessions in Mexico (Gálvez-López et al., 2010) or of 200 accessions in China (Gao et al., 2013). Additional studies performed with AFLPs in mango include testing of outcrossing rate in 'Haden' and 'Tommy Atkins' (Santos and Neto, 2011) and the analysis of phylogenetic relationships among *Mangifera* species (Eiadthong et al., 2000; Yamanaka et al., 2006).

3.4 Inter simple sequence repeats (ISSRs) or SSR anchored primers

ISSR markers consist in the amplification of an interrepeat region by a primer and electrophoretic separation of the amplicons (Zietkiewicz et al., 1994). ISSR markers have been used in mango to identify cultivars and establish geographical relationships. Eiadthong et al. (1999b) analysed 22 genotypes, 13 from Thailand and the rest from other countries, González et al. (2002) were able to detect differences between different 'Kensington Pride' trees grown in Australia, Xie et al. (2007) studied 32 accessions, mainly from China, Singh et al. (2007) analysed 12 samples from India, including some well-known cultivars such as 'Langra', 'Alphonso' or 'Mulgoa', and Tomar et al. (2011) analysed 20 Indian cultivars with 21 ISSR primers. Also, ISSRs were used to detect intracultivar variation in Indian commercial varieties (Singh et al., 2009). Damodaran et al. (2012) studied mango diversity of 29 genotypes and four wild *Mangifera* species in the Indian Andaman Islands with 27 ISSR markers and were able to distinguish monoembryonic and polyembryonic types. More recently, Rocha et al. (2014) used ISSR markers to identify zygotic and nucellar seedlings in the Brazilian 'Uba' cultivar showing that, at least in this genotype, the most vigorous seedling is not always of nucellar origin.

3.5 Microsatellites or simple sequence repeats (SSRs)

Microsatellites or simple sequence repeats (SSRs) (Tautz, 1989) are based on the presence of tandem repetitive DNA sequences flanked by specific conserved regions that allow their amplification by PCR. In recent years, SSRs have become the markers of choice for fingerprinting and diversity analyses in plants due to their codominant inheritance, high abundance, allelic diversity and easy application (Powell et al., 1996). About 200 SSR loci have been developed in mango (Chiang et al., 2012; Duval et al., 2005; Honsho et al., 2005; Ravishankar et al., 2011, 2015a; Schnell et al., 2005; Surapaneni et al., 2013; Tsai, 2014; Viruel et al., 2005), allowing significant advances on variety identification, diversity, paternity analysis, breeding programmes and germplasm conservation. Moreover, SSRs developed in mango can be successfully transferred to closely related species in the *Mangifera* genus, avoiding the need of developing SSR markers in each species of interest. Thus, SSRs developed in *M. indica* were used in *M. odorata*, *M. andamanica* King,

M. zeylanica Hook.f., *M. camptosperma* Pierre and *M. griffithii* (Ravishankar et al., 2011) and in *M. andamanica*, *M. camptosperma*, *M. odorata* and *M. griffithi* (Ravishankar et al., 2015a). SSR markers have also been very useful to perform paternity analyses such as those of the Florida breeding programme (Olano et al., 2005; Schnell et al., 2006; Viruel et al., 2005) and of others (Begun et al., 2013; Honsho et al., 2013). SSR markers have also been used to study outcrossing rate in monovarietal and multivarietal orchards (Perez et al., 2015, 2016) showing self-fertility in mango but a higher proportion of fruits resulting from outcrossing. These results indicate a preference for cross-fertilization, corroborating previous results obtained with isozyme markers (Dag et al., 1997, 1998, 2001, 2009; Degani et al., 1997) and suggesting that preferential cross-fertilization without excluding self-fertilization could be a bet-hedging strategy in mango for avoiding inbreeding depression or for ensuring reproduction.

Most of the published work with SSR markers in mango deals with molecular characterization and genetic diversity studies in different mango-growing areas such as the Caribbean (Duval et al., 2009), India (Bajpai et al., 2016; Begum et al., 2012; Ravishankar et al., 2015b; Singh and Bhat, 2009; Surapaneni et al., 2013; Vasugi et al., 2013), Myanmar (Hirano et al., 2010), Brazil (Dos Santos Ribeiro et al., 2012), Taiwan (Chiang et al., 2012; Tsai et al., 2013), Iran (Shamili et al., 2012), Australia (Dillon et al., 2013a,b), Pakistan (Azmat et al., 2016) and Kenya (Gitahi et al., 2016; Sennhenn et al., 2014). Similar to works with other molecular markers (Ravishankar et al., 2004; Abirami et al., 2008; Damodaran et al., 2012), studies with SSRs show clustering of mango varieties according to embryo type (Viruel et al., 2005). SSR markers have also produced information about the propagation methods as reported by Hirano et al. (2011) that analysed genetic variability in two traditional mango varieties in Myanmar, 'Sein Ta Lone' and 'Yin Kwe', both polyembryonic. 'Yin Kwe' mango, usually propagated by seed, showed higher genetic variability than 'Sein Ta Lone', that is clonally propagated through grafting.

EST-SSR markers are microsatellites developed from expressed sequence tags (EST), which development is increasingly easier due to the advances in next-generation RNA sequencing. EST-SSRs are more transferable between closely related species since they correspond to coding DNA and, consequently, they are located in highly conserved genomic regions (Varshney et al., 2005). Dillon et al. (2014) obtained 24,840 EST sequences from libraries prepared from different tissues of 'Kensington Pride' and 'Irwin' from which 25 EST-SSRs were extracted. The EST-SSRs obtained were transferable to other closely related species (*M. caesia* Jack, *M. foetida*, *M. laurina* Blume or *M. odorata*). Luo et al. (2015) also developed an additional set of 93 EST-SSR markers from seven mango cultivars from China.

4 Other molecular markers

In addition to the main molecular marker systems described above, additional studies have been performed in mango using other less common molecular markers. Diversity analyses have been performed with 14 sequence-tagged microsatellite sites (STMS) with 689 mango cultivars in India (Ravishankar et al., 2013). Start codon targeted (SCoT) polymorphism markers, described by Collard and Mackill (2009) and based on the short conserved regions flanking the ATG translation start codon, have been used by Luo et al. (2010, 2011, 2012) with foreign and local mango accessions from Guangxi (China). Gajera et al. (2014), also studying a collection of mango in China, found that clustering using SCoT markers correlated with physical and biochemical characteristics of fruits. Cleaved amplified polymorphic sequence (CAPS) markers developed from cloned sucrose synthase (SS) and sucrose phosphate synthase (SPS) genes have been used for the identification of true hybrids in F_1 mango progeny (Shudo et al., 2013).

Chloroplast DNA (cpDNA) markers can also be used to estimate intraspecific genetic diversity, although they are more useful at the interspecific level as intraspecific chloroplast DNA polymorphism is low. In mango, the intergenetic spacer *rpl20-rps12* was used to test genetic diversity in 19 cultivars from Pakistan (Khan and Azin, 2011).

Genomic in situ hybridization approaches (GISH) have been used by Nishiyama et al. (2006) to differentiate nine *Mangifera* species on the basis of signal intensity of hybridization probes to somatic metaphase chromosomes.

5 Next-generation sequencing technologies

Next-generation sequencing (NGS) is a high-speed, low-cost, multiply parallel sequencing technology (Schuster, 2007). NGS is replacing traditional Sanger (Sanger et al., 1977) sequencing and is revolutionizing many areas in biology due to continuously decreasing prices for whole genome and transcriptome sequencing. This technology allows genome-wide characterization and profiling of mRNAs, small RNAs, chromatin structure, DNA methylation patterns and metagenomics (Ansorge, 2009). Such information obtained has increased knowledge in evolution, ecology, domestication and breeding (Kilian and Graner, 2012). Furthermore, NGS will provide answers for the behaviour of crops under changing environmental conditions (Varshney et al., 2011). NGS can also provide such information for the main pests and diseases that affect mango. A recent example is the sequencing of the complete genome of the bacteria *Pseudomonas syringae* pv. *syringae* Van Hall, the causal agent of apical necrosis of mango trees, and the discovery of candidate genes for virulence (Martínez-García et al., 2015).

6 Genetic linkage maps

Genetic linkage maps are highly useful to detect quantitative trait loci (QTLs) for important horticultural traits and in marker assisted selection (MAS) programmes. So far, at least five genetic maps of mango are available using different molecular markers. Chunwongse et al. (2000) used 197 RFLPs and 650 AFLPs to construct a genetic map on 31 F_1 plants from a cross 'Alphonso' x 'Palmer', with 63 linkage groups for 'Alphonso', with an average of 3.3 markers per linkage group, and 59 linkage groups for 'Palmer', with an average of 4.2 markers per linkage group. Kashkush et al. (2001) used 34 AFLPs on 29 F_1 individuals from a cross 'Keitt' x 'Tommy Atkins' and obtained a linkage map of 13 groups, with between 2-5 markers. Fang et al. (2003) used 191 AFLPs on 60 F_1 individuals from a cross 'Keitt' x 'Tommy Atkins' reporting 15 linkage groups. Chunwongse et al. (2015) used 9 microsatellites and 67 RFLPs markers on 31 F_1 plants from a cross 'Alphonso' x 'Palmer', and obtained a linkage map with 29 groups. However, all those linkage maps are insufficient to develop a saturated map. More recently, Luo et al. (2016) used high-throughput sequencing to develop 318,414 specific-locus amplified fragments (SLAFs) of which 6,594 markers containing 13,844 SNPs (single-nucleotide polymorphisms) loci were used on 173 F_1 plants from a cross 'Jin-Hwang' x 'Irwin'; a linkage map consisting of 20 linkage groups with 330 SLAF markers per linkage group was constructed.

7 Other 'omics'

NGS combined with genomics, transcriptomics, proteomics or metabolomics may dramatically increase the current knowledge of mango gene function and regulation. Proteomic studies are starting to provide valuable information of the role of gene expression and translation in ripening (Andrade et al., 2012), leaf metabolism (Renuse et al., 2012) or fruit pulp and exocarp tissues during development (Fasoli and Rhigetti, 2013).

Genomic studies in mango include the characterization of the mango leaf transcriptome and chloroplast genome, from the cultivar Langra, identifying genes involved in biosynthetic pathways, plant hormone signal transduction, proteolytic enzymes and stress response genes (Azim et al., 2014). Other transcriptome studies are of fruit tissue from the cultivar Zill (Wu et al., 2014), fruit pericarp from the cultivar Shelly (Luria et al., 2014) or fruit mesocarp from the cultivar Kent (Dautt-Castro et al., 2015), expanding previous results of gene expression during fruit development and ripening of mango (Pandit et al., 2010). Recently (Kuhn et al., 2016), a thorough transcriptome analysis from six different tissues (leaves, flowers, exocarp, mesocarp, seed coat and seed) and from fruit tissues at four different developmental stages has been performed in 'Tommy Atkins' mango, representing more than 30,000 transcripts. Transcriptomic analysis has been used to identify genes involved in mango defence response to anthracnose caused by *Colletotrichum gloeosporioides* (Penz.) Penz. and Sacc. (Hong et al., 2016).

NGS can also be used to develop a large number of molecular markers such as SSRs and single-nucleotide polymorphisms (SNPs), which are variations of single nucleotides without change in the length of the DNA sequence. SNPs overcome some of the limitations of other molecular markers such as low marker abundance and the difficulty of sharing genotype data among different laboratories (Kuhn et al., 2016). Ravishankar et al. (2015a) sequenced genomic DNA from the cultivar Alphonso and identified 106,049 microsatellite repeats of which 90 were tested in 64 mango cultivars and four *Mangifera* species *(M. andamanica, M. camptosperma, M. odorata and M. griffithii*). Sherman et al. (2015) identified 332,016 SNPs and 1,903 SSRs, and used a subset of 293 SNPs to analyse genetic diversity in the Israel mango collection, with 74 mango accessions from different origins. Samples were separated into two main groups: a group including mostly Southeastern Asian accessions (Malaysia, Thailand and Indonesia) and India and a second group with Florida and Israeli mango cultivars. *M. laurina and M. odorata* grouped with the Southeastern Asian and Indian subgroup. Recently, Kuhn et al. (2016) reported the development of a high number of SNP markers after RNA analysis of 24 mango cultivars from diverse origins.

NGS has made entire genome sequencing possible for any plant species. Although the number of sequenced genomes is much higher in annual crops, the number of published whole-genome sequences in fruit tree crops is increasing rapidly (Larranaga and Hormaza, 2016). In the case of mango, although a fully assembled genome sequence has not yet been published, its relatively small genome should allow to sequence and assemble the genome in the near future, and different research groups are working in this direction. Examples include the sequencing of 'Kensington Pride' in Australia (Innes et al., 2015), 'Tommy Atkins' in the United States and Israel and a draft genome of 'Amrapali' in India (Singh et al., 2014).

8 Future trends and conclusion

NGS can generate hundreds to thousands of molecular markers, facilitating marker-assisted selection (MAS) (Ribaut and Hoisington, 1998). NGS is also revolutionizing genetic mapping and phylogenetics with the development of genotyping-by-sequencing (GBS), a technology for sequencing multiplexed samples, which combines molecular marker discovery and genotyping (He et al., 2014; Poland and Rife, 2012). GBS and similar technologies may soon become powerful approaches due to the increased availability of sequencing data, development of reference genomes and improved bioinformatic tools (Poland and Rife, 2012). All this should improve the identification of associations between genotype and phenotype for MAS, making breeding more efficient and improving parental selection for new crosses. Although currently genomic resources for breeding in mango are limited, soon, as in other crops, the main limitation will be the availability of thorough and reliable phenotypic information. This highlights the need for curation, analysis and exchange of mango germplasm among mango collections worldwide to ensure the preservation of genetically and phenotypically diverse material for breeding programmes.

9 Where to look for further information

Although no specific books have been published on mango genetics and genomics, several general books on mango can provide additional information. An example is the book edited by R.E. Litz *The Mango: Botany, Production and Uses* (CABI). Some other books on biotechnology also include chapters on mango. An example is *Biotechnology of Fruit and Nut Crops* (CABI), edited by R.E. Litz in 2005, of which a new edition is currently being edited by J.I. Hormaza, F. Pliego and R.E. Litz. Additional resources are included in the reference list of this chapter.

Several databases on genomic resources are available on the Internet:

- Plant Genome Database. Useful resources for plant comparative genomics: http://www.plantgdb.org/
- National Center for Biotechnology Information (NCBI). Access to GenBank, Blast, and additional resources: https://www.ncbi.nlm.nih.gov/
- A mango genomics workshop takes place every year in January at the International Plant and Animal Genome meeting in San Diego: http://www.intlpag.org/

Additionally, there is a mango group in the International Society for Horticultural Science (ISHS) that organizes an international congress every two years: http://www.ishs.org/mango

10 Acknowledgements

The research was supported by the Ministerio de Economía y Competitividad – European Regional Development Fund, European Union (AGL2013-43732-R, AGL2016-77267-R and Recupera2020). V. Pérez was supported by a Cabildo de La Palma- CSIC PhD scholarship.

11 References

Abirami, K., Singh, S. K., Singh, R., Mohapatra, T. and Kumar, A. R. (2008). Genetic diversity studies on polyembryonic and monoembryonic mango genotypes using molecular markers. *Indian Journal of Horticulture*, 65(3): 258–62.

Adato, A., Sharon, D., Lavi, U., Hillel, J. and Gazit, S. (1995). Application of DNA fingerprints for identification and genetic analyses of mango (*Mangifera indica*) genotypes. *Journal of the American Society for Horticultural Science*, 120(2): 259–64.

Ahmad, I., Malik, A. U., Malik, S. A., Tabassam, N., Rahman, M. and Zafar, Y. (2008). Application of DNA fingerprinting technology to estimate genetic divergence among mango cultivars-genotypes. *Acta Horticulturae*, 773: 127–32.

Andrade, J. M., Toledo T. T., Nogueira S. B., Cordenunsi B. R., Lajolo F. M. and Nascimento J. R. O. (2012). 2D-DIGE analysis of mango (*Mangifera indica* L.) fruit reveals major proteomic changes associated with ripening. *Journal of Proteomics*, 75: 3331–41.

Ansorge, W. J. (2009). Next-generation DNA sequencing techniques. *New Biotechnology*, 25(4): 195–203.

Aron, Y., Czosnek, H., Gazit, S. and Degani, C. (1997). Segregation distortion and linkage of mango isozyme loci. *HortScience*, 32(5): 918–20.

Arumuganathan, K. and Earle, E. D. (1991). Nuclear DNA content of some important plant species. *Plant Molecular Biology Reporter*, 9: 208–18.

Azim, M. K., Khan, I. A. and Zhang, Y. (2014). Characterization of mango (*Mangifera indica* L.) transcriptome and chloroplast genome. *Plant Molecular Biology*, 85(1–2): 193–208.

Azmat, M. A., Khan, A. A., Khan, I. A., Rajwana, I. A., Cheema, H. M. N. and Khan, A. S. (2016). Morphological characterization and SSR based DNA fingerprinting of elite commercial mango cultivars. *Pakistan Journal of Agricultural Sciences*, 53: 321–30.

Bajpai, A., Muthukumar, M., Ahmad, I., Ravishankar, K. V., Parthasarthy, V. A., Sthapit, B., Rao, R., Verma, J. P. and Rajan, S. (2016). Molecular and morphological diversity in locally grown non-commercial (heirloom) mango varieties of North India. *Journal of Environmental Biology*, 37(2): 221.

Bally, I. S. E., Graham, G. C. and Henry, R. J. (1996). Genetic diversity of Kensington mango in Australia. *Animal Production Science*, 36(2): 243–7.

Begum, H., Reddy, M. T., Malathi, S., Reddy, B. P., Arcahk, S., Nagaraju, J. and Siddiq, E. A. (2012). Molecular analysis for genetic distinctiveness and relationships of indigenous landraces with popular cultivars of mango (*Mangifera indica* L.) in Andhra Pradesh, India. *Asian and Australasian Journal of Plant Science and Biotechnology*, 6(1): 24–37.

Begum, H., Reddy, M. T., Malathi, S., Reddy, B. P., Narshimulu, G., Nagaraju, J. and Abubaker, E. (2013). Molecular analysis of intracultivar polymorphism of 'Panchadarakalasa' mango by microsatellite markers. *Jordan Journal of Biological Sciences*, 6(2): 127–36.

Bompard, J. M. (2009). Taxonomy and Systematics. In: Litz, R. E. ed. *The Mango: Botany, Production and Uses*, CAB International, Wallingford, Oxon.

Chiang, Y. C., Tsai, C. M., Chen, Y. K. H., Lee, S. R., Chen, C. H., Lin, Y. S. and Tsai, C. C. (2012). Development and characterization of 20 new polymorphic microsatellite markers from *Mangifera indica* (Anacardiaceae). *American Journal of Botany*, 99(3): e117–9.

Chunwongse, J., Phumichai, C., Barbrasert, C., Chunwongse, C., Sukonsawan, S. and Boonreungrawd, R. (2000). Molecular mapping of mango cultivars 'Alphonso' and 'Palmer'. *Acta Horticulturae*, 509: 193–206.

Chunwongse, C., Phumichai, C., Tongyoo, P., Juejun, N. and Chunwongse, J. (2015). Development of di-nucleotide microsatellite markers and construction of genetic linkage map in mango (*Mangifera indica* L.). *Songklanakarin Journal of Science and Technology*, 37: 119–27.

Collard, B. C. and Mackill, D. J. (2009). Start codon targeted (SCoT) polymorphism: a simple, novel DNA marker technique for generating gene-targeted markers in plants. *Plant Molecular Biology Reporter*, 27(1): 86–93.

Cordeiro, M. C. R., Pinto, A. C. D. Q., Ramos, V. H. V., Faleiro, F. G. and Fraga, L. M. S. (2006a). RAPD markers utilization and other parameters in the determination of mango hybrids genitors. *Revista Brasileira de Fruticultura*, 28(2): 164–7.

Cordeiro, M. C. R., Pinto, A. C. D. Q., Ramos, V. H. V., Faleiro, F. G. and Fraga, L. M. S. (2006b). Identification of plantlet genetic origin in polyembryonic mango (*Mangifera indica*, L.) cv. Rosinha seeds using RAPD markers. *Revista Brasileira de Fruticultura*, 28(3): 454–7.

Coroza-Almontero, C. and Espino, R. R. C. (2010). Genetic characterization of Mango (*Mangifera indica* L.) cultivars through isozyme analysis. *Philippine Agricultural Scientist*, 93: 238–44.

Dag, A., Eisenstein, D., Degani, C., El-Batsri, R., Zelig, M., Ziv, G. and Gazit, S (1997) 'Tommy Atkins' mango as pollinizer for 'Lily'. *Acta Horticulturae*, 455: 209–16.

Dag, A., Eisenstein, D., Gazit, S., El-Batsri, R. and Degani, C. (1998). Effect of pollenizer distance and selective fruitlet abscission on outcrossing rate and yield in 'Tommy Atkins' mango. *Journal of the American Society for Horticultural Science*, 123: 618–22.

Dag, A., Degani, C. and Gazit, S. (2001) In-hive pollen transfer in mango. *Acta Horticulturae*, 561: 61–5.

Dag, A., Degani, C. and Gazit, S. (2009) Gene flow in mango orchards and its impact on yield. *Acta Horticulturae*, 820: 347–50.

Damodaran, T., Kannan, R., Ahmed, I., Srivastava, R. C., Rai, R. B. and Umamaheshwari, S. (2012). Assessing genetic relationships among mango (*Mangifera indica* L.) accessions of Andaman Islands using inter simple sequence repeat markers. *New Zealand Journal of Crop and Horticultural Science*, 40(4): 229–40.

Dautt-Castro, M., Ochoa-Leyva, A., Contreras-Vergara, C. A., Pacheco-Sanchez, M. A., Casas-Flores, S., Sanchez-Flores, A., Kuhn, D. N. and Islas-Osuna, M. A. (2015). Mango (*Mangifera indica* L.) cv. Kent fruit mesocarp de novo transcriptome assembly identifies gene families important for ripening. *Frontiers in Plant Science*, 6: 62.

De Souza, V. A. B. and Lima, P. S. C. (2004). Genetic variability in mango genotypes detected by RAPD markers. *Acta Horticulturae*, 645: 303–10.

Degani, C., El-Batsri, R. and Gazit, S. (1990) Enzyme polymorphism in mango. *Journal of the American Society for Horticultural Science*, 115: 844–7.

Degani, C., Cohen, M., El-Batsri, R. and Gazit, S. (1992). PGI isozyme diversity and its genetic control in mango. *HortScience*, 27(3): 252–4.

Degani, C., Cohen, M., Reuveni, O., El-Batsri, R. and Gazit, S. (1993). Frequency and characteristics of zygotic seedlings from polyembryonic mango cultivars, determined using isozymes as genetic markers. *Acta Horticulturae*, 341: 78–85.

Degani, C., Yutko, O., El-Batsri, R. and Gazit, S. (1997). Outcrossing rate in adjacent 'Maya' and 'Tommy Atkins' blocks. *Scientia Horticulturae*, 70:25–30.

Díaz-Matallana, M., Schuler-García, I., Ruiz-García, M. and Hodson de Jaramillo, E. (2009). Analysis of diversity among six populations of Colombian mango (*Mangifera indica* L. cvar. Hilacha) using RAPDs markers. *Electronic Journal of Biotechnology*, 12(3): 1–2.

Dillon, N. L., Bally, I. S. E., Hucks, L. A., Wright, C. L., Innes, D. J. and Dietzgen, R. G. (2013a). Implementation of SSR markers in mango breeding in Australia. *Acta Horticulturae*, 992: 259–67.

Dillon, N. L., Bally, I. S. E., Wright, C. L., Hucks, L. A., Innes, D. J. and Dietzgen, R. G. (2013b). Genetic diversity of the Australian national mango genebank. *Scientia Horticulturae*, 150: 213–26.

Dillon, N. L., Innes, D. J., Bally, I. S. E., Wright, C. L., Devitt, L. C. and Dietzgen, R. G. (2014). Expressed sequence tag-simple sequence repeat (EST-SSR) marker resources for diversity analysis of mango (*Mangifera indica* L.). *Diversity*, 6(1): 72–87.

Dos Santos Ribeiro, I. C., Neto, F. P. L. and Santos, C. A. (2012). Allelic database and accession divergence of a Brazilian mango collection based on microsatellite markers. *Genetics and Molecular Research*, 11: 4564–74.

Duval, M. F., Bunel, J., Sitbon, C. and Risterucci, A. M. (2005). Development of microsatellite markers for mango (*Mangifera indica* L.). *Molecular Ecology Notes*, 5(4): 824–6.

Duval, M. F., Risterucci, A. M., Calabre, C., Le Bellec, F., Bunel, J. and Sitbon, C. (2009). Genetic diversity of Caribbean mangoes (*Mangifera indica* L.) using microsatellite markers. *Acta Horticulturae*, 820: 183–8.

Eiadthong, W., Yonemori, K., Sugiura, A., Utsunomiya, N. and Subhadrabandhu, S. (1999a). Analysis of phylogenetic relationships in *Mangifera* by restriction site analysis of an amplified region of cpDNA. *Scientia Horticulturae*, 80(3): 145–55.

Eiadthong, W., Yonemori, K., Sugiura, A., Utsunomiya, N. and Subhadrabandhu, S. (1999b). Identification of mango cultivars of Thailand and evaluation of their genetic variation using the amplified fragments by simple sequence repeat-(SSR-) anchored primers. *Scientia Horticulturae*, 82(1): 57–66.

Eiadthong, W., Yonemori, K., Kanzaki, S., Sugiura, A., Utsunomiya, N. and Subhadrabandhu, S. (2000). Amplified fragment length polymorphism analysis for studying genetic relationships among *Mangifera* species in Thailand. *Journal of the American Society for Horticultural Science*, 125: 160–4.

Faleiro, F. G., Pinto, A. C. Q., Cordeiro, M. C. R., Andrade, S. R. M., Ramos, V. H. V., Bellon, G. and Dias, J. N. (2009). Genetic variability of mango cultivars developed by Embrapa cerrados breeding program using rapd markers. *Acta Horticulturae*, 820: 177–82.

Fang, J. G., Liu, D. J. and Ma, Z. Q. (2003). Constructing mango (*Mangifera indica* L.) genetic map using markers for double heterozygous loci. *Molecular Plant Breeding*, 1: 313–19. (in Chinese)

Fasoli, E. and Righetti, P. G. (2013). The peel and pulp of mango fruit: a proteomic samba. *Biochimica et Biophysica Acta*, 1834: 2539–45.

Fitmawati, H. A. and Purwoko, B. S. (2010). Diversity of Indonesian mango (*Mangifera indica*) cultivars based on morphological and RAPD markers. *SABRAO Journal of Breeding and Genetics*, 42(2): 84–95.

Gajera, H. P., Bambharolia, R. P., Domadiya, R. K., Patel, S. V. and Golakiya, B. A. (2014). Molecular characterization and genetic variability studies associated with fruit quality of indigenous mango (*Mangifera indica* L.) cultivars. *Plant Systematics and Evolution*, 300(5): 1011–20.

Gálvez-López, D., Salvador-Figueroa, M., Becerra-Leor, E. N., González-Paz, M., Hernández-Delgado, S. and Mayek-Pérez, N. (2010). Molecular diversity and genetic relationships of mango germplasm from Chiapas, Mexico. *Agrociencia*, 44(8): 907–15.

Gan, Y. Y., Zaini, S. and Idris, A. (1981). Genetic variation in the grafted vegetatively propagated mango (*Mangifera indica*). *Petranika*, 4: 53–62.

Gao, A. P., Chen, Y. Y., Jonathan, H. C., Zhu, M., Huang, J. F., Luo, H. Y. and He, Y. H. (2013). AFLP analysis on the genetic diversity of two hundred accessions of mango in China. *Acta Horticulturae*, 992: 295–307.

Gitahi, R., Kasili, R., Kyallo, M. and Kehlenbeck, K. (2016). Diversity of threatened local mango landraces on smallholder farms in Eastern Kenya. *Forests, Trees and Livelihoods*: 1–16.

Glenn, T. C. (2011). Field guide to next-generation DNA sequencers. *Molecular Ecology Resources*, 11: 759–69.

González, A., Coulson, M. and Brettell, R. (2002). Development of DNA markers (ISSRs) in mango. *Acta Horticulturae*, 575: 139–43.

He, J., Zhao, X., Laroche, A., Lu, Z. X., Liu, H. and Li, Z. (2014). Genotyping-by-sequencing (GBS), an ultimate marker-assisted selection (MAS) tool to accelerate plant breeding. *Frontiers in Plant Science*, 5: 484.

Hirano, R., Htun Oo, T. and Watanabe, K. N. (2010). Myanmar mango landraces reveal genetic uniqueness over common cultivars from Florida, India, and Southeast Asia. *Genome*, 53(4): 321–30.

Hirano, R., Ishii, H., Oo, T. H., Gilani, S. A., Kikuchi, A. and Watanabe, K. N. (2011). Propagation management methods have altered the genetic variability of two traditional mango varieties in Myanmar, as revealed by SSR. *Plant Genetic Resources*, 9(03): 404–10.

Hong, K., Gong, D., Zhang, L., Hu, H., Jia, Z., Gu, H. and Song, K. (2016). Transcriptome characterization and expression profiles of the related defense genes in postharvest mango fruit against *Colletotrichum gloeosporioides*. *Gene*, 576(1): 275–83.

Honsho, C., Nishiyama, K., Eiadthong, W. and Yonemori, K. (2005). Isolation and characterization of new microsatellite markers in mango (*Mangifera indica*). *Molecular Ecology Notes*, 5(1): 152–4.

Honsho, C., Hazama, H., Yonemori, K., Yuji, K. and Bally, I. S. E. (2013). Genetic relationships of 'Beni-Keitt' to other mango cultivars using microsatellite markers. *Acta Horticulturae*, 992: 251–7.

Innes, D. J., Dillon, N. L., Smyth, H. E., Karan, M., Holton, T. A., Bally, I. S. E. and Dietzgen, R. G. (2015). Mangomics: information system supporting advanced mango breeding. In: Thangadurai, D. and Sangeetha, J. (eds) *Genomics and Proteomics: Principles, Technologies and Applications*. Apple Academic Press, Oakville, Canada, pp: 281–307.

International Plant Genetic Resources Institute (IPGRI). 2006. Descriptors of Mango (*Mangifera indica* L.). Rome, Italy. 44pp. *Subtropical Plant Science*, 62:18–26.

Iyer, C. P. A. and Degani, C. (1997). Classical breeding and genetics. In: Litz, R. E. (ed.) *The Mango: Botany, Production and Uses*. CAB International, Wallingford, UK.

Jeffreys, A. J. (1979). DNA sequence variants in the G gamma-, A gamma-, delta- and beta-globin genes of man. *Cell*, 18: 1–10.

Jintanawongse, S. and Changtragoon, S. (2000). Identification of cultivars and certification of hybrids in mango (*Mangifera indica* Linn.) by isoenzyme gene markers. *Acta Horticulturae*, 509: 177–84.

Karihaloo, J. L., Dwivedi, Y. K., Archak, S. and Gaikwad, A. B. (2003). Analysis of genetic diversity of Indian mango cultivars using RAPD markers. The *Journal of Horticultural Science and Biotechnology*, 78(3): 285–9.

Kashkush, K., Jinggui, F., Tomer, E., Hillel, J. and Lavi, U. (2001). Cultivar identification and genetic map of mango (*Mangifera indica*). *Euphytica*, 122(1): 129–36.

Khan, I. A. and Azim, M. K. (2011). Variations in intergenic spacer rpl20-rps12 of mango (*Mangifera indica*) chloroplast DNA: implications for cultivar identification and phylogenetic analysis. *Plant Systematics and Evolution*, 292(3–4): 249–55.

Khan, A. S., Ali, S. and Khan, I. A. (2015). Morphological and molecular characterization and evaluation of mango germplasm: An overview. *Scientia Horticulturae*, 194: 353–66.

Kilian, B. and Graner, A. (2012). NGS technologies for analyzing germplasm diversity in genebanks. Briefings in Functional Genomics, elr046.

Kostermans, A. J. G. H. and Bompard, J. M. (1993). *The Mangoes: Their Botany, Nomenclature, Horticulture and Utilization*. Academic Press, London.

Kuhn, D. N., Dillon, N. L., Innes, D. J., Wu, Le-Shin. and Mockaitis, K. (2016). Development of single nucleotide polymorphism (SNP) markers from the mango (*Mangifera indica*) transcriptome for mapping and estimation of genetic diversity. *Acta Horticulturae*, 1111: 315–22.

Kumar, H., Narayanaswamy, P., Prasad, T., Mukunda, G. K. and Sondur, S. (2001). Estimation of genetic diversity of commercial mango (*Mangifera indica* L.) cultivars using RAPD markers. *The Journal of Horticultural Science and Biotechnology*, 76(5): 529–33.

Larranaga, N. and Hormaza, J. I. (2016). Advances in genetic diversity analysis in fruit tree crops. In: Lüttge, U, Cánovas, F.M. and Matyssek R. (eds). *Progress in Botany*, Springer International Publishing, 77: 245–64.

López-Valenzuela, J. A., Martinez, O. and Paredes-Lopez, O. (1997). Geographic differentiation and embryo type identification in *Mangifera indica* L. cultivars using RAPD markers. *HortScience*, 32(6): 1105–8.

Luo, C., He, X. H., Chen, H., Ou, S. J. and Gao, M. P. (2010). Analysis of diversity and relationships among mango cultivars using Start Codon Targeted (SCoT) markers. *Biochemical Systematics and Ecology*, 38(6): 1176–84.

Luo, C., He, X. H., Chen, H., Ou, S. J., Gao, M. P., Brown, J. S., Tondo, C. T. and Schnell, R. J. (2011). Genetic diversity of mango cultivars estimated using SCoT and ISSR markers. *Biochemical Systematics and Ecology*, 39(4): 676–84.

Luo, C., He, X. H., Chen, H., Hu, Y. and Ou, S. J. (2012). Genetic relationship and diversity of *Mangifera indica* L.: revealed through SCoT analysis. *Genetic Resources and Crop Evolution*, 59(7): 1505–15.

Luo, C., Wu, H. X., Yao, Q. S., Wang, S. B. and Xu, W. T. (2015). Development of EST-SSR and TRAP markers from transcriptome sequencing data of the mango. *Genetics and Molecular Research*, 14(3): 7914–19.

Luo, C., Shu, B., Yao, Q., Wu, H., Xu, W. and Wang, S. (2016). Construction of a high-density genetic map based on large-scale marker development in mango using Specific-Locus Amplified Fragment Sequencing (SLAF-seq). *Frontiers in Plant Science*, 7: 1310.

Luria, N., Sela, N., Yaari, M., Feygenberg, O., Kobiler, I., Lers, A. and Prusky, D. (2014). De-novo assembly of mango fruit peel transcriptome reveals mechanisms of mango response to hot water treatment. *BMC Genomics*, 15(1): 1.

Mansour, H., Mekki, L. E. and Hussein, M. A. (2014). Assessment of genetic diversity and relationships among Egyptian mango (*Mangifera indica* L.) cultivares grown in Suez Canal and Sinai Region using RAPD markers. *Pakistan Journal of Biological Sciences*, 17(1): 56.

Martinez, E. D. C., Andrade-Rodríguez, M., Rodríguez, M. R. and Monter, A. V. (2012). Identification of zygotic and nucellar seedlings in polyembryonic mango cultivars. *Pesquisa Agropecuária Brasileira*, 47(11): 1629–36.

Martínez-García, P. M., Rodríguez-Palenzuela, P., Arrebola, E., Carrión, V. J., Gutiérrez-Barranquero, J. A., Pérez-García, A., Ramos, C., Cazorla, F. M. and de Vicente, A. (2015). Bioinformatics analysis of the complete genome sequence of the mango tree pathogen *Pseudomonas syringae* pv. *syringae* UMAF0158 reveals traits relevant to virulence and epiphytic lifestyle. *PLoS One*, 10(8): e0136101.

Mukherjee, S. K. (1950). Mango: its allopolyploid nature. *Nature*, 166: 196–7.

Mukherjee, S. K. (1972). Origin of mango (*Mangifera indica*). *Economic Botany*, 26(3): 260–4.

Mukherjee, S. K. (1977). Improvement of mango, grapes and guava. In: Nijjar, G. S. (ed.) *Fruit Breeding in India; Symposium on Fruit Crop Improvement*, pp: 15–20.

Mukherjee, S. K. and Litz, R. E. (2009). Introduction: botany and importance. In: Litz, R. E. (ed.) *The Mango: Botany, Production and Uses*. CAB International: 1–18, Wallingford, Oxon.

Mullis, K., Faloona, F., Scharf, S., Saiki, R., Horn, G. and Erlich, H. (1986). Specific enzymatic amplification of DNA *in vitro*: the polymerase chain reaction. *Cold Spring Harb Symp Quant Biol*, 51: 263–73.

Nishiyama, K., Choi, Y. A., Honsho, C., Eiadthong, W. and Yonemori, K. (2006). Application of genomic in situ hybridization for phylogenetic study between *Mangifera indica* L. and eight wild species of *Mangifera*. *Scientia Horticulturae*, 110(1): 114–17.

Olano, C. T., Schnell, R. J., Quintanilla, W. E. and Campbell, R. J. (2005). Pedigree analysis of Florida mango cultivars. *Proceedings of the Florida State Horticultural Society*, 118: 192–7).

Pandit, S. S., Kulkarni, R. S., Giri, A. P., Köllner, T. G., Degenhardt, J., Gershenzon, J. and Gupta, V. S. (2010). Expression profiling of various genes during the fruit development and ripening of mango. *Plant Physiology and Biochemistry*, 48(6): 426–33.

Perez, V., Hormaza, J. I. and Herrero, M. (2015). Reproductive biology of mango (*Mangifera indica*) in a Mediterranean climate. *Acta Horticulturae*, 1075: 143–7.

Perez, V., Herrero, M. and Hormaza, J. I. (2016). Self-fertility and preferential cross-fertilization in mango (*Mangifera indica*). *Scientia Horticulturae*, 213: 373–8.

Poland, J. A. and Rife, T. W. (2012). Genotyping-by-sequencing for plant breeding and genetics. *The Plant Genome*, 5(3): 92–102.

Powell, W., Machray, G. C. and Provan, J. (1996). Polymorphism revealed by simple sequence repeats. *Trends in Plant Science*, 1(7): 215–22.

Rajwana, I. A., Khan, I. A., Malik, A. U., Saleem, B. A., Khan, A. S., Ziaf, K., Anwar, R. and Amin, M. (2011). Morphological and biochemical markers for varietal characterization and quality assessment of potential indigenous mango (*Mangifera indica*) germplasm. *International Journal of Agriculture and Biology*, 13: 151–8.

Ramessur, A. D. and Ranghoo-Sanmukhiya, V. M. (2011). RAPD marker-assisted identification of genetic diversity among mango (*Mangifera indica*) varieties in Mauritius. *International Journal of Agriculture and Biology*, 13(2): 167–73.

Ravishankar, K. V., Anand, L. and Dinesh, M. R. (2000). Assessment of genetic relatedness among mango cultivars of India using RAPD markers. The *Journal of Horticultural Science and Biotechnology*, 75(2): 198–201.

Ravishankar, P., Chandrashekara, S. A., Sreedhara, M. R., Dinesh, L. and Saiprasad G. V. S. (2004). Diverse genetic bases of Indian polyembryonic and monoembryonic mango (*Mangifera indica* L) cultivars. *Current Science*, 85: 870–1.

Ravishankar, K. V., Mani, B. H. R., Anand, L. and Dinesh, M. R. (2011). Development of new microsatellite markers from Mango (*Mangifera indica*) and cross-species amplification. *American Journal of Botany*, e96–9.

Ravishankar, K. V., Dinesh, M. R., Mani, B. H., Padmakar, B. and Vasugi, C. (2013). Assessment of genetic diversity of mango (*Mangifera indica* L.) cultivars from Indian peninsula using sequence tagged microsatellite site (STMS) markers. *Acta Horticulturae*, 992: 269–76.

Ravishankar, K. V., Dinesh, M. R., Nischita, P. and Sandya, B. S. (2015a). Development and characterization of microsatellite markers in mango (*Mangifera indica*) using next-generation sequencing technology and their transferability across species. *Molecular Breeding*, 35(3): 1–13.

Ravishankar, K. V., Bommisetty, P., Bajpai, A., Srivastava, N., Mani, B. H., Vasugi, C., Rajan, S. and Dinesh, M. R. (2015b). Genetic diversity and population structure analysis of mango (*Mangifera indica*) cultivars assessed by microsatellite markers. *Trees*, 29(3): 775–83.

Renuse, S., Harshaa, H. C., Kumar, P., Acharya, P. K., Sharma, J., Goel, R., Kumar, G. S. S., Raju, R Prasad, T. S. K., Slotta, T. and Pandey, A. (2012). Proteomic analysis of an unsequenced plant – *Mangifera indica*. *Journal of Proteomics*, 75: 5793–6.

Ribaut, J. M. and Hoisington, D. (1998). Marker-assisted selection: new tools and strategies. *Trends in Plant Science*, 3(6): 236–9.

Rocha, A., Salomão, T. M. F., Siqueira, D. L. D., Cruz, C. D. and Salomão, L. C. C. (2014). Identification of 'Ubá'mango tree zygotic and nucellar seedlings using ISSR markers. *Revista Ceres*, 61(5): 597–604.

Roy, S. C. and Chattopadhyay, A. (2011). Evaluation of genetic diversity in mango germplasm resources using RAPD markers and characterization of cultivar Guti based on 18S rRNA gene sequence. *The Indian Journal of Genetics and Plant Breeding*, 71(3): 254–61.

Samal, K. C., Jena, R. C., Swain, S. S., Das, B. K. and Chand, P. K. (2012). Evaluation of genetic diversity among commercial cultivars, hybrids and local mango (*Mangifera indica* L.) genotypes of India using cumulative RAPD and ISSR markers. *Euphytica*, 185(2): 195–213.

Sanger, F., Nicklen, S. and Coulson, A. R. (1977). DNA sequencing with chain-terminating inhibitors. *Proceedings of the National Academy of Sciences*, 74(12): 5463–7.

Santos, C. A. F. and Neto, F. P. L. (2011). Outcrossing rate between 'Haden' and 'Tommy Atkins' mangoes estimated using microsatellite and AFLP markers. *Pesquisa Agropecuária Brasileira*, 46(8): 899–904.

Santos, C. A. F., Neto, F. P. L., Rodrigues, M. A. and Costa, J. G. D. (2008). Similaridade genética de acessos de mangueira de diferentes origens geográficas avaliadas por marcadores AFLP. *Revista Brasileira de Fruticultura*, 30(3): 736–40.

Scandalios, J. G. (1969). Genetic control of multiple molecular forms of enzymes in plants: a review. *Biochemical Genetics*, 3(1): 37–79.

Schnell, R. J. and Knight Jr, R. J. (1992). Frequency of zygotic seedlings from 5 polyembryonic mango rootstocks. *HortScience*, 27: 174–6.

Schnell, R. J. and Knight Jr., R. J. (1993). Genetic relationships among *Mangifera* spp. based on rapd markers. *Acta Horticulturae*, 341: 86–92.

Schnell, R. J., Ronning, C. M. and Knight Jr, R. J. (1995). Identification of cultivars and validation of genetic relationships in *Mangifera indica* L. using RAPD markers. *Theoretical and Applied Genetics*, 90(2): 269–74.

Schnell, R. J., Olano, C. T., Quintanilla, W. E. and Meerow, A. W. (2005). Isolation and characterization of 15 microsatellite loci from mango (*Mangifera indica* L.) and cross – species amplification in closely related taxa. *Molecular Ecology Notes*, 5(3): 625–7.

Schnell, R., Brown, J. S., Olano, C., Meerow, A., Campbell, R. and Kuhn, D. (2006). Mango genetic diversity analysis and pedigree inferences for Florida cultivars using microsatellite markers. *HortScience*, 41(4): 993.

Schuster, S. C. (2007). Next-generation sequencing transforms today's biology. *Nature*, 200(8): 16–18.

Sennhenn, A., Prinz, K., Gebauer, J., Whitbread, A., Jamnadass, R. and Kehlenbeck, K. (2014). Identification of mango (*Mangifera indica* L.) landraces from Eastern and Central Kenya using a morphological and molecular approach. *Genetic Resources and Crop Evolution*, 61(1): 7–22.

Shamili, M., Fatahi, R. and Hormaza, J. I. (2012). Characterization and evaluation of genetic diversity of Iranian mango (*Mangifera indica* L., Anacardiaceae) genotypes using microsatellites. *Scientia Horticulturae*, 148: 230–4.

Sherman, A., Rubinstein, M., Eshed, R., Benita, M., Ish-Shalom, M., Sharabi-Schwager, M., Rozen, A., Saada, D., Cohen, Y. and Ophir, R. (2015). Mango (*Mangifera indica* L.) germplasm diversity based on single nucleotide polymorphisms derived from the transcriptome. *BMC Plant Biology*, 15(1): 277.

Shudo, A., Tarora, K., Makishi, Y., Ichi, R., Takahashi, K., Matsumura, M., Shimabuku, S., Matsuda, N., Nakasone, S. and Urasaki, N. (2013). Development of CAPS markers and their application in breeding for mango, *Mangifera indica* L. *Euphytica*, 190(3): 345–55.

Singh, S. and Bhat, K. V. (2009). Molecular characterization and analysis of geographical differentiation of Indian mango (*Mangifera indica* L.) germplasm. *Acta Horticulturae*, 839: 599–606.

Singh, S., Karihaloo, J. L. and Gaikwad, A. B. (2007). DNA Fingerprinting of some Mango (*Mangifera indica* L) cultivars using anchored-ISSR markers. *Journal of Plant Biochemistry and Biotechnology*, 16(2): 113–17.

Singh, S., Gaikwad, A. B. and Karihaloo, J. L. (2009). Morphological and molecular analysis of intracultivar variation in Indian mango (*Mangifera indica* L.) cultivars. *Acta Horticulturae*, 829: 205–12.

Singh, N. K., Mahato, A. K., Sharma, N., Gaikwad, K., Srivastava, M., Tiwari, K., Dogra, V., Rawal, H. C., Jayaswal, P., Singh, A., Rai, V., Mithra, A., Bajpai, A., Dinesh, M. R., Ravishankar, K. V., Rajan, S., Rai, A., Singh, A. K. and Sharma T. R (2014). A draft genome of the king of fruit, mango (*Mangifera indica L.*). Plant and Animal Genome Conference XXII, P045.

Souza, I. G. B., Valente, S. E. S., Britto, F. B., de Souza, V. A. B. and Lima, P. S. C. (2011). RAPD analysis of the genetic diversity of mango (*Mangifera indica*) germplasm in Brazil. *Genetics and Molecular Research*, 10(4): 3080–9.

Srivastava, A. P., Chandra, R., Saxena, S., Rajan, S., Ranade, S. A. and Prasad, V. (2007). A PCR-based assessment of genetic diversity, and parentage analysis among commercial mango cultivars and hybrids. *The Journal of Horticultural Science and Biotechnology*, 82(6): 951–9.

Srivastava, N., Bajpai, A., Chandra, R., Rajan, S., Muthukumar, M. and Srivastava, M. K. (2012). Comparison of PCR based marker systems for genetic analysis in different cultivars of mango. *Journal of Environmental Biology*, 33: 159–66.

Surapaneni, M., Vemireddy, L. R., Begum, H., Reddy, B. P., Neetasri, C., Nagaraju, J., Anwar, S. Y and Siddiq, E. A. (2013). Population structure and genetic analysis of different utility types of mango (*Mangifera indica* L.) germplasm of Andhra Pradesh state of India using microsatellite markers. *Plant Systematics and Evolution*, 299(7): 1215–29.

Tautz, D. (1989). Hypervariability of simple sequences as a general source for polymorphic DNA markers. *Nucleic Acids Research*, 17(16): 6463–71.

Tomar, R. S., Gajera, H. P., Viradiya, R. R., Patel, S. V. and Golakiya, B. A. (2011). Phylogenetic relationship among mango (*Mangifera indica* L.) landraces of Saurashtra based on DNA fingerprinting. *Journal of Horticulture and Forestry*, 3: 379–85.

Tsai, C. C. (2014). Identification and characterization of 16 polymorphic microsatellite markers from *Mangifera indica* l. (Anacardiaceae). *Acta Horticulturae*, 1048: 187–92.

Tsai, C. C., Chen, Y. K. H., Chen, C. H., Weng, I. S., Tsai, C. M., Lee, S. R., Lin, Y. S. and Chiang, Y. C. (2013). Cultivar identification and genetic relationship of mango (*Mangifera indica*) in Taiwan using 37 SSR markers. *Scientia Horticulturae*, 164: 196–201.

Varshney, R. K., Graner, A. and Sorrells, M. E. (2005). Genic microsatellite markers in plants: features and applications. *Trends in Biotechnology*, 23(1): 48–55.

Varshney, R. K., Bansal, K. C., Aggarwal, P. K., Datta, S. K. and Craufurd, P. Q. (2011). Agricultural biotechnology for crop improvement in a variable climate: hope or hype?. *Trends in Plant Science*, 16(7): 363–71.

Vasugi, C., Dinesh, M. R., Ravishankar, K. V. and Padmakar, B. (2013). Morphological and molecular characterization – nomenclature ambiguity in Indian mangoes. *Acta Horticulturae*, 992: 331–9.

Viruel, M. A., Escribano, P., Barbieri, M., Ferri, M. and Hormaza, J. I. (2005). Fingerprinting, embryo type and geographic differentiation in mango (*Mangifera indic*a L., Anacardiaceae) with microsatellites. *Molecular Breeding*, 15(4): 383–93.

Vos, P., Hogers, R., Bleeker, M., Reijans, M., Van de Lee, T., Hornes, M., Frijters, A., Pot, J., Peleman, J., Kuiper, M and Zabeau, M. (1995). AFLP: a new technique for DNA fingerprinting. *Nucleic Acids Research*, 23(21): 4407–14.

Williams, J. G., Kubelik, A. R., Livak, K. J., Rafalski, J. A. and Tingey, S. V. (1990). DNA polymorphisms amplified by arbitrary primers are useful as genetic markers. *Nucleic Acids Research*, 18(22): 6531–5.

Wu, H. X., Jia, H. M., Ma, X. W., Wang, S. B., Yao, Q. S., Xu, W. T., Zhou, Y. G, Gao, Z. S. and Zhan, R. L. (2014). Transcriptome and proteomic analysis of mango (*Mangifera indica* Linn) fruits. *Journal of Proteomics*, 105: 19–30.

Xie, J. H., Ma, W. H. and Liu, C. M. (2007). Analysis of genetic relationships among mango germplasm using ISSR marker. *Acta Horticulturae*, 763: 185–90.

Yamanaka, N., Hasran, M., Xu, D. H., Tsunematsu, H., Idris, S. and Ban, T. (2006). Genetic relationship and diversity of four *Mangifera* species revealed through AFLP analysis. *Genetic Resources and Crop Evolution*, 53: 949–54.

Zietkiewicz, E., Rafalski, A. and Labuda, D. (1994). Genome fingerprinting by simple sequence repeat (SSR)-anchored polymerase chain reaction amplification. *Genomics,* 20(2): 176–83.

Chapter 2

The genetic diversity of mangoes

Noris Ledesma, Fairchild Tropical Botanic Garden, USA

1 Introduction

The genus *Mangifera* consists of 69 species restricted in their native range to tropical Asia, Indonesia and Malaysia. Wild members of the *Mangifera* genus occur in Borneo, Malaysia, Singapore, Indonesia, Brunei, the Philippines, Papua New Guinea, Micronesia, Sri Lanka, Bangladesh, Myanmar, Thailand, Cambodia, Vietnam, Laos, southern China and India. The greatest diversity, with approximately 28 species, occurs in western Malaysia, especially in peninsular Malaysia, Borneo and Sumatra, a region considered to be the centre of diversity of the genus. It is thought that the mango started its spread from its home in the Far East to Tropical Africa, the islands of the Indian Ocean and the West Indies and Tropical America (Popenoe, 1924; Fairchild, 1951).

Mangifera indica L. and the commercial cultivars of the mango originated predominantly in India through seedling selections of the variants due to recombination and segregation of characters in the progenies. Early records show that India has been cultivating mangoes since 4000 BC. Mangoes were domesticated in India, and spread to East Asia around 400 BC. By the tenth century AD, mangoes were cultivated in East Africa, followed by the Philippines in the fifteenth century. Portuguese traders brought mango seeds along their route, spreading the trees to their African and Brazilian colonies in the sixteenth century. The Portuguese are given the credit for bringing the mango to America. It is believed to have been first planted at Bahia, Brazil, in the early 1700s. We can assume then that the American colonies were introduced to mangoes in the seventeenth century (Popenoe, 1924). Mangoes were first introduced into Florida in 1833 in Cape Sable; however, mango seedlings had already been in existence in South Florida. In the same year, Henry Perrine introduced mango plants from Mexico to Miami, but the trees survived only until his death. In 1888, grafted varieties were introduced from India. Of the thirty-five mango

http://dx.doi.org/10.19103/AS.2017.0026.03

trees sent to South Florida, just one variety survived, the 'Mulgoba'. A new generation of mangoes was developed around 1910. Since then suitable seedling progenies with much improved traits have been developed in Florida, with the Haden and other cultivars exhibiting bigger-sized fruits and more attractive colours than the parent Mulgoba with dull colour (Popenoe, 1924). The majority of mango production continues using ancient cultivars, except in recent years with expansion into new growing regions of the New World, China and parts of Africa with new cultivars (Mukherjee and Litz, 2009).

Agricultural practices are currently undergoing another revolution, as integrated pest and disease management replaces the earlier reliance on agrichemicals, and emerging fields within biotechnology begin to impact cultivar development and the increased high-density planting. The search continues for additional cultivars offering greater manageability, smaller tree size, and superior quality and horticultural traits.

A mango cultivar collection was established by Fairchild Tropical Botanic Garden in Miami, Florida, in the early 1980s. At present, the programme consists of well over 600 mango cultivars from various mango-growing regions of the world and 25 *Mangifera* species. The cultivars are evaluated throughout the mango-growing season for their performance, that is, vegetative vigour, susceptibility to pests and diseases, yield, and fruit quality.

2 Description of the principal mango cultivars

Appendix 1 provides a list of the principal mango cultivars, which are selected on the basis of their local and global importance. The description of cultivars grown under South Florida conditions is based on evaluations conducted by Fairchild Tropical Botanic Garden in Miami. The data reflect some 30 years of research. Description information includes origin, information on tree form and vigour (low, moderate or high).

The cultivars are in alphabetic order with the description of the fruit, including colour, shape, flesh characteristics; tree size; origin; and other relevant characteristics of the cultivar. Mango fruit are green, yellow or red, with or without colour blushes and blankets of overlaid white or yellow dots. Aromas are subtle to intense, and range from fruity to resinous. Flesh texture and fibre content are highly variable, and the range of flavour among cultivars defies description.

The mango reaches the size and bright colour characteristic when left to mature on the tree itself with adequate lighting conditions. Mango flavours can be powerful and balanced with the perfect sweetness. Aromas can differ among cultivars. Mango colour, shape and firmness can vary depending on environmental conditions. Fruit can change from orange to full red. It can be equally sophisticated, wrapped in soft pastels, earthy greens and saffron yellows.

There is a considerable variation in tree size among mango cultivars. Rootstocks that restrict tree size are not available. The growth potential of each cultivar – along with pruning, watering and fertilizing techniques – can restrict growth and maintain proper fruiting. Different cultivars have shown some natural disease tolerance. Cultivar information is focused on the most important mango diseases: powdery mildew and anthracnose.

Mangoes generally ripen during the warm season. The following descriptions are grouped as early, middle and late season. The calendar date of ripening for each cultivar can vary up to a month or more in different years depending on the bloom date and weather conditions during fruit development. Some cultivars can have multiple blooms under tropical conditions, resulting in multiple crops. The season descriptions of early, middle and late harvest refer to the beginning of physiological maturity of the varieties in the Northern Hemisphere.

Yield in predominant cultivars including 'Tommy Atkins' and 'Keitt' is significantly high in average of 90–136 kg/tree (200 to 300 lb/tree) in a wide variety of locations and climates for large-scale commercial production. Compared with lower-yielding cultivars such as 'Edward' and 'Nam Doc Mai' may be grown for their superior flavour, but a realistic yield would be 23–45 kg/tree 50–100 lb/tree (Campbell et al., 2002).

The mango is highly versatile, and this versatility is accentuated by cultivar selection. Some are used in immature stage for cooking. Other cultivar sections provide flavour for preserves, juices or flesh product. While far from complete, they demonstrate the diversity and considerable potential of the world's genetic resources.

3 Photographs of the principal mango cultivars

Appendix 1 is accompanied by Appendix 2, which includes photographs of the principal mango cultivars. Since the photographs were taken at different maturation stages, the colourations and shades may not always correspond with what is typical of the variety.

4 Acknowledgements

For their help in reviewing portions of this chapter and contributing vast qualities of information on mango cultivar attributes: the author is profoundly grateful to Dr. Richard J. Campbell, Director of Horticulture of Fairchild Tropical Botanic Garden, Coral Gables, FL, USA. Thank you to Dr. Lue Ping for his contribution with some of the Chinese varieties including in this chapter. Thank you to Westfalia Fruit for some of the pictures.

5 References

Campbell, R. J. (1992). *A Guide to Mangos in Florida*. Fairchild Tropical Botanic Garden, Miami, Florida, USA.

Campbell, R. J., Ledesma, N. and Campbell, C.W. (2002). *Tropical Mangos: How to Grow the World's Most Delicious Fruit*. Fairchild Tropical Botanic Garden, Miami, Florida, USA.

Campbell, R. J. and Ledesma, N. (2003). *Fairchild's Quick Guide to Mangos of the World*. Fairchild Tropical Botanic Garden, Miami, Florida, USA.

Campbell, R. J. and Ledesma, N. (2009). *Trends in Mango Production and Cultivars Wordwide*. Fairchild Tropical Botanic Garden, Coral Gables. American Society for Tropical Horticulture.

Campbell, R. J. and Ledesma, N. (2010). *Locations and Cultivars for Indian Mango Production in the Americas*. International Society for Horticultural Science, Proceedings of the Ninth International Mango Symposium, Number 992, China, pp. 59–62.

Fairchild, D. (1951). *Mango studies*. Florida Mango Forum.

Knight, R. J., Campbell, R. J. and Maguire, I. (2009). Important mango cultivars and their descriptions. *The Mango: Botany, Production and Uses*, 2nd Edition. CABI.

Lavi, U., Kaufman, D. and Sharon, D. (1997). 'Shelly': A new mango cultivar. *Hort Science* 32(1):138.

Mukherjee, S. K. and Litz, R. E. (2009). Introduction, botany and importance. *The Mango Botany, Production and Uses*, 2nd Edition. CABI.

Popenoe, W. (1924). *Manual of Tropical and Subtropical Fruits*. The Macmillan Company.

6 Appendix 1: list of the principal mango cultivars

		Characterization of the ripe fruit				
Cultivar	Tree size	Size	Weight gr.	Shape	Skin colour	Flesh colour
Alphonso	Large	Medium	283	Oblong	Yellow	Yellow
Ah Ping	Large	Large	348	Oblong	Red	Orange
Alampur Baneshan	Small	Large	453	Oblong oblique	Yellow	Yellow
Anderson	Large	Large	1500	Long slender	Yellow with dark red blush	Yellow
Angie	Small	Large	500	Oblong oblique	Yellow and pink blush	Yellow
Aroemanis	Medium	Small	250	Oblong	Yellow	Yellow
Ataulfo	Large	Small	250	Sigmoid	Yellow	Yellow
Azúcar	Medium	Small	200	Oval	Yellow and red	Yellow
Bangalora	Large	Large	792	Sigmoid-long	Yellow	Yellow
Banilejo	Small	Small	230	Oblong	Yellow with pink blush	Yellow
Baptiste	Small	Medium	320	Oval	Yellow	Orange
Beverly	Medium	Large	520	Oval	Yellow	Yellow
Bombay	Large	Small	311	Rounded	Green-yellow	Orange
Borsha	Medium	Medium	368	Oblong	Green-yellow	Yellow
Brahm Kai Meu	Medium	Large	360	Sigmoid-long	Green-yellow	Orange
Brander	Small	Large	680	Oblique	Orange	Orange
Brooks	Medium	Large	550	Oblong	Green-yellow	Yellow
Cairo	Large	Medium	368	Oval	Yellow and pink blush	Yellow
Cambodiana	Large	Medium	280	Oblong	Green-yellow	Yellow
Carabao	Large	Medium	350	Long and slender	Yellow	Yellow

Fibre	Fruiting season	Origin	Additional information
Fibreless	Middle	India	Named after the General Alfonso De Albuquerque; considered one of the best mangoes in the world
Fibreless	Early	Hawaii, USA	Mapulehu, Molokai, Hawaii; eye-catching colour
Fibreless	Middle	India	Central India state of Andhra Pradesh; The bloom and fruit are highly susceptible to disease
Fibreless	Middle	Florida, USA	Significant fruit splitting occurs in years with erratic rainfall; perfect for pickling and chutneys
Fibreless	Early	Florida, USA	Selected by Fairchild Tropical Botanic Garden; Fruit with good potential. Grown commercially in South Florida
Fibreless	Early	Indonesia	Selected in Jakarta; bloom and fruit are tolerant of fungal disease
Fibreless	Middle	Mexico	Disease susceptibility and modest yield; dominate the Mexican mango market
Fibre	Middle	Colombia	Commercially grown in Colombia; Used fresh fruit and juice processing
Fibreless	Middle	India	It also call as 'Totapuri', 'Sandersha', or 'Kallamai; Used fresh fruit and juice processing
Some fibre	Middle	Dominican Republic	The tree and fruit have some disease tolerance; Used fresh fruit and juice processing
Fibreless	Middle	Haiti	Tree has a compact in growth habit; susceptible to fungal diseases
Fibreless	Late	Florida, USA	Selected by L. Zill; Excellent flavour
Fibreless	Middle	Jamaica	The tree, blooms and fruit are highly susceptible to disease; production is light, highly demanded by people from the Caribbean
Fibreless	Middle	India	The bloom and fruit are highly susceptible to disease; Excellent quality fresh fruit
Fibreless	Middle	Thailand	The bloom and fruit are tolerant of disease; consistent, heavy production in the hot tropics
Fibreless	Middle	South Pacific	Blooms and fruit have considerable disease tolerance; tree is compact, forming a dense canopy with consistent and heavy production
Fibre	Late	Florida, USA	The blooms and fruit have good tolerance of disease; Used juice processing
Fibreless	Early	Egypt	Production is consistent and heavy; The bloom and fruit are susceptible to disease
Fibreless	Early	Vietnam	Considerable tolerance to disease; there are often multiple crops
Fibreless	Early	Philippines	The tree grows vigorously and is difficult to maintain; Important commercial crop in the Philippines

Cultivar	Tree size	Characterization of the ripe fruit				
		Size	Weight gr.	Shape	Skin colour	Flesh colour
Carrie	Medium	Medium	380	Oval	Green-yellow	Orange
Chokanan	Small	Small	340	Sigmoid-long	Yellow	Yellow
Chousa	Large	Medium	400	Oval	Yellow	Orange
Cogshall	Small	Large	480	Oblong	Red	Yellow
Cushman	Large	Large	850	Rounded	Yellow	Yellow/orange
Dot	Medium	Medium	380	Oval	Yellow/pink	Yellow
Duncan	Medium	Large	550	Oblong	Yellow	Orange
Diplomatico	Large	Medium	340	Oblong oblique	Yellow/orange	Yellow
Early Gold	Medium	Medium	368	Oblong	Yellow/pink	Yellow
Edward	Large	Large	560	Ovate to oblong	Yellow/pink	Yellow
Esmeralda	Small	Large	500	Sigmoid-long	Yellow/green	Yellow
Fairchild	Medium	Small	283	Oblong	Yellow	Yellow
Fernandin	Large	Small	225	Ovoid	Red	Yellow
Florigon	Large	Medium	380	Oblong	Yellow	Yellow
Fukuda	Medium	Large	509	Oval	Yellow/orange	Yellow
Gaylour	Large	Large	500	Oblong	Yellow	Yellow
Glenn	Small	Medium	500	Ovate to oblong	Yellow/pink	Yellow
Golek	Large	Medium	226	Oblong	Yellow	Green/orange
Gouveia	Medium	Large	280	Oblong	Red	Yellow
Graham	Medium	Large	425	Oval	Yellow	Orange
Haden	Large	Large	600	Oval	Red	Orange
Hatcher	Medium	Large	680	Oval	Yellow/red	Yellow

Fibre	Fruiting season	Origin	Additional information
Fibreless	Middle	Florida, USA	Good disease tolerance; named after Mrs. Carrie Zill
Fibreless	Middle	Thailand	The blooms and fruit have good tolerance of disease; adapted to the humid tropics
Fibreless	Middle	Pakistan	Exported to the Americas, Europe and Japan; adapted to the subtropics
Fibreless	Middle	Florida, USA	The blooms and fruit have good tolerance of disease; excellent fruit quality
Fibreless	Middle	Florida, USA	Selected by Mr. E. L. Cushman; unique appearance
Fibreless	Middle	Florida, USA	Selected by L. Zill; excellent flavour
Fibreless	Middle	Florida, USA	Selected by Mr. David Sturrock of Palm Beach; The bloom and fruit are tolerant to disease
Some fibre	Middle	Mexico	The blooms and fruit have good tolerance of disease; The fruit are generally consumed ripe
Fibreless	Early	Florida, USA	Selected by Mr. Frank Adams in Pine Island, Florida; production is inconsistent if grown in humid conditions
Fibreless	Middle	Florida, USA	Developed by Mr. Edward Simmonds. Excellent flavour; The blooms and fruit have good tolerance of disease
Some fibre	Middle	Mexico	Consistent and heavy producer; Multicropping
Fibreless	Middle	Panama	Highly tolerant of disease; introduced to USA by David Fairchild
Fibreless	Middle	India	The blooms and fruit have good tolerance of disease; tendency towards alternate bearing
Fibreless	Early	Florida, USA	Selected by Mr. John Kaiser of Ft Lauderdale; The blooms and fruit have good tolerance of disease
Some fibre	Middle	Hawaii, USA	The blooms and fruit are susceptible to disease; excellent flavour
Fibreless	Middle	Egypt	The bloom and fruit are susceptible to disease; The tree is vigorous and open in growth habit
Fibreless	Early	Florida, USA	Selected by Mr. Roscoe E. Glenn; The blooms and fruit have good tolerance of disease
Some fibre	Middle	Indonesia	Heavy and consistent production; The blooms and fruit have good tolerance of disease
Fibreless	Middle	Hawaii, USA	Pirie seedling; The bloom and fruit are susceptible to disease
Fibreless	Late	Trinidad and Tobago	Adapted well in humid climates; fruit is soft when ripe
Some fibre	Early	Florida, USA	Selected by Mrs. Florence Haden, Coconut Grove, FL; Mulgoba seedling. Excessively vigorous
Fibreless	Middle	Florida, USA	Selected by John Hatcher and Lawrence Zill; good disease tolerance and storage characteristics

Cultivar	Tree size	Characterization of the ripe fruit				
		Size	Weight gr.	Shape	Skin colour	Flesh colour
Heidi	Medium	Large	410	Oval	Red	Yellow
Hodson	Large	Large	450	Oblong	Yellow/red	Orange
Imam Pasand	Small	Large	500	Oval	Yellow	Yellow
Itamaraca	Medium	Small	113	Flattened ovoid	Yellow/green	Yellow
Ivory	Large	Medium	500	Sigmoid-long	Yellow	Yellow
Jakarta	Medium	Large	750	Round-oblong	Red-orange	Yellow
Jean Ellen	Small	Small	280	Ovoid long	Yellow	Yellow
Jehangir	Medium	Medium	454	Oval	Green/yellow	Yellow
Jubilee	Medium	Large	420	Oval	Purple	Orange
Julie	Small	Small	255	Oval-flat	Green/red	Orange
Keitt	Large	Large	1000	Oval	Green/pink	Yellow
Kensington Pride	Large	Large	500	Oval	Yellow/red	Yellow
Kent	Large	Large	700	Oval	Red	Orange
Kesar	Large	Medium	340	Oblong-long	Green/yellow	Orange
Lily	Medium	Large	450	Oblong	Red	Yellow
Madame Francis	Large	Large	450	Oblong sigmoid	Yellow	Yellow
Malindi	Medium	Large	560	Oblong	Red	Yellow
Mallika	Medium	Large	500	Oblong	Yellow/green	Yellow
Mamita	Small	Medium	280	Oblong	Yellow	Yellow
Manzanillo	Medium	Large	450	Oblong	Red	Orange
Nam Doc Mai	Medium	Large	520	Long	Yellow	Yellow
Neelum	Small	Medium	300	Oval	Yellow	Orange
Number 11	Medium	Small	250	Ovoid	Yellow	Yellow

Fibre	Fruiting season	Origin	Additional information
Fibreless	Middle	South Africa	Dense canopy with good production; The blooms and fruit have good tolerance of disease
Some fibre	Middle	Florida, USA	The blooms and fruit have good tolerance of disease; fruit is beautiful
Fibreless	Middle	India	From district of Andhra State, India; The tree is small and consistent in production
Fibre	Middle	Brazil	The fruit shape is unique; The bloom and fruit are susceptible to disease
Fibreless	Middle	Thailand	Disease resistance
Some fibre	Early	Florida, USA	Grown in the Caribbean. The bloom and fruit are susceptible to disease
Fibre	Early	Florida, USA	Selected by Fairchild Tropical Botanic Garden; Multicrop
Fibreless	Middle	India	The bloom are susceptible to disease; Excellent flavour
Some fibre	Middle	Florida, USA	Selected by Mr. Ed Mitchell; fruit are tolerant of disease
Fibreless	Middle	Jamaica	The tree is a natural dwarf; Fruit is juiced with strong pleasant flavour
Some fibre	Late	Florida, USA	Selected by Mrs. J.N. Keitt; Productive, heavy producer. Fruit are tolerant of disease
Fibreless	Early	Australia	Tree is vigorous and upright; The bloom and fruit are susceptible to disease
Fibreless	Late	Florida, USA	Selected by Mr. Leith Kent; good disease tolerance and storage characteristics
Fibreless	Middle	Gujarat, India	Exported to USA and Europe; fruit is intensely sweet and rich
Some fibre	Middle	Florida, USA	Selected by the Carmichael family; resistance to disease is excellent
Fibre	Middle	Haiti	Exported to USA and Europe; preforms well under semi-arid Caribbean climate
Some fibre	Middle	Kenya	Consistent production; fruit is susceptible to disease
Fibreless	Middle	India	Hybrid: Neelum x Dasheri; consistent production, resistant to diseases
Fibre	Middle	Puerto Rico	The production is consistent and heavy with good disease tolerance
Some fibre	Late	Mexico	The tree is vigorous and upright; fruiting consistent and heavy
Fibreless	Middle	Thailand	Tree has compact and upright growth habit; Considerable disease tolerant
Fibreless	Late	India	Genetic dwarf tree; Very late season
Fibre	Middle	Jamaica	Commercially used as rootstock in the Caribbean; fruit is susceptible to disease

		Characterization of the ripe fruit				
Cultivar	Tree size	Size	Weight gr.	Shape	Skin colour	Flesh colour
Okrong Tong	Large	Small	220	Long-sigmoid	Yellow	Yellow/green
Oro	Small	Large	560	Long-sigmoid	Green/red	Orange
Osteen	Large	Large	680	Oblong	Yellow/red	Yellow
Pairi	Large	Medium	380	Oblong	Green/red	Yellow
Palmer	Large	Large	700	Oblong	Red	Orange
Parvin	Large	Large	560	Ovate to oblong	Red	Yellow
Pedda Rasam	Large	Large	600	Ovate to oblong	Yellow	Yellow
Phimsen Man	Medium	Large	480	Long-sigmoid	Green/yellow	Yellow
Prieto	Medium	Small	250	Oblong	Green/red	Orange
Rapoza	Medium	Large	720	Ovoid	Red	Orange
Rataul	Large	Medium	396	Oblong	Yellow	Yellow
Rosigold	Small	Medium	320	Oblong	Pink/yellow	Orange
Royal Special	Medium	Small	283	Round	Yellow	Yellow
Sabre	Large	Medium	320	Long	Red/yellow	Orange
San Felipe	Medium	Large	520	Oval	Red/yellow	Orange
Sensation	Large	Medium	300	Oval	Purple	Yellow
Step	Small	Large	500	Long-sigmoid	Green/pink	Yellow
Sunset	Large	Medium	280	Oval	Red/yellow	Orange
Toledo	Small	Small	105	Oval	Green/yellow	Yellow
Tommy Atkins	Large	Large	650	Oval-oblong	Red	Yellow
Tong Dam	Large	Large	420	Oblong	Green	Yellow
Torbet	Small	Medium	330	Round	Red/yellow	Orange
Turpentine	Large	Small	170	Oval	Yellow	Yellow
Tyler	Large	Large	450	Oval	Yellow	Yellow
Valencia Pride	Large	Large	800	Long	Yellow/red	Yellow

Fibre	Fruiting season	Origin	Additional information
Fibre	Middle	Thailand	Good disease tolerance and production; used for cooking in Thailand
Some fibre	Early	Mexico	Dwarf tree and highly productive; the blooms are tolerant to disease
Fibreless	Middle/Late	Florida, USA	Haden seedling; Disease tolerance is excellent
Fibreless	Early	India	The flavour is very good; The bloom and fruit are tolerant to disease
Fibreless	Middle/Late	Florida, USA	The tree is very vigorous; fruit are moderately susceptible to disease.
Fibreless	Middle/Late	Florida, USA	Propensity for the production of seedless fruit; Disease tolerance is excellent
Fibre	Middle	India	The tree is vigorous and open in growth habit; The bloom and fruit are susceptible to disease
Fibreless	Early	Thailand	Eaten green and used for cooking; The blooms and fruit have good tolerance of disease
Fibre	Middle	Cuba	Tree is vigorous and highly susceptible to disease; very productive
Fibreless	Middle	Hawaii, USA	Beautiful fruit, firm and sweet; fruit has good storage and long distance transport characteristics
Fibreless	Middle	India	Fruit is used for pickles and preserves; The blooms and fruit have good tolerance of disease
Fibreless	Early	Florida, USA	Adapted well in tropical climates; very early fruiting season
Some fibre	Middle	India	Heavy producer; disease tolerant
Fibre	Early	Africa	The tree is vigorous and upright; Used as a rootstock
Fibreless	Middle	Cuba	Consistent and productive; fruit is susceptible to disease
Fibreless	Middle	Florida, USA	Heavy producer, good shipping and handling; well adapted to cool subtropical climates
Fibreless	Middle	Hawaii, USA	Resistance to disease is excellent; fruit are highly uniform, consistent and heavy production
Some fibre	Middle	Florida, USA	Vigorous tree; considered heavy producer
Fibre	Middle	Cuba	Small and productive; fruit is susceptible to diseases
Some fibre	Middle	Florida, USA	Heavy producer, good shipping and handling; resistance to disease is excellent
Fibreless	Early	Thailand	Remains green when ripe; Heavy producer
Fibreless	Middle	Florida, USA	Selected by Nellie Torbet; fruit shaped like an apple
Fibre	Early	Caribbean	Used as a rootstock in the Americans and the Caribbean Islands; common mango of the Americas
Fibreless	Middle	Africa	Unique appearance; good eating quality
Some fibre	Middle	Florida, USA	Haden seedling; commercially grown in the Americas

		Characterization of the ripe fruit				
Cultivar	Tree size	Size	Weight gr.	Shape	Skin colour	Flesh colour
White Pirie	Medium	Medium	311	Oval	Yellow/orange	Yellow
Winters	Medium	Small	340	Oval	Red	Yellow
Zebda	Medium	Large	450	Oblong-cylindrical	Green	Orange
Hong Xiangya	Medium	Large	485	Long-sigmoid	Red	Orange-yellow
Gui re-mang #82	Medium	Medium	275	Oblong	Light green	Yellow
Gui re-mang #3	Medium	Medium	250–400	Regular oval	Yellow-orange	Orange-yellow
Gui re-mang #10	Medium	Medium-large	350–550	Oval-with Nose	Green-orange	Orange-yellow
Gui re-mang #120	Medium	Medium-large	270–500	Regular oval	Orange-Red	Yellow
Jin-Hwang	Large	Large	1050	Oblong	Green/yellow	Orange
Kaohsiung No.3 (Shia Sheue)	Large	Medium	500	Oval	Green/yellow	Orange
Kaohsiung No.4 (Sweety)	Medium	Medium	350	Oval	Yellow/red	Orange
Taichung No. 1	Large	Large	600	Oval	Yellow/red	Yellow
Tainung No.1	Medium	Small	254	Oval	Yellow/red	Orange
Calypso®	Medium	Medium-large	350–500	Round	Blush colour – Pink	Yellow
R2E2	Large	Large	600–1000	Round ovate	Green-yellow	Yellow
Shelly	Small	Medium	500	Rounded/unsymmetrical	Red/yellow	Yellow
Vallenato	Large	Medium	280–320	Oval	Red/purple	Yellow
Van Dyke	Large	Medium	280–500	Oval, base rounded	Yellow with red blush	Orange-yellow
Jumbo Kesar	Medium	Medium-large	450–600	Rounded/unsymmetrical	Green/yellow with pink blush	Orange

Fibre	Fruiting season	Origin	Additional information
Some fibre	Early	Hawaii, USA	Selected by Oahu by a Mr. Tong Chew in Makiki, Hawaii; The bloom and fruit are susceptible to disease
Fibreless	Early	Florida, USA	Exceptional flavour; Tree is vigorous and upright
Fibreless	Middle	Egypt	Flavour is exceptional; The bloom and fruit are tolerant to disease
Some fibre	Middle-later	Guangxi, China	Poly, selected from seedlings of Xiangya #26; regular bearer, flesh 75.5% dry matter 15%
Some little	Middle-later	Guangxi, China	Mono, selected from seedlings of Indian's #901; regular bearer, flesh 73%, dry matter 20.1%, sugar 17%
Fibreless	Late	Guangxi, China	Poly, seedling of Hong Xiangya; thick skin
Some fibre	Late	Guangxi, China	Poly, seedling of Hong Xiangya
Fibreless	Middle-Later	Guangxi, China	Mono, seedling of Hong Xiangya
Fibreless	Middle	Taiwan	Selected by Mr. Jin-Hwang, Hwang; Tree is vigorous and upright
Fibre	Middle	Taiwan	Selected by Mrs. Sheue-Ru, Lee; Excellent flavour
Some fibre	Middle	Taiwan	Selected by Mrs. Sheue-Ru, Lee; fruit is beautiful
Some fibre	Middle	Taiwan	Selected by Mr. Chih-Sheng, Chang and Ching-Chang, Shiesh; Tree is vigorous
Fibreless	Early	Taiwan	The flavour is very good; Pulp aroma is strong
Fibreless	Middle	Australia (Trademark)	Monoembryonic, progeny of Kensington Pride and Sensation known also as a B74; moderately susceptible to anthracnose, and may also be affected by stem end rots (*Botryosphaeria* spp., and *Lasiodiplodia theobromae*)
Fibreless	Middle	Australia	Progeny of Kent, polyembryonic; internal fruit breakdown has been reported in trees growing on sandy soils
Fibreless	Late	Israel (Trademark)	Progeny Tommy Atkins and Keitt, known as BD-14-63; Mild flavour, very productive
Some fibre	Early	Colombia	The skin is uncommonly smooth and waxy, exceptional storage life; It flowers naturally in the tropics
Fibreless	Middle	Florida, USA	It was planted in commercial scale in the 1950s and 1960s; regular heavy producer, resistant to anthracnose, good colour and storage characteristics.
Fibreless	Medium	India	Yields are moderate and consistent. The flavour is intensely sweet; Fruit is juiced with strong pleasant flavour

7 Appendix 2: photos of principal mango cultivars

Figure 1 Alphonso.

Figure 2 Ah Ping.

Figure 3 Alampur Baneshan.

Figure 4 Anderson.

Figure 5 Angie.

Figure 6 Aroemanis.

Figure 7 Ataulfo.

Figure 8 Azúcar.

Figure 9 Bangalora (also known as Sandersha).

Figure 10 Banilejo.

Figure 11 Baptiste.

Figure 12 Beverly.

Figure 13 Bombay.

Figure 14 Borsha.

Figure 15 Brahm Kai Meu.

Figure 16 Brander.

Figure 17 Brooks.

Figure 18 Cairo.

Figure 19 Cambodiana.

Figure 20 Carabao.

Figure 21 Carrie.

Figure 22 Chokanan.

Figure 23 Chousa.

Figure 24 Cogshall.

Figure 25 Cushman.

Figure 26 Dot.

Figure 27 Duncan.

Figure 28 Diplomatico.

Figure 29 Early Gold.

Figure 30 Edward.

Figure 31 Esmeralda.

Figure 32 Fairchild.

Figure 33 Fernandin.

Figure 34 Florigon.

Figure 35 Fukuda.

Figure 36 Gaylour.

Figure 37 Glenn.

Figure 38 Golek.

Figure 39 Gouveia.

Figure 40 Graham.

Figure 41 Haden.

Figure 42 Hatcher.

Figure 43 Heidi.

Figure 44 Hodson.

Figure 45 Imam Pasand.

Figure 46 Itamaraca.

Figure 47 Ivory.

Figure 48 Jakarta.

Figure 49 Jean Ellen.

Figure 50 Jehangir.

Figure 51 Jubilee.

Figure 52 Jumbo Kesar.

Figure 53 Julie.

Figure 54 Keitt.

Figure 55 Kensington Pride.

Figure 56 Kent.

Figure 57 Kesar.

Figure 58 Lily.

Figure 59 Madame Francis.

Figure 60 Malindi.

Figure 61 Mallika.

Figure 62 Mamita.

Figure 63 Manzanillo.

Figure 64 Nam Doc Mai.

Figure 65 Neelum.

Figure 66 Number 11.

Figure 67 Okrong Tong.

Figure 68 Oro.

Figure 69 Osteen.

Figure 70 Pairi.

Figure 75 Prieto.

Figure 76 Rapoza.

Figure 77 Rataul.

Figure 78 Rosigold.

Figure 79 Royal Special.

Figure 80 Sabre.

Figure 81 San Felipe.

Figure 82 Sensation.

Figure 83 Step.

Figure 84 Sunset.

Figure 85 Toledo.

Figure 86 Tommy Atkins.

Figure 87 Tong Dam.

Figure 88 Torbet.

Figure 89 Turpentine.

Figure 90 Tyler.

Figure 91 Valencia Pride.

Figure 92 Vallenato.

Figure 93 Van Dyke.

Figure 94 White Pirie.

Figure 95 Winters.

Figure 96 Zebda.

Figure 97 Hong Xiangya.

Figure 98 Gui re-mang #82.

Figure 99 Gui re-mang #3.

Figure 100 Gui re-mang #10.

Figure 101 Gui re-mang #120.

Figure 102 Jin-Hwang.

Figure 103 Kaohsiung No.3 (Shia Sheue).

Figure 104 Kaohsiung No.4 (Sweety).

Figure 105 Taichung No. 1.

Figure 106 Tainung No.1.

Chapter 3

Advances in understanding mango tree growth and canopy development

Frédéric Normand, CIRAD, France; and Pierre-Éric Lauri, INRA, France

1 Introduction

One may wonder why a whole chapter is dedicated to mango tree growth and canopy development. The reader might expect that a book devoted to a fruit crop talks about the final product. It might discuss the fruit (beginning with flowering and concluding with quality characteristics such as colour, taste or nutritional content), or might review varieties and orchard management to obtain a profitable and sustainable crop. The subject of vegetative growth is often neglected or only mentioned in terms of tree vigour and the cultivation practices required to manage tree growth.

However, the above-ground vegetative part of the fruit tree is of great importance. It is the perennial part of the tree which produces leaves, flowers and fruits. It fulfils essential functions, such as the conduction of vital elements (water, nutrients, carbohydrates and hormones), the storage of carbohydrates or nitrogen and the support of the whole tree and its fruit production. But this vegetative part can also be a disadvantage for fruit production, for example with excessive vegetative growth at the expense of flowering and fruiting.

It is therefore important to better understand the vegetative part of the tree, how it grows, how it works, what are the differences between cultivars, and what are the interactions between vegetative growth, flowering and fruiting, in order to manage tree

http://dx.doi.org/10.19103/AS.2017.0026.06

and canopy more efficiently. The objective of this chapter is to summarize information on these different topics for the mango tree. Based on current knowledge and the most recent research results, it describes the architectural model of the mango tree and the basic entity of its structure, the growth unit. Morphology and the growth and development of the growth unit are reviewed. The relationships between the growth unit and canopy development are then outlined, along with the effects of environmental and endogenous factors on tree phenology. Finally, the interactions between vegetative growth and flowering and fruiting are described, leading to a more complete, if complex, picture of the mango tree canopy.

2 Mango tree architecture

The architecture of a plant is commonly defined by the nature and structural arrangement of plant parts (Hallé et al., 1978; Barthélémy and Caraglio, 2007). Aerial plant parts can be categorized according to their nature (e.g. leaves, buds, flowers, fruits, etc.) and their scale [e.g. metamers (leaves and associated internodes), growth units, axes, etc.]. The structural arrangement of these plant parts refers to their topology, that is, the connection between them, and to geometry, that is, their dimensions and position in space, which together define the form of the plant. Plant architecture is established over time and is the result of the balance between endogenous growth processes and exogenous constraints such as environment.

The architectural model of a plant is the output of its endogenous growth strategy (Hallé and Oldeman, 1970). It is defined by the combination of characteristics related to four simple morphological features: the growth pattern, the branching pattern, the morphological differentiation of axes and the position of flowering (Barthélémy and Caraglio, 2007). Twenty-three architectural models have been described in nature up to now, each named after a famous botanist.

The mango tree follows the architectural model of Scarrone (Hallé et al., 1978; Goguey, 1995, 1997). The main characteristics of this model for the mango tree are (Fig. 1):

- rhythmic growth: vegetative buds are active for a short period, providing new portions of an axis, and then remain quiescent until the next active period. The new portion of axis, which can develop from apical or lateral buds, is called a growth unit. It is defined as the portion of an axis that develops during an uninterrupted period of growth (Hallé and Martin, 1968). The opposite of rhythmic growth is continuous growth, when the apical bud of an axis is continuously active, producing new leaves and internodes without interruption.
- terminal flowering: mango inflorescences result from the transformation of the apical bud of the terminal growth units. They therefore appear at the extremity of the woody structure of the tree. This is important because terminal flowering changes the nature of the apical bud, from vegetative to reproductive, leading to the cessation of further axis growth. Growth is then only possible through branching, the development of lateral buds.
- monopodial and sympodial branching: monopodial branching relates to the growth and branching of an axis with an active apical bud (Barthélémy and Caraglio, 2007).

Figure 1 Mango terminal flowering, cultivar José (a); growth units (GU 1–6) succession and monopodial and sympodial branching, cultivar Kent (b). The development of new growth units at the distal end of each growth unit is named acrotony. Leaves of growth units 1 and 2 have been removed for a better visibility.

This occurs during the vegetative growth season on growth units that have not flowered. When the apical bud is transformed into an inflorescence, the development of new axes from lateral buds equates to sympodial branching. This is typically the case for vegetative growth stemming from growth units that have flowered.

- acrotonic branching: acrotony, mesotony and basitony indicate the zone along an axis where branching occurs preferentially – the distal zone, the median zone and the basal zone, respectively (Barthélémy and Caraglio, 2007). The mango tree has an acrotonic branching pattern. Branching results from the activity of the distal-most lateral buds of the terminal growth units. The development of new axes, therefore, appears at the extremity of the parent growth unit. More rarely, new axes can also appear at the distal end of the penultimate growth unit, according to the acrotonic branching pattern.

Figure 2 Reiteration process on a Kensington Pride branch after severe pruning.

Figure 3 Branch of a mango tree where the successive growth units are clearly visible, delimited by ring-like marks of growth arrest of each apical bud (white arrows), cultivar Kent.

The growth patterns described earlier correspond to sequential growth, which is the natural sequence of growth and flowering at the periphery of the tree. New growth units can also appear in old woody structures, after dedifferentiation of dormant buds, producing new axes and vegetative structures as a result of growth and branching processes. It is rarely spontaneous in cultivated mango trees, and occurs mainly as a response to damage to the woody framework, for example, after severe pruning (Fig. 2). This particular sort of growth is called reiteration and results in a reiterated complex (Barthélémy and Caraglio, 2007).

Rhythmic growth begins from seed germination or from grafting for grafted trees. A mango tree is therefore a succession of growth units, connected through the activity of the apical bud as it forms an axis or through the activity of lateral buds that form ramifications. Growth units remain visible for several years since they are separated by a clear ring-like mark that corresponds to the scars of apical bud scales, indicating the growth arrest of the apical bud (Fig. 3). Moreover, acrotony and terminal flowering lead to a concentration of vegetative growth, flowering and fruiting at the distal part of the terminal growth units at the periphery of the canopy. Terminal growth units are therefore essential in the mango tree, from a structural, temporal and functional point of view, as explained later. The following section is dedicated to these units.

3 Morphology of the mango growth unit

This section is devoted to the growth unit. It describes the growth unit's morphology and the factors affecting this morphology. To illustrate, the chapter uses the kinship terms devised by Dambreville et al. (2013a). When a growth unit produces new growth units, the former is called the mother growth unit and the latter are called daughter growth units.

3.1 Growth unit morphology

A mango growth unit is a succession of nodes, each composed of a leaf and an axillary bud, and internodes of various lengths. Different types of leafy organs follow one another from the base to the apex of the growth unit (Goguey, 1995):

- Scaly cataphylls are the scales protecting the apical bud (Fig. 4). They fall rapidly after bud burst, leaving thin scars. The scars of scaly cataphylls typically indicate the transition between two growth units (Hallé and Martin, 1968).
- Leafy cataphylls are leaves with a small round lamina (Fig. 4). Their lifespan is few days.
- Photosynthetic leaves are the persistent large leaves of the growth unit. Mango leaves are glabrous, simple and entire. Lamina margins are more or less wavy. Their characteristics, shape, colour and dimensions vary with the cultivar. Leaf shape varies from lanceolate to elliptic, with the apex varying from acuminate to acute. The petiole base is rounded, leaving a typical badge-shaped scar when the leaf falls. Leaf size is highly variable, even within a tree. Lamina length can vary from a few centimetres to 40 cm, and lamina width from 1 to 7 cm (Goguey, 1995).
- Aborted leaves are very small leaves around the apical bud. They have a short lifespan and are not always present.

Figure 4 Close up of a mango apical bud, cultivar Kent (a), and of the basal section of an apical growth unit, cultivar José (b). The apical bud is sheltered by scaly cataphylls (SC) and surrounded by lateral buds (Lb) at each leaf axil (only petioles are visible). The base of the growth unit shows scars of SC from the former apical bud and leafy cataphylls (LC) below the first photosynthetic leaves (only petioles are visible). Two badge-shaped leaf scars are visible on the mother growth unit.

Mango phyllotaxy is alternate, with a 2/5 spiral pattern. The angle between two consecutive leaves is 144°, and five consecutive leaves are included in two gyres. Leaves are not distributed regularly along the stem. Internodes between scaly cataphylls are very short, less than 1 mm. Their length increases regularly and is greatest in the middle of the growth unit. It then decreases and internodes at the distal part of the growth unit are very short – less than 1 mm – forming a rosette of leaves (Goguey, 1995).

Immediate or sylleptic branching, that is, the concomitant development of lateral growth units from an elongating growth unit, does not occur in the mango tree (Goguey, 1995). Lateral buds remain quiescent after growth unit elongation. The mango growth unit ends with a highly visible, cone-shaped apical bud. On the contrary, quiescent lateral buds are hard to see, as they are within the leaf axil. A short, light green line is the most frequent sign of the presence of the quiescent lateral bud. When a growth unit flowers, apical, and possibly lateral, buds transform into inflorescence(s). Once the inflorescence disappears, a large, round, flat scar replaces the apical bud, and circular crater-like scars replace the lateral buds. Further vegetative growth is then only possible from the remaining lateral buds of the growth unit.

At the time of bud burst, all the small and folded leaves of the growth unit are visible. However, one may wonder if some of these leaves, and the nodes supporting them, are neoformed, that is, formed at or after bud burst, as opposed to being the preformed nodes existing in the quiescent bud (Hallé and Martin, 1968). Goguey (1995) carried out dissection experiments and found that the number of leaf primordia in the apical bud increases just before bud burst, clearly indicating neoformation. The neoformed distal part of the growth unit stem is short, composed, in his study, of about four leaves with a small leaf area and three aborted leaves. Recent results suggest that at least the two distal-most leaves are connected to neoformed nodes (Dambreville et al., 2013b; see Section 3.2). The characteristic feature of these leaves is that they have a smaller area than the other leaves of the growth unit (Taylor, 1970; Dambreville et al., 2013b).

A growth unit can be characterized by several structural and temporal traits. Structural traits describe growth unit topology, morphology, fate and position in the sequence of tree development. Temporal traits are the different dates associated with growth unit life and activity. As will be explained later, these traits are important because they are related to growth unit activity, vegetative growth, flowering and fruiting.

Growth unit topology is described as the position, or origin, of the growth unit with respect to its mother growth unit. It can be in an apical position, that is, stemming from the apical bud of the mother growth unit, or in a lateral position, that is, stemming from a lateral bud of the mother growth unit (Fig. 5).

The position of the growth unit in the sequence of tree development indicates whether the growth unit appears as sequential growth, that is, on terminal mother growth units, or as the first growth unit of a reiterated complex, that is, stemming from a dormant bud in the old woody structure. In the latter case, the growth unit appears in a lateral position. In this chapter, we consider only sequential growth units, which are more numerous and important in a mango tree canopy.

Growth unit morphology can be described as the stem length and diameter; the number of leaves; and the length, width and surface area of leaves. After the extension of the stem and the leaves, these traits remain stable, except for the stem diameter and the number of leaves. Stem diameter increases with further vegetative growth (Normand et al., 2008; Normand and Lauri, 2012) or fruit growth (Lauri et al., 2010; Capelli, 2017) as a result of secondary growth. Stem diameter must therefore be recorded after the complete extension of the growth unit and before further vegetative growth. The mango tree is an evergreen tree, but the number of leaves within a growth unit decreases with time, because of damage by pests and diseases or as a result of shading due to adjacent

Figure 5 Apical growth unit (GU) stemming from the apical bud, and two lateral growth units stemming from lateral buds of the mother growth unit, cultivar Kent.

vegetative growth. The average lifespan of a mango leaf is three years (Holdsworth, 1963) and old growth units are leafless.

The fate of a growth unit describes growth unit activity at a given time in terms of vegetative or reproductive growth through the functioning of apical and/or lateral buds (Dambreville et al., 2013a). Growth unit fate is:

- quiescent when all buds on a growth unit are quiescent,
- vegetative when at least one bud bursts and produces a daughter growth unit,
- flowering when at least one bud produces an inflorescence,
- fruiting when a flowering growth unit bears one or several fruits.

Growth unit fate evolves with time, particularly as long as the growth unit is in a terminal position. Typically, it is quiescent after complete extension, and then it can be flowering, fruiting and/or finally vegetative. At this stage, the growth unit is no longer in a terminal position and its vegetative and reproductive activity is null, except in the case of reiteration or rare flowering on an aged structure.

The temporal traits that describe a growth unit are its date of birth, that is, the date when the bud bursts on the mother growth unit, and the dates related to its reproductive and vegetative activity, such as the date of flowering, the date of fruit maturity or the date of vegetative growth. These dates correspond to events that must be clearly specified. For example, considering that inflorescence development lasts from three to four weeks from the emergence of the inflorescence to the last open flower, when is the flowering date of a given growth unit? The best method to determine this precisely is to define these events by a particular phenological stage (see Section 4.2), especially a phenological stage of short duration, such as inflorescence or vegetative bud burst.

3.2 Factors affecting growth unit morphology

Several endogenous and exogenous factors affect leaves or stem morphology, or both. Stem length, the number of leaves per growth unit and the individual leaf area or growth unit leaf area can vary enormously. These traits vary at different scales: between cultivars, between trees, between growth units within a tree, and even between leaves within a growth unit. It is therefore important to identify the factors that affect these traits. First, because flowering and fruiting are related to vegetative traits (see Section 6.2), knowing the factors that affect vegetative traits can suggest tree or canopy management practices to improve and to regulate yield across years. Second, this knowledge is helpful for experimental purposes, in order to better control the sources of variability in the data related to growth unit morphology.

The cultivar is an important factor that affects stem length, the number of leaves and the individual leaf area or growth unit leaf area (Whiley et al., 1989; Normand et al., 2008, 2009a; Shaban, 2009; Dambreville et al., 2013b; Ramírez et al., 2014). The cultivar effect relates to the tree's vigour, provided that temperature and growth conditions are favourable (Whiley et al., 1989). For example, compared to the low-vigour cultivars Irwin, Cogshall and José, the vigorous cultivar Kensington Pride has longer growth units, with more leaves and a larger individual leaf area, leading to a larger leaf area at the scale of the growth unit (Normand et al., 2009a).

Large tree fruit load may negatively affect stem length, the number of leaves and the growth unit leaf area of different cultivars (Shaban, 2009). In this two-year study in

Egypt (Shaban, 2009), the fruit load effect was combined with the effect of the period of vegetative growth. Three periods of vegetative growth occurred: in spring during flowering, in summer during or after harvest and in autumn. The stem length, the number of leaves and the growth unit leaf area were larger for growth units developed during summer than for those developed in autumn. Those developed in spring had an intermediate size. The negative fruit load effect suggests competition between fruit growth and vegetative growth, even after harvest in autumn. The largest growth units developed during or just after harvest in summer, whatever the tree fruit load, which suggests a positive effect of temperature on leaf initiation during this warm season (see the following section).

Growth unit dimorphism has been evidenced in four mango cultivars, in relation to the position of the growth unit – apical or lateral – with respect to the mother growth unit (Fig. 5; Normand et al., 2009a). This dimorphism is more pronounced for leaf traits than for stem length. Apical and lateral growth units are morphologically distinct, and the differences on most of the traits are more significant than the differences among cultivars, denoting the importance of the growth unit position on its morphology (Normand et al., 2009a).

The effect of the growth unit position on stem length is cultivar-dependent. Apical growth units have significantly longer stems than lateral growth units for Cogshall and Kensington Pride, but not for Irwin and José (Normand et al., 2009a). To illustrate this difference, in a dataset containing 72 mature growth units sampled on seven trees of the cultivar Cogshall, mean stem length was 17.7 $\pm$ 4.2 cm (n = 30) for apical growth units and 13.5 $\pm$ 3.7 cm (n = 42) for lateral growth units (Normand, unpublished data). The difference was highly significant ($P < 0.001$). The above standard deviations denote a large variability of stem length among apical growth units and among lateral growth units. This suggests that other factors, currently unspecified and likely to be endogenous and exogenous, affect this trait.

The number of leaves, individual leaf area and the growth unit's total leaf dry mass and leaf area were significantly higher in apical than in lateral growth units. In particular, the higher total leaf area in apical growth units is the result of a higher number of leaves as well as a higher individual leaf area (Normand et al., 2009a). A consequence of these differences is that the axialization index (Lauri and Normand, 2017), the ratio of stem length to leaf dry mass in this study, was significantly lower in apical than in lateral growth units.

Individual leaf size varies according to the order of the leaves along the growth unit stem (Fig. 6; Taylor, 1970; Dambreville et al., 2013b). Leaf size is relatively constant among the basal and median leaves of the growth unit and decreases progressively up to the most distal leaf. This general pattern may vary among cultivars (Taylor, 1970; Dambreville et al., 2013b). In the cultivar Cogshall, the individual area of the distal and subdistal leaves represents on average 37.9% and 52.1%, respectively, of the mean leaf area of the other leaves of the growth unit (Normand, unpublished results). The smaller leaf area of the most distal leaves is probably related to their neoformed origin. One hypothesis is that at the time of bud burst, the preformed basal and median leaves are more developed and take up more resources, which limits the resources that are available for cell division in the neoformed, more distal leaves (Dambreville et al., 2013b). As leaf size in mango is determined by the number of cells (Taylor, 1970), this would result in smaller distal leaves.

The daily mean temperature during growth, hereafter called mean temperature, may affect stem length or final leaf size and the number of leaves per growth unit, but it depends on the cultivar, the leaf position along the stem and the temperature range (Whiley et al., 1989; Dambreville et al., 2013b). In an experiment using ten cultivars in

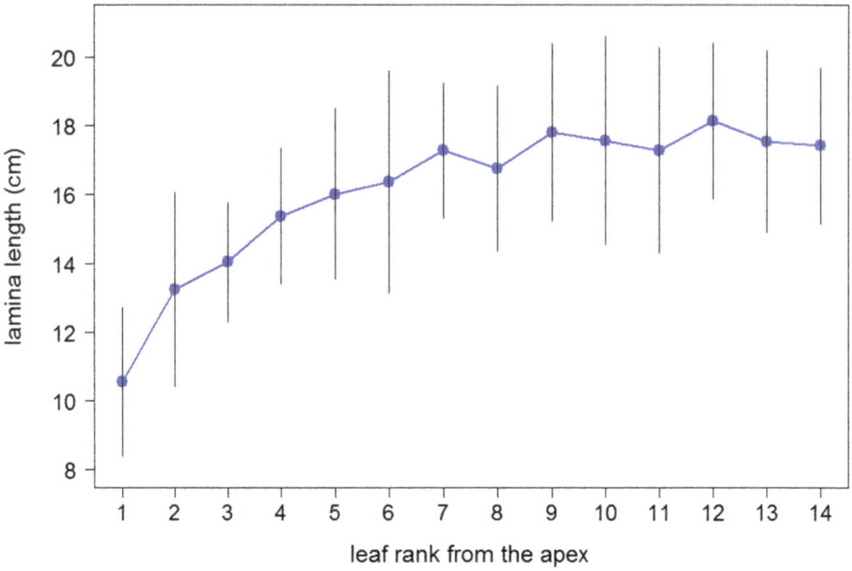

Figure 6 Distribution (mean ± sd, n = 20) of leaf lamina length along a mango growth unit with 14 leaves, cultivar Cogshall. Rank 1 corresponds to the most distal leaf, close to the apex (Dambreville, unpublished data).

controlled day/night temperatures, from 15/10 to 30/25°C, across 20 weeks, Whiley et al. (1989) showed that the number of leaves per growth unit drops dramatically as the temperature decreases. On average, across the ten cultivars, growth units bear 13.6 leaves at 30/25°C, 9.7 leaves at 25/20°C, 7.1 leaves at 20/15°C, and 1.2 leaves at 15/10°C (for the cultivars that produced growth units within this temperature range). Stem length and leaf area per growth unit also decrease as the temperature decreases. There is a significant variability among cultivars as to how their traits respond to temperature, in particular at the two lower temperatures in this range.

Dambreville et al. (2013b) studied the effect of temperature during growth unit growth, from bud break to final leaf size, in natural conditions. Their results denote the direct effects of temperature on growth unit morphology during the growth phase and not on morphogenesis at the bud stage. Daily mean temperatures were less even and in a narrower range (19.3–28.5°C) than in Whiley et al. (1989). Leaf area tends to increase with mean temperature, but only for the cultivar José and not for Cogshall. When considering leaf position along the stem, this relationship is highly significant for the distal and subdistal leaves, but not for the proximal leaf of the cultivar José. The same trend is observed in the cultivar Cogshall, but is not significant. This positive effect of current temperature on the size of distal leaves, but not on the size of proximal leaves, also supports the hypothesis of neoformation of the distal leaves (Dambreville et al., 2013b).

These two studies show that temperature has a positive effect on leaf initiation, during morphogenesis at the bud stage, as well as on the first stages of leaf development, which is likely to be during the period of cell division. A consequence is that the individual area of preformed leaves depends on temperature during bud development (leaf initiation), and the individual area of neoformed leaves depends on temperature during growth of the growth unit.

Other differences in growth unit and leaf size that are not explained by the previous endogenous and environmental factors can be observed on mango trees. A well-known example is the larger individual leaf area in the first growth unit of the reiterated complex following severe pruning, compared to sequential growth. The light environment also seems to affect leaf size, with larger leaves forming in the shade rather than in sunlight (e.g. Charbonnier et al., 2017). The availability of resources (nitrogen, water, etc.), determined by the environment and cultivation practices, can also affect growth unit morphology, with larger growth units and leaf areas developing when these resources are not limiting. However, these differences, and the factors generating them, are not documented for the mango tree.

4 Growth and development of the mango growth unit

Growth and development are distinct but closely related processes. They describe what happens between bud burst and the mature growth unit. Growth describes how a growth unit increases in mass or size over time, and development describes the visible changes the growth unit undergoes. This section reviews current knowledge related to growth unit growth and development, and their relationships.

4.1 Stem and leaves growth

Growth is defined as an irreversible increase in plant or organ dimensions over time. These dimensions can be length, width, area, diameter, volume or mass (Guinochet, 1965; Dambreville et al., 2015). The growth pattern of a plant or an organ generally follows a sigmoid curve and is modelled by a logistic function. This curve is composed of three consecutive phases: a first phase where growth is exponential, a second phase where growth is linear, and a third phase, often called a plateau, where growth slows down and ceases as the final size is reached (Goudriaan and Van Laar, 1994). The growth in length of growth unit stem and in area of the individual leaf follow this typical sigmoid pattern (Taylor, 1970; Dambreville et al., 2013b, 2015; Fig. 7).

Leaf growth and the underlying processes have been studied by Taylor (1970). Mango leaves grow faster in length than in width. The absolute growth rates in length and width are not uniform along leaf length and width, and the centre of the lamina is the most actively growing area. Duration of growth is, however, similar for the different parts of the leaf lamina, from the base to the tip.

The duration of leaf growth is affected by both endogenous and exogenous factors. It depends on the cultivar and temperature (Dambreville et al., 2013b). The position of the leaf along the stem also seems to affect leaf growth duration, although this effect is small. Leaf growth duration increases across the most basal leaves (leaves 1–5) and then remains constant for the upper leaves (leaves 6–14) (Taylor, 1970). Dambreville et al. (2013b) recorded small differences in growth duration between lower and upper leaves for two cultivars (10.7 and 11.6 days for Cogshall, 12.2 and 13.1 days for José, respectively), but they were not significant. This is likely because these values are means of data recorded over a range of temperatures varying from 19.3 to 28.5°C.

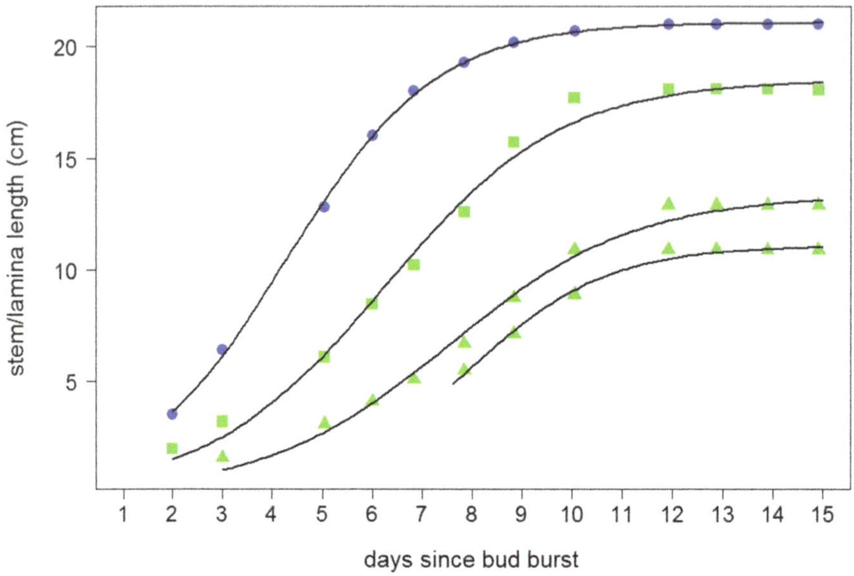

Figure 7 Growth in length of a Cogshall growth unit stem (blue circle) and three of its leaves: a proximal leaf (green square) and the two most distal leaves (green triangle). Adjusted logistic models are represented for each organ with a black line (Dambreville, unpublished data).

Leaf growth lasts about 14 days (in Taylor's (1970) experiment), with 8–9 days of exponential growth before slowing down and cessation. Growth is mainly related to cell division during the first seven days, and then to cell expansion. The mean area of epidermal cells doubles between days 7 and 14, changing from 141 to 309 μm^2, whereas their total number remains constant, 32.3 to 34.1×10^6 $leaf^{-1}$. After the final leaf area is reached, leaf thickness continues to increase up to day 35, mainly in relation to the mesophyll development. Contrary to leaf area, leaf dry mass increases linearly with time, at least during 14 days. The faster growth in area than in dry mass leads to a minimum value of dry mass per leaf area of 4–5 mg cm^{-2} at day 14 before it increases linearly up to 20 mg cm^{-2} at six weeks. These results suggest that leaf growth in area and in dry mass are at least partially independent (Taylor, 1970).

Leaf area can be easily and accurately estimated from the measurement of leaf length and width. Prasada Rao et al. (1978) compared different linear models with leaf length, leaf width and the product of both as predictors for mature leaves in six Indian cultivars. They concluded that the most accurate model was that using the product of leaf length and leaf width as predictor. Taylor (1970) and Dambreville et al. (2013b) proposed a similar model that can be used for both growing and mature leaves. When the y-intercept of the linear model between the leaf area and the product of leaf length and leaf width is zero, which is generally the case (Taylor, 1970; Prasada Rao et al., 1978; Dambreville et al., 2013b), the slope of the model represents the ratio between the actual leaf area and the area of the rectangle in which the leaf is inscribed. Despite different leaf shapes and sizes, this slope is about constant, ranging from 0.692 to 0.740, among six cultivars from the cultivars discussed in the three cited studies.

The duration of growth is significantly shorter for the stem than for the leaves, between 7.3 and 11.4 days for Cogshall and José, respectively. It is significantly affected by temperature (Dambreville et al., 2013b). Durations of growth of the stem and of its leaves are closely and positively related, indicating a strong temporal dependency within the growth unit. This dependency seems to result from a constraint imposed by temperature on growth durations (Dambreville et al., 2013b).

4.2 Growth unit development

Plant or organ development is defined as a series of identifiable events resulting in a qualitative (e.g. germination, bud burst, flower opening) or quantitative (e.g. number of leaves, number of flowers) change in plant or organ structure (Bonhomme, 2000). Vegetative and reproductive developments are generally distinguished, where the former relates to vegetative changes (e.g. the development of new growth units) and the latter relates to reproductive changes (e.g. the development of inflorescences).

Developmental stages, which are also called phenological stages, have been defined to characterize these changes in development. They correspond to a period during which the plant or organ shows a precise combination of morphological traits. The accurate definition of phenological stages requires traits with visible discontinuities, such as bud burst or petal fall, rather than traits displaying continuous changes, such as leaf colour, texture or orientation, because, in this latter case, the distinction between two successive stages may not be obvious. However, traits with visible discontinuities are not numerous, so precise phenological stages need to be defined by combining several traits.

Phenological stages have been described for mango growth units, inflorescences and fruits, according to two main methods. The first method is based on observation of several traits of these organs. These traits are combined to define the phenological stages (Fig. 8). The phenological stages are named by letters (Aubert and Lossois, 1972), abbreviations (Ramírez et al., 2014), numbers (Oosthuyse (1991) for inflorescences only) or a combination of letters and numbers (Dambreville et al., 2015). The number of stages depends on the precision with which the stages are described. The phenological stages described for mango growth unit development are based on bud, stem and leaf traits. Their number ranges from five (Aubert and Lossois, 1972) to eight (Ramírez et al., 2014) and nine (Dambreville et al., 2015). However, most of the stages, from the growth unit with dormant buds to the mature daughter growth unit, are common among these different scales. The phenological stages proposed by Aubert and Lossois (1972) describe more specifically bud burst and the beginning of stem elongation and do not detail the later stages of growth unit development, as done by Ramírez et al. (2014) and Dambreville et al. (2015).

The second method is an adaptation of the Biologische Bundesanstalt, Bundessortenamt und Chemische Industrie (BBCH) scale to the mango (Hernández Delgado et al., 2011). The BBCH scale has been developed to uniformly code similar phenological stages of mono- and dicotyledonous plant species (Bleiholder et al., 1989; Meier et al., 2009). In the general BBCH scale, plant development is divided into ten principal stages, from germination/bud break to senescence/beginning of dormancy. These stages are themselves split into ten secondary stages, which are defined as the percentage of the final organ size attained; they are not based on morphological traits, as in the previous method. When necessary, ten mesostages can also be inserted between the principal and the secondary stages. The stages are then coded with a two- or three-digit code.

Figure 8 Nine vegetative phenological stages defined by Dambreville et al. (2015) from the resting bud (stage A) to the mature growth unit (stage H). Detailed description of each stage is in the original publication.

Hernández Delgado et al. (2011) split the vegetative and reproductive development of the mango into seven principal stages, the other three not being applicable for the mango tree. For vegetative development, they proposed phenological stages for bud, leaves and stem development during the first and the second flush. Rajan et al. (2011) validated the BBCH scale for mango on the Indian cultivar Totapuri and proposed minor adaptations.

The main factor affecting the time needed to complete development of a plant or an organ is temperature (Arnold, 1959; Bonhomme, 2000; Tardieu, 2013). In a particular range of temperature, the rate of development (i.e. the inverse of the development duration expressed in d^{-1}) is linearly related to temperature. Thermal time models, which are widely used in plant modelling (e.g. Brisson and Delécolle, 1991; Cave et al., 2013) and/or in predicting harvest dates (e.g. Malézieux et al., 1994; Dufault, 1997), are built on this property. They integrate temperature into the timescale by assuming that, above a certain threshold temperature θ, the daily increment of development, expressed in degree.day (°C.d), is the difference between the daily mean temperature and θ. The plant or the organ, here the mango growth unit, accumulates these daily increments of development until development is completed. Thermal time models have two parameters: the threshold temperature θ (in °C) and the thermal time sum to complete development (in °C.d). The threshold temperature is a statistical parameter and it differs from the physiological base temperature of a plant/organ, under which the development is null (Arnold, 1959; Bonhomme, 2000; de Parcevaux and Hubert, 2007).

Relationships between development and temperature have been poorly investigated for the mango tree. They mainly concern fruit development (Mosqueda-Vázquez and Ireta-Ojeda, 1993; Shinde et al., 2001; Normand and Léchaudel, 2006). Thermal time models have also been developed for growth unit development of two cultivars, Cogshall and José, from bud burst to the end of stage G (Fig. 8) when the leaves have matured (Table 1).

The results in Table 1 illustrate that the time needed to complete growth unit development depends on the cultivar. The threshold temperatures of the two cultivars are significantly different, indicating that they react differently to temperature. For a same daily mean temperature above the threshold temperature, for example 25°C, José growth units accumulate daily more development (11.5°C.d) than do Cogshall growth units (11.0°C.d). But the development of José growth units requires a higher thermal time sum, meaning that the average duration of development is longer than that for Cogshall.

4.3 Combining growth and development

Growth and development are distinct processes that are generally considered separately, although they are related (Guinochet, 1965; Bonhomme, 2000; Dambreville et al., 2015). The mango growth unit is an interesting illustration. The growth and development of the stem and leaves occur concurrently, suggesting that the two processes are closely related. An integrated approach, then, appears necessary to better understand growth and development, and their interaction.

To investigate this question, Dambreville et al. (2015) defined six growth stages for the leaves and the stem by splitting their growth into a sequence of stages based on changes in the absolute growth rate. These growth stages, determined by calculating the measured data, match well with the growth unit phenological stages, as defined above (Fig. 8). This clear correspondence led to the definition of 'developmental growth stages'. These integrative stages describe both the growth and the development of the growth unit and

Table 1 Threshold temperature, thermal time sum and average duration for the development of growth units of two mango cultivars, Cogshall and José (Dambreville et al., 2015; Normand, unpublished results)

Cultivar	Threshold temperature (°C)	Thermal time sum (°C.d)	Average development duration (d)
Cogshall	14.0	364.2	31.5
José	13.5	410.1	36.9

open up the possibility of deducing growth dynamics from developmental observations. They have also been found in the inflorescence (Dambreville et al., 2015).

Developmental growth stages illustrate the coordination between growth and development at the growth unit scale. For example, stage 3 of leaf growth, which is characterized by the maximal absolute growth rate in leaf area, corresponds to the phenological stage F, where the leaves are limp and hang down (Fig. 8; Dambreville et al., 2015). It corresponds to the time when cells are expanding in the lamina, and when there is a decrease of dry mass and an increase of water content per unit leaf area (Taylor, 1970). During this period, growth in dry mass is less important than growth in leaf area, and limpness appears to be more related to a minimum in dry mass per unit area than a lack of lamina turgor (Taylor, 1970). Stage F is also characterized by a large angle between the stem and the petiole (~135° downward). This large angle, associated to leaf limpness, likely acts to protect the leaves from excessive radiation, and therefore temperature, during this important period of maximal absolute growth rate (Dambreville et al., 2015).

While phenological stages are defined at the growth unit scale, and are therefore similar for the stem and the leaves, there is evidence of asynchronisms during the growth of the growth unit. The first asynchronism is between the initial growth of the stem and of the leaves (Fig. 7). Stem length increases first, followed by an increase of leaf area (Taylor, 1970; Dambreville et al., 2015). Two non-exclusive hypotheses are proposed to explain this delay. First, the stem must grow first in order to support the growth of the leaves and to supply them with water and carbohydrates. Second, at the beginning of growth unit extension, the leaves are still small and in a distal position with respect to the stem base. They therefore probably attract less carbohydrate than the growing stem (Dambreville et al., 2015). The second asynchronism is between the lower and upper leaf growth (Fig. 7). Although all the leaves are visible at bud burst, leaf growth begins earlier in the lower leaves than in the upper leaves (Taylor, 1970; Dambreville et al., 2015), with a clear cultivar effect (Dambreville et al., 2015). In contrast to the beginning of growth, the growth arrest is more synchronous between the stem and the leaves (Dambreville et al., 2015).

5 From the growth unit to the current-year branch

Terminal growth units can produce daughter growth units during the period of vegetative growth. These daughter growth units can themselves produce new growth units after a rest period (rhythmic growth) during the same period of vegetative growth, and so on. Consequently, branched systems are produced during a period of vegetative growth and contribute to canopy development. This section proposes a unified framework – the growing cycle – to describe the main phenological periods on a mango tree and to study their interactions. The environmental and endogenous factors affecting vegetative phenology are then reviewed. Finally, a simple method to estimate leaf area or dry mass of annual vegetative growth is presented.

5.1 The growing cycle

The review of studies describing vegetative and reproductive growth and their interactions (see Section 6) shows two problems which can make the text confusing for a non-specialist.

Nevertheless, this topic is important in understanding mango phenology and production, so we need to tackle these problems.

The first problem is that the main phenological periods (vegetative growth, rest, flowering and fruit growth) are generally considered on an annual or seasonal basis, or are separated into vegetative and reproductive phenology. However, results show close interactions between vegetative and reproductive growth, which are developed in the last section of this chapter. To better comprehend these interactions, we propose to use the concept of 'the growing cycle' (Dambreville et al., 2013a, 2015; Capelli et al., 2016).

The growing cycle is a coherent, temporal sequence of the main phenological periods from the beginning of vegetative growth to the harvest of fruits produced by this vegetative growth. It implicitly considers the effects of vegetative growth on flowering and fruiting within the growing cycle (see Section 6.2). This sequence comprises vegetative growth, rest, flowering and fruiting, from fruit set to fruit maturity (Fig. 9). The duration of the growing cycle extends beyond one year. For an early cultivar, such as Cogshall, which is grown on Réunion Island, it lasts about 18 months, from August of year n to February of year n + 2 (Fig. 9; Capelli et al., 2016). Consequently, two consecutive growing cycles overlap during the flowering and fruiting periods on a given tree, which is the end of a growing cycle and the beginning of the period of vegetative growth of the following growing cycle. This overlap represents the interaction between the two cycles, that is, the effects of reproduction on vegetative growth (Fig. 9).

The pattern of the growing cycle depends on environmental factors, which to a large extent drive the main phenological periods (see Section 5.2). The pattern illustrated here relates to mango trees in the subtropics, where there is a marked alternation of the cool dry season and the hot rainy season. It can be adapted to the case of a mild tropical climate where there are two dry and two rainy seasons per year (Ramírez et al., 2014). Each growing cycle then starts with the onset of vegetative growth at the beginning of each rainy season. Growing cycles then occur closer together and overlap for a longer period of time. In the warm tropics, where the period of vegetative growth is continuous and flowering is sporadic, depending on the phenology of each terminal growth unit (Chacko, 1991; Davenport, 2009), the growing cycle needs to be described for each growth unit, which is not convenient.

The second problem encountered is that different names are used to denote the growth unit: shoot, flush, growth flush, intercalary unit, terminal and stem terminal. Similarly, a vegetative growth event within the period of vegetative growth may be called a flush or a growth cycle. A mango specialist can probably picture what these terms represent, but this multiplicity is in general confusing. For example, the term 'flush' can be used in a same publication to indicate an event of vegetative growth as well as the result of vegetative growth, that is, the growth unit.

Since 'growth unit' is the actual botanical term for the entity that issues from the activity of a vegetative bud between two quiescent phases, we propose to use this term to standardize the vocabulary. We will also use the term 'current-year branch' to mean the branch complex that develops during a period of vegetative growth. It is composed of an axis issued from a single bud, in an apical or a lateral position, including one or more growth units and possibly bearing side axes (Normand et al., 2008, 2009a, 2012). A terminal growth unit can therefore produce none, one or several current-year branches during a growing cycle, depending on the number of buds that burst (Fig. 9).

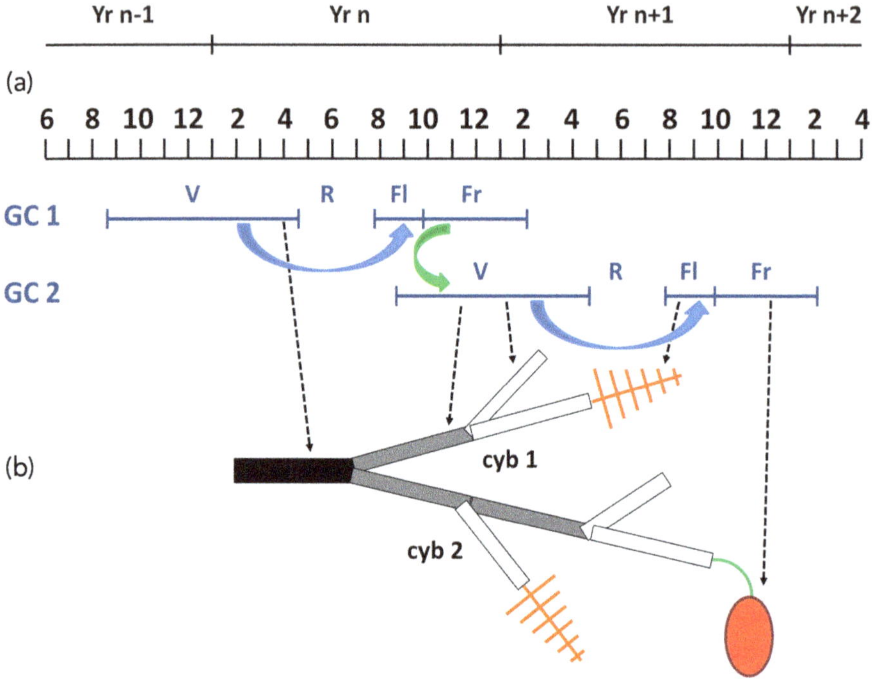

Figure 9 Diagrammatic representation of two growing cycles (GC 1 and GC 2) corresponding to the cultivar Cogshall on Réunion Island, Southern hemisphere, (a) and of the succession of growth units during two growing cycles (b). The phenological periods are vegetative growth (V), rest (R), flowering (Fl) and fruiting (Fr). The blue arrows represent the within-cycle interactions between vegetative growth and reproduction, and the green arrow represents the interaction between reproduction during a cycle and vegetative growth during the following cycle. The terminal growth unit of cycle 1 is black. It produces two current-year branches (cyb) during growing cycle 2. The first one (cyb 1) is made up of three growth units, one non-terminal (in grey) and two terminal (in white) growth units. The second one (cyb 2) is made up of five growth units, two non-terminal and three terminal growth units. During the reproductive season of growing cycle 2, three of the terminal growth units flowered (fish-bone-like symbols) and one of them set a fruit (red ellipse). Adapted from Capelli et al. (2016).

5.2 Environmental factors affecting vegetative phenology

Mango tree phenology and its relationship with the environment have been recognized as a cornerstone to improving cultivation practices because of the profound influence of the environment, in particular temperature and water availability (Whiley, 1986; Chacko, 1991), on vegetative growth and flowering (Cull, 1991; Crane et al., 2009). Mango vegetative growth is unlikely to be photoperiodic since vegetative growth can occur during seasons with contrasted day length (Holdsworth, 1963). More recent studies have shown that mango phenology is also determined to a large extent by endogenous factors (see Section 5.3). These determining endogenous factors must also be considered if cultivation practices are to be improved (Davenport, 2006; Ramírez et al., 2014).

Temperature is one of the most important environmental factors since it affects both vegetative and reproductive growth. This latter aspect, however, is beyond the scope of this chapter. The range of optimum temperature for vegetative growth varies from 19.5–28.8°C (Chacko, 1986) to 24–27°C (Whiley, 1986). Vegetative growth is stimulated by temperature. At 15/10°C, only four cultivars among ten (Alphonso, Glenn, Irwin and Haden) produced new growth units during a 20-week experiment, with a mean value of 0.48 growth units per plant (Whiley et al., 1989). At 30/25°C, the ten studied cultivars exhibited vegetative growth and produced on average 3.21 successive growth units per plant. This average value hides variability among the cultivars (e.g. 2.0 successive growth units for Irwin and 4.7 for Kensington Pride), clearly indicating a genotype × temperature interaction for vegetative growth. Whiley et al.'s (1989) results suggest that monoembryonic cultivars are able to produce vegetative growth at a lower temperature than polyembryonic cultivars. Across the ten studied cultivars, the mean daily temperature for zero growth was 15°C (Whiley et al., 1989).

The effects of water availability on mango vegetative growth have not been studied as much because vegetative growth generally occurs during the rainy season. Research on the water status of mango trees is mainly related to floral induction and fruit growth. Triggering bud burst by artificial watering showed contrasting results: bud burst occurs on aged, mature terminal growth units, but not on young growth units (Holdsworth, 1963). On the other hand, defoliation of terminal growth units induces bud burst and vegetative growth. Holdsworth (1963) suggested that the initial signal for bud burst is endogenous, and is likely to be related to the auxin efflux from the leaves. Bud burst and growth unit extension are then dependent on water availability. Low soil water content and high vapour pressure deficit during the beginning of the dry season induce a check in vegetative growth (Chacko, 1991). In subtropical areas and tropical highlands, cool temperatures at this period reinforce the effect of low water availability in preventing vegetative growth.

The mango tree is mainly cultivated in environments ranging from subtropical areas, which are characterized by cool dry winters and hot rainy summers, to tropical and equatorial areas which possibly have two dry and two rainy seasons per year. In these latter areas, the climate is generally warm throughout the year and cooler temperatures can be achieved at elevation. The close relationship between mango vegetative phenology and temperature and water availability means the phenology differs in subtropical and tropical areas (Chacko, 1986; Cull, 1991; Whiley, 1993; Ramírez et al., 2014).

In subtropical areas, vegetative phenology has a rather regular pattern from year to year, driven by temperature and water availability, provided by rain or irrigation (Cull, 1991; Robert and Wolstenholme, 1992, Normand et al., 2009a). Vegetative growth begins at the end of flowering, when the temperature is rising. Most vegetative growth occurs after the harvest, during the hot and rainy season or under irrigation. It stops at the end of summer or in autumn, at the beginning of the cool and dry season, usually due to falling temperatures and rain or irrigation cessation. This rest period is necessary for floral induction (Davenport, 2009).

In general, new growth units appear intermittently, and not continuously, at the scale of the tree and the orchard, when environmental conditions are favourable. Each period of vegetative growth is called a flush, which lasts between two and several weeks (e.g. Scholefield et al., 1986). The number of flushes and their position within the period of vegetative growth depend on the cultivar (Shaban, 2009) and the growing conditions, mainly temperature, water availability, intensity of flowering and fruit load. During each flush, only a proportion of terminal growth units within a canopy produce daughter growth units (Scholefield et al., 1986; Issarakraisila et al., 1997; Shaban, 2009; Ramírez et al., 2014).

For example, in Northern Australia near Darwin (lat. 12°S), in a warm tropical climate, four flushes are recorded (during flowering, at harvest, after harvest and before the following flowering) on the cultivar Kensington Pride during a cycle with a light crop (61 kg $tree^{-1}$). Four flushes are also recorded during the following cycle with a heavy crop (174 kg $tree^{-1}$), but the flushes are weaker and no flush occurs during harvest (Scholefield et al., 1986). In contrast, in south-west Australia near Perth (lat. 31°S), in a Mediterranean climate with continuous irrigation, the same cultivar shows a continuous period of growth unit development, from the end of flowering until the harvest, instead of intermittent flushes. This period corresponds to temperatures above 15°C (Issarakraisila et al., 1997), confirming in the field the results of Whiley et al. (1989).

In warm tropical areas without a marked dry season, vegetative growth occurs continuously on different growth units of the tree canopy, leading to large but unproductive trees (Chacko, 1986). A dry season may prevent bud burst, even in a warm climate, until the following rainfall. In a mild tropical area (22°C mean annual temperature) with two rainy and two dry seasons per year, vegetative growth is synchronous with the onset of each rainy season (Ramírez et al., 2014). The trees therefore have two main seasons for vegetative growth per year, but only a proportion of terminal growth units within a tree canopy produce daughter growth units in each season. If the rainy season is late because the first rain is a long time coming, trees without irrigation stay in rest.

5.3 Endogenous factors affecting vegetative phenology

Irregular flushing of the trees within an orchard, or vegetative growth occurring on only a proportion of growth units within a canopy, suggests that endogenous factors, at the scale of the tree and the growth unit, are crucial in determining vegetative phenology. The most important endogenous factor identified is related to reproduction at the scale of the terminal growth unit and of the tree (Scholefield et al., 1986; Issarakraisila et al., 1997; Shaban, 2009; Dambreville et al., 2013a). This point is presented in the next section, which is dedicated to the interactions between vegetative growth and reproduction. Here, we focus on the other endogenous factors affecting the occurrence and timing of vegetative growth.

The pattern of vegetative growth is affected by the age of the tree (Holdsworth, 1963; Chacko, 1986). Young seedlings or grafted unproductive trees tend to flush regularly and continuously, independently of the environmental conditions. On the contrary, old productive trees generally produce one to a few vegetative flushes per year. These flushes are partial because they concern only a proportion of the terminal growth units of the canopy. This phenomenon is related to the ontogenetic development of the tree (Barthélémy and Caraglio, 2007), which evolves from a vegetative state that establishes the tree structure to a reproductive state where vegetative growth is limited.

The position, apical or lateral, of the growth unit affects its probability of burst and the number of daughter growth units (Normand et al., 2009a; Dambreville et al., 2013a). In general, apical growth units have a higher probability of burst. However, this conclusion needs to be put into perspective. It is true for four cultivars (Cogshall, Irwin, José and Kensington Pride) when the position of the growth unit is the only factor used to explain the probability of burst (Normand et al., 2009a). But when other factors (the growth unit's burst date and the number of sister growth units) are used together with the position of the growth unit to explain the probability of burst, while the result remains true for Cogshall, the position of the growth unit has no significant effect for the three other cultivars (Dambreville et al., 2013a). This is probably related to the significant effect of the growth unit's burst date in the four cultivars (see the following section). Growth units in an apical position in general produce more daughter growth units than those in a lateral position (Normand et al., 2009a; Dambreville et al., 2013a).

A growth unit without an apical bud produces more daughter growth units than that with an apical bud (Normand et al., 2009a). This is related to the removal of apical dominance, exerted by the apical bud, allowing the outbreak of several lateral buds. As a consequence, growth units which have flowered produce more daughter growth units, when they burst, than vegetative growth units (Normand et al., 2009b; Capelli et al., 2016). Moreover, vegetative growth units produce one apical daughter growth unit and possibly lateral growth units, whereas growth units without an apical bud produce only lateral daughter growth units.

The probability of burst of a growth unit is also correlated with its own date of burst. The earlier a growth unit appears during the period of vegetative growth, the higher its probability to produce daughter growth units (Issarakraisila et al., 1997; Normand et al., 2009a; Dambreville et al., 2013a). This relationship is consistent with the fact that, in the subtropical conditions where the studies were carried out, the period of vegetative growth, defined by favourable environmental conditions, is limited and growth units that appear late do not have the time to produce daughter growth units before the rest period.

In relation to the previous point, an endogenous growth rhythm has been evidenced: within the season favourable for vegetative growth, the burst date of daughter growth units is positively correlated to the burst date of the mother growth unit (Dambreville et al., 2013a). The mean delay observed in the field is about two months (Anwar et al., 2011; Dambreville et al., 2013a). However, this delay depends on the temperature: the higher the temperature, the shorter the delay (Whiley et al., 1989).

Issarakraisila et al. (1997) defined the vigour of a current-year branch ('shoot vigour') as the number of vegetative flushes during which growth units making up the branch are produced. The vigour of the current-year branch positively influences the probability of burst and the number of the daughter growth units produced during the following period of vegetative growth. For example, terminal growth units of a current-year branch made up of one vegetative flush in the previous season have a 0.27 probability to burst, and the average number of daughter growth units for those which burst is 1.55. These values are 0.50 and 2.45, respectively, for terminal growth units of a current-year branch made up of three vegetative flushes.

The structural and temporal characteristics of vegetative growth mentioned earlier are also affected by the cultivar. The pattern of vegetative growth, in terms of the number and date of vegetative flushes and the number of growth units produced, varies among cultivars (Chacko, 1986; Shaban, 2009; Nafees et al., 2010). This cultivar effect is related to at least two aspects.

The first aspect is the variability in architecture traits and development among cultivars. For example, some cultivars produce fewer daughter growth units than others in similar conditions (Normand et al., 2009a,b). In particular, terminal growth units on young trees of the Irwin and Tommy Atkins cultivars produce on average 1.18 and 1.43 daughter growth units, respectively, compared to Kensington Pride (2.37) and Nam Doc Maï (2.94) (Normand et al., 2009b). Similarly, the probability of vegetative burst may vary among cultivars (Shaban, 2009; Nafees et al., 2010), or the effect of other factors, such as the position of the growth unit on the probability of burst, may vary among cultivars (Normand et al., 2009a; Dambreville et al., 2013a; Capelli et al., 2016).

The second aspect is the interaction of the cultivar and the environment, in particular temperature. The effect of temperature on vegetative growth differs among cultivars (Whiley et al., 1989). As a consequence, different cultivars may show different vegetative growth patterns in a given environment, according to the genotype × environment interaction. The same cultivars in a different (e.g. warmer) environment are likely to behave differently (see e.g. Scholefield et al. (1986) and Issarakraisila et al. (1997) for the differences in the vegetative growth pattern of Kensington Pride in contrasted environments).

5.4 Assessment of current-year branch leaf area and dry mass

The characterization of vegetative growth at different scales within the tree is important to study the interactions between vegetative growth and reproduction, or to evaluate the effects of treatments or cultivation practices on vegetative growth. The current-year branches are generally described by the number and the date of appearance of the constitutive growth units (Scholefield et al., 1986; Whiley et al., 1989; Issarakraisila et al., 1991, 1997; Shaban, 2009; Nafees et al., 2010; Anwar et al., 2011, Dambreville et al., 2013a). Leaf area is also sometimes recorded by measuring the leaf area of individual leaves (Whiley et al., 1989; Shaban, 2009). This is relevant for a sample of current-year branches, but not relevant at a larger scale, such as a scaffold branch or a whole tree. However, the non-destructive quantification of leaf area or dry mass of vegetative growth at different scales is pertinent because of the ecophysiological meaning of these variables.

Normand and Lauri (2012) proposed predictive allometric models to non-destructively estimate the leaf area, stem dry mass and branch (leaves + stem) dry mass of mango current-year branches from a simple measurement of the basal diameter of the branch stem. The models assume a power relationship between each of these variables – regarded as the dependent variable – and the basal cross-sectional area of the branch, calculated from the branch basal diameter, assuming a circular section – regarded as the independent variable. The power relationships are linearized by log-transformation of both dependent and independent variables:

$$\log(X) = \alpha \log(Y) + \beta \tag{1}$$

where X is the dependent variable (the leaf area, stem dry mass or branch dry mass), Y is the basal cross-sectional area of the branch, and α and β are parameters, respectively, the slope and the y-intercept of the linearized relationship.

Linearization is a convenient way to estimate the two parameters α and β and to analyse the model properties or to test for a cultivar or a year effect on model parameters. The variable of interest is then calculated by back-transformation:

$$X = e^{(\alpha \log(Y) + \beta')} \tag{2}$$

where β' is the corrected y-intercept (see Normand and Lauri (2012) for details).

Models have been created with set parameters calculated for the leaf area, stem dry mass and branch dry mass for four cultivars, and validated with independent data from different years and sites. The 12 adjusted linearized relationships are in general very satisfactory ($r^2 > 0.88$) and validations are good. For each variable of interest, a cultivar

Table 2 Slope (α) and corrected y-intercept (β') of the cultivar-specific and the common models to estimate, according to equation 2, leaf area, stem dry mass or branch dry mass of a current-year branch from its basal cross-sectional area (Y in mm²). From Normand and Lauri (2012)

Variable of interest	Cogshall		Irwin		José		Kensington Pride		Common model	
	α	β'	α	β'	α	β'	α	β'	α	β'
Leaf area (dm²)	0.09	−2.15	1.23	−3.01	1.12	−2.41	1.11	−2.46	1.13	−2.47
Stem dry mass (g)	1.54	−4.78	1.64	−5.50	1.76	−5.36	1.67	−5.38	1.66	−5.27
Branch dry mass (g)	1.25	−2.06	1.40	−3.02	1.37	−2.56	1.34	−2.72	1.33	−2.51

effect was detected on the y-intercept of the linearized relationship, and not on the slope. The cultivar-specific models can be used for more accurate predictions than the common models (Table 2).

For a different cultivar or in a different environment, it is advisable to estimate specific models according to the procedures for data acquisition and parameter calculation described in Normand and Lauri (2012).

The basal diameter of each current-year branch is measured with a calliper during the vegetative rest period, once the last developed growth units are fully expanded and mature (Fig. 10). This measurement is easy and rapid to make. The variable of interest, for example leaf area, of each current-year branch is then estimated using the corresponding model. These data can then be aggregated at different scales where the leaf area is of interest (e.g. at the terminal growth unit, the scaffold branch or the whole tree scale; see, e.g. Capelli et al., 2016; Normand et al., 2016).

To go further, the allometric relationships between the size (leaf area or dry mass) of a branch and its basal cross-sectional area refer to the distribution of dry mass between leaves and stem within a current-year branch. This distribution is shaped by the two main functions of a branch: the conduction of sap to the leaves and the support for the leaves and the branch itself (Preston and Ackerly, 2003; Taneda and Tateno, 2004). For the mango tree, wood density and hydraulic conductivity of current-year branches, two characteristics related to conduction and support, differ significantly among cultivars (Normand et al., 2008). This variability induces a different allocation of dry mass within the branch, and between the branch components, leaves and stem, for a given basal cross-sectional area (Fig. 11). The significant differences among cultivars in the y-intercept of the linearized relationships are related to these functional differences.

6 Interactions between vegetative growth and reproduction

Several studies describe vegetative and reproductive growth and their interactions (Scholefield et al., 1986; Issarakraisila et al., 1991, 1997; Muhammad et al., 1999; Shaban, 2009; Nafees et al., 2010; Anwar et al., 2011; Dambreville et al., 2013a; Ramírez et al.,

Figure 10 Terminal growth unit that flowered and produced in lateral position in two current-year branches of contrasted size. In the close up, triangles indicate where the basal diameter of each current-year branch must be measured in order to estimate the leaf area and dry mass with allometric relationships.

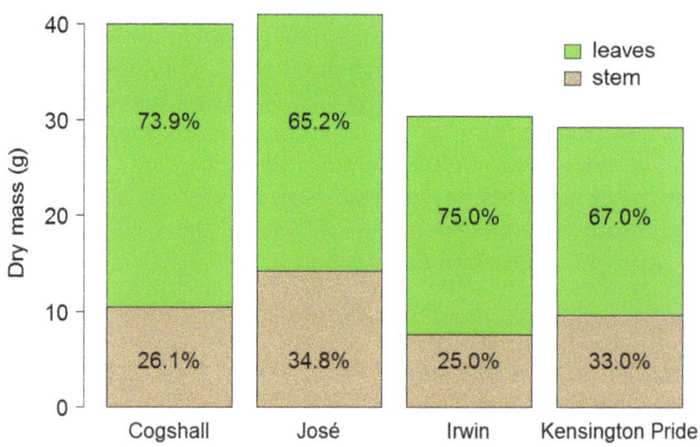

Figure 11 Total dry mass and its distribution among leaves and stem in a current-year branch with a 100 mm^2 basal cross-sectional area (about 11.3 mm in diameter) for four mango cultivars. Percentages of leaves and stem dry mass are indicated for each cultivar. Data calculated from Normand et al. (2008).

2014; Capelli et al., 2016; Normand et al., 2016). They show that vegetative growth and reproductive growth are closely related, from the scale of the growth unit to that of the whole tree. These interactions must be considered in two ways: the effects of reproduction on vegetative growth and the effects of vegetative growth on flowering and fruiting. These topics are discussed in the following sections.

6.1 Effects of reproduction on vegetative growth

In general, reproduction has a negative effect on the subsequent vegetative growth, at the tree, scaffold branch and terminal growth unit scales. This is called the cost of reproduction on the subsequent vegetative growth (Obeso, 2002; Capelli et al., 2016; Normand et al., 2016). At the tree scale, heavy fruit load reduces vegetative growth (Scholefield et al., 1986; Issarakraisila et al., 1997; Shaban, 2009; Capelli et al., 2016; Normand et al., 2016), and this effect is cultivar-dependent (Normand et al., 2016). In a study by Normand et al. (2016), the leaf area produced during a growing cycle was, on average, 44% larger on thinned trees (fruitlets of the previous cycle were removed at fruit set) than on unthinned trees of the cultivar Cogshall. The difference was 180% for Kensington Pride. More precisely, a negative linear relationship was evidenced between the leaf area produced during a growing cycle and the fruit load during the previous cycle, at the tree and at the scaffold branch scales. The slope was more negative for Kensington Pride than for Cogshall, indicating, as above, the higher susceptibility of Kensington Pride to the negative effect of fruit load on vegetative growth (Normand et al., 2016).

Moreover, heavy fruit load delays vegetative growth when it occurs (Issarakraisila et al., 1997; Shaban, 2009; Dambreville et al., 2013a). Vegetative flushes are rare before harvest on trees with a heavy crop, and they mainly occur after harvest (Scholefield et al., 1986; Shaban, 2009). In areas with cool temperature or for late cultivars, this could be a trigger for alternate bearing. If the temperature drops soon after harvest, vegetative growth either does not occur or occurs late, limiting flowering during the next cycle (Issarakraisila et al., 1997).

These negative effects at the tree scale are observed also at the growth unit scale. One of the most important factors affecting vegetative growth at this scale is the fate of the terminal growth unit during the previous cycle, that is, quiescent, flowering or fruiting. The fate of the terminal growth unit represents its investment in reproduction during the period of flowering and fruiting. This investment is null for a quiescent growth unit which does not flower, moderate for a flowering growth unit which flowers but does not set fruit, and high for a fruiting growth unit which flowers and produces one or several fruits. The fate impacts quantitatively and temporally on the beginning of vegetative growth.

The reproductive growth units, and in particular the fruiting ones, have a lower probability of vegetative bud burst than growth units that did not flower (Issarakraisila et al., 1991; Dambreville et al., 2013a; Capelli et al., 2016). For example, Issarakraisila et al. (1991) showed that the percentages of quiescent, flowering and fruiting growth units of Kensington Pride exhibiting vegetative growth during a cycle were 49.1%, 36.0% and 3.7%, respectively.

This lower probability of bearing further vegetative growth on reproductive growth units results from two processes. First, vegetative growth is prevented in flowering and fruiting growth units as long as inflorescence and fruits are present (Issarakraisila et al., 1991, 1997; Muhammad et al., 1999; Shaban, 2009; Ramírez et al., 2014). Second, when vegetative growth occurs after harvest, it is delayed on growth units that bore fruits (Muhammad et al., 1999; Ramírez et al., 2014). As a consequence, new growth units appearing at the end of flowering and during fruit growth are produced by quiescent growth units or by flowering growth units which did not set fruit. Recent research (Capelli, 2017) suggests that the physiological mechanisms involved are (1) bud burst inhibition by auxin efflux from the inflorescence or the growing fruit, and (2) no, or delayed, recovery of depleted starch content in fruiting growth units after harvest, leading to a delay in bud burst.

The consequence is that, at the end of the vegetative growth period, a reproductive growth unit produces significantly less leaf area than a quiescent growth unit (Capelli et al., 2016; Normand et al., 2016). This result depends on the cultivar and, interestingly, the more regular bearing cultivar Irwin does not show this negative consequence. Irwin even shows a larger leaf area produced by reproductive growth units, contrary to the more irregular bearing cultivars such as José (Capelli et al., 2016). Several compensatory mechanisms (Obeso, 2002, Capelli et al., 2016) can reduce the negative effects of reproduction on vegetative growth. First, reproductive growth units are not subjected to apical dominance because of the loss of the apical bud, and more lateral buds can burst in a reproductive growth unit than in a quiescent growth unit. Second, photosynthesis is enhanced close to growing fruits (Urban et al., 2004), leading to a slower local depletion of carbohydrates. Third, a proportion of quiescent growth units, which did not flower, does not burst either, and may die as the canopy ages (Issarakraisila et al., 1997).

The mango tree is characterized by within- and between-tree phenological asynchronisms. They are defined by the presence of a particular phenological stage (e.g. burst of a growth unit) at different periods and/or on different growth units within a tree or between trees within an orchard. The reproductive asynchronisms generate real agronomic drawbacks. They maintain stages that are susceptible to pests and diseases (e.g. flowering) over a long period in the orchard, and they lead to a spread harvest, with logistics issues and fruit quality and maturity problems.

Within-tree asynchronisms are mainly determined by the fate of the terminal growth units during the period of reproduction since vegetative growth is delayed on reproductive growth units (Dambreville et al., 2013a). This delay is then maintained by endogenous growth rhythms, as presented in the previous section. At the tree scale, fruit load delays vegetative growth and is a trigger of between-tree asynchronisms.

6.2 Effects of vegetative growth on flowering and fruiting

The effects of vegetative growth on flowering and fruiting have been studied mainly at the growth unit scale. The temporal and structural characteristics of the growth unit and the current-year branch affect the probability of flowering and fruiting. They are reviewed in this section.

The temporal characteristic is related to the requirement of a rest period for the growth unit to favour flowering (Chacko, 1986; Davenport, 2009; Ramírez et al., 2010, 2014). Growth units which appear early in the period of vegetative growth, or even during the previous season, and stay quiescent up to the reproductive period have a higher probability of flowering (Scholefield et al., 1986). Different minimum durations for the rest period have been reported, probably because of differences in the cultivars' requirements and environmental conditions. Chacko (1986) indicated a minimum duration of two months. In a mild tropical climate, the minimum age to flower is 5–5.5 months for Keitt and six months for Tommy Atkins (Ramírez et al., 2010, 2014).

Dambreville et al. (2013a) showed an average delay of approximately seven months between the date of burst of a growth unit and its date of flowering in four cultivars (Cogshall, Irwin, José and Kensington Pride). As a consequence, the asynchronisms in vegetative growth to a large extent explain the asynchronisms in flowering. This result also supports the idea that, regardless of the underlying mechanisms, there is a negative relationship between the age of the growth unit and the requirement for cool temperatures to induce flowering (Davenport, 2009).

The structural characteristics are the position of the growth unit and the vigour of the current-year shoot. The apical or lateral position of the growth unit affects its probability of flowering and fruiting (Normand, 2009a). Apical growth units tend to flower and fruit more than lateral growth units. For the cultivar Cogshall, the effect of the growth unit position is in fact related to the diameter of the growth unit. The relationship between growth unit diameter and the probability of flowering is quadratic. Its pattern differs between apical and lateral growth units in terms of the maximum probability of flowering (0.89 vs 0.75, respectively), the growth unit diameter corresponding to this maximum (6.8 mm vs 4.5 mm, respectively) and the curvature demonstrated in the quadratic relationship. On the other hand, apical and lateral growth units follow the same linear relationship between growth unit diameter and the probability of fruiting. Apical growth units have a larger diameter than lateral growth units, leading to a higher average probability of fruiting for the former. These contrasted relationships clearly indicate that different mechanisms underlie flowering and fruiting.

The position of the growth unit also affects the number of inflorescences produced and the date of flowering (Dambreville et al., 2013a). Among the four cultivars Cogshall, Irwin, José and Kensington Pride, apical growth units produced more inflorescences than lateral growth units, just as they also produce a larger number of daughter growth units. The second effect is cultivar-specific. Apical growth units tend to flower earlier than lateral growth units for Irwin and José, but later for Cogshall and Kensington Pride.

Similar to its effect on vegetative growth (see Section 5.3), the vigour of the current-year branch positively influences the probability of flowering and fruiting of the terminal growth units as well as the number of fruits produced (Issarakraisila et al., 1997). This vigour effect is even more important if vegetative growth occurs during the same growing cycle, that is, a few months before flowering. Indeed, the same effect was observed to a lesser extent on current-year branches that appeared during the previous growing cycle, that is, more than one year before flowering. Dambreville et al. (2013a) confirmed the positive effect of current-year branch vigour on flowering in three cultivars, including Kensington Pride from the study by Issarakraisila et al. (1997). However, no effect was detected for the cultivar José, indicating that this effect is cultivar-dependent.

Independent of the specific effects of these structural and temporal traits on flowering and fruiting, Capelli et al. (2016) showed a clear relationship between vegetative growth and fruit production during the same growing cycle. The number of fruits produced is positively and linearly related to vegetative growth, expressed in leaf area or in the number of terminal growth units, established before flowering. This relationship holds for four cultivars and two growing cycles at the scale of the growth unit. The slope of the relationship differs among cultivars. Irwin shows a higher slope for the two growing cycles, indicating that a given increase in vegetative growth, for example, leaf area, leads to a larger increase in fruit production than for the other cultivars. The slope of the relationship is therefore a mark of the efficiency of vegetative growth to produce fruits.

At the scale of the scaffold branch, the relationship is significant only for some cultivars and/or growing cycles (Capelli et al., 2016). Oosthuyse and Jacobs (1995) also found evidence of a linear relationship between the number of terminal growth units and the number of fruits produced at the scaffold branch scale for the Kent and Sensation cultivars. The quality of the relationship differs, however, between the two cultivars, confirming a cultivar-dependent effect, as is the case for so many of the other relationships presented in this chapter.

6.3 Complexity of a mango tree canopy

During vegetative rest, a mango tree canopy seems to be a homogeneous set of mature terminal growth units ready for flowering. In fact, this is not true. Each terminal growth unit is characterized by its age, diameter, leaf area, and apical or lateral position, and the vigour (in the sense of Issarakraisila et al., 1997, see Section 5.3) of the current-year branch supporting it. Even in the case of an important and synchronous vegetative flush, most of these traits differ among terminal growth units. The mango tree canopy must therefore be considered as a collection of terminal growth units, each with specific traits conferring a specific ability to flower, to set fruit, to grow vegetatively or to remain quiescent during the reproductive period.

This diversity originates during the vegetative growth period. The occurrence and timing of vegetative growth at the scale of the terminal growth unit depend mainly on the reproductive fate of the growth unit during the previous growing cycle. Consequently, flowering and fruiting during one cycle affect flowering and fruiting during the following cycle through their effects on vegetative growth. These relationships are generally negative (Capelli et al., 2016), indicating that these natural structural and temporal interactions within a mango tree canopy may play a role in irregular bearing.

However, this picture at the growth unit scale is more complex for several reasons. First, terminal growth units are not independent, and the relationships described above are also affected by other factors at the branch, scaffold branch or whole-tree scale. These hierarchical entities have their own structural characteristics (e.g. wood volume, leaf area or root system size), and each year they produce a different quantity of fruit. The level of fruit production affects their physiology, in particular their carbohydrate content, and therefore their ability to grow and to flower. The effects of factors at the tree scale on vegetative growth demonstrate this higher level of complexity. However, despite these fundamental rules, a number of the studies presented in this chapter were only carried out at the scale of the growth unit on several trees. So, although these studies demonstrate the relevance of the factors and phenomena presented here, more generally, it would be useful for future studies to consider different hierarchical scales in their experimental designs in order to better decipher the observed phenomena.

The second reason is that these relationships are cultivar-dependent, as shown in this chapter. Consequently, the effect of a particular factor on vegetative growth may vary from being significant to non-significant, or from being positive to negative, among different cultivars in a given environment (see e.g. Dambreville et al., 2013a or Capelli et al., 2016).

The third reason, the environment, is closely related to the second since the observed relationships result from the interactions between the genetic background of the cultivar and the environment (see e.g. Scholefield et al., 1986 and Issarakraisila et al., 1997). As discussed in this chapter, temperature and rainfall regimes have a significant impact on mango tree phenology.

The fourth reason is that most of the cited references deal with the natural vegetative or reproductive development of the growth units, without disturbance by cultivation practices during the studies. However, several practices, such as pruning, trellising, fertilizing and irrigation, may affect the vegetative and reproductive growth patterns (e.g. Davenport, 2006; Ramírez et al., 2010). These practices are important levers to improving mango production. It is essential to comprehend their effects on vegetative and reproductive growth, in order to get the best out of the tree and to improve yield in harmony, rather than in conflict, with its growing cycle (e.g. Cull, 1991).

7 Conclusion

This chapter presents the architectural model of the mango tree, the major structural unit of the mango tree, that is, the growth unit, the transition from the growth unit to the current-year branch and the interactions between vegetative and reproductive growths. It also shows that different environmental and endogenous factors quantitatively and temporally affect vegetative growth at different scales, from individual leaf size to the whole-tree leaf area established during a growing cycle. Finally, the chapter demonstrates that flowering and fruiting depend on previous vegetative growth at different scales, which relies on the reproductive behaviour during the previous growing cycle. This leads to complex interactions across cycles.

These data show that the design of cultivation practices to achieve more sustainable mango production must not only focus on flowering and fruiting, but must also consider vegetative growth and its relationships with reproduction. However, knowledge is lacking on these aspects. For example, the effects of cultivation practices are mostly considered on fruit production and rarely on vegetative growth. This opens up a new field for research into improving canopy management strategies with the objective of increasing orchard productivity (Menzel and Le Lagadec, 2017). This chapter also shows that while several phenomena and the effects of some factors are well described, the underlying mechanisms are mostly unknown. This represents a gap in understanding tree behaviour and in improving cultivation practices. Further efforts are needed to close this gap.

Modelling is an interesting tool for synthesizing the current state of our knowledge and can be used to assist our understanding of tree behaviour in the quest to improve yield and fruit quality. As a first step, modelling can be based on the known relationships within the tree and on the effects of identified environmental and endogenous factors (Boudon et al., 2017). Mechanistic processes can then be added as soon as new results are obtained. A model simulating mango yield and fruit quality would also be a useful tool in assessing the future effects of climate change on mango production (Normand et al., 2015).

8 Where to look for further information

The better understanding of mango tree functioning and the improvement of mango production with appropriate cultivation practices require future basic and applied research on mango vegetative growth. Five specific points can be considered.

First, up-scale the observations from the growth unit to the whole tree. Most of the temporal and structural relationships presented in the chapter have been evidenced at the growth unit scale. However, the tree is not a collection of independent growth units and it would be interesting to investigate the relationships at larger scales, that is, the scaffold branch and the whole tree, to apprehend the regulatory mechanisms at these scales and integrate this knowledge to design cultivation practices.

Second, understand the mechanisms underlying the relationships described in the chapter, in particular the interactions between vegetative growth and reproduction which are involved in yield and irregular bearing. Most of the relationships show differences among cultivars and it is important to decipher the cultivar effects in these mechanisms.

Third, improve mango yield through an appropriate control of vegetative growth, on the basis of the previous results. The design of innovative practices for canopy and orchard management must take into account the genetic and environmental effects on vegetative growth and reproduction. It seems utopian to design a unique practice for all cultivars and environmental conditions.

Fourth, model mango yield and fruit quality with an explicit consideration of vegetative growth and development. The development of such a model can fulfil several objectives. First, it synthesizes the knowledge on the complex interactions between vegetative growth and reproduction, and the effects of environment. Second, it can be used to explore in silico the response of the mango tree to different climatic and cultivation scenarios, and contribute to the design of cultivation practices (point 3). The knowledge of mechanisms underlying the observed relationships (points 1 and 2) can help in building more mechanistic models.

Fifth, integrate records on vegetative growth in experiments. An objective of the chapter is to make practitioners and researchers fully aware of the importance of vegetative growth in mango fruit production. It is therefore important to consider and characterize vegetative growth in experiments, which are generally dedicated to mango yield, in order to better explain the results and the effects of the tested factors. These data could also contribute to the progress of knowledge on vegetative growth

To our knowledge, few organizations are involved in the specific research on mango tree growth and canopy development. The French Agricultural Research Centre for International Development (CIRAD; http://www.cirad.fr/en/) carries out, on Réunion Island, a research programme on the mango tree including the study of the interactions between vegetative growth and reproduction, as well as the effects of plant material (cultivar, rootstock) and cultivation practices (pruning, irrigation) on vegetative growth, yield and fruit quality. A model of mango yield and fruit quality is developed in collaboration with the CIRAD-French Institute for Research in Computer Science and Automation (INRIA) team Virtual Plants in Montpellier, France.

The Queensland Department of Agriculture, Fisheries and Forestry, Australia, (QDAFF; https://www.daf.qld.gov.au/) has developed, in Mareeba, a research programme on understanding mango tree canopy architecture, physiology and productivity. It includes the study of the effects of plant material (scion genotype, rootstock) and cultivation practices (planting density, training system, pruning) on vegetative growth and yield. A model of mango yield is developed in collaboration with the University of Queensland at Brisbane.

More information on mango tree growth and canopy development can be found in scientific journals. The mango working group of the International Society for Horticultural Science (ISHS; http://www.ishs.org/) organizes an International Mango Symposium every 2–4 years. Manuscripts corresponding to the oral presentations and the posters are published after reviewing in a volume of *Acta Horticulturae* (http://www.actahort.org/). These volumes are mines of information about all aspects of mango cultivation, and in particular on vegetative growth and canopy development.

9 References

Anwar, R., Ahmad, S., Rajwana, I. A., Khan, A. S., Memon, N. and Nafees, M., (2011), Phenological growth patterns and floral malformation of mango (*Mangifera indica* L.) tree under subtropical climate. *Pak. J. Agric. Sci.*, 48, 109–15.

Arnold, C. Y., (1959), The determination and significance of the base temperature in a linear heat unit system. *Proc. Am. Soc. Hortic. Sci.*, 74, 430–45.

Aubert, B. and Lossois, P., (1972), Considérations sur la phénologie des espèces fruitières arbustives. *Fruits*, 27, 269–86.

Barthélémy, D. and Caraglio, Y., (2007), Plant architecture: A dynamic, multilevel and comprehensive approach to plant form, structure and ontogeny. *Ann. Bot.*, 99, 375–407.

Bleiholder, H., van den Boom, T., Langelüddeke, P. and Stauss, R., (1989), Einheitlich Codierung der phänologischen Stadien bei Kultur- und Schadpflanzen. *Gesunde Pflanzen*, 41, 381–4.

Bonhomme, R., (2000), Bases and limits to using 'degree. day' units. *Eur. J. Agron.*, 13, 1–10.

Boudon, F., Jestin, A., Fernique, P., Lauri, P.-É., Dambreville, A., Guédon, Y. and Normand, F., (2017), The role of structural and temporal factors in the architectural development of the mango tree: Evidences from simulation. *Acta Hortic.*, 1160, 83–90.

Brisson, N. and Delécolle, R., (1991), Développement et modèles de simulation de culture. *Agronomie*, 12, 253–63.

Capelli, M., (2017), Décrypter l'irrégularité de production des fruitiers tropicaux via l'analyse des coûts de la reproduction. Le cas du manguier (*Mangifera indica* L.). PhD thesis, Université de Montpellier, France.

Capelli, M., Lauri, P.-É. and Normand, F., (2016), Deciphering the costs of reproduction in mango – vegetative growth matters. *Front. Plant Sci.*, 7:1531, doi:10.3389/fpls.2016.01531

Cave, R. L., Hammer, G. L., McLean, G., Birch, C. J., Erwin, J. E. and Johnston, M. E., (2013), Modelling temperature, photoperiod and vernalization responses of *Brunonia australis* (Goodeniaceae) and *Calandrinia* sp. (Portulacaceae) to predict flowering time. *Ann. Bot.*, 111, 629–39.

Chacko, E. K., (1986), Physiology of vegetative and reproductive growth in mango (*Mangifera indica* L.) trees. In: *Proceedings of First Australian Mango Research Workshop*, CSIRO, Melbourne, Australia, pp. 54–70.

Chacko, E. K., (1991), Mango flowering – Still an enigma!. *Acta Hortic.*, 291, 12–21.

Charbonnier, F., Roupsard, O., le Maire, G., Guillemot, J., Casanoves, F., Lacointe, A., Vaast, P., Allinne, C., Audebert, L., Cambou, A., Clément-Vidal, A., Defrenet, E., Duursma, R. A., Jarri, L., Jourdan, C., Khac, E., Leandro, P., Medlyn, B. E., Saint-André, L., Thaler, P., Van den Meersche, K., Barquero Aguilar, A., Lehner, P. and Dreyer, E., (2017), Increased light-use efficiency sustains net primary productivity of shaded coffee plants in agroforestry system. *Plant Cell Envir.*, 40(8), 1592–608.

Crane, J. H., Salazar-García, S., Lin, T. S., de Queiroz Pinto, A. C. and Shü, Z.-H., (2009), Crop production: Management. In: *The Mango, Botany, Production and Uses*, 2nd Edition, R. E. Litz (Ed.), CAB International, Wallingford, UK, pp. 432–83.

Cull, B. W., (1991), Mango crop management. *Acta Hortic.*, 291, 154–73.

Dambreville, A., Lauri, P.-É., Trottier, C., Guédon, Y. and Normand, F., (2013a), Deciphering structural and temporal interplays during the architectural development of mango trees. *J. Exp. Bot.*, 64, 2467–80.

Dambreville, A., Normand, F. and Lauri, P.-É., (2013b), Plant growth co-ordination *in natura*: A unique temperature controlled law among vegetative and reproductive organs in mango. *Funct. Plant Biol.*, 40, 280–91.

Dambreville, A., Lauri, P.-É., Normand, F. and Guédon, Y., (2015), Analysing growth and development of plants jointly using developmental growth stages. *Ann. Bot.*, 115, 93–105.
Davenport, T. L., (2006), Pruning strategies to maximize tropical mango production from the time of planting to restoration of old orchards. *HortScience*, 41, 544–8.
Davenport, T. L., (2009), Reproductive physiology. In: *The mango, Botany, Production and Uses*, 2nd Edition, R. E. Litz (Ed.), CAB International, Wallingford, UK, pp. 97–169.
de Parcevaux, S. and Hubert, L., (2007), *Bioclimatologie: Concepts et applications. Collection Synthèses (INRA)*, Editions Quae, Versailles, France.
Dufault, R. J., (1997), Determining heat unit requirements for broccoli harvest in coastal South Carolina. *J. Am. Soc. Hortic. Sci.*, 122(2), 169–74.
Goguey, T., (1995), Approche architecturale des mécanismes de la croissance aérienne et de la floraison du manguier. PhD thesis, Université Montpellier II, France.
Goguey, T., (1997), Architectural approach of the mechanisms of canopy growth and flowering of mango trees. *Acta Hortic.*, 455, 124–31.
Goudriaan, J. and Van Laar, H. H., (1994), *Modelling potential crop growth processes*. Springer, Dordrecht, the Netherlands.
Guinochet, M., (1965), *Notions fondamentales de botanique générale*. Masson, Paris, France.
Hallé, F. and Martin, R., (1968), Étude de la croissance rythmique chez l'hévéa (*Hevea brasiliensis* Müll.-Arg., Euphorbiacées, Crotonoïdées). *Adansonia*, 8, 475–503.
Hallé, F. and Oldeman, R. A., (1970), *Essai sur l'architecture et la dynamique de croissance des arbres tropicaux*. Masson, Paris, France.
Hallé, F., Oldeman, R. A. and Tomlinson, P. B., (1978), *Tropical trees and forests. An architectural analysis*. Springer-Verlag, Berlin, Germany.
Hernández Delgado, P. M., Aranguren, M., Reig, C., Fernández Galván, D., Mesejo, C., Martínez Fuentes, A., Galán Saúco, V. and Agusti, M., (2011), Phenological growth stages of mango (*Mangifera indica* L.) according to the BBCH scale. *Sci. Hortic.*, 130, 536–40.
Holdsworth, M., (1963), Intermittent growth of the mango tree. *J. West Afr. Sci. Ass.*, 7(2), 163–71.
Issarakraisila, M., Considine, J. A. and Turner, D. W., (1991), Pattern of vegetative and reproductive growth of mango trees in a warm temperate region of Western Australia. *Acta Hortic.*, 291, 188–97.
Issarakraisila, M., Considine, J. A. and Turner, D. W., (1997), Vegetative and reproductive growth aspects of mango growing in a Mediterranean climate in Western Australia. *Acta Hortic.*, 455, 56–63.
Lauri, P.-É. and Normand, F., (2017), Are leaves only involved in flowering? Bridging the gap between structural botany and functional morphology. *Tree Physiol.*, 37(9), 1137–9.
Lauri, P.-É., Kelner, J. J., Trottier, C. and Costes, E., (2010), Insights into secondary growth in perennial plants: Its unequal spatial and temporal dynamics in the apple (*Malus domestica*) is driven by architectural position and fruit load. *Ann. Bot.*, 105, 607–16.
Malézieux, E., Zhang, J., Sinclair, E. R. and Bartholomew, D. P., (1994), Predicting pineapple harvest date in different environments, using a computer simulation model. *Agron. J.*, 86, 609–17.
Meier, U., Bleiholder H., Buhr, L., Feller, C., Hack, H., Heß, M., Lancashire, P. D., Schnock, U., Stauß, R., van den Boom, T., Weber, E. and Zwerger, P., (2009), The BBCH system to coding the phenological growth stages of plants: History and publications. *Nachrichtenblatt des Deutschen Pflanzenschutzdienstes*, 61, 41–52.
Menzel, C. M. and Le Lagadec, M. D., (2017), Can the productivity of mango orchards be increased by using high-density plantings? *Sci. Hortic.*, 219, 222–63.
Mosqueda-Vázquez, R. and Ireta-Ojeda, A., (1993), Degree-days and base temperature required for inflorescence and fruit development in mango 'Manila'. *Acta Hortic.*, 341, 232–9.
Muhammad, F., Ibrahim, M. and Pervez, M. A., (1999), Vegetative and reproductive growth pattern of mango (*Mangifera indica* L.). *Int. J. Agric. Biol.*, 1(3), 97–9.
Nafees, M., Anwar, R., Jameel, M., Aslam, M. N., Ahmad, S., Akhtar, F. Z. and Memon, N., (2010), Flushing pattern of mango (*Mangifera indica* L.) cultivars in response to pruning of panicles and its effect on carry over effect of floral malformation. *Pak. J. Agric. Sci.*, 47, 13–18.
Normand, F. and Léchaudel, M., (2006), Toward a better interpretation and use of thermal time models. *Acta Hortic.*, 707, 159–65.

Normand, F., Bissery, C., Damour, G. and Lauri, P.-É., (2008), Hydraulic and mechanical stem properties affect leaf-stem allometry in mango cultivars. *New Phytol.*, 178(3), 590–602.
Normand, F., Pambo Bello, A. K., Trottier, C. and Lauri, P-É., (2009a), Is axis position within tree architecture a determinant of axis morphology, branching, flowering and fruiting? An essay in mango. *Ann. Bot.*, 103, 1325–36.
Normand, F., Magne, C. and Lauri, P.-É., (2009b), Effect of flowering and fruiting on branching of young mango trees: An architectural approach. *Acta Hortic.*, 820, 221–7.
Normand, F. and Lauri, P.-É., (2012), Assessing models to predict vegetative growth of mango at the current-year branch scale. *Am. J. Bot.*, 99(3), 425–37.
Normand, F., Lauri, P.-É. and Legave, J-M., (2015), Climate change and its probable impact on mango production and cultivation. *Acta Hortic.*, 1075, 21–32.
Normand, F., Capelli, M. and Lauri, P-É., (2016), The costs of reproduction in plants: A novel approach to study irregular bearing of fruit crops. *Acta Hortic.*, 1130, 207–13.
Obeso, R., (2002), The costs of reproduction in plants. *New Phytol.*, 155, 321–48.
Oosthuyse, S. A., (1991), Stages of development of the mango panicle. *S. Afr. Mango Growers Assoc. Yearb.*, 11, 59–61.
Oosthuyse, S. A. and Jacobs, G., (1995), Relationship between branching frequency, and growth, cropping and structural strength of 2-year-old mango trees. *Sci. Hortic.*, 64, 85–93.
Prasada Rao, G. S., Khan, B. H. and Chadha, K. L., (1978), Comparison of methods of estimating leaf-surface area through leaf characteristics in some cultivars of *Mangifera indica*. *Sci. Hortic.*, 8, 841–8.
Preston, K. A. and Ackerly, D. D., (2003), Hydraulic architecture and the evolution of shoot allometry in contrasting climates. *Am. J. Bot.*, 90, 1502–12.
Rajan, S., Tiwari, D., Singh, V. K., Saxena, P., Singh, S., Reddy, Y. T.N., Upreti, K. K., Burondkar, M. M., Bhagwan, A. and Kennedy, R., (2011), Application of extended BBCH scale for phenological studies in mango (*Mangifera indica* L.). *J. Appl. Hortic.*, 13(2),108–14.
Ramírez, F., Davenport, T. L., Fischer, G. and Pinzón, J. C. A., (2010), The stem age required for floral induction of synchronized mango trees in the tropics. *HortScience*, 45, 1453–8.
Ramírez, F., Davenport, T. L., Fischer, G., Pinzón, J. C. A. and Ulrichs, C., (2014), Mango trees have no distinct phenology: The case of mangoes in the tropics. *Sci. Hortic.*, 168, 258–66.
Robert, J. P. and Wolstenholme, B. N., (1992), Phenological cycles, carbohydrate status, and CPPU spray trials for four mango cultivars – Current research in the Nkwalini Valley. *S. Afr. Mango Growers Assoc. Yearb.*, 12, 9–13.
Shaban, A. E. A., (2009), Vegetative growth cycles of some mango cultivars in relation to flowering and fruiting. *World J. Agric. Sci.*, 5(6), 751–9.
Scholefield, P. B., Oag, D. R. and Sedgley, M., (1986), The relationship between vegetative and reproductive development in the mango in Northern Australia. *Aust. J. Agric. Res.*, 37, 425–33.
Shinde, A. K., Burondkar, M. M., Bhingarde, R. T., Waghmare, G. M., Rangwala, A. D. and Wagh, R. G., (2001), Heat unit requirement for fruit maturity in mango varieties. *Indian J. Plant Physiol.*, 6, 194–6.
Taneda, H. and Tateno, M., (2004), The criteria for biomass partitioning of the current shoot: Water transport versus mechanical support. *Am. J. Bot.*, 91, 1949–59.
Tardieu, F., (2013), Plant response to environmental conditions: Assessing potential production, water demand, and negative effects of water deficit. *Front. Plant Physiol.*, 4, 17, doi:10.3389/fphys.2013.00017.
Taylor, F. J., (1970), Some aspects of the development of mango (*Mangifera indica* L.) leaves. I. leaf area, dry weight and water content. *New Phytol.*, 69, 377–94.
Urban, L., Léchaudel, M. and Lu, P., (2004), Effect of fruit load and girdling on leaf photosynthesis in *Mangifera indica* L. *J. Exp. Bot.*, 55, 2075–85.
Whiley, A. W., (1986), Crop management – A review. In: *Proceedings of First Australian Mango Research Workshop*, CSIRO, Melbourne, Australia, pp. 186–95.
Whiley, A. W., (1993), Environmental effects on phenology and physiology of mango – A review. *Acta Hortic.*, 341, 168–76.
Whiley, A. W., Rasmussen, T. S., Saranah, J. B. and Wolstenholme, B. N., (1989), Effect of temperature on growth, dry matter production and starch accumulation in ten mango (*Mangifera indica* L.) cultivars. *J. Hort. Sci.*, 64(6), 753–65.

Chapter 4

Advances in understanding flowering, pollination and fruit development in mangoes

Maria Hilda Pérez-Barraza and Jorge Alberto Osuna-García, Instituto Nacional de Investigaciones Forestales Agrícolas y Pecuarias, Mexico

1 Introduction

The number of flowers produced in mango is determined during the year before flowering through the processes of induction, initiation and floral differentiation. The number of fruits and their production will depend on the success of pollination, fecundation and fruit set processes. This chapter addresses topics related to these processes, starting from the vegetative shoot formation which results in flowering, and later to pollination and fruit development. It has been compiled using information from global research work including that carried out by the author in Mexico.

It appears that floral induction is stimulated by low temperatures in subtropical conditions. The shoot age and a period of dormancy are the determinant factors in tropical conditions. The FT protein has been identified as being responsible for floral induction and it is believed that this protein could also be the floral promoter in fruits such as mango.

Reports show that gibberellins (GA) can inhibit or delay the process in mango. Other GA synthesis inhibitors like prohexadione-calcium (P-Ca) are mostly focused on temperate fruits. There is information with respect to management practices such as pruning and application of nitrates to modify flowering and harvest.

http://dx.doi.org/10.19103/AS.2017.0026.07

There is evidence that pollination is affected by internal factors such as the presence of imperfect flowers, heterostilia, effective pollination period, low pollen viability and incompatibility. Some external factors affecting pollination are temperature, wind, rain and the presence of pollinators. Fecundation, hormones and nutrition are also important in fruit set and growth. The chapter seeks to provide a clear and comprehensible perspective on these processes and to assist in solving the recurring problems of irregular flowering and low fruit set which are caused by climate change among other factors.

2 Vegetative shoot development

During the adult stage, vegetative shoots are developed under favourable environmental conditions: usually warm temperatures (30/25°C, day/night) and high humidity in the ground and environment due to the presence of rain or irrigation in areas of low precipitation (300–500 mm annually).

Growth begins with the development stimulus of leaf primordia meristems in apical buds which remained dormant, or in lateral or axillary buds where shoots have been pruned or where the apical bud does not exist due to earlier flowering (Whiley et al., 1989; Núñez-Elisea and Davenport, 1995; Davenport, 2000).

These meristems are small structures programmed to produce leaves or bracts. They have a central apical dome which will give rise to the stem. A group of cells spread over its periphery will form the specific organ primordia which result in leaves and buds. Some central cells are not destined to form primordia so they form the axis internodes. Vegetative meristems are composed of three cell layers forming the tunica, L1, L2 and L3. These cell layers differ from each other by their location in the meristem and by their cellular division pattern (Fig. 1). L1 superficial cells divide anticlinally inside the meristem resulting in the epidermis. L2 and L3 generate leaf and stem cell tissues. In the apical dome, the L2 cell division is also anticlinal, but during tissue formation, cell division takes place in another plane. By contrast, L3 cells are divided in all planes; they disappear during the formation of tissues such as leaves and stem (Atwell et al., 2003).

Davenport (2009) observed that following the stimulus of leaf primordia meristems, a genetic signalling cascade occurs for leaf development and formation in each node. It appears that the signal to primordia development in leaves is known as a vegetative promoter rather than a floral inhibitor. In species like *Arabidopsis*, molecular studies have identified a gene controlling the vegetative meristem: the *terminal flower* (*TFL*) which is responsible for shoot formation. Its expression inhibits the development of inflorescence (Atwell et al., 2003), indicating that

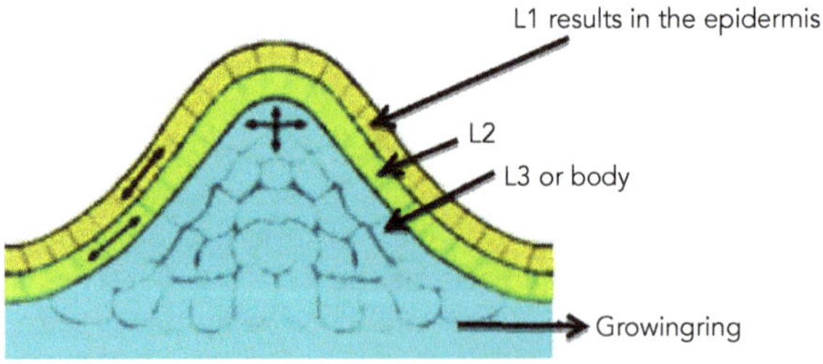

Figure 1 Vegetative meristem showing the three cell layers that form the tunica. Taken from http://datateca.unad.edu.co

an homologue gene from *Arabidopsis* in mango is probably involved in vegetative growth. Consequently its identification, expression factors and conditions are of interest.

New vegetative shoots in mango are yellowish, light green, brown or reddish, depending on the cultivar. However, regardless of cultivar, the leaves of a mature shoot are dark green and their maturation occurs from approximately two to three months (Davenport, 2009; Pérez-Barraza and Vázquez-Valdivia, 2006). Emerged vegetative shoots, once mature, can develop leaf primordia again in the apical bud if the inductive conditions are still present in the environment (Núñez-Elisea et al., 1996). These are known as vegetative flushes and they increase in number as the temperature rises, occurring several times during a year (Whiley et al., 1989; Davenport, 2007) The vigour of the shoot will depend on fruit load, pruning and nutrition (Davenport, 2000). In several mango cultivars, trees produce a high percentage of vegetative shoots and there are three to four flushes during spring, summer and autumn during an 'off' year (low production). In an 'on' year there are only two flushes. The main cycle, in both conditions, occurs during the summer, where the shoot length, the number of leaves and the foliar area in general are higher than in the spring and autumn cycles. The summer vegetative cycle appears to give a higher flowering percentage than the other cycles (Shaban, 2009). Nevertheless, under Mexico's Nayarit conditions, the autumn flush contributes better to flowering (Pérez-Barraza and Vázquez-Valdivia, 2006). Under Australian conditions, Scholefield et al. (1986) reported that in the 'Kensington' cultivar the main vegetative flush occurs between March and May (before flowering), which is the end of spring–summer in the Southern Hemisphere. According to these authors, the fruit load determines whether there are two or more flushes. However, this will also depend on the cultivar, environment and crop management.

The kind of generated growth (vegetative, mixed or floral) in a terminal bud of a mature shoot will depend mostly on environmental stimulus, especially the temperature (Whiley et al., 1989; Davenport, 2007). Warm temperatures and high humidity give rise to new vegetative shoots, while cool temperatures (15–18°C) result in reproductive shoots (inflorescences). However, it has been observed that the presence of high and low temperatures during the beginning of floral differentiation leads to mixed shoots of leaves and flowers (Fig. 2) or transition shoots from reproductive to vegetative or vice

Figure 2 (a) Mixed shoot with leaves (l) and flowers (f) presence and (b) reproductive shoot or inflorescence.

versa (Núñez-Elisea et al., 1996; Davenport, 2007). The production of mixed shoots has also been observed with the use of gibberellins to delay flowering (Vázquez-Valdivia et al., 2009b; Pérez and Vázquez, 2006).

The flush development time and the age of the shoot are important because both vegetative and reproductive development only happen in mature flushes, although the development time and their synchronisation can be achieved by pruning at the appropriate time to intensify flowering (Yeshitela et al., 2005a; Vázquez et al., 2009a).

3 Induction, initiation and floral differentiation

3.1 Induction or floral stimulus

Induction takes place when an environmental stimulus incites flowering. In fruit trees, this process is totally different to annual or biennial plants such as *Arabidopsis thaliana*. In this species, there are four routes to floral induction where photoperiod, vernalisation and gibberellins are important factors, as well as the autonomous route. These routes stimulate or activate the gene expression related to the process, including the flowering locus T (FT) gene in leaves and the subsequent transportation of FT protein through the phloem to the meristem, where it induces the gene expression of the apical meristem identity and consequently, the flowering (Boss et al., 2004; Giakountis and Coupland, 2008; Liu et al., 2013). FT protein is considered to be the 'florigen', long mentioned hypothetically as being responsible for flowering in many other species. It is thought that a similar mechanism could be involved in floral induction in fruit trees (Brunner and Nilsson, 2004).

There is evidence that where mango are grown under subtropical conditions, the signal or stimulus to floral induction are low temperatures of 15/10°C (day/night) in some cultivars, while others require 20/15°C (Núñez Elisea and Davenport, 1994; Osuna-Enciso et al., 2000; Whiley et al., 1989) Shu and Sheen (1987) found that floral induction in 'Haden' occurred with temperatures of 19/13°C while surprisingly, in others 30/20°C was required (Sukhvibul et al., 2000). By contrast, Whiley et al. (1989) reported floral induction inhibition with temperatures of 20/15°C in some cultivars.

The duration of temperatures needed for floral induction varies from four days (Reece et al., 1949) to two weeks (Shu and Sheen, 1987) in the cultivar 'Haden' and up to 35 days in 'Tommy Atkins' and 'Keitt' (Yeshitela et al., 2004).

Pandey (1989) noted that in tropical conditions, low temperatures are rare or non-existent in some seasons and it is unclear what the stimulus for the induction is. On the other hand, several authors agree that the last flush of vegetative growth age is the dominant factor regulating induction. Shoots must be in a state of dormancy long enough, (normally from seven weeks to five months), to induce flowering in the absence of low temperatures. (Davenport, 2000; Nuñez-Elisea and Davenport, 1995; Ramírez and Davenport, 2010). However, in 'Tommy Atkins', induction occurred under Nayarit's tropical conditions with daytime temperatures of between 18 and 20°C (Pérez and Vázquez, 2006). Whereas floral induction in 'Ataulfo' mango buds occurs with temperatures of 27/17°C (day/night) in shoots between two and four months old. The application of GA inhibitors in warm temperatures did not promote floral initiation and differentiation in the year of application. However, initiation does occur in the same year when the temperature is low. This indicates that under tropical conditions, age and temperature are important (Pérez-Barraza, 2015).

Some reports suggest that mature leaves are capable of sensing environmental stimulus, producing large amounts of florigen under inductive conditions and transporting it to the apical meristem from a 100 cm distance in subtropical conditions and 52 cm in tropical conditions (Davenport and Ying, 2004; Davenport et al., 2006; Ramírez and Davenport, 2010). Studies on the number of leaves needed for translocation to take place have found that in the 'Tommy Atkins' cultivar, branches where shoots have four leaves each was sufficient, while in 'Keitt', fewer were required (Ramírez et al., 2010). Under tropical conditions, mature leaves in 'Tommy Atkins' were eliminated a month before differentiation began (September 30th), causing the production of vegetative shoots instead of floral shoots. By contrast, defoliated shoots in the first differentiation stages (December 1st) produced inflorescences (Pérez and Vázquez, 2006), suggesting the leaves had received and sent the stimulus to the apical bud. This confirmed the importance of these kind of leaves in floral initiation and differentiation of this cultivar. It was also shown that the signal to flowering can be transmitted by a graft, proving that the stimulus is transported via phloem (Kulkarni, 1991).

3.2 Initiation and floral differentiation

When the meristem reaches the reproductive adult stage, it is committed to blooming. In this phase, floral initiation (the transition from vegetative to reproductive) occurs with the enlargement of the main axis and the presence of leaf primordia in the secondary axis of the meristem. This is the first evidence of the floral differentiation process in mango and presents first in the axillary meristems furthest from the apex (Scholefield et al., 1986; Osuna-Enciso et al., 2000; Pérez-Barraza et al., 2009; Pérez-Barraza, 2015).

Floral differentiation (FD) is the process whereby formation of floral primordia takes place in a reproductive meristem which has been initiated. With regard to inflorescence, the formation of flower parts occurs after the secondary and tertiary axis has been formed. The process comprises morphologic changes in the apex structure to form the flower parts (Osuna-Enciso et al., 2000). The beginning of FD in mango is characterised by a slight lengthening of the principal axis and the formation of meristematic protuberances in the primordial leaf axils. There are four main defined stages of FD, where the last stage is characterised by a lengthening of the tertiary axis of the inflorescence, the formation of reproductive meristems and the development of floral primordia, sepals, petals and stamens (Ravishankar et al., 1979; Osuna-Enciso et al., 2000; Pérez-Barraza et al., 2009; Palanichamy et al., 2012; Pérez-Barraza, 2015).

The time at which FD occurs may vary depending on species, cultivar, environment and crop management. In mango, as has been mentioned, this normally occurs at the end of autumn and the beginning of winter and lasts four to six weeks, depending on the cultivar (Ravishankar et al., 1979; Osuna-Enciso et al., 2000; Davenport, 2007, Pérez-Barraza et al., 2009; Palanichamy et al., 2012; Pérez-Barraza, 2015).

In India, the beginning of FD may occur in October ('Amatrapali', 'Mallika', 'Alphonso' 'Totapuri'); in Mexico it takes place during November ('Tommy Atkins') or February ('Manila') and ends in November, December and March in the same cultivars respectively (Ravishankar et al., 1979; Osuna-Enciso et al., 2000; Pérez-Barraza et al., 2009; Palanichamy et al., 2012). In the 'Ataulfo' cultivar, growth regulators PBZ (2500 ppm, applied to the soil) and P-Ca (1500 ppm sprayed on the foliage) reduced the time necessary for the differentiation process by 28 days. From the principal axis lengthening and the appearance of meristematic protuberances in bract axils (initiation) to the appearance of floral organs, the process took 39 days in the control. The start of the process was advanced by four to six weeks compared to the control tree with no regulators applied (Pérez-Barraza, 2015).

3.3 Flowering

Once the differentiation process has finished, the inflorescences emerge from the apical or lateral bud and continue to grow until anthesis occurs at the end of winter or beginning of spring (Wilkie et al., 2008). Flowering begins in zones near the equator, moving both north and south in mango-producing countries around the world. However, climate change greatly affects this process in both the tropics and subtropics, leading to wide variations in season and flowering quantity. This has led to irregular flowering or total failure to flower as well as to differences in the quantity of hermaphrodite and masculine flowers (Liu et al., 2012; Rajan, 2012; Scaven and Rafferty, 2013; Kumar et al., 2014). In Nayarit, Mexico, warm autumns with high humidity due to intermittent rain have been associated with irregular flowering in 'Ataulfo' and even a failure to flower in 'Tommy Atkins', 'Kent' and 'Keitt' (Perez-Barraza, 2013, unpublished data).

The amount of inflorescence determines the quantity and quality of fruit production. A low inflorescence production (irregular flowering) results in low fruit yield. This occurs in 'off' or alternate years and is caused by multiple factors, though principally by the previous year's production and climate conditions. An abundance of flowers and fruit set results in high fruit production. However, the quality can be affected by the small fruit size produced biennially. Therefore, flowering control and a balance between flowering and vegetative growth are needed to reach regular production every year (Stover, 2000; Samach and Smith, 2013).

3.4 Factors affecting floral induction and differentiation

Floral induction in fruit trees such as mango is affected by exogenous and endogenous factors. It is also affected by horticultural and hormonal factors which may stimulate or inhibit the process (Bangerth, 2009).

Within the exogenous factors stimulating the process, light duration or photoperiod has a considerable effect in the floral induction of annual plants (due to variation in daylight hours). However, most fruit trees do not respond to photoperiod, even though light intensity has an important impact on the process: for example, shade significantly reduces floral induction and differentiation (Jackson and Palmer, 1977; Tromp, 1984).

Low temperatures or vernalisation stimulates floral induction in *Arabidopsis* and other plants by silencing the *FLC* gene, which is a floral induction repressor in this species. The silencing is due to a histone mutilation of *FLC* which reduces its repressive effect in downstream genes, allowing early flowering (Bangerth, 2009). In some fruit trees such as mango, low temperatures induce flowering, indicating that modifications in the histone could be considered, although to date, this has not been elucidated (Chen and Coleman, 2006).

Endogenous factors such as fruit overload and vegetative growth inhibit the process due to the large quantity of gibberellins present in the tree .The rootstock (vigorous/weak) as well as the distance from the bud to the signal capitation site (leaves) may also inhibit or assist the process (Bangerth, 2009).

It was previously believed that carbohydrates played an important role in apical bud formation in fruit trees. Bernier (1988) and Bernier et al. (1993) concluded that sucrose levels found in apical buds are a vital part of the floral induction theory and that both starch and soluble carbohydrates are used during inflorescence development. However, studies show some differences on this topic. Chacko (1991) reported that flowering in mango is stimulated by the presence of carbohydrates and by some conditions which promote flowering such as water stress, low temperatures, high evaporation, flooding and ringing. However, in a study of ten mango cultivars, Whiley et al. (1989) detected an accumulation of carbohydrates in stems resting before flowering. Ravishankar and Rao (1982) found that insoluble carbohydrates decreased during bud differentiation time, together with the non-reductive sugars, resulting in a significant increase in reductive sugar levels. By contrast, Mishra and Dhillon (1978) did not find that carbohydrates and sugars significantly influenced the differentiation process. Phavaphutanon et al. (2000) maintain that carbohydrate reserves in reproductive shoots may be essential for advanced flowering caused by paclobutrazol in mango.

The author suggests that gibberellins and carbohydrates are related to the floral induction process affected by fruit overload. Even though there is some evidence, the exact role performed by carbohydrates in floral induction of mango has not been clarified (Suryanarayana, 1978; Chacko and Ananthanarayanan, 1982; Davenport and Nuñez-Elisea, 1987; Pongsomboon et al., 1997).

It is known that GA inhibits floral induction in mango, promoting vegetative growth instead of reproductive growth, although the effect depends on the phenological stage at which it is applied, the dose and cultivar. Davenport (2007) mentioned that GA could be the promoter inducing vegetative growth. Núñez-Elisea and Davenport (1998) pointed out that the application of exogenous GA can regulate flowering indirectly by delaying flowering under inductive conditions (low temperatures). Previous studies in Mexico showed that 50 mg L^{-1} of AG_3 inhibited flowering in 'Tommy Atkins' by 76%. In 'Ataulfo' the inhibition was up to 94%, causing a delay in flowering of four to six weeks respectively (Vázquez and Pérez, 2006; Pérez et al., 2008; Váquez-Valdivia et al., 2009b).

Many authors report low GA activity in floral initiation. Low AG_3 content has been found in shoots close to flowering, and high levels in shoots that remain vegetative. (Chen, 1987; Pongsomboon et al., 1997; Tongumpai et al., 1991; Abdel-Rahim et al., 2011). Apparently, high levels of cytokinins are found in shoots on the point of flowering (Ulger et al., 2004; Abdel-Rahim et al., 2011).

On the other hand, there is evidence that some growth regulators reduce vegetative growth and enhance flowering as in the case of GA inhibitors. There are different kinds of products which act as inhibitors:

- inhibitors that block the cyclases copalyl-diphosphate synthase and ent-Kaurene synthase in the first stage of the GA metabolic pathway, such as Chlormequat chloride;
- inhibitors that block monooxigenases, inhibiting the ent-kaurene oxidation to ent-kaurenoic acid such as paclobutrazol (PBZ) and uniconazole (Rademacher, 2000), avoiding the GA synthesis such as GA_4, GA_3, GA_7 and GA_1 in leaves and apical buds (Abdel-Rahim et al., 2011; Upreti et al., 2013) and
- compounds with a similar structure to 2-oxoglutaric acid such as prohexadione-calcium (P-Ca) that particularly block 3β-hydroxylation inhibiting active GA formation by forming inactive precursors (Rademacher et al., 1998; Rademacher, 2000).

PBZ is also able to affect other hormones by reducing the abscisic acid, ethylene and indolacetic acid levels and increasing the cytokinins level (Berova and Zlatev, 2000; Cárdenas and Rojas, 2003; Abdel-Rahim et al., 2011; Upreti et al., 2013).

Positive results in flowering have been obtained by applying PBZ in many cultivars such as 'Tommy Atkins' (Cárdenas and Rojas, 2003; Rodríguez et al., 2007) 'Kensington Pride' (Blaikie and Kulkarni, 2002; Winston, 1992) 'Dashehri' (Singh and Bhattacherjee, 2005) and 'Keitt' (Yeshitela et al., 2004). Some authors produced evidence that PBZ is able to induce flowering in the tropics without the presence of low temperatures. In 'Keitt' and 'Tommy Atkins', two weeks at inductive temperatures (10/15°C) PBZ reduced the days required for flowering, indicating that the regulator complemented the need for low temperatures in floral induction (Yeshitela et al., 2004, 30). In 'Khiew Sawoey mango, 30% of the apical buds developed floral primordia 91 days after PBZ application (6 g·tree^{-1} of a.i.). Ninety to 100% of the treated shoots produced flowers between 104 and 112 days after application. None of the control buds developed flowers (Tongumpai et al., 1996). In Mexico, the use of PBZ has been effective in the 'Manila' cultivar. (Pérez-Barraza et al., 2011; Rebolledo et al., 2008) 'Ataulfo' (Vázquez et al., 2009a) and Tommy Atkins (Medina-Urrutia, 1995). P-Ca has been studied especially in temperate species. It has obtained positive results in apples by reducing shoot length and increasing fruit production (Cline et al., 2008; Kiessling-Davison et al., 2008; Ramírez et al., 2006). In subtropical fruit trees such as 'Eureka' lemon, as well as in 'Hass' avocado (Garner et al., 2010) a 30% decrease in shoot length was achieved by applying 250 mg L^{-1} of P-Ca, however these doses had no effect on the 'Navel' orange.

There are a few reports of P-Ca decreasing vegetative growth and increasing flowering in tropical fruit trees such as mango. In the 'Kent' mango, the combination of P-Ca (1.5 g a.i.·plant^{-1}) with PBZ (3.0 g a.i.·plant^{-1}) applied 40 days after pruning was effective in reducing vegetative growth by 67%, but there was no effect on flowering or production (Do Carlo-Mouco et al., 2011). Flowering in 'BaladiAbuZaid' and 'BaladiBurai' mangoes was advanced by 60 days with PBZ and 40 days with P-Ca applied after harvest (July 18th) (Abdel Rahim et al., 2011). Studies in Mexico showed that 1500 mg L-1 of P-Ca sprayed 30 days after pruning in 'Ataulfo' mango was enough to match or even exceed in some cases the effect PBZ had on intensity and flowering time, achieving abundant (>80%) and anticipated flowering (35 days compared to the control). In an 'off' year, the effect on flowering was greater with any of the bio-regulators. This indicates that P-Ca could be a possible substitute to PBZ in reducing vegetative growth, increasing and advancing flowering and fruit production in the 'Ataulfo' cultivar (Pérez-Barraza et al., 2016); nevertheless, more studies and other regulators are required to determine a substitute for PBZ.

Horticultural factors such as balanced nutrition (especially nitrogen), partial elimination of young fruits (fruit thinning), ringing, branch bending, moderate pruning and nitrate applications stimulate the flowering process by allowing the meristem to receive the message and progress from vegetative to reproductive (Bangerth, 2009). Yeshitela (2005b) increased flowering by pruning in 'Keitt' and 'Tommy Atkins' mangoes by manipulating the vegetative shoot development time and synchronising vegetative flushes in the tree top. Similarly, Srilatha and Reddy (2015) advanced flowering by up to 20 days in 'Rapuri' mango by pruning and PBZ application.

The use of water stress as a technique for floral manipulation has given contradictory results (Chaikiattiyos et al., 1994; Núnez-Elisea and Davenport, 1994).

Potassium nitrate (KNO_3) and other nitrates (ammonium, calcium) have been used for floral induction in many mango-producing regions around the world (Mosqueda and De los Santos, 1982; Sargent and Leal, 1989; Tongumpai et al., 1989; Goguey, 1993; Ravishankar et al., 1993). However, the results were not conclusive in many environments (Davenport and Núnez-Elisea, 1997). It was previously believed that these products induced flowering

(Barba, 1974) but it has been observed that depending on the time of application, some shoots respond with vegetative growth and others by flowering, indicating that KNO_3 does not influence the flowering process but stimulates bud break, be it vegetative or floral (Núñez-Elisea, 1985; Salazar-García et al., 2000; Pérez-Barraza and Vázquez-Valdivia, 2006). It is therefore believed that mature leaves respond to the floral stimulus and low temperatures and that these factors rather than nitrates are responsible for the initiation of the reproductive process in apical buds (Tongumpai et al., 1989; Núñez-Elisea, 1994).

Many studies agree that the response to nitrates depends on the cultivar, climate conditions, geographic location, physiologic maturity of the plant, shoot age and previous harvest. This is the reason for inconsistent results (Winston and Wright, 1986; Fierro and Ulloa, 1991; Goguey, 1993; Machado and Sao Jose, 2000). Recent studies still show contrasting results as to the effect of these products. An increase in hermaphrodite flowers and in inflorescence length was demonstrated in 'Langra' mango with KNO_3 (3%) Apparently an advance in flowering was achieved with this product, though no data was produced (Azam et al., 2007). 'Tommy Atkins' and 'Amrapali' showed an advance in flowering of between 45 and 17 days with KNO_3 4% + 1 g of Urea and KNO_3 4% respectively (Yeshitela et al., 2004; Sarker and Rahim, 2013).

4 Genes related to the flowering process

Identification, quantification and expression of genes related to flowering have become important tools in understanding this process.

Competent adult meristems can produce reproductive structures, but the flowering process really depends on environmental stimulus and is controlled by a large group of genes called 'meristem identity genes'. These genes specify floral identity as *APETALA1* and *LEAFY*, even though there are other genes such as *CAULIFLOWER*, *FRUITFULL* and *AP2*. The meristem identity genes activate the 'identity genes of floral organs' in discrete regions of the flower which activate 'organ construction genes' downstream (A, B, C and E Model in *Arabidopsis*) the latter specify the different tissue and cell types that are part of floral organs (Krizek and Meyerowitz, 1996; Ruiz-García et al., 1997; Ditta et al., 2004; Liu et al., 2009).

Studies in *Arabidopsis* have been the basis for molecular studies in fruit trees. The studies in subtropical fruit trees normally include a relation of gene expression with floral initiation and development, environment and management. The levels of *FT* homologous gene in 'Satsuma' tangerine were raised with low temperatures during the floral induction time (Nishikawa et al., 2007) and in sweet orange, there is evidence that the level of the *LFY* and AP1 homologous rise during and after floral induction with low temperatures (Pillitteri et al., 2004). It was observed that in the sweet orange tree 'Salustiana' cultivar, gibberellins inhibited flowering by repressing the *FLOWERING LOCUS T* (*CiFT*) homologous genes from Citrus, meanwhile, PBZ increases flowering by stimulating their expression (Muñoz-Fambuena et al., 2012). These reports suggest that *FT*, *AP1* and *LFY*, which are key to promoting flowering in *Arabidopsis*, are functionally conserved in these fruits.

There is little information about this in mango. Davenport et al. (2006) isolated and characterised a gene similar to *CONSTANT* in mango (*MiCOL*), *MiCOL* was identical to *MdCOL2*, *MiCOL1* from apple trees and *AtCO* from *Arabidopsis*. However, the *FT* gene could not be identified. Recently, a gene similar to *FLOWERING LOCUS T* (*FT*) was identified and characterised in matures leaves from 'Irwin' mango. The *MiFT* sequence

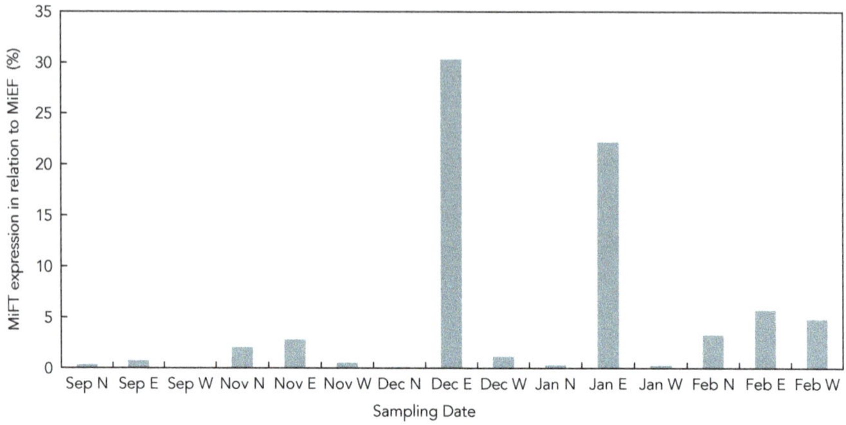

Figure 3 *FLOWERING LOCUS T* (*MiFT*) gene expression in relation to *MiEF* in mature leaves of 'Ataulfo' mango, treated with PBZ and with or without shading. N, shaded north side; E, shaded east side in the morning and W, shaded west side in the afternoon.

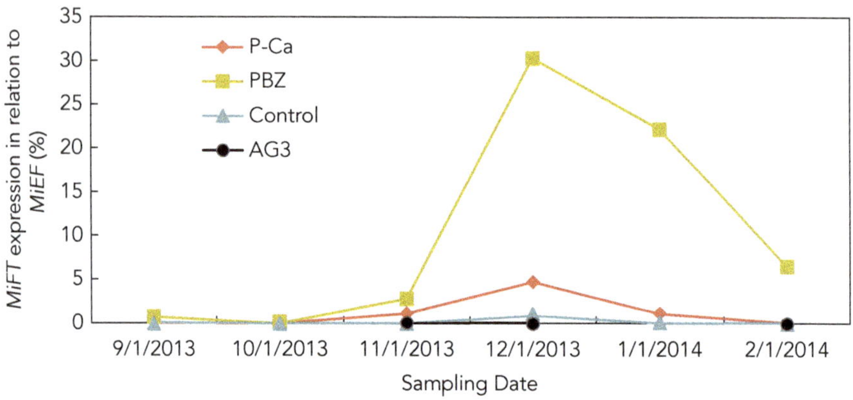

Figure 4 *FLOWERING LOCUS T* (*MiFT*) gene expression in relation to *MiEF* in mature leaves of 'Ataulfo' mango, treated with growth regulators (PBZ, P-Ca and AG3) and without regulator (control).

showed a high similarity to the *FT* gene from *Arabidopsis* and its expression increased in leaves under floral induction conditions (temperatures of 15°C), and in trees with low fruit production. The application of 250 mg L^{-1} of AG_3 completely inhibited flowering and the *MiFT* expression in both high and low production. *APETALA1* (*MAP1*) was identified and characterised in the apical buds of reproductive shoots under the same conditions.

In this study, the authors concluded that *MiFT* is the key factor to flowering in mango (Nakagawa et al., 2012). In Mexico, it was found that in mature leaves of 'Ataulfo' mango, both light and the application of bio-regulators which inhibit synthesis of gibberellins (PBZ and P-Ca; 2500 and 1500 ppm respectively) increased expression of the *MiFT* gene (Figs. 3 and 4) which controls floral induction in mango and therefore favoured flowering.

Conversely, application of 250 ppm of gibberellins inhibited the *MiFT* expression and hence the flowering (Fig. 4) (Pérez-Barraza, 2015).

5 Pollination and fertilisation

Pollination is the transfer of pollen grain to flower stigma, while fertilisation is the process by which the pollen grain germinates into stigma, develops the pollen tube and fuses with the egg cells. This last process is considered a stimulating event in fruit set where seed development starts, hormones are produced and metabolites and nutrients are attracted.

5.1 Floral biology

The mango tree is a polygamous and monoic plant, having both male and hermaphrodite flowers in the plant itself (Fig. 5a, b and c), their distribution or location in the inflorescence depending on the cultivar. The flowers emerge in a pyramidal inflorescence which normally develops from an apical bud (Galán-Saúco, 1999; Mukherjee, 2009). The number of flowers in an inflorescence varies from 1000 to 6000 according to the cultivar and tree position. For example, 'Tommy Atkins' showed a greater number of flowers per inflorescence than 'Kent' and 'Keitt' (Ram, 1992; Mukherjee, 2009; Abourayya et al., 2011) and 'Langra' had a greater number of hermaphrodite flowers on the west and southeast side of the tree than 'Anwar Ratual' (Asif et al., 2002). The flower size is small, between 5 and 10 mm, and the number of male flowers in an inflorescence is greater than hermaphrodite flowers, although the number of these is mainly determined by ambient temperature (Galán-Saúco, 1999; Tharanathan, et al., 2006; Mukherjee, 2009; Ding and Darduri, 2013).

The flowers have four to five stamens, of which only one or two are fertile, the others being sterile with a small filament close to a floral disc (Mukherjee, 2009). Studies performed in the 'Chok Anan' cultivar showed hermaphrodite flowers with a longer style than the stamen filament, indicating a degree of self-incompatibility, thus requiring outcrossing pollination (Ding and Darduri, 2013). The flower opens in the morning, and the sigma is immediately

Figure 5 'Ataulfo' and 'Tommy Atkins' mango hermaphrodite flower, (a) and (b) respectively, with their stamen (St) and carpel (Ca). 'Ataulfo' male flower (c) with a single fertile stamen (St) and floral disc (Fd).

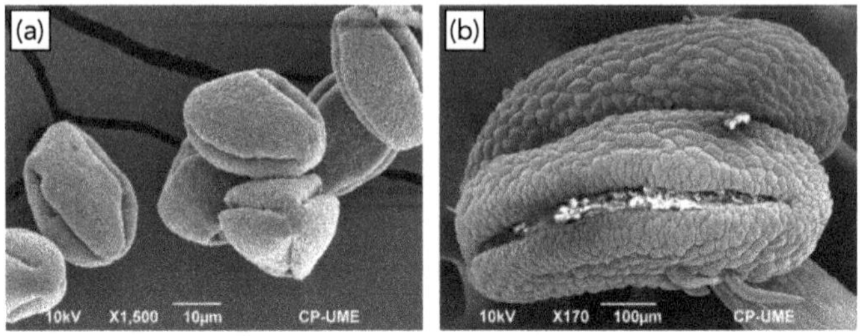

Figure 6 'Ataulfo' mango pollen grain (a) and anthers dehiscence starting and release of pollen grain (b).

receptive. The maximum window is between 11.00 and 12.00 am and the pollen is released from 8.00 am until noon (Gehrke-Vélez, 2008). Medeiros et al. (2008) suggest that mango flowers exhibit dichogamy because anther dehiscence occurs 24 hours after anthesis.

The pollen grain (Fig. 6a) may have two or three nuclei. In the latter case, one of the three nuclei lost viability quickly. The viability of cultivars varies. Khan and Perveen (2009) demonstrated that the pollen of cultivars 'Langra', 'Chaunsa' and 'Dasheri' had a higher viability with 70% germination rates when stored for four weeks at −30°C. Gehrke et al. (2011) showed that 'Ataulfo' had 70–85% viability and 14.5 and 1.75% germination in hermaphrodite and male flowers, respectively.

5.2 Factors affecting pollination

Outcrossing pollination in fruit trees can be a prerequisite for fertilisation and is affected by internal and external factors (Peña, 2003). Highlighted among internal factors are the presence of imperfect flowers (one sex; male or female) heterostilia (flowers with long stigma and short stamen, or vice versa) a restricted effective pollination period, low viability of the pollen grains and incompatibility. The most important external factors are temperature, wind, rain and the presence of pollinators (Wani et al., 2010).

The cultivar 'Ataulfo' showed a higher rate of pollen tube growth (58 $\mu m\ h^{-1}$) which decreases 12 hours after germination to 12.8 $\mu m\ h^{-1}$, indicating the possibility that growth is restricted by an environmental or physiological factor, causing inadequate fertilisation and/or embryo abortion (Gehrke et al., 2011). Rani et al. (2013) showed that the presence of anthers with fused lobes and curved stigmas with low receptivity were the factors responsible for the restriction of pollen grain germination and pollen tube growth in the 'Baramasi' cultivar.

Temperature is one of the most important external factors in pollination and fertilisation. It affects several developmental phases in plants, beginning in post-pollination (development of reproductive organs), progamous phase (from pollination to fertilisation) and during embryo development. Temperature stress may lead to an asynchronous development in pollen-pistil-ovule, so reducing fertilisation. However, independent of successful fertilisation, the persistence of temperature stress does not guarantee a good fruit set, impeding embryo development (Hedhly, 2011).

Low temperatures (20/10°C; day/night) decreased the percentage of hermaphrodite flowers in polyembryonic cultivars but increased the percentage in monoembryonic ones. 'Kensington' produced flowers with short styles (0.62 mm) and small stigmata. 'Nam Dok Mai' and 'Irwin' cultivars produced flowers with deformed or fused ovaries. These abnormalities, caused by low temperatures, contributed to a low fruit set in subtropical areas (Sukhvibul et al., 1999, 2000).

Dag et al. (2000) reported that 'Kent' mango had poor pollination at the beginning of the flowering period with temperatures of 20/10°C day/night, affecting the development of the reproductive organs. At the end of the period, pollination improved with temperatures of between 25 and 15°C (day/night). An increased number of germinated pollen grains on the stigma surface were found in the 'Haden' cultivar at temperatures of 25–30°C. Pollination was increased by the application of boron at anthesis (Wet et al., 1989).

5.3 Pollination by insects

In some mango cultivars, pollen transfer depends partly or totally on vectors such as wind and insects. There is disagreement over mango pollen transfer as some studies had previously shown that mango cultivars were pollinated by insects rather than by wind, while other studies showed that wind and gravity were also important (Waite, 2002). In fact, it is generally accepted that mango cultivars are mainly pollinated by insects. Sung et al. (2006) found 126 different insects including visitors or pollinators in mango. Several of these insects belong to the order Diptera and Hymenoptera, with housefly and bees (*Apis cerana and A. mellifera*) being the best pollinators. Carvalheiro et al. (2012) showed that the presence of small areas of native flowers close to mango orchards combined with the moderate use of insecticides increased production due to better insect pollination.

Several mango studies show the importance of insects for pollination, If insect control is necessary then the use of insecticides should be moderate to avoid damaging beneficial pollinating insects.

6 Fruit set and growth

The fruit set is considered an inductive stage which begins with pollination of the flower and finishes with initiation of the ovary growth and its permanency in the plant. These processes are key because they determine the production potential of any cultivar. The mango fruit is a drupe, the pericarp comprising exocarp, mesocarp and endocarp that includes the seed. The growth follows a sigmoid curve behaviour (Fig. 7) and is divided into three stages (Ram, 1983; Mukherjee, 2009; Tharanathan et al., 2006; Pérez-Barraza et al., 2015):

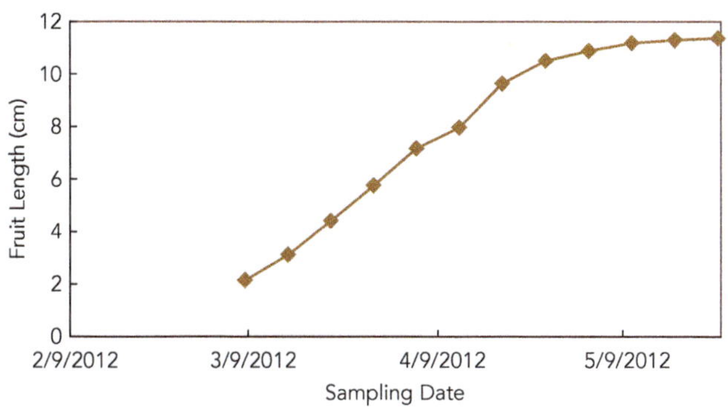

Figure 7 Growth pattern of 'Ataulfo' mango fruits, developing under tropical conditions in Nayarit, Mexico.

- Stage I–14 days after flowering (DAF) is characterised by slow growth but high cellular division in the pericarp after pollination and fertilisation.
- Stage II, 14–42 DAF fast growth occurs with cellular expansion and the start of seed development.
- Stage III 42–77 DAF the fruit increases rapidly in cell size (cellular expansion) until it reaches its final size (Ram, 1992; Ruan et al., 2012). In this last stage the embryo reaches the torpedo phase and continues its development until maturity so that the fruit achieves its maximum growth (physiological maturity) (Varoquaux et al., 2000).

6.1 Parthenocarpy

The absence of seed in mango fruits is known as stenospermocarpy (Davenport, 2009), a kind of parthenocarpy where there is pollination and fecundation but the newly fertilised embryo is aborted (Sedgley and Griffin, 1989; Pérez-Barraza, 2015).

Production of stenospermocarpic fruit occurs in many situations in the subtropics (Galán-Sáuco, 2009). Under Mexico's Nayarit conditions, this occurs in 'Ataulfo' mango with increasing frequency, decreasing yield and orchard productivity. Even though the lack of seed may seem attractive in fruits normally having many seeds, such as citrus, tomato and papaya, or in those fruits that have a large seed such as mango (Baker et al., 1973; Varoquaux et al., 2000), seed development promotes cellular expansion via hormone synthesis or other indeterminate compounds, resulting in larger fruit (Gillaspy et al., 1993).

According to several authors (Sukhvibul et al., 2005; Galán Sáuco, 2009; Irenaeus and Mitra, 2014), the main causes of seedless fruits in mango are extremes of temperature during flowering time and the early stages of fruit development. Sukhvibul et al. (2005) demonstrated that exposure to low temperatures (20/10°C; day/night) three days after manual pollination caused embryo abortion and an increased percentage of stenospermocarpic fruits (nubbin) in 'Irwin', 'Kensington' and 'Nam Dok Mai'. They noted that polyembryonic cultivars were more susceptible to low temperatures than were the monoembryonic plants. In Mexico, stenospermocarpy has been associated with temperatures under 10°C and above 35°C. It has been observed that below this temperature embryo abortion occurs after the initial development stages, in 3–5 mm length fruits and even in advanced stages of fruit development (Fig. 8a,b and c) (Pérez-Barraza, 2015).

In parthenocarpic tomato fruits, there is some evidence that auxins and GA are required for fruit development (Bunger-Kibler and Bangerth, 1982; Serrani et al., 2007) even though

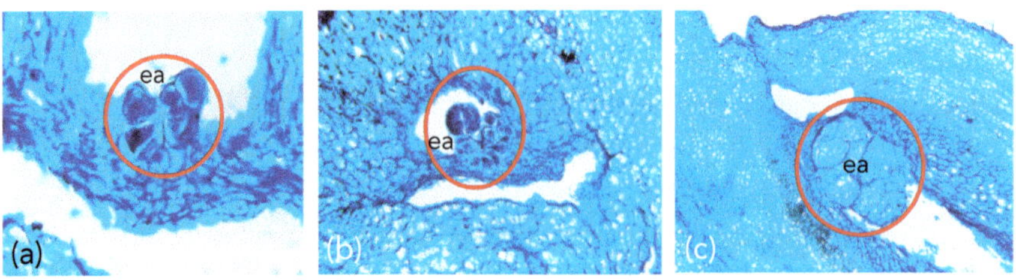

Figure 8 Embryo abortion presence (ea) in 3–5 mm length 'Ataulfo' mango (a), 1–2 cm (b) and 4–5 cm (c) under Nayarit, Mexico conditions. Note the presence of many embryos (c).

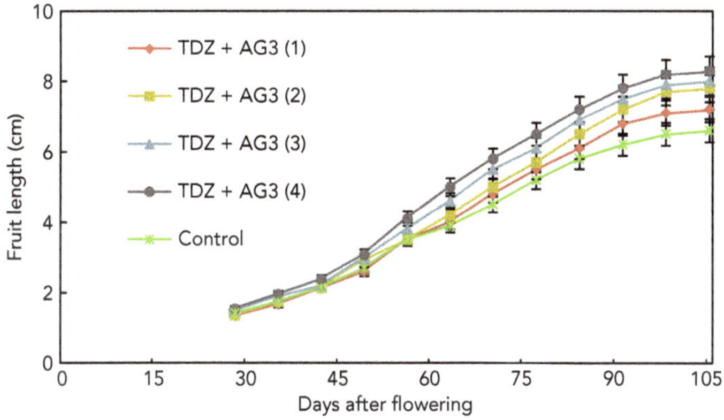

Figure 9 Parthenocarpic fruits length of 'Ataulfo' mango with and without growth regulator. TDZ simple application in anthesis, and one to four applications of AG_3 at 15, 15 + 30, 15 + 30 + 45 and 15 + 30 + 45 + 60 DAF.

Table 1 Fresh weight of parthenocarpic fruit (PF) and fruit with seed (SF), and obtained yield by regulators growth effect in 'Ataulfo' mango trees

	Fresh Weight (g)		Yield
Treatment	PF	SF	($Kg{\cdot}tree^{-1}$)
1	105.3a	259.5b	48.7a[1]
2	104.8a	277.9ab	45.0a
3	126.0a	279.9ab	46.9a
4	116.9a	299.5ab	57.5a
5	66.2b	315.7a	41.2b

[1] Means with the same letter in a column are equal according to Tukey test with a $P \leq 0.05$

there are other reports showing that auxins act before or independently from GA and are able to replace pollination and fertilisation (Serrani et al., 2008). Some studies show that cytokines regulate cellular division so it is possible their association with fruit growth increases the number of cells. A correlation has been found between cytokine content and cellular division in tomato (Srivastava and Handa, 2005; Matsuo et al., 2012). A high concentration of cytokines has also been demonstrated in developing tomato seeds, suggesting an important role in embryo/seed development and fruit growth (Pandolfini, 2009). Jun-hu et al. (2013) showed that 10–30 DAF, parthenocarpic mango fruits have lower GA and auxin content and higher abscisic acid (ABA) content than seeded fruits. This probably indicates that lower hormone content in the pericarp, coupled with an absence of seed in parthenocarpic fruits, results in smaller fruit because the seed is an important source of GA and cytokine synthesis.

It is interesting to note that the application of 50 mg L^{-1} of cytokines to the 'Ataulfo' mango in full bloom (Fig. 9, Table 1) increased the size of seedless fruits which was reflected in better yield (Table 1); while seeded fruits size were not affected (Pérez-Barraza et al., 2015).

6.2 Fruit drop

In all mango cultivars, fruit drop happens throughout the whole period of fruit development but especially during the three or four weeks after pollination. Some related factors should be noted. Most fruits fall during the first weeks after flowering; this being attributed to failure of the pollination–fertilisation processes.

Hormones play an important role in the second drop flush (60–75 DAF) due to the lack of cytokines and the high levels of inhibitors that apparently match embryo development failures. The third drop flush, just before harvest, (90–120 DAF), is attributed to nutrition and photosynthesis (Ram, 1992; Singh et al., 2005). The main fruit drop occurred between 10 and 25 days after anthesis (DAA) in the 'Ataulfo' mango under Mexico's Nayarit conditions, the second flush occurred at 45 DAA, and the third between 70 and 90 DAA (Pérez-Barraza, 2015). Even though the number of dropped fruits is small, the cost is high because it happens near to harvest (Fig. 10) which concurs with the findings of Ram (1992) and Singh et al. (2005).

6.3 Factors affecting fruit set

Among the factors involved in fruit set and growth; pollination and fertilisation, hormones and nutrition, play an important role.

There are some contradictory reports in mango about the hormone content during growth (Jun-hu et al., 2013; Ram, 1992). There is low auxin content in pericarp during the first 14 DAF, this rises 28–42 DAF and then decreases and remains at low levels during Stage III (Ram, 1992). In contrast to Ram, Jun-hu et al. (2013) found a high auxin content in the pericarp of pollinated fruits of 'the Tainung-1' and 'Jinhuang' cultivars 10 days after fruit set (DAFS), decreasing between days 20 and 30 and followed by a small increase between days 40 and 60. The content in the embryo is high at 30 DAFS and decreases as it grows, but the content here is higher during the whole growth period than it is in the pericarp. A high GA content is reported during seed growth (20–42 days, Stage II) decreasing later and remaining low during fruit maturation (Stage III).The pericarp content is unreported (Ram, 1992). However, Jun-hu et al. (2013) found a high GA content in the

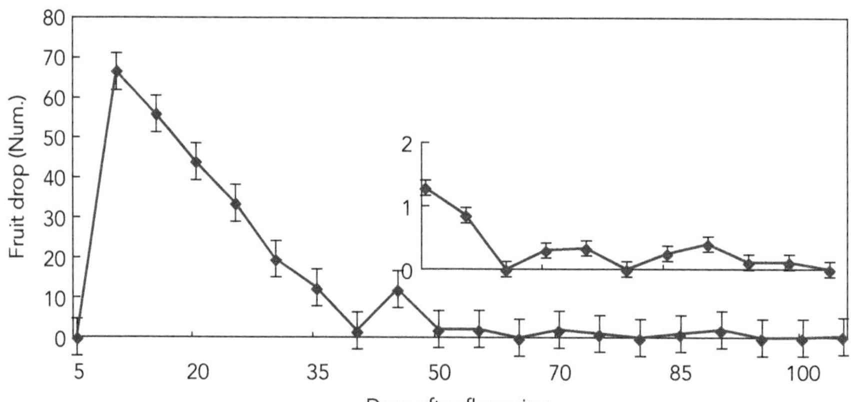

Figure 10 Fruit drop per inflorescence in 'Ataulfo' mango, from 15 DAF to harvest. Nayarit, Mexico.

pericarp between 10 and 20 DAFS, probably in Stage II; and a slight increase in Stage III. There is a high GA content in embryos at 30–40 DAFS, decreasing at 50–60 days even when there is a higher content when compared to the pericarp during its growth. The GA content in seeds reported by Ram could be the trigger for the beginning of fruit growth, also the results of Jun-hu et al. (2013) could indicate that the GA content in pericarps is also involved in fruit growth due to pericarp cell elongation.

Cytokines are present in seed and pericarp at a high level between 14 and 21 DAF with a dry weight of 0.001 and 0.01 µg g^{-1} respectively (Ram, 1992) and they decrease at day 35–0.0001 µg g^{-1} (Stage II). The cytokine content increases in both organs (seed and pericarp) at 42–72 days and then decreases, remaining at low levels after 82 days (Stage III). Jun-hu et al. (2013) found a high cytokine content in the pericarp 10 DAF (40 ng g^{-1} in dry weight approximately) and low content between days 20 and 30 (10–20 ng g^{-1}). This agrees with the results reported by Ram (1992), with the exception of the low content found at 60 days as Ram found a high content at the same stage. The content in embryos is higher than in pulp between days 30 and 60 (50–60 ng g^{-1}). According to both authors, cytokines are present in Stage I, indicating the key role of these hormones in cellular division, resulting in a better fruit set. Meanwhile the high content in Stage II (Ram, 1992) could be related to the presence of GA at the same stage, suggesting that both GA and cytokinins participate in fruit growth.

However, inhibitors like ABA are found at high concentrations in pericarp from 7 to 21 DAF (end of Stage I and beginning of Stage II). This later decreases and remains at low levels during fruit maturation, Stage III (Ram, 1992). Jun-hu et al. (2013) found a high content at 10 DAF, decreasing at days 20–30 with a slight increase at day 60. In the case of the embryo, between 30 and 40 DAF, the content is high, decreasing later with its development. In spite of this, the content is higher than in the pulp. The high content of inhibitors at the end of Stage I and II could be related to fruit drop, causing abscission zones (Ram, 1992).

Schaffer et al. (2009) observed that a water deficit during the first four to six weeks after anthesis can affect retention to the tree. It seems that water stress is associated with ABA accumulation, resulting in fruit drop (Nakano et al., 2002).

Fruits accumulate carbohydrates during development, usually as starch or sugars such as sucrose, glucose and fructose. Developing fruits attract photosynthates. If these increase, the demand rises, resulting in an increased photosynthetic capacity of mature leaves near the fruit. In contrast, a small number of fruits causes photosynthate accumulation in leaves because of reduced demand by fruits, so decreasing photosynthetic activity in the mature leaves (Fisher et al., 2012; Hansen, 1982; Urban et al., 2003, 2004). Simao et al. (2008) pointed out that the starch content in an immature fruit is approximately 70%, developing into soluble sugars during maturation.

In regard to nutrition, it is known that availability of major and minor elements is required for an adequate fruit set and for their growth and maturation. Nutrient content per fruit is higher during the cellular division stage and there is some evidence that suitable N, P and K concentrations in this stage as well as Ca and Mg after flowering gives rise to better fruit set and size in 'Dashehari' mango (Pathak and Pandey, 1977). It appears that the nutrient with the greatest impact on production is N. P is required to maintain fruit set and K contributes to reinforcing trees against different types of stress such as drought, low temperatures, pests and diseases. Zn and B are effective in increasing the proportion of hermaphrodite flowers, inflorescence size, improving pollination and obtaining better fruit quality. So to avoid excessive fruit drop, a nutritional programme must be adopted to match the requirements of the different phenological stages of the crop (Samra and Arora, 1997; Singh and Rajput, 1977; Singh, 1977; Rath et al., 1980).

6.4 Use of hormones to increase fruit size

One of the most studied aspects of increasing fruit set and size has been the use of endogen growth regulators. Several studies with GA have shown a positive effect on mango fruit set. According to Singh (2009), spraying mango with AG_{13} before flowering was effective for initial fruit set but GA_4 (10 ppm) was better for the final fruit set, the conclusion being that GA application is important to ensure fruit set in zones with low temperatures during flowering. Chutichudet et al. (2006) found a fruit set increase in 'Srisaket 007' mango with GA_3 application at 50 ppm in Stage I. Similar results were obtained with GA application in the 'Irwin' cultivar, improving parthenocarpic fruit set (Ogata et al., 2010). Hormone application in Stage I has been effective in increasing fruit size by stimulating cellular division and elongation. Fruit growth in mango was achieved by the application of GA_3 in the 'Srisaket 007' cultivar (Chutichudet et al., 2006) and in 'Irwin' parthenocarpic fruits (Ogata et al., 2010).

7 Conclusion

In Mexico, initial studies on inducing flowering by the use of nitrates were performed in 'Manila' mango under the tropical conditions of Veracruz State. This technology was transferred to the main producing areas of the country, mainly at Chiapas, Guerrero, Oaxaca, Michoacán and Colima states, resulting in positive advances in flowering in various cultivars, with the exception of 'Tommy Atkins'. Michoacán is now the only state bringing forward flowering in the 'Haden' cultivar through the use of nitrate applications in the soil. Although the yield obtained by advanced flowering is lower than the normal harvest, the price is five times higher, thus increasing crop productivity. In Nayarit, technologies to advance flowering were generated using potassium or ammonium nitrate, in 'Haden', 'Ataulfo' and 'Manila' cultivars, but the results proved erratic when the technology was validated with farmers. However, the application of growth regulators such as PBZ that inhibit gibberellin synthesis have been very promising for abundant and early flowering in 'Ataulfo', 'Manila' and 'Tommy Atkins' cultivars, even without inductive conditions during FD and without the application of nitrates. Pruning after harvest has been effective in synchronising vegetative shoot age and flowering. PBZ and pruning have been used successfully in the states of Oaxaca and Guerrero.

There are also technologies to delay flowering in 'Ataulfo' mango. This is achieved by applying gibberellin (GA_3) to foliage shortly before floral initiation, allowing the producer to achieve greater profitability in cultivation by increasing the product price through harvesting in off-season.

Scientists are now looking for alternatives to PBZ for controlling flowering. Prohexadione-calcium (P-Ca), another gibberellin synthesis inhibitor, appears to be the favoured choice. However, more studies are required to understand and validate its effect. The use of growth regulators such as thidiazuron (TDZ) and gibberellic acid (GA_3) in combined applications has been effective in increasing fruit size when applied to the seedless fruit of the 'Ataulfo' mango, achieving increases in size up to 21% in length and 37–48% in fresh weight, thus increasing the productivity of this cultivar.

8 Future trends

In Mexico the study of growth regulator alternatives to PBZ is necessary for modifying blooming. The results obtained for delayed flowering by GA_3 requires further studies in different environments and cultivars as well as evaluation of alternatives such as late and severe pruning after harvest, and the pruning of inflorescences. These studies must now be extended to different environments and cultivars, especially 'Tommy Atkins', 'Kent', 'Haden', 'Keitt' and 'Ataulfo'. The expression and quantification of genes related to flowering and factors which modify its expression should be studied to better understand the flowering process in mango if the problem of irregular flowering is to be solved.

9 Where to look for further information

It is possible to search information in the following books, scientific journals and Symposia and scientific meetings:

Books:

- *The Mango: Botany, Production and Uses*. 2009. Richard E. Litz (Editor)
- *El cultivo del Mango*. 2009. Víctor Galán Sauco. MundiPrensa Libros Second edition.
- *Plant Physiology and Development* by L. Taiz and E. Zeiger. Sixth Edition. 2015
- Flowers: Floral organs and floral organ identity mutants. In *Arabidopsis: An Atlas of Morphology and Development* (ed. J. Bowman), pp. 162–163, 244, New York: Springer-Verlag.

Journals: Acta Horticulturae, Scientia Horticulturae, Annals of Botany, Journal of Experimental Botany, Hort Technology, New Zealand Journal of Crop and Horticultural Science, Rev. Bras. Frutic., Jaboticabal – SP, Annu. Rev. Plant Physiol. Plant Mol. Biol., Journal of Experimental Botany, Annu. Rev. Plant Biol., Journal of American Science, Plant physiol., J. Hort. Sci

Symposia and scientific meetings:

- International Mango Symposium. International Society for Horticultural Science (ISHS).
- International Symposium on Growth Regulators in Fruit Production ISHS
- National and International Congress of Horticultural Sciences. Mexican Society of Horticultural Sciences. Nayarit, Mexico. October, 2017.
- Meetings for diffusion of technology by National Mango Board, Research Program and generic promotion (USDA)

10 References

Abdel Rahim, A. O. S., O. M. Elamin and F. K. Bangerth. 2011. Effects of growth retardants, paclobutrazol (PBZ) and prohexadione-Ca on floral induction of regular bearing mango (*Mangifera indica* L.) cultivars during off season. *ARPN J. Agric. Biol. Sci.* 6: 18. ISSN: 1990–6145.

Abourayya, M. S., N. E. Kassim, M. H. El-Sheikh and A. M. Rakha. 2011. Comparative study between inflorescences characteristics, pollen viability, germination and dimensions of Tommy Atkins, Kent and Keitt mango cultivars. *Life Sci. J.* 81: 100–105.

Asif, M., M. Usman, M. Jafar-Jaskani and M. Mumtaz-Khan. 2002. Comparative study of flower sex ratio in different cultivars of mango (*Mangifera indica* L.). *Int. J. Agric. Biol.* 4: 220–2.

Atwell, B. J., P. E. Kriedemann and C. G. N. Turnbull. 2003. *Plants in Action: Adaptation in Nature, Performance in Cultivation.* First Edition. Reprints 2003. Macmillan Education Australia Pty Ltd: Melbourne, Australia, p. 650.

Azam, M., F. M. Tahir, R. Anwar, M. A. Pervez and S. Rehman. 2007. Effect of gibberellic acid and potassium nitrate spray on panicle physiology of mango (*Mangifera indica* L.). Proceedings International Symposium on Prospect of Horticultural Industry in Pakistan. 28th to 30th March. Institute of Horticultural Science, University of Agriculture, Faisalabad, pp. 126–30.

Baker, L. R., J. W. Scott and J. E. Wilson. 1973. Seedless pickles – a new concept. *Farm Sci.* 227: 1–12.

Bangerth, K. F. 2009. Floral induction in mature, perennial angiosperm fruit trees: Similarities and discrepancies with annual/biennial plants and the involvement of plant hormones. *Sci. Hort.* 122: 153–63.

Barba, R. C. 1974. Induction of flowering of the mango by chemical spray. Crop Science Society of the Philippines. Proceedings of the 5th. Scientific meeting. University of Nueva Caceres, Naga City, pp. 154–60.

Bernier, G. 1988. The control of floral evocation and morphogenesis. *Ann. Rev. of Plant Physiol.* 39: 175–219.

Bernier, G., Havelange, A., Houssa, C., Petitjean, A. and Lejeune, P. 1993. Physiological signal that induce flowering. *Plant Cell* 5: 1147–55.

Berova, M. and Z. Zlatev. 2000. Physiological response and yield of paclobutrazol treated tomato plants. *Plant Growth Regul.* 30: 117–23.

Blaikie, S. J. and V. Kulkarni. 2002. Manipulating flowering in mango, cv. Kensington Pride. *Acta Hortic.* 575: 791–6.

Boss, P. K., R. M. Bastow, J. S. Mylne and C. Dean. 2004. Multiple pathways in the decision to flower: Enabling, promoting, and resetting. *Plant Cell* 16: S18–S31.

Brunner, A. M. and O. Nilsson. 2004. Revisiting tree maturation and floral initiation in the poplar functional genomics era. *New Phytol.* 164: 43–51.

Bunger-Kibler, S. and F. Bangerth. 1982. Relationship between cell number, cell size and fruit size of seeded fruits of tomato (*Lycopersiconesculentum* Mill.), and those induced parthenocarpically by the application of plant growth regulators. *Plant Growth Regul.* 1: 143–54.

Cárdenas, K. and E. Rojas. 2003. Efecto del paclobutrazol y nitratos de potasio y calcio sobre el desarrollo del mango 'Tommy Atkins'. *BioAgro* 15: 83–90.

Carvalheiro, L. G., C. L. Seymour, S. W. Nicolson and R. Veldtman. 2012. Creating patches of native flowers facilitates crop pollination in large agricultural fields: Mango as a case study. *J. Appl. Ecol.* 49: 1373–83.

Chacko, E. K. 1991. Mango flowering-Still an enigma. *Acta Hort.* 291: 12.

Chacko, E. K. and T. V. Ananthanarayanan. 1982. Accumulation of reserve substances in Mangifera indica L. during flower initiation. *Z. Pflanzenphysiol.* 106: 281–5.

Chaikiattiyos, S., Menzel, C. M. and Rasmussen, T. S. 1994. Floral induction in tropical fruit trees: Effects of temperature and water supply. *J. Hortic. Sci.* 69, 397–415.

Chen, W. S. 1987. Endogenous growth substances in relation to shoot growth and flower bud development of mango. *J. Am. Soc. Hortic. Sci.* 112: 360–3.

Chen, K. Y. and Coleman, G., 2006. Type II MADS-box genes associated with poplar apical bud development and dormancy. Ann. Meeting Amer. Soc. Plant Biologists. Abstr. p. 11.

Chutichudet, B., P. Chutichudet and T. Chanaboon. 2006. Effect of gibberellins (GA_3) on fruit yield and quality of Kaew mango (*Mangifera indica* L.) cv. Srisaket 007 in Northeast Thailand. *Pak. J. Biol. Sci.* 9: 1542–6.

Cline, J. A., C. G. Embree, J. Hebb and D. S. Nichols. 2008. Performance of prohexadione-calcium on shoot growth and fruit quality of apple - Effect of spray surfactants. *Can. J. Plant Sci.* 88: 165–74.

Dag, A., D. Eisenstein and S. Gazit. 2000. Effect of temperature regime on pollen and the effective pollination of `Kent' mango in Israel. *Sci. Hort.* 86: 1–11.

Davenport T. L. and R. Nuñez-Elisea. 1997. *Reproductive Physiology in the Mango.* Edited by R. E. Litz. CAB, International New York: USA, pp. 69–146.

Davenport, T. L. 2000. Processes Influencing floral initiation and bloom: The role of phytohormones in a conceptual flowering model. *HortTechnology* 10: 733–9.

Davenport, T. L. and Z. Ying. 2004. Translocation of the mango florigenic promoter. *HortScience* 39: 887.

Davenport, T. L., T. Zhang and Z. Ying. 2006. Isolation of genes potentially regulating mango flowering. Proceedings 33rd PGRSA Annual Meeting.109.

Davenport, T. L. 2007. Reproductive physiology of mango. *Braz. J. Plant Physiol.* 19: 363–76.

Davenport, T. L. 2009. Reproductive physiology. In Litz, R. E. (Ed.), *The Mango: Botany, Production and Uses.* 2nd. Edition. CAB International: USA, pp. 97–169.

Ding, P. and K. B. Darduri. 2013. Morphology of Chok Anan mango flower grown in Malaysia. *Afr. J. Agric. Res.* 18: 1877–80.

Ditta, G., A. Pinyopich, P. Robles, S. Pelaz and M. F. Yanofsky. 2004. The SEP4 Gene of *Arabidopsis thaliana* functions in floral organ and meristem identity. *Curr. Biol.* 14: 1935–40.

Do Carmo-Mouco, M. A., E. O. Ono and J. D. Rodriguez. 2011. Controle do crescimento vegetativo e floração de mangueiras cv. Kent com reguladores de crescimento vegetal. *Rev. Bras. Frutic., Jaboticabal – SP* 33: 1043–7.

Fierro, C. A. and M. Ulloa. 1991. A developmental reference stage for flower induction response to potassium nitrate. *Acta Hortic.* 291, 71–7.

Fischer, G., P. J. Almanza-Merchán and F. Ramírez. 2012. Source-sink relationships in fruit species: A review. *Revista Colombiana de Ciencias Hortícolas* 6: 238–53.

Galán-Saúco, V. 1999. *El Cultivo del Mango.* Ediciones Mundi Prensa: Madrid, España, p. 298.

Galán- Saúco, V. 2009. *El cultivo del Mango* Ediciones Mundi Prensa (2nd ed.): Madrid, España p. 340.

Garner, L. C., Y. Zheng, T. Khuong and C. J. Lovatt. 2010. Prohexadione-Calcium affects shoot growth of evergreen subtropical woody perennials differently than deciduous temperate zone woody perennials–Is it a case of apples and oranges? *Acta Hortic.* 884: 249–56.

Gehrke-Vélez, M. R. 2008. Reflexiones sobre problemas de biología reproductiva del mango 'Ataúlfo' en el Soconusco, Chiapas. *Tecnología en Marcha* 21: 174–83.

Gehrke-Vélez, M. R., A. Castillo-Vera, C. Ruiz-Bello and J. L. Moreno-Martínez. 2011. Viabilidad y germinación del polen en mango (*Mangifera indica* L.) cv. 'Ataulfo'. *Interciencia* 36: 378–85.

Giakountis, A. and G. Coupland. 2008. Phloem transport of flowering signals. *Curr. Opin. Plant Biol.* 11: 687–94.

Gillaspy, G, H. Ben-David and W. Gruissem. 1993. Fruits: A developmental perspective. *Plant Cell* 5: 1439–51.

Goguey T. 1993. Study of the effects of three flower-inducing substances on 'Kent' and 'Zill' mango. *Acta Hortic.* 341, 216–24.

Hansen, P. 1982. Assimilation and carbohydrate utilization in apple. Proc. 21st Int. *Hort. Cong.* 1: 257–68.

Hedhly, A. 2011. Sensitivity of flowering plant gametophytes to temperature fluctuations. *Environ. Exper. Bot.* 74: 9–16.

Jackson, I. E. and I. W. Palmer. 1977. Effects of shade on the growth of apple trees. II. Effects on components on yield. *J. Hortic. Sci.* 52: 253–66.

Jun-hu, H., M. Feng-wang and Ch. Ye-yuan. 2013. Fruit development and endogenous hormone concentrations in mango fruit with aborted and non-aborted embryos. *Acta Hortic.* 992: 139–47.

Khan, S. A. and A. Perveen. 2009. Pollen germination capacity of three mango cultivars (*Mangifera indica* L., Aanacardiaceae) from Pakistan. *Pak. J. Bot.* 41: 1009–12.

Kiessling-Davison, C. M., J. E. Magaña-Magaña, A. Segovia-Lerma, A. J. Obando-Rodríguez and V. H. Villareal-Ramírez. 2008. Prohexadiona de calcio como regulador de crecimiento en el manzano (*Malus domestica* Borkh.) 'Golden Delicious', Ciudad Cuauhtémoc, Chihuahua, Mexico. *TecnoCiencia Chihuahua* 1: 7–12.

Krizek, B. A. and E. M. Meyerowitz. 1996. The *Arabidopsis* homeotic genes APETALA3 and PISTILLATA are sufficient to provide the B class organ identity function. *Development* 122: 11–22.
Kulkarni, V. J. 1991. Physiology of flowering in mango studied by grafting. *Acta Hortic.* 291: 95–104.
Kumar, M., V. Ponnuswami, P. J. Kumar and S. Saraswathy. 2014. Influence of season affecting flowering and physiological parameters in mango. *Sci. Res. Essays.* 9: 1.
Liu, C. H., W. Xi, L. Shen, C. Tan and H. Yu. 2009. Regulation of Floral Patterning by Flowering Time Genes. *Dev. Cell* 16: 711–22.
Liu, Y., J. Mu, K. J. Kiklas, G. Li and S. Sun. 2012. Global Warming reduces plant reproductive output for temperate multi-inflorescence species on the Tibetan plateau. *New. Physiol.* 195: 427–36.
Liu, L., Y. Zhu, L. Shen and H. Yu. 2013. Emerging insights into florigen transport. *Curr. Opin. Plant Biol.* 16: 607–13.
Machado, E. M. and A. R. Sao Jose. 2000. Effect of different intervals of potassium nitrate spraying on flowering and production of mango trees *(Mangifera indica* L.) cv. Tommy Atkins. *Acta Hortic.* 509, 581–5.
Matsuo, S., K. Kikuchi, M. Fukuda, I. Honda and S. Imanishi. 2012. Roles and regulation of cytokinins in tomato fruit development. *J. Exp. Bot.* 63: 695–709.
Medeiros de Siqueira, K. M., L. H. Pieda de Kiill, C. Feitosa-Martins, I. Borges Lemos, S. Pitombeira-Monteiro and E. de Araújo Feitoza. 2008. Estudo comparativo da polinização de *Mangifera indica* L. em cultivo convencional e orgânico na região do Vale do Submédio do São Francisco. *Rev. Bras. Frutic.* 30: 303–10.
Medina-Urrutia, V. M. 1995. Effect of paclobutrazol and severe pruning on mango trees cv Tommy Atkins at high densities. *Proc. Fla. State Hort. Soc.* 108: 364–8.
Mishra, K. A. and B. S. Dhillon. 1978. Carbohydrates and mineral composition of leaves in relation to fruit bud differentiation in 'Langra' mango. *Ind. Agric. Sci.* 48: 46–50.
Mosqueda V. R. and F. De los Santos R. 1982. Aspersiones de nitrato de potasio para adelantar e inducir la floración del mango cv. Manila en México. Proceedings of the Tropical Region. 29th Congress American Society of Horticultural Science. Campinas-SP-Brasil. Vol. 25, pp.311–15.
Mukherjee, S. 2009. Introduction: Botany and importance. In Litz, R. (Ed.), *The Mango Botany, Production and Uses.* CAB International: New York, USA, pp. 1–22.
Muñoz-Fambuena, N., C. Mesejo, M. C. González-Mas, D. J. Iglesias, E. Primo-Millo and M. Agustı. 2012. Gibberellic Acid Reduces Flowering Intensity in Sweet Orange [*Citrus sinensis* (L.) Osbeck] by repressing CiFT Gene Expression. *J. Plant Growth Regul.* 3: 529–36.
Nakagawa, M., C. Honsho, S. Kanzaki, K. Shimizu and N. Utsunomiya. 2012. Isolation and expression analysis of FLOWERING LOCUS T-like and gibberellin metabolism genes in biennial-bearing mango trees. *Sci. Hort.* 139: 108–17.
Nakano, R., S. Inoue, Y. Kubo and A. Inaba. 2002. Water stress-induced ethylene in the calyx triggers autocatalytic ethylene production and fruit softening in 'Tonewase' persimmon grown in a heated plastic-house. *Postharvest Biol. Technol.* 25: 293–300.
Nishikawa, F., T. Endo, T. Shimada, H. Fuji, T. Shimizu, M. Omura and Y. Ikoma. 2007. Increased CiFT abundance in the stem correlates with floral induction by low temperature in Satsuma mandarin (Citrus unshiu Marc.). *J. Exp. Bot.* 58: 3915–27.
Núñez-Elisea, R. 1985. Flowering and fruit set of a monoembryonic and polyembryonic mango as influenced by potassium nitrate sprays and shoot decapitation. *Proc. Fla. State Hortic. Soc.*, 98: 179–83.
Núñez-Elisea, R. 1994. Environmental, developmental and bioregulator control of flowering in mango (Mangifera indica L.). Ph. D. Dissertation. University of Florida, pp. 129–49.
Núñez-Elisea, R. and T. L. Davenport. 1994. Flowering of mango trees in containers as influenced by seasonal temperature and water stress. *Sci. Hort.* 58: 57–66.
Núñez-Elisea, R. and T. L. Davenport. 1995. Effect of leaf age, duration of cool temperature treatment, and photoperiod on bud dormancy release and floral initiation in mango. *Sci. Hort.* 62: 63–73.
Núñez-Elisea, R., T. L. Davenport and M. L. Caldeira. 1996. Control of bud morphogenesis in mango (Mangifera indica L.) by girdling,defoliation and temperature modification. *J. Hort. Sci.* 71: 25–40.

Núnez-Elisea, R. and T. L. Davenport. 1998. Gibberellin and temperature effects on dormancy release and shoot morphogenesis of mango (Mangifera indica L.). *Sci. Hort.* 77: 11–21.

Ogata, T., H. Tamura, K. Hamada and K. Hasegawa. 2010. Effect of gibberellin on setting and growth of non-pollinated parthenocarpic fruit in mango. *Acta Hortic.* 884: 597–604.

Osuna-Enciso, T., E. M. Engleman, A. E. Becerril R., R. Mosqueda V., M. Soto H. and A. Castillo M. 2000. Iniciación y diferenciación floral en mango 'Manila'. *Agrociencia* 34: 573–81.

Palanichamy, V., N. N. Reddy, S. Babu, E. Selvaraj, Aranganathan and B. Mitra. 2012. Determination of time period of fruit bud differentiation and the associated histological and biochemical changes in mango Hybrids. *Res. J. Pharm. Biol. Chem. Sci.* 3: 271–90.

Pandey, R. M. 1989. Physiology of flowering in mango. *Acta Hortic.* 231: 361–80.

Pandolfini, T. 2009. Seedless fruit production by hormonal regulation of fruit set. *Nutrients* 1: 168–77.

Pathak, R. A. and R. M. Pandey. 1977. A note on the status of mineral content of inflorescence and fruits at different stages of their growth in mango cv. Dusehri. *Indian J. Plant Physiol.* 20:41–3.

Peña, J. E. 2003. Insectos polinizadores de frutales tropicales: No solo las abejas llevan la miel al panal. *Manejo Integrado de Plagas y Agroecología (Costa Rica)* 69: 6–20.

Pérez-Barraza, M. H. and V. Vázquez-Valdivia. 2006. Fenología y Fisiología Reproductiva del Mango. En Vázquez-Valdivia and Pérez-Barraza (Eds), *El Cultivo del Mango: Principios y Tecnología de Producción*, Libro Técnico Núm. 1. INIFAP-CIRPAC.CESIX: Nayarit, México, pp. 149–85.

Pérez B. M. H., and V. Vázquez V. 2006. Defoliación de brotes apicales y su efecto en la diferenciación floral del mango Tommy Atkins. *Revista Fitotecnia Mexicana* 29: 197–202.

Pérez-Barraza, M. H., V. Vázquez V. and J. A. Osuna G. 2008. Uso de giberelinas para modificar crecimiento vegetativo y floración en mango Tommy Atkins y 'Ataulfo'. *Revista Chapingo Serie Horticultura* 14: 169–75.

Pérez-Barraza, M. H., V. Vázquez-Valdivia. and J. A. Osuna-García. 2009. Floral bud development of 'Tommy Atkis' mango under tropical condition in Nayarit, Mexico. *Acta Hortic.* 820: 197–204.

Pérez-Barraza, M. H., Osuna-García, J. A., Sánchez-Lucio, R. and Vázquez-Valdivia, V. 2011. El paclobutrazol como promotor de la floración en mango 'Manila', aun sin condiciones ambientales inductivas. *Revista Chapingo Serie Horticultura* 17(1):47–52.

Pérez-Barraza, M. H., T. Osuna-Enciso, M. A. Gutiérrez-Espinosa, M. de J. Santiago-Cruz, E. Avitia-García and R. Cano-Medrano. 2015. Thidiazuron y ácido giberélico en amarre y tamaño de frutos de mango 'ataulfo' polinizados y partenocárpicos. *Interciencia* 40: 677–83.

Pérez-Barraza, M. H. 2015. Regulación del proceso de floración y desarrollo de frutos de mango 'Ataulfo' mediante biorreguladores de crecimiento. Tesis de Doctarado en Ciencias. Colegio de Postgraduados, Montecillo, Texcoco Edo. de México, México, p. 148.

Pérez-Barraza, M. H., T. Osuna-Enciso, E. Avitia-García, M. A. Gutiérrez-Espinosa, M. de J. Santiago-Cruz, H. Ramírez and R. Cano-Medrano. 2016. Prohexadiona de calcio reduce crecimiento vegetativo e incrementa brotación floral en mango 'Ataulfo'. *Revista Mexicana de Ciencias Agrícolas* 7: 263–76.

Phavaphutanon, L., K. Krisanapook, A. Pichakum and K. Jutamanee. 2000. Changes of total non-structural carbohydrates within shoots of 'Nam Dok Mai' mango after paclobutrazol application. *Acta Hort.* 509: 559–65.

Pillitteri, L. J., C. J. Lovatt and L. L. Walling. 2004. Isolation and characterization of a TERMINAL FLOWER homolog and its correlation with juvenility in citrus. *Plant Physiol.* 135: 1540–51.

Pongsomboon, W., S. Subhadrabandhu and R. A. Stephenson. 1997. Some aspects of the ecophysiology of flowering intensity of mango (*Mangifera indica* L.) cv. Nam Dok Mai in a semi-tropical monsoon Asian climate. *Sci. Hort.* 70: 45–56.

Rademacher, W. 2000. Growth retardants: Effects on gibberellin biosynthesis and other metabolic pathways. *Annu. Rev. Plant Physiol. Plant Mol. Biol.* 51: 501–31.

Rademacher, W., J. B. Speakman, R. R. Evans, J. R. Evans, S. Roemmelt, S. Michalek, A. Lux-Endrich, D. Treutter, T. Iturriagagoitia-Bueno and P. John. 1998. Prohexadione-Ca: A new plant growth regulator for apple with interesting biochemical features. *HortScience* 33: 540.

Rajan, S. 2012. Phenological responses to temperature and rainfall: A case study of mango. In Sthapit, B. R., Rao, R. V., Sthapit, S. R. (Eds), *Tropical Fruit Tree Species and Climate Change*. Bioversity International: New Delhi, India, pp. 71–96. ISBN: 978-92-9043909-7.

Ram, S. 1992. Naturally occurring hormones of mango and their role in growth and drop of the fruit. *Acta Hortic*. 321: 400–11.

Ramirez, F. and T. L. Davenport. 2010. Mango (*Mangifera indica* L.) flowering physiology. *Sci. Horti.* 126: 65–72.

Ramírez, F., T. L. Davenport and G. Fischer. 2010. The number of leaves required for floral induction and translocation of the florigenic promoter in mango (*Mangifera indica* L.) in a tropical climate. *Sci. Hort.* 123: 443–53.

Ramírez, H., S. Alonso and A. Benavides. 2006. Prohexadione-Ca modifies growth and endogenous hormones in the shoot apex in apple trees. *Acta Hortic.* 727: 117–24.

Rani, V., M. W. Ansari, A. Shukla, N. Tuteja and G. Bains. 2013. Fused lobed anther and hooked stigma affect pollination, fertilization and fruit set in mango. *Plant Signal. Behav.* 8: 23167.

Rath, S., L. R. Singh, B. P. Singh and D. B. Singh. 1980. Effect of boron and zinc on physico-chemical composition of mango fruits. *Punjab Hort. J.* 20: 30–5.

Ravishankar, H., M. M. Rao and K. M. Bojappa. 1979. Fruit bud differentiation in mango 'Alphonso' and 'Totapuri' under mild tropical rainy conditions. *Sci. Hort.*, 10: 95–9.

Ravishankar, H. and M. M. Rao 1982. Studies on changes in carbohydrate fractions and minerals in Alphonso mango shoot at pre, during and post-fruit bud differentiation stages. *J. Maharashtra Agric. Univ.* 7: 2.

Ravishankar, H., U. G. Nalawadi and N. H. Hulamani. 1993. Use of growth regulators to manipulate alternate bearing rhythm in mango (*Mangifera indica* L.). *Karnataka J. of Agric. Sci.* 6:7–12.

Rebolledo, M. A., A. L. Del Angel P. and J. Rey M. 2008. Effect of paclobutrazol and KNO_3 over flowering and fruit quality in two cultivars of mango Manila. *Interciencia* 33: 518–22.

Reece, P. C., J. R. Furr and W. C. Cooper. 1949. Further studies of floral induction in the 'Haden' mango (*Mangifera indica* L.). *˘mer. J. Bot.* 36:734–40.

Rodriguez, K., M. Aranguren and E. Farrés. 2007. Efecto del paclobutrazol en el desarrollo vegetativo e inicio de la floración en dos cultivares de mango (Mangifera indica l.). Memorias del II Simposio Internacional de Fruticultura Tropical y Subtropical. La Habana, Cuba. s/p.

Ruan Y. L., J. W. Patrick, M. Bouzayen, S. Osorio and A. R. Fernie. (2012). Molecular regulation of seed and fruit set. *Trends Plant Sci.* 17, 656–65.

Ruiz-García, L., F. Madueño, M. Wilkinson, G. Haughn, J. Salinas and J. M. Martínez-Zapatera. 1997. Different roles of flowering-time genes in the activation of floral initiation Genes in *Arabidopsis*. *Plant Cell* 9: 1921–34.

Salazar-García, S., M. H. Pérez-Barraza and V. Vázquez-Valdivia. 2000. Effect of ammonium nitrate sprays on flowering and harvest time of 'Manila', 'Ataulfo', and 'Tommy Atkins' mango in Nayarit, México. *Acta Hort.* 509:573–80.

Samach, A. and H. M. Smith. 2013. Constraints to obtaining consistent annual yields in perennials. II: Environment and fruit load affect induction of flowering. *Plant Sci.* 207: 168–76.

Samra, J. S. and Y. K. Arora. 1997. Mineral nutrition. In Litz, R. E. (Ed.), *The Mango: Botany, Production and Uses*. CAB International: Wallingford, UK, pp. 175–201.

Sargent, E. and F. Leal. 1989. Inducción floral en mango con KNO3. *Revista de la Facultad de Agronomía*. 15: 17–32.

Sarker, B. C. and M. A. Rahim. 2013. Yield and quality of mango (*Mangifera indica* L.) as influenced by foliar application of potassium nitrate and urea. *Bangladesh J. Agril. Res.* 38: 145–54.

Scaven, V. L. and N. E. Rafferty. 2013. Physiological effects of climate warming on flowering plants and insect pollinators and potential consequences for their interactions. *Curr Zool.* 59(3): 418–26.

Schaffer, B., L. Urban, P. Lu and A. W. Whiley. 2009. Ecophysiology. In Litz, R. E. (Ed.), *The Mango: Botany, Production and Uses*. CAB International: Cambridge, UK, pp. 210–30.

Scholefield, P. B., D. R. Oag and M. Sedgley. 1986. The relationship between vegetative and reproductive development in the mango in northern Australia. *Aust. J. Agric. Res.* 37: 425–33.

Sedgley, M. and A. R. Griffin. 1989. *Sexual Reproduction of Tree Crops*. Academic Press Inc.: Sa Diego, CA, p. 363. ISBN: 0-12-634470-1.

Serrani, J. C., M. Fos, A. Atares and J. L. García-Martínez. 2007. Effect of gibberellin and auxin on parthenocarpic fruit growth induction in the cv. micro-tom of tomato. *J. Plant Growth Regul.* 26: 211–21.

Serrani, J. C., O. Ruiz-Rivero, M. Fos and J. L.García-Martínez. 2008. Auxin-induced fruit-set in tomato is mediated in part by gibberellins. *Plant J.* 56: 922–34.

Shaban, A. E. A. 2009. Vegetative growth cycles of some mango cultivars in relation to flowering and fruiting. *W. J. Agric. Sci.* 5: 751–9.

Shu, Z. H. and T. F. Sheen. 1987. Floral induction in axillary buds of mango (*Mangifera indica* L) as affected by temperature. *Sci. Hort.* 31: 81–7.

Simao, R. A., A. P. Silva, F. H. Peroni, J. R. do Nascimento, R. P. Louro, F. M. Lajolo and B. R. Cordenunsi. 2008. Mango starch degradation. I. A microscopic view of the granule during ripening. *J. Agric. Food Chem.* 56: 7410–15.

Singh, R. R. 1977. Effect of various concentrations of boron on growth characters and chemical composition of leaves of mango (*Mangifera indica* L.) cv. Langra. *Bangladesh Hort.* 5: 30–4.

Singh, R. R. and C. B. S. Rajput. 1977. Effect of various concentrations of zinc on vegetative growth characters, flowering, fruiting and physico-chemical composition of fruits in mango (*Mangifera indica* L.) cultivar Chausa. *Haryana J. Hort. Sci.* 6: 10–14.

Singh, V. K. and A. K. Bhattacherjee. 2005. Genotypic response of mango yield to persistence of paclobutrazol in soil. Sci. Hort. 106: 53–9.

Singh, Z., U. Mallik and T. L. Davenport. 2005. Fruit drop in mango. In Janick, J. (Ed.), *Horticultural Reviews, Volume 31*. John Wiley & Sons: USA, pp. 11–153

Sing, Z. 2009. Gibberellin type and time of application influence fruit set and retention in mango. *Acta Hortic.* 820: 407–12.

Srilatha, V. and Y.T.N. Reddy. 2015. Pruning and Paclobutrazol Induced Flowering and Changes in Phenols and Flavonoids of Mango (*Mangifera indica L.*) cv. Raspuri. *J. Eng. Comput. Appl. Sci.* 4: 43–7.

Srivastava, A. and A. K. Handa. 2005. Hormonal regulation of tomato fruit development: A molecular perspective. *J. Plant Growth Regul.* 24: 67–82.

Stover, E. 2000. Relationship of Flowering Intensity and Cropping in Fruit Species. *HotTechnology.* 10: 729–32.

Sukhvibul, N., A. W. Whiley, M. K. Smith, S. E. Hetherington and V. Vithanage. 1999. Effect of temperature on inflorescence and floral development in four mango (*Mangifera indica* L.) cultivars. *Sci. Hort.* 82: 67–84.

Sukhvibul, N., S. E. Hetherington, V. Vithanage, A. W. Whiley and M. K. Smith. 2000. Effect of temperature on inflorescence development and floral biology of mango (*Mangifera indica* L.). *Acta Hortic.* 509: 601–7.

Sukhvibul, N., A. W. Whiley and M. K. Smith. 2005. Effect of temperature on seed and fruit development in three mango (*Mangifera indica* L.) cultivars. *Sci. Hort.* 105: 467–74.

Sung, I. H., M. Y. Lin, C. H. Chang, A. S. Cheng, W. S. Chen and K. K. Ho. 2006. Pollinators and their behaviors on mango flowers in Southern Taiwan. *Formos. Entomol.* 26: 161–70.

Suryanarayana, V. 1978. Seasonal changes in sugar, starch, nitrogen and C:N ratio in relation to flowering in mango. *Plant Biochem. J.* 5: 108–17.

Tharanathan, R. N., M. H. Yashoda and T. N. Prabha. 2006. Mango (*Mangifera indica* L.), 'The king of fruits': An overview. *Food Rev. Int.* 22: 95–123.

Tongumpai, P., N. Hongsbhanich and C. H. Voon. 1989. Cultar – for flowering regulation of mango in Thailand. *Acta Hort.* 239: 375–8.

Tongumpai, P., K. Jutamanee and S. Subhadrabandu. 1991. Effect of paclobutrazol on flowering of mango cv. Khiew Sawoey. *Acta Hortic.* 291: 67–70.

Tongumpai, P., K. Chantakulchan, S. Subhadrabandhu and R. Ogata. 1996. Foliar application of paclobutrazol on flowering of mango. *Acta Hortic.* 455: 175–9.

Tromp, J. 1984. Flower bud formation in apple as affected by air and root temperature, air humidity, light intensity, and day length. *Acta Hortic.* 149: 39–47.

Ulger, S., S. Sonmez, M. Karkacier, N. Ertoy, O. Akdesir and M. Aksu. 2004. Determination of endogenous hormones, sugars and mineral nutrition levels during the induction, initiation and differentiation stage and their effects on flower formation in olive. *Plant Growth Regul.* 42: 89–95.

Upreti, K. K., Y. T. N. Reddy, S. R. Shivu Prasadb, G. V. Bindua, H. L. Jayarama and S. Rajan. 2013. Hormonal changes in response to paclobutrazol induced early flowering in mango cv. Totapuri. *Sci. Hort.* 150: 414–18.

Urban, L., X. Le Roux, H. Sinoquet, S. Jaffuel and M. Jannoyer. 2003. A biochemical model of photosynthesis for mango leaves: Evidence for the effect of fruit on photosynthesis capacity of nearby leaves. *Tree Physiol.* 23: 289–300.

Urban, L., M. Léchaudel and P. Lu. 2004. Effect of fruit load and girdling on leaf photosynthesis in *Mangifera indica* L. *J. Exp. Bot.* 55: 2075–85.

Varoquaux, F., R. Blanvillain, M. Delseny, and P. Gallois. 2000. Less is better: New approaches for seedless fruit production. *Trends Biotechnol.* 18: 233–42.

Vázquez V. V. and M. H. Pérez B. 2006. Dosis y épocas de aplicación de ácido giberélico en la floración y la cosecha del mango 'Ataulfo'. *Revista Fitotecnia Mexicana* 29: 197–202.

Vázquez, V. V., M. H. Pérez B., J. A. Osuna G. and M. A. Urías L. 2009. Intensidad de poda sobre el vigor, producción y peso del fruto, del mango 'Ataulfo'. *Revista CHapingo Serie Horticultura* 15: 127–32.

Vázquez-Valdivia, V., M. H. Pérez-Barraza and J. A. Osuna-García. 2009. Effect of gibberellic acid on flowering of 'Ataulfo' mangos. *Acta Hortic.* 820: 413–17.

Waite, G. 2002. Pests and Pollinators of mango. In Peña, J., Sharp, J. and Wysoki, M. (Eds), *Tropical Fruit Pests y Pollinators.* CAB International: Wallingford, UK, pp. 103–30.

Wani, I. A., M. Y. Bhat, A. A. Lone and M. Y. Mir. 2010. Unfruitfulness in fruit crops: Causes and remedies. *Afr. J. Agric. Res.* 5: 3581–9.

Wet, E., P. J. Robbertse and H. T. Groeneveld. 1989. The influence of temperature and boron on pollen germination in *Mangifera indica* L. *S. Afr. J. Plant Soil* 6: 228–34.

Whiley, A. W., T. S. Rasmussen, J. B. Saranah and B. N. Wolstenholme. 1989. Effect of temperature on growth, dry matter production and starch accumulation in ten mango (*Mangifera indica* L.) cultivars. *J. Hortic. Sci.* 64: 753–65.

Wilkie, J. D., Sedgley, M. and Olesen, T. 2008. Regulation of initiation in horticultural trees. *J. Exp. Bot.* 59:3215–28.

Winston, E. C. and R. M. Wright. 1986. Mango flower induction: Ethephon, potassium nitrate and cincturing. Proc. -1st Aust. Mango Res. Workshop. CSIRO, Melbourne, pp. 202–10

Winston, E. C. 1992. Evaluation of paclobutrazol on growth, flowering and yield of mango cv. Kensington Pride. *Aust. J. Exp. Agric.* 32: 97–104.

Yeshitela, T., P. J. Robbertse and P. J. C. Stassen. 2004. Effects of various inductive periods and chemicals on flowering and negative growth of 'Tommy Atkis' and 'Keitt' mango (*Mangifera indica* L.) cultivars. *N. Z. J. Hortic. Sci.* 31: 209–15.

Yeshitela, T., P. J. Robbertse and P. J. C. Stassen. 2005a. Effects of pruning on flowering, yield and fruit quality in mango (*Mangifera indica*). *Aust. J. Exp. Agric.* 45: 1325–30.

Yeshitela, T., P. J. Robbertse and P. J. C. Stassen. 2005b. Potassium nitrate and urea sprays affect flowering and yields of 'Tommy Atkins' (Mangifera indica) mango in Ethiopia. *S. Afr. J. Plant Soil*, 22: 28–32.

Chapter 5

Mango cultivation practices in the tropics: good agricultural practices to maximize sustainable yields

Sisir Mitra, International Society for Horticultural Science, India; and A. Bhagwan, Fruit Research Station, India

1 Introduction

Mango (*Mangifera indica* L.) originated in the Indo-Burma (Myanmar) region (Pandey and Dinesh, 2010). From its tropical centre of origin, the fruit gradually moved to subtropical countries and is now considered a pan-tropical fruit. Mango is grown in more than 100 countries of the world in different agroclimatic conditions and is commercially an important fruit crop for many countries in the tropics and subtropics. From ancient times this delicious fruit crop has achieved great cultural, socio-economic and religious significance in the Indian subcontinent. Mango is the national fruit of India, Pakistan and the Philippines and the mango tree is the national tree of Bangladesh. India has the richest collection of mango varieties.

Almost eighty per cent of the world's mango production comes from the following countries: India (40%), China (11%), Indonesia (6%), Pakistan (6%), Thailand (6%), Mexico (5%), Brazil (3%) and Bangladesh (2%) (Menzel and Le Lagadec, 2017). Asia accounts for nearly 78% of global mango production, whilst the Americas and Africa account for approximately 12.6% and 10.8%, respectively. Approximately 3–4% of world production is traded internationally, whilst the rest is traded and consumed within the countries of production (Mitra, 2016). India's major marketing season is April to July, while harvesting continues for eight to ten months in Brazil, Colombia, Kenya and Venezuela. The season is also quite long in Burkina Faso, Costa Rica, Indonesia, Jamaica, Mexico, Nicaragua and Puerto Rico (Mitra, 2016).

http://dx.doi.org/10.19103/AS.2017.0026.08

Mango is well adapted to tropical and subtropical climates between 25°N and 25°S of the equator and from sea level up to a height of 1200 m (Morton, 1987); however, generally mango is not grown commercially in areas above 600 m. A rainfall of about 125 cm, falling mostly during monsoon, is considered very suitable for mango production. In the subtropics mango is commercially cultivated even at 36°N (Galán Sauco, 2008); the cold months ensure greater floral induction, but late frost is a major risk that destroys the tender parts of the trees and flowers or young fruits. In the tropics, the mango grows almost everywhere but, for good commercial fruit production, a prominent dry season lasting more than three months is necessary. However, flowering is rather erratic in the tropics due to variable temperature, uneven rainfall distribution and high humidity. Mango growth is optimum in seasonally wet/dry climate zones of the lowland tropics. A dry and/or cool season causes uniform floral initiation and tends to synchronize bloom and harvest. Mango is best adapted to hot, dry areas that receive less than 400 mm of rainfall annually, but supplemental irrigation is desirable for highest yields in those areas. Dry weather during the flowering period is best for fruit production. High winds can knock fruits off trees or cause scarring, since the fruits hang on long, pendulous floral branches at the periphery of the canopy. Mango trees should be protected from strong winds, but windbreaks that shade or compete with them should be avoided. Anthracnose disease often destroys both flowers and developing fruits in humid, high-rainfall areas.

1.1 Factors affecting the productivity of mango in the tropics

1. Lack of quality planting material.
2. Majority of the mango orchards are old and senile with low productivity.
3. Imbalanced application of fertilizers and insufficient addition of organic manures.
4. Prevalence of micronutrient deficiency, especially zinc, magnesium and boron.
5. Poor water management practices in the orchard at the time of fruit development.
6. Lack of integrated pest management (IPM) and integrated disease management (IDM) package of practices for sustainable mango production.

1.2 Constraints on mango production in the tropics

1. Large tracts of saline and sodic soils.
2. Effects of climate change on flowering behaviour and productivity.
3. Depleting water table, making the mango production system unsustainable.
4. Orchard management problems in old and senile mango orchards.
5. Lack of standardized canopy management protocol specific to tropical mango production, including high-density orcharding.
6. Incidence of pest and diseases in the changing climatic scenario.

Major problems that are experienced while exporting fresh fruits from the tropics include low productivity (cost competitiveness) as compared to global standards, prevalence of low level of pre- and post-harvest technologies, fruit fly incidence, international quality standards and distortion in market channels. It is time to effect a paradigm shift so that the future development in this sector brings a better balance between production and other subsystems, *viz.* preharvest technologies, post-harvest processing, quality management, export infrastructure, supply chain, market information and marketing strategies. One of the major constraints in the export of mangoes from many tropical mango-growing countries is its non-compliance with Good Agricultural Practices (GAPs).

The concept of GAP has evolved in the context of a rapidly changing and globalizing food economy and as a result of the concerns and commitments of a wide range of stakeholders about food production and security, food safety and quality and the environmental sustainability of agriculture. These stakeholders include governments, food processing and retailing industries, farmers and consumers who seek to meet specific objectives of food security, food quality, production efficiency, livelihoods and environmental benefits in both the medium and long term. GAP offers a means to help reach those objectives. GLOBALGAP (initially EUREPGAP) is an organization established in 1997 and put into practice by the largest retail supermarket owners in Europe who came together in order to make sure that the agricultural products they place on their shelves are safe and not harmful or hazardous to human health.

Over the next ten years it extended to the whole European continent and then to the rest of the world. With the impulse of globalization, a growing number of producers and retailers from all over the world began to adhere to this initiative, which came to be known as GLOBALGAP in 2007, becoming the benchmark of GAP at the international level. Today, it has become a global standard valid worldwide. Approximately 70–80% of the big European retail supermarkets and producers are either members or are registered in GLOBALGAP (EUREPGAP) (www.globalgap.org/es).

GAP is a collection of principles applied for on-farm production and the post-production process, which results in the production of safe and healthy food and non-food agricultural products, taking into account economic, social and environmental sustainability. GAP arose because of global initiatives to improve food safety and quality from the Food and Agricultural Organization (FAO) of the United Nations and the WTO (World Trade Organization) which assist in fair trade between countries (Protacio, 2013).

2 Constraints and strategies: soil

2.1 Salinity and sodicity

Salinity and especially sodicity are the soil conditions that mainly occur under arid and semi-arid regions. The area where mango cultivation occurs in some places (e.g. Punjab, India) is fast declining due to an increase in soluble salts in soil. High concentration of all three sodium salts, *viz.* chloride, sulphate and carbonate, has detrimental effects on the leaves and plant parts of mango. Scorching of old mango leaves, which first appears on the tips and margins of leaves, was due to sodium accumulation, either by soil or irrigation water sodicity (Gazit and Kadman, 1983).

Mango is considered sensitive to saline conditions (Maas, 1986), leading to scorched leaf tips and margins, leaf curling and in severe cases reduced growth, abscission of leaves and death of trees (Jindal et al., 1976). To overcome the effects of soil salinity, the best method is the use of salt-resistant rootstocks. Kadman et al. (1976) demonstrated the saline-tolerant effect of the rootstock 13/1 which is widely used in Israel. The 13/1 rootstock has increased the productivity of mango cultivation in Israel where the soil is very infertile. Such rootstocks were also developed in the Canary Islands of Spain. Gomera-1 and Gomera-3 rootstocks were extensively studied for their resistance to saline conditions (Galán Saúco et al., 1988; Galán Saúco, 1990, 2008). Further, Gomera-1 rootstock was also found to be superior than Gomera-3 in terms of higher micronutrient uptake under saline conditions in mango cv Osteen (Duran Zuazo et al., 2004). In India, certain polyembryonic rootstocks, such as Kurukan, Moovandan and Nekkare, were found to be tolerant to saline conditions. Extensive research on their effects in mitigating the effects of salinity needs to be conducted for evolving a commercial rootstock specific to tropical Indian conditions. As many of the mango orchards in India are grown under irrigation, mainly by drip systems, analyses on the mango rootstocks and the influence of salt on nutrient uptake are also needed. In addition to salinity, mango is affected by alkalinity and drainage as well.

Alkalinity is injurious especially to young plants, which exhibit symptoms of burning. In hilly terrains, in the soft rocky areas, performance of trees is highly erratic, and deep black 'cotton' soils (so-called because they are used to grow cotton) are generally unsuitable. Trees growing in light sand must be fertilized periodically for satisfactory growth and fruit production. Mango trees tolerate some flooding or wet conditions only for a short period, and prolonged stagnation leads to leaf wilting, desiccation, stem dieback, reduced growth and tree death. Trees under water-stagnated conditions remain unhealthy and chlorotic. Vegetation, flowering and fruiting will be adversely affected. Hence, moderately sloping sites are preferred to prevent water logging.

2.2 Suitable soil conditions

Mangoes are well adapted to many soil types (light sandy loams to red and clay soils). Though mango can be cultivated in a wide range of soils, its most profitable cultivation is limited to certain groups, and it is most successful under red loam soil conditions. For best performance, it prefers deep, well-drained soils of loamy texture (Singh, 1967) and can be grown from alluvial to lateritic soils including red soils, medium black soils and deep red loam (Pandey and Dinesh, 2010). Like most other fruit crops, it prefers a slightly acidic soil. Soils beyond a pH of 7.5 (Majumdar and Sharma, 1990) are not suitable for mango cultivation, while those with a pH of 5.5–7.5 is preferred. Deep rich soils give the best production and fruit quality. Deep soils without impermeable layers permit the development of deep taproot, which helps with drought tolerance and wind resistance. Well-drained deep red sandy loams add colour and lustre to the skin of fruits (Pandey and Dinesh, 2010). Loamy, silt loam, sandy loam soils are positively correlated with productivity of mango in the tropics, while clay and clay loam soils are negatively correlated (Najmus Saadat, 2016). Depth of soil should be at least 2 m with uniform texture in order for the production to be higher. In the rainy season, the water table should be below 2 m from the ground level. After an orchard has been established, it is virtually impossible to correct soil limitations (physical or chemical) in many instances, but chemical adjustment of pH, phosphorus and other elements can be done by means of stock applications during soil preparations.

For sustainable cultivation of mango in tropical conditions where soil temperature greatly influences the nutrient uptake, constant replenishment of the soil with organic matter is essential. In tropical conditions, this is addressed by addition of organic carbon in the form of various organic sources such as farmyard manure, poultry manure and sheep manure. In addition, to make the nutrient available to the plant, research on the use of biofertilizers such as *Azotobacter*, *Pseudomonas fluorescens* and *Azospirillum* is being carried out. Preliminary information has revealed that the addition of 250 g of *Azotobacter* and 250 g of *Azospirillum* for a ten-year-old tree reduced the use of inorganic fertilizer by half in realizing the same yield potential from 100% of the recommended dose of fertilizer (Kundu et al., 2011). Further, addition of bio fertilizers was found to improve soil conditions, which in the long run helps in the sustainability of mango cultivation in erratic climatic conditions.

3 Constraints and strategies: climate

Though adapted to varied climatic conditions, mango is highly productive in the tropics and lowlands of the subtropics. Areas with bright sunny days and relatively low humidity during the flowering period are ideal for mango cultivation. Tropical mango production is constrained by cloudy weather and increased humidity, which encourages greater incidence of pests and diseases. This also interferes with the activity of pollinating insects, thereby significantly affecting the fruit set.

3.1 Altitude

Altitude has a definite role at the time of mango flowering. When altitude is 1000 m above sea level in the tropics, growth and productivity of the crop are reduced. It has been observed that for an increase of each 400 ft of altitude and each degree of latitude south or north of the tropics, flowering is delayed by four days (Hopkins, 1938).

3.2 Temperature

Temperature is probably the most important environmental factor to consider when selecting mango cultivars for particular sites. The mean temperature range for optimum growth of mango is 24–30°C (Whiley et al., 1989). Mature trees can withstand temperatures from −3.9°C for a few hours with injury to leaves and small branches. Young trees, however, die at −1.7°C to 1.1°C (Carmichael, 1958). Flowers and small fruits may get damaged if the temperature falls below 4.4°C for a few hours. Low temperature adversely affects plant growth, besides decreasing the percentage of perfect flowers (Young, 1955).

3.3 Effects of climate change on mango productivity

Climatic changes, especially temperature, during the flowering and fruit set period have been attributed to erratic flowering and poor fruit set in mango cv. Banganpalli, a commercial cultivar of southern India (Bhagwan et al., 2011). The flowering and fruit set in mango are mostly influenced by the temperature during flowering (Davenport, 2007). Floral induction of mango occurs during bud dormancy in cool temperatures around 15°C, and warm temperatures around 30°C prevent floral initiation of induced buds (Nunez-Elisea and Davenport, 1995). A certain number of days during which the temperatures are below 13°C are required for optimum flowering in mango cv. Alphonso (Rao, 1998). A night temperature of less than 15°C for three to four weeks is necessary for mango flowering to occur, and that above 14°C is needed for proper fruit set (Davenport, 2003). Mango cv. Keitt was found sensitive towards low temperature for flower induction (Yeshtela et al., 2004).

Analysis of temperature data from tropical regions of India showed that the shift of low night temperature of less than 15°C from November–December to January–February, and the subsequent sudden rise in temperature of above 35°C during February progressively delayed flowering and subsequently exposed the emergence of the flower panicle to high day temperatures, resulting in fewer perfect flowers, poor fruit set and decreased productivity of mango cv. Banganpalli. The peculiar climatic conditions prevailing in southern parts of India make it possible for the production of the off-season mango crop during October, in addition to the main season crop in May. However, the stimulus of this off-season is weak, leading to a poor off-season crop. The lack of a proper crop model for mango prevents prediction of the effects of climate change on mango tree development and production (Normand et al., 2015). Such a crop model would pave the way to the development of mitigation strategies such as cultivar and rootstock selection and improvement of cultivation practices.

Recommendations for mango growers in the tropics in fruiting branch formation include pruning of autumnal shoots, properly timing floral enforcing and maintaining flowers and fruitlets. This has largely increased mango yield and reduced the phenomenon of 'off years' (Chen, 2013). Synchronization of the vegetative growth of tree canopies in an orchard is a requisite first step in the flowering management programme. Synchronous growth throughout each tree allows all the stems in the canopy to be in the same physiological stage of maturity so that management can be most efficiently achieved uniformly throughout a section of orchard. Synchronized growth is best accomplished by tip pruning all the stems on the tree (Davenport, 2011), making it the first step in an annual flowering programme. It not only causes a uniform flush of growth throughout the canopy, but also eliminates growth- and flower-inhibiting factors in stems derived from the previous season's flowering and fruiting panicles (Davenport, 2006). Tip pruning also stimulates lateral shoot development, forming five to ten times the original number of productive stems, thus increasing potential yield.

It is essential that the trees have adequate water at the time of pruning to facilitate a rapid flushing response as even mild water stress during the tropical dry season discourages shoot initiation. A deeper level of pruning to shape trees, such as cutting branches larger than 2 cm in diameter, usually results in a second vegetative flush occurring about three months after the first one. It is essential that there be only one flush of vegetative growth that occurs about one month following the synchronizing prune (Davenport, 2006).

The next step in a mango flowering management programme is to decide whether to use paclobutrazol or uniconazole. Paclobutrazol and uniconazole are gibberellins-biosynthesis-inhibiting triazole plant growth retardants; they reduce the levels of gibberellins which are thought to be a vegetative promoter (Davenport, 2011). Both products are effective for assisting floral induction, with uniconazole being more effective than paclobutrazol (Davenport, 1994). Either product should be applied after the onset of regrowth following pruning (one to one-and-a-half months after prune date) depending upon cultivar. Although the correct dosage depends on cultivars, paclobutrazol should be applied in a soil drench containing 1–1.5 g of active ingredient per metre of canopy diameter (Nartvaranant et al., 2000), if applied during the rainy season, while half that amount is needed if applied during the dry season (Davenport, 2011).

It takes at least 90 days for either product to exert an effect on trees (Nartvaranant et al., 2000). Earlier initiation of flushes results in the formation of vegetative shoots, so it is still important to avoid growing conditions that promote initiation of frequent flushes when utilizing paclobutrazol. Another advantage of paclobutrazol is that it increases the proportion of tree shoots with flowers from 25 to 50% of the canopy to 80–100%. Increased yields result because of more flowering trees and the full canopy bears fruit instead of 25–50% of the canopy (Portacio, 2013).

4 Constraints and strategies: orchard management

One of the major constraints of mango production in the tropics is the presence of old and senile orchards. The traditional spacing of mango, which varies from eight to ten metres between the rows and plants, has resulted in the huge growth and intermingling of tree canopies. This intermingling of trees has resulted in less light penetration and humidity build-up, leading to low productivity as well as increased incidence of disease. The high growth rate of mango in the tropics has resulted in huge trees, which cause problems for many orchard management practices, especially spraying and harvesting. Spraying of the trees becomes difficult in old orchards due to lack of proper equipment, leading to heavy build-up of pest and diseases, ultimately leading to low productivity. The traditional old orchard also poses a problem when harvesting the fruits, leading to high post-harvest losses. Even though the damage is not visible during harvesting, it later develops due to various diseases such as anthracnose or black spot. Technologies such as centre opening and rejuvenation of old mango orchards are being commercially implemented by tropical mango farmers. A recent trend in mango cultivation is high-density plantation for which there are many advantages, such as low maintenance costs, low canopy trees, higher productivity and early returns. The success in high-density orcharding depends largely on canopy management and input management.

4.1 Canopy management

Canopy management is the manipulation of tree canopies to optimize the production of quality fruits. Tree training and pruning affect the quantity of sunlight intercepted by trees, as tree shape determines the proportion of leaf area to radiation. An ideal training strategy centres around the arrangement of plant parts, especially to develop a better plant architecture that optimizes the utilization of sunlight and promotes productivity.

Light is essential for the growth and development of mango. The green leaves harvest the sunlight to produce carbohydrates and sugars, which are transported to the sites where they are needed – buds, flowers and fruits. Better light penetration into the tree canopy improves tree growth, productivity, yield and fruit quality. The density and orientation of mango planting also impact light penetration in an orchard. Generally, in close plantings, quicker shading becomes a problem. An east-to-west row orientation results in more shading as compared to the west and south orientation of trees. Strong bearing branches tend to produce larger fruits. The initial objective should be to build up a strong and balanced framework of trees, which then equips them with appropriate fruiting. Obviously, pruning in the early years has to provide a strong and stocky framework with well-spaced limbs or any other desired shape.

Some of the basic principles of canopy management in mango are:

1 Maximum utilization of light.
2 Avoidance of built-up microclimate congenial for diseases and pest infestation.
3 Convenience in carrying out cultivation practices.
4 Maximization of productivity with quality fruit production.

Tree canopy management, especially size control, has become a priority for reducing production costs and increasing fruit yield and quality. Like temperate fruits, where tree management technologies have been developed and refined for over a century, similar tools and experiences can be applied with a few modifications in mango.

Tree management techniques specific to mango have been developed and are being used in different parts of the world. These can be adopted after certain modifications in different mango-growing regions. Early height control and tree canopy management are some important tree management techniques. Similarly, the problem of large tree size can be tackled by topping and hedging, because large and crowded trees pose many disadvantages. Appropriate height, topping and hedging and cutting angles, as well as time and frequency of hedging determined for mango – which are common practices in Israel, USA, Australia and South Africa – can be used by other countries for increased efficiency and production. Shaping the mango tree immediately after planting is important for maintaining desirable plant height at first branching, so that proper clearance for harvesting equipment is possible.

When the plants attain the necessary volume, the tree should be shaped into an informal pyramidal shape (keeping the top narrower than the bottom). This can be done by cutting back one or more nodes. Plants should be pruned to attain balance between vegetative activity and regular yield. The pruning method should be followed with the objectives of developing a compact plant, avoiding branch breaking and favouring plant balance. Maximum plant height should not exceed 60% of the row spacing and the canopy should not project to more than 45% of row spacing, to avoid plant interlacing.

The time of pruning is an important factor influencing the flowering and yield the following year. Regular bearing cultivars/hybrids, which flower on new shoots, can be pruned after harvest. Early cultivars can also be pruned in a similar way, while for late cultivars, complete pruning after harvest will result in a severe decrease in yield. Renewal pruning is also possible in mango where 25–30% of the shoots are cut back yearly and others remain to bear the crop.

4.2 New mango orchards

Heading back of plants should be carried out when they attain the age of one year. Heading back should be done with sharp secateurs to give a sharp and smooth cut. The height of heading back should be 60–100 cm from the ground, resulting in the emergence of new shoots. To develop an ideal open canopy, thinning of excessive shoots is needed. Thinning should be done in such a manner as to retain four well-distributed shoots in all directions, which will develop as primary branches. If the crotch angle of retained shoots is smaller, bending should be carried out at this stage to increase the crotch angle of newly developed shoots and should be done with a jute rope (use of nylon or poly threads should be avoided). Second cutting is required when these shoots attain maturity, determined by a colour change from green to brown. Generally, this stage comes after seven to eight months of shoot growth. This cutting also induces new growth during the ensuing summer. Again, thinning of excessive shoots should be carried out to ensure two to three shoots per primary branch. These shoots develop as secondary branches, and this initial training results in an open and spreading canopy of trees.

4.3 Mango-bearing orchards

Normally, the height of a planned mango orchard should not exceed 5 m. Over this height, harvesting and other cultivation operations become difficult. To manage the canopy of mango-bearing trees, the upright branches of trees should be identified and thinned out, in order to increase their productivity. Only one or two upright-growing branches from the centre of the tree should be removed, to significantly reduce tree height and to increase the availability of light inside the canopy for better photosynthesis. This process opens up the centre of the tree, allowing more light to pass inside the tree canopy. Cutting of upright-growing branches should be done at the beginning of summer from the base of their origin.

During the removal of branches, the first cut should be made on the lower side of the branch to give a smooth cut and avoid bark splitting. This method of pruning is commercially followed in southern India. Branches with a wide crotch angle should be protected as they are more productive. In mango-bearing trees, not more than 25% biomass should be removed at a time for better canopy management; otherwise it results in excessive vegetative growth. Under a high-density planting system, 10–15% of the biomass should be removed annually to increase light penetration inside the canopy. Removal of 10–15% biomass should include crisscross branches, dead wood and diseased shoots. Pruning of shoots is essential in plants that touch their canopy with other trees. In these plants, 20 cm of pruning every alternate year gives better yield in the bearing trees.

4.4 Rejuvenation of orchards

Old and senile orchards that have not been under a tree size control maintenance programme may need to be drastically pruned to establish a productive and manageable canopy. This type of drastic pruning is sometimes called 'skeletonizing' or 'hat-racking' and usually consists of removing most of the tree canopy back to either just main scaffold limbs or to the main trunk. This rejuvenating pruning process can be done in the whole orchard at once, or may be done over a two- to three-year period, pruning one-half or one-third of the orchard at a time. Generally, trees will re-establish moderate fruit production within three to four years. Proper tree management after pruning, including regular thinning of new shoots, nutrition and pest and disease management, is necessary for successful management of rejuvenation.

4.5 High-density orchards

Unlike in temperate fruit crops, the concept of high-density planting has gained momentum in mango. With the development of dwarf mango hybrids such as 'Amrapali' (Dashehari × Neelum) in India and by regular canopy management after harvest, it is possible to make high-density orchards at 2.5 m × 2.5 m or 3 m × 2.5 m. Although high-density planting has been standardized in India for this cultivar (2.5 m × 2.5 m), after a profitable yield for up to 14–15 years, it shows progressive decline in yield due to overcrowding of canopies (Sharma et al., 2001). The fruit production and quality depends on several factor prevailing during their growth and development. Amongst these, pruning is an important operation for obtaining quality yield from the fruiting trees, which involves judicious removal of vegetative parts. An unpruned tree can become very large, inhibiting light penetration inside the canopy. As a result, leaf sprout is decreased, photosynthetic activity remains low and high incidence of pests and diseases occurs due to high relative humidity (Lal and Misra, 2007).

Sunlight not only influences the flowering and fruit set, but also enhances quality and colour development of fruits (Hampson et al., 2002). For this reason, fruits in the top of the tree always have better quality than fruits in the lower, shaded part of the canopy (Crisosto et al., 1997).

Nowadays, due to consumer concern for quality, the concept and priorities of fruit production are changing all over the world. Both intrinsic and extrinsic attributes are integral parts of fruit quality. A previous study conducted on pruning in the mango tree in relation to better light penetration, fruit set and yield in pruned trees (Sharma and Singh, 2006) mainly focused on vegetative growth and fruiting behaviour without much attention to fruit quality (shape, size, total soluble solids (TSS), acidity and aroma). Ram et al. (2013) reported that pruning of 'Amrapali' mango trees resulted in a significant increase in fruit weight and other quality attributes such as better TSS and sugars as compared with fruits harvested from unpruned trees. Pruning protocols have been standardized in Israel where the plants are planted at 5 × 1.5 m spacing and also in South Africa, which has resulted in an early higher yield. However, pruning practices have not been standardized for such high densities in tropical India. For the success of high-density planting and ultra-high-density planting in India, standardization of pruning practices specific to tropical conditions is necessary.

5 Constraints and strategies: irrigation and nutrition

5.1 Irrigation in relation to productivity

For mango production to be successful, proper irrigation management needs to be in place. The most important decision in this aspect is to know when to irrigate and how much. In order to maintain an optimal water status and prevent trees from a significant loss of productivity under limited water availability, water relations of mango trees (water status, water requirement, physiological response to water deficit, etc.) need to be understood (Lu, 2013).

Irrigation promotes vegetative growth under favourable climatic conditions. It is, therefore, a commercial practice that, in tropical situations, should be discouraged during the fruit bud differentiation period. In addition to a low night temperature requirement for flowering, stress in the form of withholding irrigation two months before flower bud differentiation is necessary for good flowering in mango grown under tropical conditions. Irrigation during this period promotes vegetative bud instead of reproductive development, which ultimately adversely affects the fruit yield. Irrigation starts when the panicle emerges with an increase in the temperature. Irrigation during flowering enhances the fruit set and reduces fruit drop in mango grown under tropical conditions. Deficit irrigation during this period is highly detrimental to yield, resulting in poor fruit set and dropping of young fruitlets.

The mango tree needs continuous irrigation throughout the period of fruit development. During this period, under a hot and dry climate, irrigation checks the drop of immature fruits. Moisture deficit in soil can cause early maturity of fruits, so regular irrigation during fruit development and the maturity period improves the quality of fruits. Climate change aberrations in the tropics, which result in the erratic distribution of rainfall and drought–like environments, have impacted mango productivity. The depleting ground water concomitant with low water use efficiency practices in orchards has resulted in the declining productivity of mango. Water status in trees is commonly studied by measuring leaf water potential, but excessive latex exudation in mango prohibits the reliable measurement of leaf water potential.

Alternative methods, such as xylem sap flow in the tree trunk, microvariation (shrinking and expansion) of branch diameter (microdendrometry), leaf gas exchange and other techniques, have been used to study mango water relations (Lu, 2013). A low-cost farmer–friendly tool for irrigators – 'FullStop' wetting front detectors – was developed by CSIRO in Australia. 'FullStop' is a simple device buried in the ground in the rooting zone, which will instruct the irrigators on when to switch it off. This system has great potential as an aid to irrigation decision-making (Stirzaker et al., 2004). In Israel, the irrigation schedule was pulsed three to 12 times a day, resulting in increased water use efficiency and more yield per quantum of water used. Such standardization of irrigation scheduling is needed as a way of dealing with climatic variation.

5.2 Nutrition

One of the most important constraints in the productivity of mango in tropics is nutrition management. Nutrition of perennial fruit trees such as mango has not been given due attention in the past; however, in recent years, attention is being paid towards nutrition for increasing the productivity of mango. Mango responds well to nitrogen (N), phosphorus (P), potassium (K), calcium, magnesium and micronutrients, *viz.* zinc, boron, copper and molybdenum. Presence of any of these nutrients may directly or indirectly affect the health of the tree. Rao and Mukherjee (1989) recorded positive correlations between yield, leaf and soil N and P. In leaf, N concentrations are generally low, and many orchards belong to the deficient (%N) and severely deficient (%N) categories. Soil deficiencies of micronutrients are also common in mango.

It is important to realize that the existence of visual deficiency symptoms indicates an advanced stage of inadequate nutrition. Historically, nutrient deficiency is realized only based on visual symptoms without correlating with leaf and soil analyses. The non–quantitative approach is compelled to develop techniques of tissue and soil analyses for determining nutritional requirements on a proactive as well as scientific basis. In Israel, soil nutrition is well replenished with daily fertigation and the complete protocol for mango nutrition has been standardized with increased productivity to the tune of 40 t/ha. In fact, irrigation scheduling and fertigation go hand in hand in realizing such productivity levels in Israel. Over-application of nitrogen will stimulate vigorous growth, which will reduce productivity, and when applied at the wrong phonological stage, has the effect of decreasing the internal quality of the fruit (Galan Saúco, 2009).

To be able to apply the correct amounts and have feedback on previous application, it is important to have leaf samples analysed annually. Leaf analysis is usually done by collecting young mature leaves from fruit-bearing twigs, while in some countries (e.g. Australia) the leaves are collected just prior to panicle emergence for analysis (Meurant et al., 1978). Measurement of leaf N more regularly than annually or in individual trees provides an improved understanding of leaf N fluctuations throughout the phonological cycle and enables growers to monitor the effects of regular management practices such as fertilizer application, irrigation and pruning on leaf N concentrations (Bally and Still, 2013). The SPAD-502 chlorophyll metre can be used to rapidly assess mango leaf N content in the field (Bally and Still, 2013).

Mango productivity is also constrained by micronutrient deficiency, especially in tropical conditions where soil microbes play an important role. Foliar spraying of micronutrients is a common practice to overcome micronutrient deficiencies in order to improve fruit quality. Nutrients are generally made more quickly available to the plants by the foliar application than the soil application (Silberbush, 2002), and foliar application of micronutrients may be 6–20 times more effective than soil application in tropical mango cultivation (Liew, 1988). Spraying of boric acid to provide boron requirement increases vegetative flushes, number of inflorescences, fruit setting percentage and improved fruit quality of mango cv. Langra (Rajput et al., 1976). Foliar spraying of zinc or boron each at the rate of 0.8% on 13-year-old mango cv. Langra at full bloom stage was reported to improve total sugars, TSS and ascorbic acid content of fruits (Rath et al., 1980).

Syamal and Mishra (1989) observed that the ascorbic acid content of fruit increased by increasing the concentration of N, P and K alone or in combination with micronutrients as a foliar spray on 17-year-old Langra trees. Kumar and Chakrabarti (1992) reported an increase in sugar content and decrease in acidity of fruit by spraying 1% $ZnSO_4$ to 30-year-old Dushehari trees. Unbalanced fertilization, deficiencies of micronutrients and poor tree management are responsible for low productivity and poor fruit quality (Ahmad and Rashid, 2003). With an increase in high-density planting in the tropics, the supply of nutrients to the plant through fertigation and scheduling irrigation needs to be established for tropical mango production.

6 Conclusion

Mango is an important fruit for the world and the crop is produced in more than 100 countries. Mango production provides an excellent livelihood for growers and workers engaged in its production and allied activities, *viz.* transportation and distribution, trade, storage, nursery growing, mango processing, etc. There are, however, many production problems that require urgent attention for sustainable production.

The main problems are the irregular fruit-bearing nature of most of the favoured mango cultivars of the world. This low productivity is due to the effects of climate during flowering, poor harvesting methods and inadequate canopy, nutrition, irrigation and IPM and IDM methods followed by most of the growing countries. All these negatively affect the productivity of mango.

In mitigation, technologies must be generated to evolve tolerant rootstocks for abiotic stress, irrigation scheduling and fertigation under tropical conditions. These technologies will improve the sustainable productivity of mango in the long run. Further, canopy management protocol along with fertigation has to be standardized for high-density and ultra-high-density planting of mango.

7 References

Ahmad, N. and Rashid, M. (2003). *Fertilizers and their Use in Pakistan. Extension Bulletin*(3rd Ed.). NFDC, Islamabad, pp. 3–4.

Bally, I. S. E. and Still, L. A. (2013). Rapid monitoring of nitrogen in mango trees. *Acta Horticulturae*, 992: 107–13.

Bhagwan, A., Vanajalatha, K., Prabhakar Reddy, I., Sarkar, S. K. and Girwani, A. (2011). Standardization of dosage and time of soil application of Cultar on flowering and yield of Mango cv. Banganpalli. *Paper Presented at the Global Conference on Augmenting Production and Utilization of Mango: Biotic and Abiotic Stresses*, held on 21–24th June 2011, Lucknow, Uttar Pradesh, India.

Carmichael, W. W. (1958). Observations of cold damage to mangoes in Dade County and the lower west coast. *Proceedings of the Florida State Horticultural Society*, 90: 254.

Chen, Q. B. (2013). Perspectives on mango industry in Mainland China. *Acta Horticulturae*, 992: 25–35.

Crisosto, C. H., Johnson, R. S., Day, K. R. and DeJong, T., (1997). Orchard factors affecting postharvest stone fruit quality. *HortScience*, 32: 820–3.

Davenport, T. L. (1994). Potential problems with use of uniconazole on mango (*Mangifera indica* L.). *PGRSA Quarterly*, 22: 143–53.

Davenport, T. L. (2003). Management of flowering in three tropical and subtropical fruit tree species, *HortSci.*, 38:1331–5.

Davenport, T. L. (2006). Pruning strategies to maximize tropical mango production from time of planting to restoration of old orchards. *HortScience*, 41: 544–8.

Davenport, T. L. (2007). Reproductive physiology of mango, *Brazilian J. Pl. Physiol.*, 19:363–76.

Davenport, T. L. (2011). Mango reproductive physiology: challenges and opportunities. In: *Souvenir, Global Conference on Augmenting Production and Utilization of Mango: Biotic and Abiotic Stresses*, 21–24 June 2011. Central Institute for Subtropical Horticulture, Lucknow, India, pp. 58–71.

Duran Zuazo, V. H., Martínez-Raya, A., Aguilar Ruiz, J. and Franco Tarifa, D. (2004). Impact of salinity on macro and micro nutrient uptake in mango (*Mangifera indica* L. cv. Osteen) with different rootstocks. *Spanish Journal of Agricultural Research*, 2(1): 121–33.

Galán Saúco, V., Acuña Barreto, J. F., Fernández Galván, D., Socorro Monzón, A. R. and Herrera Rodríguez, L. 1988. Tolerancia de distintos patrones de mango (*Mangifera indica* L.) a la salinidad. *Actas de Horticultura*, 1:83–9.

Galán Sauco, V. (1990). Los frutales tropicales en los subtrópicos I. In: Galán, V. S. (Ed.), *Mango*. Mundi-Prensa, Madrid, Spain, pp. 54–94.

Galán Saúco, V. 2008. *El cultivo del mango* (2nd ed.). MundiPrensa, Madrid, 340p.

Galán Saúco, V. 2009. Physiological disorders. In: Litz, R. (Ed.), *The Mango, Botany, Production and Uses* (2nd ed.). CAB Internacional, pp. 303–16.

Gazit, S. and Kadman, A. (1983). *Selection of Mango Rootstocks for Calcareous Soils and Saline Irrigation Water in the Arava*, Special publication 222. Institute of Horticulture, The Volcani Center, Scientific activities, Israel, 93p.

Hampson, C. R., Quamme, H. A. and Brownlee, R. T. (2002). Canopy growth, yield, and fruit quality 'Royal Gala' apple trees grown for 8 years in five tree training systems. *HortScience*, 37: 627–31

Hopkins, A. D. (1938) *Bioclimatics: A Science of Life and Climate Relations*, US Dept. of Agric. Pub., 280p.

Jindal, P. C., Singh, J. P. and Gupta, O. P. (1976). Studies on salt tolerance in mango injurious effect of salts on young mango seedlings. *Progressive Horticulture*, 8: 65–74.

Kadman, A., Gazit, S. and Ziv, G. (1976). Selection of mango rootstock for adverse water and soil conditions in arid areas. *Acta Horticulturae*, 57: 81–8.

Kumar, R. and Chakrabarti, D. K. (1992). Assessment of loss in yield of mango (*Mangifera indica* L.) caused by mango malformation. *Indian Journal of Agricultural Science*, 67: 130–1.

Kundu, S., Dutta, P., Mishra, J., Rashmi, K. and Ghosh, B. (2011). Influence of biofertilizer and inorganic fertilizer in pruned mango orchard cv. Amrapali. *Journal of Crop and Weed*, 7(2): 100–3.

Lal, B. and Mishra, D. (2007). Effect of pruning on growth and bearing behaviour of mango cv. Chausa. *Indian Journal of Horticulture*, 64: 268–70.

Liew, C. S. (1988). Foliar fertilizers from Uniroyal and their potential in Pakistan. In: *Proceedings of Seminar on Micronutrients in Soils and Crops in Pakistan*, 277p.

Lu, P. (2013) Whole-tree water relations and irrigation scheduling for mango. *Acta Horticulturae.*, 992: 115–22.

Maas, E. V. (1986). Salt tolerance of plants. In: *Applied Agricultural Research*, Volume 1. Springer-Verlag, NY, USA, pp. 12–26.

Majumdar, P. K. and Sharma, D. K. (1990). Mango. In: Bose, T. K. and Mitra, S. K. (Ed.), *Fruits: Tropical and Subtropical*. Naya Prokash, Calcutta, pp. 1–61.

Menzel, C. M. and Le Lagadec, M. D. (2017) Can the productivity of mango orchards be increased by high-density plantings? *Scientia Horticulturae*, 218: 222–63.

Meurant, N., Holmes, R., MacLeod, N., Fullelove, G., Bally, I. and Kernot, I. (1978). *Mango Information Kit*. Queensland Dept. Primary Indust. and Fisheries, Australia.

Mitra, S. K. (2016). Mango production in the world-present situation and future prospect. *Acta Horticulturae*, 1111: 287–95.

Morton, J. (1987). *Mango: In Fruits of Warm Climates*, Miami, Florida, pp. 221–39.

Najmus Saadat, Md, (2016). Impact of soil on productivity of mango crop in Malda district, west Bengal. *American International Journal of Research in Humanities, Arts and Social Sciences*, 14(2): 121–6.

Nartvaranant, P., Subhadrabandhu, S. and Tongumpai, P. (2000). Practical aspects in producing off season mango in Thailand. *Acta Horticulturae*, 509: 661–8.

Normand, F., Lauri, P. E. and Legave, J. M. (2015) Climate change and its probable effects on mango production and cultivation. *Acta Horticulturae*, 1075: 21–32.

Nunez Elisea, R. and Davenport, T. L. (1995). Effect of leaf age, duration of cool temperature treatment and photoperiod on bud dormancy release and floral initiation in mango. *Scientia Horticulturae*, 62: 63–73.

Pandey, S. N. and Dinesh, M. R. (2010). *Mango*. Directorate of Information and Publications of Agriculture, ICAR, New Delhi.

Protacio, C. M. (2013). Towards a Good Agricultural Practice (GAP)- compliant mango production system in the Philippines. *Acta Horticulturae*, 992: 69–73.

Rajput, C. B. S., Singh, B. P. and Mishra, H. P. (1976). Effect of foliar application of boron on mango. *Scientia Hort.*, 5: 311–3 .

Ram, A., Patel, V. B., Barman, K. and Pal, R. K. (2013). Pruning affects fruit yield and postharvest quality in mango (*Mangifera indica* L) cv Amrapali. *Fruits*, 68: 367–80.

Rao, M. M. (1998). Role of low night temperature days in fruit bud differentiation in mango trees under mild tropical rainy conditions. *Karnataka J.Agric Sci.*, 11(4): 1142–4.

Rao, D. P. and Mukherjee, S. K. (1989). Nutrient status in leaf and soil of some cultivars of mango in relation to yield. *Acta Horticulturae*, 231: 286–9.

Rath, S., Singh, R. L. and Singh, D. B. (1980). Effect of boron and Zinc sprays on the physic-chemical composition of mango fruits. *Punjab Horticulture Journal*, 2: 33–5.

Sharma, R. R. and Singh, R. (2006). Effect of pruning intensity on light penetration and leaf physiology in Amrapali mango trees under high density planting. *Tropical Science*, 46: 16–19 .

Sharma, R. R., Singh, C. N., Saxena, S. K., Pandey, S. N. and Chhonkar, O. P. (2001). Cluster planting favours malformation and influences yield and fruit quality in mango. *Annals of Agricultural Research*, 22: 48–51.

Silberbush, L. F. (2002). Response of maize to foliar vs. Soil application of nitrogen, phosphorus and potassium fertilizers. *Journal of Plant Nutrition*, 25: 2333–42.

Singh, K. K. (1967). Climate and cultivation. In: Kurup, C. R. (Ed.), *The Mango: A Handbook*. ICAR, New Delhi, pp. 70–98.

Stirzaker, R., Etherington, R., Lu, P., Thompson, T. and Weilke, J. (2004). *The Farmer's Road to Clean and Green Horticulture: Evaluation of the FullStop Wetting Front Detector On-farm (2001–2003)*. RIDC, Australia, 74p.

Syamal, M. M. and Mishra, A. K. (1989). Effect of NPK on growth, flowering and quality of mango. *Acta Horticulturae*, 231: 276–81.

Whiley, A. W., Rasmussen, T. S., Saranah, J. B. and Wolstenholme, B. N. (1989). Effect of temperature on growth, dry matter production and starch accumulation in ten mango (*Mangifera indica* L.) cultivars. *Journal of Horticultural Sciences*, 64: 753–6.

Yeshitela, T. P., Robbertse, J. and Stassen, P. J. C. (2004). Effect of various inductive periods and chemicals on flowering and vegetative growth of 'Tommy Atkins' and 'Keitt' mango (*Mangifera indica*) cultivars. *New Zealand Journal of Crop and Horticultural Science*, 32: 209–15.

Young, T. W. (1955). Influence of temperature on growth of mango pollen. *Florida State Horticultural Society*, 68: 308–10.

Chapter 6

Mango cultivation practices for the subtropics

Víctor Galán Saúco, Instituto Canario de Investigaciones Agrarias, Spain

1 Introduction

Although there is no clear definition of the subtropics, this term can refer to the areas of the world that are free of frost, extending from above the 23° North (Tropic of Cancer) and below 23° South of the Equator (Tropic of Capricorn), which roughly reaches to around 35° of latitude in both hemispheres. But these boundaries are not very rigid and there are areas inside the tropics, especially at altitude, which may be considered subtropical. In addition, there are areas at higher latitude, particularly at sea level and with insular climates, which may exhibit climatic conditions that are similar to those of the tropics. This is the reason why some consider that the tropics can be extended to 30° North and South of the Equator (Nakasone and Paull, 1988).

Climatologically speaking, the subtropics have hotter summers and cooler winters than the tropics. Since tropical crops require a climate with average mean temperatures higher than 10 °C for the cooler months (Watson and Moncour, 1985), the isotherm of 10 °C average temperature for the coldest month has been considered by Nakasone and Paull (*op. cit.*) as the limit for distinguishing between tropical and subtropical areas.

The most prevalent climatic characteristics in the subtropics include (a) wide temperature fluctuations between day and night and between summer and winter, (b) high and low temperature extremes in summer and winter, respectively, and (c) low annual rainfall which generally is poorly distributed. Humidity is also generally lower than in the tropics, and day length differences become greater with increased latitude (Galán Saúco et al., 2012). These widely different climatic conditions explain the difficulties in cultivating tropical

http://dx.doi.org/10.19103/AS.2017.0026.09

fruits in the subtropics. However, there are specific advantages in the subtropics relating to the limitations of low temperatures and humidity on pathogen activity which will favour organic cultivation with the inherent added value. In addition, moderate low winter temperatures in the mild subtropics have a beneficial effect on flower induction, which is of special relevance in the case of mango (Albrigo and Galán Saúco, 2004).

The mango is well adapted to hot tropical conditions, but the ideal temperature conditions for its sustainable growth and cultivation are not much different from those occurring in the mild subtropics and include the following (Galán Saúco, 2008):

- moderate cold winter (minimum temperatures around 10 °C–15 °C) to induce abundant flowering;
- relatively warm spring (minimum temperatures above 15 °C) to favour good fruit set;
- warm summer and autumn seasons to get good fruit development and good vegetative growth after harvesting and a climate with small differences between night and day temperatures.

The mango is commercially planted both in the tropics and in the subtropics from 37° North latitude in Sicily (Italy) to 33° South latitude in South Africa. The incidence of frost is a limiting factor for mango cultivation and sets the geographic boundary for mango plantings. Young mango trees can die when exposed for several hours to temperatures below –4 °C, with death of adult trees occurring at −6 °C and young shoots being burnt at temperatures below 0 °C (Campbell et al., 1977). Leaf chlorosis may occur both at temperatures below 10 °C or above 40 °C, although the mango can withstand temperatures close to 50 °C during the period of fruit development if irrigated regularly (Majumder and Sharma, 1985; Donadio, 1980).

Although the most important mango-producing countries are located in the tropics, the cultivation of mango in the subtropics has increased much in the last few decades as a result of improved cultural techniques, especially those related to control of growth and flowering, and, on a minor scale, by obtaining better adapted cultivars and by protected cultivation. Furthermore, the highest yields for this crop are obtained in Israel, a subtropical country (see Table 1). Climatic and genetic factors, as well as good cultural practices that allow sustainability of growing mangoes are behind the successful cultivation of mangoes in the subtropics, and these will be discussed in this chapter.

2 Tropical versus subtropical mango cultivation: climatic considerations

Because of the tropical origin of mangoes – located between Assam (India) and Myanmar (Kosterman and Bompard, 1993) – this plant is better adapted to the tropical areas. Experiments conducted by subjecting mangoes to continuous day/night temperatures of 30 °C/25 °C or 20 °C/15 °C (Whiley et al., 1989; Núñez-Elisea and Davenport, 1994; Núñez-Elisea et al., 1996) indicate clear differences on growth behaviour, with their magnitude variable with cultivars, and these can be summarized as follows:

1. The interval dormancy growth lasts for an average of 6 weeks in warmer conditions (30 °C/25 °C) versus 20 weeks in cooler conditions (20 °C/15 °C).

Table 1 Main mango-producing countries in the world and yield (www.fao.org)

	Production (x 10^3 t)				Yield (t/ha)	
Country	2010	2011	2012	2013	2012	2013
India	15,020	15,188	15,250	18,002	6.63	7.20
China	4,351	4,519	4,400	4,450	9.56	9.56
Thailand	2,550	3,277	2,650	3,142	8.33	8.27
Indonesia	1,287	2,131	2,376	2,059	10.20	0.50
Mexico	1,633	1,827	1,761	1,902	9.00	9.56
Pakistan	1,845	1,888	1,950	1,659	11.21	9.68
Brazil	1,190	1,250	1,178	1,163	16.04	6.53
Bangladesh	1,048	889	945	950	7.62	7.66
Nigeria	843	801	860	850	6.51	6.54
Egypt	505	598	786	835	10.21	9.09
Philippines	790	795	783	831	3.97	4.23
Israel (*)	30	34	36	22	28.20	22.20

(*) Israel is not a major producer of mangoes at global level but is an important exporter and also one of the main mango-producing countries in the subtropics.

2 The average number of leaves/vegetative flux also varies in favour of warmer conditions (13.6 leaves vs 7.1).
3 Leaf size is also an important factor, with bigger leaves (up to 300%) seen in warmer conditions.
4 The plants exposed to 30 °C/25 °C temperature initiate vegetative buds in contrast to those exposed to 20 °C/15 °C temperature, which initiate flower buds.

As seen above, paradoxically, the warmer climatic conditions of the tropics are less favourable for flower induction than those of the subtropics and, in addition, the higher temperature and humidity occurring in some tropical regions, particularly in the humid tropics, may cause a severe reduction of yield (Galán Saúco, 2008). On the contrary, it is a well-known fact, repeatedly cited in the mango literature, that the stress caused by the moderately low winter temperatures occurring in the subtropics induces flowering of the receptive mango terminals (Davenport, 2009). This explains the profuse and homogeneous flowering of mango trees in the subtropics, in contrast to the phenomenon of erratic behaviour (Fig. 1) that is common to many parts of the tropics (Verheij, 1986; Goguey, 1997).

Despite the positive benefit for annual flowering and potential for increasing yield, many problems may also derive from it. Precocious flowering may occur in young grafted trees which, particularly if fruiting is allowed, will weaken the trees by depleting starch reserves, and can lead to retarded growth and premature ageing (see Fig. 2). For this reason, flowers, when produced, must be removed during at least the first two years after grafting seedlings. Several strategies, commented below, can be adopted to avoid this

Figure 1 Erratic behaviour in mango.

problem, including adequate location of mango nurseries in the subtropics, establishing them in warm areas or even under greenhouse.

Other disadvantages also indicated for the cultivation of mangoes caused by the low winter temperatures of the subtropics mentioned in previous works (Galán Saúco et al., 2012; Galán Saúco, 2015) include the following:

1 reduced pollen germination
2 slow growth of the pollen tube
3 embryo abortion and higher incidence of parthenocarpic fruits (Fig. 3a and b)
4 ovary malformation, reducing the viability of hermaphrodite flowers at temperatures between 7 and 10 °C
5 increase in the ratio of male/hermaphrodite flowers when winter temperatures during the inductive period are around 5 °C, which may cause a reduction in fruit set
6 alternate bearing in late cultivars
7 death of young plants or plant parts caused by temperatures close or below 0 °C
8 yellowing (chlorosis) of leaves caused by temperatures close to or below 10 °C
9 higher incidence of bacteriosis, both *Xanthomonas* and *Pseudomonas*

This explains why late flowering cultivars produce better than early flower cultivars in the subtropics when no flower manipulation is made (Galán Saúco, 2008). Delaying annual flowering in adult trees of most cultivars, to avoid flowering during periods of low winter temperatures, is of great importance for commercial cultivation of mango in the subtropics to tackle the first six problems mentioned above. The other problems can be addressed through appropriate nutrition and irrigation.

Figure 2 Prematurely aged 'Sensation' tree.

Figure 3 (a) Embryo abortion in mango (normal and aborted fruits) and (b) embryo abortion in mango.

Work done by several researchers has shown different procedures for delaying flowers till the onset of good temperatures such as a) pre-flowering treatments with gibberellic acid (Tomer, 1983); b) chemical or hand removal of inflorescences (several authors cited by Galán Saúco 2009); c) allowing natural destruction of early inflorescences (letting powdery mildew take care of them) (Galán Saúco, 2009); d) pre-bloom shoot tipping (Torres et al., 2009); and e) young flush thinning (Jannoyer and Lauri, 2009).

The longer the length of the cooler season, the more it is necessary to delay flowering, but late removal of inflorescences can induce vegetative growth instead of new inflorescences and, especially for late season cultivars, there might not be sufficient time to generate new vegetative flushes after harvesting. Trees may, in this case, enter the phenomenon of alternate bearing, or even erratic behaviour, which also occurs in the tropics as a consequence of insufficient flower induction due to lack of cool temperature (Verheij, 1986). An interesting strategy for late cultivars in the subtropics is explained in Section 5.

Not all are disadvantages derived from the low winter temperatures. The shortening of the juvenile life of mango indicated before as a consequence of the stress caused by the low winter temperatures can, however, be of great advantage, if properly managed, for the establishment of high-density plantings that are being increasingly used nowadays, especially in the subtropics (Oosthuyse, 2005 and 2009; Banik et al., 2013). Even under normal planting distances (see Sections 4 and 5), the lower number of flushes produced under subtropical conditions makes it easier to control mango growth facilitating all cultural practices, especially harvesting and spray programmes. On the other hand, the longer period of low winter temperatures occurring in the subtropics, compared with the tropics, allows the emission of a second wave of flowering, subterminal or axilar, once the first terminal has died or has been removed (Fig. 4), an almost unique case in tropical fruits, which may allow this second flowering coinciding with the rising of the temperatures occurring in spring. This will ensure good fruit set and avoid embryo abortion problems of susceptible cultivars.

It should be also mentioned that, although the general predicted climate changes occurring in the world moving towards an increase in temperature may suit, more or less, mango production by increasing vegetative growth and fruit set in the subtropical areas, this will pose further challenges for mango production in the tropical areas. In fact, as indicated by Normand et al. (2015), when considering the impact of climatic change on mango production and cultivation in the tropics, in the case of South Asia, for example, the climate evolution in the future may bring a warmer climate during the flowering season plus a warmer and wetter climate during the season of vegetative dormancy, leading probably to a lower flower induction. On the other hand, the climate can become hot and wet during fruit and vegetative growth which, although it will promote good and quick fruit growth and vegetative growth after harvest, may also increase pests and disease incidence, thus affecting fruit quality. This together with a projected increase in hurricanes and floods in the coastal areas will lead to a probable reallocation of mango cultivation towards the actual subtropical climate areas in the North or in the tropical areas at higher altitude. In the case of the Caribbean, the probable consequences of the expected impact of the climatic change may see a shift in mango cultivation towards the windward coast of the islands or in the altitudes to find wetter and cooler places, respectively. A similar trend may also be seen in other tropical areas of the world in Latin America or Africa where mangoes are cultivated. As also indicated by researchers, the genetic diversity of this species and its large ecological adaptability will, however, mean that the actual tropical zones of the world will continue to be important areas for mango production, but the current breeding and selection programmes should pay more attention to the future impact of climatic changes.

3 Exploiting genetic variation among mango cultivars

Both polyembryonic cultivars (i.e. 'Manila' in Mexico, 'Carabao' in the Philippines, Nam Doc Mai' and others in Thailand) and monoembryonic cultivars (i.e. 'Alphonso', 'Dasheri' and many others in India, as well as different cultivars in other tropical countries) produce high yield in the tropics. However, the monoembryonic cultivars selected in Florida in

Figure 4 Secondary flowering in mango.

the twentieth century, namely 'Tommy Atkins', 'Haden', 'Kent', 'Keitt' and 'Edwards' (Campbell and Ledesma, 2013a), are the most planted cultivars in the American tropics and also in the subtropics and even in some tropical areas in other parts of the world. Other Floridian cultivars like 'Osteen', which accounts for more than 85% of the around 5000 ha of mango planted in Spain, 'Cogshall' in Réunion Island and 'Lippens' in the dry coastal areas of the Canary Islands are also important, 'which indicates the wide adaptability of these cultivars'.

Some polyembryonic cultivars, that is, 'Gomera-1' in the Canary Islands (Galán Saúco, 2008) or 'Kensington' in Australia but also in Sicily (author's own observations), perform well under those subtropical conditions; 'Kensington' also performs well in tropical Australia. The adaptability of polyembryonic mangoes to subtropical conditions can be termed surprising because they have largely evolved in the consistently hot humid tropics of SouthEast Asia (Schaffer et al., 2009; Iyer and Schnell, 2009) under environmental conditions far from those prevailing in subtropical areas. In any case, the main mango breeding programmes in India, Thailand, Israel, South Africa and Brazil have focused most of their efforts on monoembryonic mangoes (Gazit, 1998; Galán Saúco, 2008), with the exception of Australia, which has obtained both mono- and polyembryonic interesting cultivars (Bally et al., 2009; Bally, 2013) and have given rise to several good new cultivars (i.e. Shelly in Israel or Calypso in Australia), well adapted to different environments and with excellent yields. Although these cultivars have not yet become worldwide planted, some of them – 'Shelly' in Europe or 'Calypso' in the United States – are starting to be known and marketed outside their country of origin (Galán Saúco, 2016b). Different Indian cultivars such as 'Langra', 'Dashehari', 'Amrapali' and 'Alphonso' have also been found well adapted to subtropical conditions in India (Singh et al., 2015) and the last in this list, as well as 'Mallika', seems to be well adapted to the subtropical areas of America with cool dry winters (Campbell and Ledesma, 2013a).

The exploitation of the great genetic potential of several species of the genus *Mangifera*, many of them genetically and graft compatible with the mango (Campbell, 2004; Campbell and Ledesma, 2013b), and some with cool tolerance, like *Mangifera sylvatica* (Bompard and Schnell, 1997), as well as the role that polyembryonic rootstocks can play in the adaptation of mango to different adverse edapho-climatic environments, are still two pendant subjects for mango cultivation both for the tropics and for the subtropics.

4 Cultural techniques: planting density, spacing and out-of-season production

Until recent times the mango has been considered a very rustic crop well adapted to many different edaphic and climatic conditions. In their places of origin, cultural techniques such as weeding, pest and disease control and removal of dead, badly conformed or broken branches have been kept to a minimum. It was only until its cultivation started in subtropical areas that special attention was paid to its cultural techniques. Israel in the northern subtropics was where modern techniques of control of irrigation, nutrition and flowering were developed and practised, while South Africa, in the southern subtropical area of the world, developed sound pruning techniques and control of orchard density to the point of making high-density plantings a successful venture (Oosthuyse, 1997, 2005 and 2009).

4.1 Planting density and spacing

There is a clear trend today in most fruit species to increase planting density. This is also the case for mangoes where modern high-density plantings at 4 × 2 m or even 3 × 1 m, oriented in North–South direction, are being increasingly utilized in the subtropics where normal planting distances were 5–7 × 3–4 m (Fig. 5a), and it is even closer when planted in terraces (Fig. 5b) (Galán Saúco, 2008).

Mangoes under greenhouse in the subtropics are cultivated at similar densities as in the open air in the Canary Islands (Fig. 6a), in Málaga, Spain, in trellis systems at planting distances of 2.5 × 2 m with 3 horizontal wires, the first at around 1.00 m above the ground and the highest at 2.5 m, or in palmetto systems like in El Algarve, Portugal (Fig. 6b), at 2.0–3.0 × 1.5–2.5 m (Galán Saúco, 2015). Cultivation in greenhouses is also practised in Japan and will be treated as a separate chapter in this book (Chapter 7).

Figure 5 Higher planting density in the subtropics.

The short juvenile phase, the reduced number of vegetative flushes per year and the regularity of flowering of mangoes in the subtropics allow closer distances between mango plants than in the tropics, facilitating high-density plantings, already established in several subtropical and even in some tropical countries. Such examples can be seen in Mexico and South Africa, with the cultivars 'Tommy Atkins', 'Kent', 'Keitt', 'Osteen', 'Heidi' and 'Irwin', or in Egypt with cultivars 'Eweisse', 'Saddik' and 'Sinnara' (Oosthuyse, 2009) or in India with 'Amrapali' (Majumder et al., 1982). When adequately trained and pruned to obtain maximum light interception, mangoes cultivated at high density (4 × 2 m) result in early production, higher yields and better control of pests (Oosthuyse, 2005). It is out of the scope of this chapter to discuss in depth either high-density plantings or cultivation of mangoes under greenhouses since two full chapters of this book will be devoted specifically to these two subjects.

4.2 Out-of-season production

Production of fruits in the tropics is limited normally to spring–summer time, except for multi-flowering cultivars. Under subtropical conditions, the period of harvesting in a particular location may be extended to late autumn and even in subtropical island climates like in the Canary Islands; the harvesting period may extend from the end of June under greenhouse (end of July under open-air conditions) with early cultivars such as 'Edwards' till February next year with late cultivars such as 'Keitt' (Galán Saúco, 2014). In addition, as indicated before, the removal of the normal flowering in cultivars that are not sensitive to embryo abortion (i.e. 'Lippens'), or the removal of inflorescences produced during winter and spring time in multi-flowering cultivars, may also result in out-of-season production for a particular cultivar (Galán Saúco, 2008).

Furthermore, the temperature conditions prevalent in the subtropics are not appropriate, due to the longer winter period, for the procedures used in the tropics to obtain out-of-season production which involves synchronizing canopy growth and stimulation of quick abundant vegetative flushing after harvest; stopping growth, usually with paclobutrazol or with water stress, to get flower induction; and, later, flower stimulation with chemicals such as potassium nitrate and other chemical products (Tongumpai et al., 1997, Albuquerque et al., 2002). This, together with the already mentioned longer period of harvesting in the subtropics, explains why these practices to obtain out-of-season production have not

Figure 6 (Top) Mangoes in greenhouses – Portugal (with the author). (Bottom) Mangoes in greenhouses in Canary Islands (D. F. Galvan, ICIA researcher passed away in June 2015).

been studied for the subtropics, except for some preliminary studies with early cultivars under greenhouse cultivation (Hernández Delgado et al., 2007).

5 Cultural techniques: control of growth and flowering

The lower number of vegetative flushes produced annually in the subtropics allows an easier control of growth, both laterally and in height than in the tropics. Training methods are, in any case, one of the most important cultural practices for the future performance of a mango planting with especial relevance in the subtropics.

The main objectives of training in mango are the following:

- to avoid too early flowering in the subtropics (not necessary in the tropics).
- to obtain quickly three to four branches at an appropriate distance from the ground, which serve as the main branches of the mango tree.
- to form quickly a well-ramified branch structure, for which it is necessary during the first two years after planting to remove any inflorescence produced and to pinch any mature vegetative terminal produced in spring-summer.
- to form a canopy that allows good light penetration, wind resistance, facility for application of chemical treatments and less sensibility to pest and diseases.

Control of growth and flowering is of special relevance in the subtropics to avoid early flowering in grafted young trees, which, due to cool winter temperature, can even occur at the nursery stage, weakening the tree by depleting reserves, particularly if fruiting is allowed (see Fig. 7). This explains the importance of an appropriate location for the nursery which should be established either at a warmer place or under a climatized greenhouse. Otherwise it may be necessary to spray regularly young grafted trees with gibberellic acid (GA_3) from the beginning of autumn till mid-spring at 100 ppm, the frequency and number of sprays (approximately one each 2 months) depending on cultivar and climatic condition. Alternatively, if flowers were produced, hand-pruning them immediately below the terminal node when fruits reach pea size (stage fruit set 1 of Aubert and Lossois (1972) or stage 701 of the BBCH scale is necessary (Hernández Delgado et al., 2011)) (Galán Saúco, 2008) (see Fig. 8).

Because of lower subtropical temperatures plants remain longer in the nursery than under tropical climates, which makes it necessary to keep an appropriate spacing between them to ensure good light penetration, avoiding the production of thin weak shoots, as well as the use of appropriate nursery bags. The bags should be deep enough (ideally a minimum of 35.5 cm with a width around 15 cm), preferably open totally or partially at the bottom, and placed on a structure that avoids direct contact with the soil to permit

Figure 7 Precocious flowering in a grafted plant at the nursery stage.

Figure 8 Stage 701 of the BBCH scale (pea size stage).

air pruning and to avoid root deformation. As in any place where mango is cultivated, but particularly for modern cultivation systems in the subtropics, it is also essential to avoid stem deformations for which careful selection in the nursery of healthy plants with leaves of green colour is mandatory. This includes the elimination of too compact plants with shorter internodes and especially those with a rosette of small leaves at the apex or even those that have flowered in the nursery, clear symptoms of having experienced a stress situation that may influence further development.

Early flowering can also be avoided in subtropical climates by field planting polyembryonic seedlings of rootstock and grafting two years later once the plant has grown vigorously (Galán Saúco and Fernández Galván, 1987). This may be a good procedure to reduce costs of plant material in the case of high-density plantings, but careful management practices have to be adopted to ensure uniformity and to avoid losses of grafted plants. There is also the need for specialized grafting labour in the farm (Galán Saúco, 2016a).

It is out of the scope of this chapter to discuss training and pruning procedures in depth as these have already been described in detail in several papers (Oosthuyse, 1997; Galán Saúco, 2008; Crane et al., 2009) and particularly in the subtropics by Galán Saúco (2016a). But it is important to mention that when adequately trained, flowering and fruiting can be allowed in the subtropics at the third year after planting, depending on climate and cultivars. In general, the warmer the summer–autumn season, the more and longer flushes are produced and the earlier can the mango be left to enter production (Galán Saúco, 1990). Cultivars like 'Heidi' and 'Zill', which usually produce additional vegetative shoots concurrently with fruit growth, may even be allowed to enter production at the second year, in the case of well-developed trees (Oosthuyse, 1995). In those very productive cultivars, like 'Lippens' in the Canary Islands, which produces in racemes (Galán Saúco and Fernández Galván, 1987), two to three fruits per panicle may be allowed to mature even at the second year, provided that around half of the fruiting terminals are eliminated by cutting them at least 5 cm below the terminal node when the fruits reach pea size. Shoots produced later in the season on the cut terminals develop generally vigorous vegetative growth, instead of new inflorescences. In the case of late cultivars, such as 'Keitt', and especially when most terminal set fruits, it is more recommendable under subtropical conditions to wait until even the fourth year before allowing them to fruit in order to allow the tree to reach a good size before entering production.

Because of the relatively low winter subtropical temperatures, it is of utmost importance to make the annual flowering coincide with the onset of spring temperatures over 15.5 °C to avoid especially the above-mentioned embryo abortion problems. As indicated by Galán Saúco (2008), this can be done by either delaying flowering or removing the first wave of flowering appearing during winter. Delaying annual flowering can be done through winter sprays of gibberellic acid, as indicated above for the nursery site, plus cutting back (2 cm) of all latent terminals or through selective pruning as it is done in Réunion Island with cultivar 'Cogshall' by removal of young yellow vegetative shoots as soon as they appear at each flush period. This delays flowering and also harvest (Jannoyer and Lauri, 2006). Similar results regarding delaying flowering have been obtained in different cultivars in Málaga (Spain) after tipping (15–25 cm) terminal shoots about 6–8 weeks before full bloom, but yields are not consistently good for all cultivars (Torres et al., 2006), indicating again the need of local trials before recommending this practice in a particular location.

Elimination of the first annual flowering can be done with chemical products, such as ethrel (800–1600 ppm), hydrogen cyanamide (0.4–0.75%), cyclohexamide (0.25g/l), dinoseb (0.5 ml/l) and pentachlorophenol (5.0 g/l) (several authors cited by Galán Saúco, 2009). A more sustainable and ecological approach will be the removal of inflorescences by hand, cutting them immediately above the terminal node and at the stage of full open flower [(stage E of Aubert and Lossois (1972) or 615 of the BBCH scale (Hernández Delgado et al., 2011)] (see Fig. 9). It has also been recommended in the dry subtropics letting powdery mildew naturally destroy the first flowering wave (Galán Saúco, 2008). However, this technique can be applied only when uniform terminal flowering and no young vegetative flushes have been produced simultaneously with the annual flowering.

There may be, however, some exception regarding the stage at which the first flowering should be eliminated. In the event that the terminal flowering is produced very early in the season, it may be better to wait even until the stage of pea size of fruits as there may still be cool temperature conditions for induction of a second wave of flowering; but to avoid any risks, only hand removal is recommended in this case. Some very productive cultivars that are not sensitive to embryo abortion, such as 'Lippens', may produce good fruit set even under mild subtropical winter temperature conditions (Fig. 10), but removal of the first annual flowering can still be useful when, for marketing reasons, some delay in harvesting may be required or simply to synchronize flowering, as it has been done in South Africa for cultivar 'Sensation' (Oosthuyse and Jacobs, 1997b). Multi-flowering cultivars such as 'Neelum' and 'Royal Special' that have not yet been studied under subtropical conditions, and that are able to produce out-of-season flowers in India with temperatures around 40 °C (Kulkarni, 2004), may also be able to produce out-of-season fruits by delaying flower

Figure 9 Stage 615 BBCH scale (full open flower stage).

Figure 10 Heavy fruit set. Cultivar Lippens.

Figure 11 Flowering at different months in 'Nam Doc Mai' in the Canary Islands.

removal until late in the year. Although this has to be experimentally proven as it may cause depletion of tree reserves. Observations made by the author in the subtropical conditions of Tenerife, Canary Islands makes it clear that at least the cultivar 'Nam Doc Mai' flowers in different times of the year and can produce some fruits out of season during the winter (Fig. 11).

The delay in the annual mango flowering will, of course, increase cultivation costs and, as indicated before, may cause alternate bearing phenomena in 'Keitt' and other late season cultivars in the subtropics since, the consequent delay in harvesting will reduce the number of summer days available for shoot growth after harvest. Quick pruning of fruited terminals immediately after harvest, together with good irrigation, fertilization and sanitation practices, is very important but, since this may not be sufficient to avoid alternance, it has been recommended that the annual after-harvest pruning in these cultivars be limited only to 75% of the terminals, leaving the pruning of the remaining 25% for spring and giving preference to winter pruning of those terminals that have not produced fruits (Galán Saúco, 2008). A different approach is being followed in Málaga (Torres et al., 2009), where in the 'on' year forcing the alternance on the cultivar 'Keitt' is recommended by cutting back 15–25 cm with a 'machete' all the terminals at the beginning of spring time (around mid-March), also eliminating any eventual fruit produced. This practice allows the build-up of carbohydrate reserves during the following 'off' year. The high production in the 'on' year reduces individual fruit weight and increases the colour (because of the fewer leaves produced and more light interception) of 'Keitt' mangoes, giving rise to a biennial production of an excellent crop of fruits with the appropriate size and colour demanded by the market (Galán Saúco, 2015).

6 Conclusion

Mango is successfully cultivated not only in several subtropical countries of the world including places such as Israel, where the highest yield of the world is obtained. This chapter reviews the differences between mango cultivation in tropical and subtropical climates, especially referring to climatic and genetic considerations and cultural techniques other than irrigation and nutrition. It indicates the importance of flower manipulation to make this coincide with appropriate temperatures for fruit set and the importance of cultural techniques such as appropriate plant density including high-density planting and pruning to obtain sustainable high yields in the subtropics both for open air and greenhouse cultivation.

More research is needed in relation to the use of *Mangifera* species for breeding and as rootstocks for mangoes and also in the use of polyembryonic rootstocks more adapted to the climatic and edaphic condition of the subtropics as these two subjects may substantially contribute not only to improve yield but also to sustainable cultivation of mango in the subtropics. Of especial interest will be to find rootstocks with greater ability to absorb nutrients, especially calcium, which may help to the control of the Internal Fruit Breakdown, this being of great interest not only for the subtropics but also for the tropics (Galán Saúco, 2009).

The impact of climate change may be even an advantage for mango cultivated in the subtropics, because of the increasing temperature in winter (Normand et al., 2015). However, salinity and water shortage problems may become worse in future. This will require rootstock research as well as the kind of water deficit research being undertaken in Israel (Levin et al., 2015). Changing climatic conditions may also introduce new pests and diseases into the subtropics which will require new phytopathological research. Finally, there need to continuing efforts to find new cultivars which are less susceptible to internal fruit breakdown and embryo-abortion problems, thus avoiding the need either to delay flowering or to remove the first wave of flowers appearing during winter.

7 Where to look for further information

Research in the topics discussed in this chapter is currently done in several countries in the subtropics including Australia, Israel, South Africa, Spain and USA with the main research institutions listed below:

- Australia - Queensland Department of Agriculture, Horticulture and Forestry Science (www.daf.qld.gov.au)
- Israel - Agricultural Research Organization, Volcani Center (www.agri.gov.il)
- South Africa - ARC Institute for Tropical and Subtropical Crops (www.arc.agric.za)
  HortResearch. www.hortresearchsa.co.

- Spain - Instituto de Hortofruticultura Subtropical y Mediterránea. Estación Experimental 'La Mayora' (www.ihsm.uma-csic.es)
  Instituto Canario de Investigaciones Agrarias (www.icia.es)
- USA - University of Florida's Institute of Food and Agricultural Sciences (UF/IFAS) (trec.ifas.ufl.edu/tropical-fruits-crop-faculty
- Fairchild Tropical Botanical Garden (www.fairchildgarden.org)

8 References

Albrigo, G. L. and Galán Saúco, V. (2004). Flower Bud Induction, Flowering and Fruit-set of Some Tropical and Subtropical Fruit Tree Crops with Special Reference to Citrus. *Acta Horticulturae*, 632, 81–90.

Albuquerque, J. A. S., Medina, V. D. and Mouco, M. A. C. (2002). Induçao Floral. In Genú, P. J. C. and Pinto. A. C. Q. (Eds), *A Cultura da Mangueira*. Embrapa Informaçao Tecnológica. Brasilia, 259–76.

Aubert, B. and Lossois, P. (1972). Considérations sur la Phènologie des Espèces Fruitières Arbustives. *Fruits*, 27 (4), 269–86.

Bally, I. S. E. (2013). New Mango Hybrids from Australia. *Acta Horticulturae*, 975, 55–61.

Bally, I. S. E., Lu, P., Johnson, P., Muller, W. J. and González, A. (2009). Past, Current and Future Approaches to Mango Genetic Improvement in Australia. *Acta Horticulturae*, 820, 153–63.

Banik, B. C., Maity, P. K., Hasan, M. A. and Ghosh, S. N. (2013). High Density Orchard Systems for 'Himsagar' Mango in the New Alluvial Zone of West Bengal. *Acta Horticulturae*, 975, 241–6.

Bompard, J. M. and Schnell, R. (1997). Taxonomy and Systematics. In: Litz, R. E. (Ed.), *The mango, Botany, Production and Uses*. CAB International. Wallingford and Oxon, 21–48.

Campbell, R. J. (2004). Graft Compatibility between *Mangifera* Species and *Mangifera indica* 'Turpentine' Rootstock and their Subsequent Horticultural Traits. *Acta Horticulturae*, 645, 311–13.

Campbell, R. J. and Ledesma, N. (2013a). The Changing Face of Cultivars for the Western Hemisphere. *Acta Horticulturae*, 992, 55–8. https://doi.org/10.17660/actahortic.2013.992.4.

Campbell, R. J. and Ledesma, N. (2013b). Update of New *Mangifera* Species in Florida, USA. *Acta Horticulturae*, 992, 95–8. https://doi.org/10.17660/actahortic.2013.992.10.

Campbell, C. W., Knight, R. J. and Zareski, N. L. (1977). Freeze Damage to Tropical Fruits in Southern Florida in 1977. *Proceedings of the Florida State Horticultural Society*, 90, 254–7.

Crane, J. H., Salazar García, S., Lin, T. S., de Queiroz Pinto, A. C. and Ziu, Z. H. (2009). Crop Production Management. In Litz, R. E. (Ed.), *The Mango, Botany, Production and Uses*, 2nd edn. CABI International. Wallingford and Oxon, 432–83.

Donadio, L. C. (1980). Cultura da mangueira. Livroceres. Piracicaba. Brazil.

Galán Saúco, V. (1990). Los Frutales Tropicales en los Subtrópicos. I. Aguacate, Mango, Litchi y Longan. Mundi-Prensa. Madrid.

Galán Saúco, V. (2008). *El Cultivo del Mango*, 2nd edn. Mundi-Prensa. Madrid.

Galán Saúco, V. 2009. Physiological Disorders. In Litz, R. (Ed.) *The Mango, Botany, Production and Uses* (2nd edition. CAB International), 303–16.

Galán Saúco, V. (2014). Mango, Litchi y Pitaya. In Hueso Martín, J. J. and Cuevas González, J. (coordinators). *La Fruticultura del Siglo XXI en España Cajamar.* Caja Rural. España, 350–80.

Galán Saúco, V. (2015). Ventajas y Desventajas del cultivo del Mango (*Mangifera indica* L.) en Zonas Subtropicales y Potencial del Cultivo bajo invernadero. *Acta Horticulturae* 1075, 167–78. https://doi.org/10.17660/actahortic.2015.1075.19.

Galán Saúco, V. (2016a). Mango Culture and Production. Control of Growth and Development in the Subtropics. In Al Saqri, E. A. (Ed.), *The Mango Tree Encyclopedy*. Vol. 1. Royal Court of Oman. Muscat. In print.

Galán Saúco, V. (2016b). Trends in Mango World Production and Marketing. Invited lecture at the XIth International Mango Symposium. Darwin. Australia. 28 September 2015 to 02 October 2015.
Galán Saúco, V. and Fernández Galván, D. (1987). El cultivo del Mango en Canarias. Cuadernos de divulgación n° 1. Consejería de Agricultura y Pesca de Canarias.
Galán Saúco, V. Herrero, M. and Hormaza, J. I. (2014). Chapter 5. Tropical and Subtropical Fruits. In Dixon, G. R. and Aldous, D.E. (Eds), *Horticulture: Plants for People and Places*. Springer Science-Business Media. Dordrech, 123–57.
Galán Saúco, V., Robinson, J. C. Tomer, E. and Daniells, J. (2012). Current Situation and Challenges of Cultivating Banana and other Tropical Fruits in the Subtropics. *Acta Horticulturae*, 928, 19–30. https://doi.org/10.17660/actahortic.2012.928.1.
Guzmán-Estrada, C. (2006). Destruction of Flowering (Blossoming) in Mango to Delay Harvest. 8th International ISHS Mango Symposium. Sun City. South Africa. 5–10 February 2006. Book of Abstract, additional not paged document.
Hernández Delgado, P. M., Fernández Galván, D., Galán Saúco, V., López Hernández, D. and Tomer, E. (2007). Comunicaciones: Poster Estudios Preliminares Sobre Tratamientos para el Adelanto de la Época de Floración y Recolección de Mango en Invernadero. In Fruticultura II. Contenido Científico. Resúmenes XI Congreso SECH. Agriculturas Minoritarias. 24–27 Abril 2007. Albacete, 47.
Hernández Delgado, P. M., Aranguren, M., Reig, C. P., Fernández Galván, D., Mesejo, C., Martínez Fuentes, A., Galán Saúco, V. and Agusti, M. (2011). Phenological Growth Stages of Mango (*Mangifera indica* L.) according to the BBCH Scale. *Scientia Horticulturae*, 130, 536–40. https://doi.org/10.1016/j.scienta.2011.07.027.
Iyer, C. P. A. and Schnell, R. J. (2009).Breeding and Genetics. In Litz, R. E. (Ed.), *The Mango, Botany, Production and Uses*, 2nd edn. CAB International. Wallingford and Oxon, 67–96.
Jannoyer, M. and Lauri, P.-E. (2009). Young Flush Thinning in Mango (cv. Cogshall) Controls Canopy Density and Production. *Acta Horticulturae*, 820, 395–402.
Kosterman, A. J. G. H. and Bompard, J. M. (1993). *The Mangoes, their Botany, Nomenclature, Horticulture and Utilization*. Academic Press. London.
Levin, A.G., Naor, A., Noy, M., Love, C., Gal, Y. and. Peres, M. 2015. Mango Response to Deficit Irrigation at Different Phenological Periods. *Acta Horticulturae*, 1075, 103–13.
Majumder, P. K. and Sharma, D. K. (1985). Mango. In Bose, T. K. and Mitra, S. K. (Eds), *Fruits: Tropical and Subtropical*. Naya Prokash. Calcutta, 1–62.
Majumder, P. K., Sharma, D. K. and Singh, R. N. (1982). A Study of High Density Orcharding in Mango (*Mangifera indica*) var. Amrapali. *Punjab Horticultural Journal*, 22, 123–7.
Nakasone, H. Y. and Paull, R. F. (1998). *Tropical Fruits*. CAB International. New York.
Normand, F., Lauri, P.-É and Legave, J.-M. (2015). Climate Change and its Probable Effects on Mango Production and Cultivation. *Acta Horticulturae*, 1075, 21–31.
Núñez-Elisea, R. and Davenport, T.-L. (1994). Flowering of Mango Trees in Containers as Influenced by Seasonal Temperature and Water Stress. *Scientia Horticulturae*, 58, 57–66. https://doi.org/10.1016/0304-4238(94)90127-9.
Núñez-Elisea, R., Davenport, T. L. and Caldeira, M. L. (1996). Control of Bud Morphogenesis in Mango (*Mangifera indica* L.) by Girdling, Defoliation and Temperature Modification. *Journal of Horticultural Science*, 71, 25–40. https://doi.org/10.1080/14620316.1996.11515379.
Oosthuyse, S. A. (1995). Relationship between Branching Frequency, and Growth, Cropping and Structural Strength of Two-year-old Mango Trees. *Scientia Horticulturae*, 64, 85–93.
Oosthuyse, S. A. (1997). Some Principles Pertaining to Mango Pruning and the Adopted Practices of Pruning Mango Trees in South Africa. *Acta Horticulturae*, 455, 413–21.
Oosthuyse, S. A. (2005). *Greatly Enhanced Ease in Growing Mangoes on Small Trees*. SQM Europe N. V. Belgium. www.sqm.com.
Oosthuyse, S. A. (2009). Management of a Tommy Atkins, Ultra-high Density Orchard and Recognised Benefits associated with Small Tree Mango Orchards. *Acta Horticulturae*, 820, 325–38.
Oosthuyse, S. A. and Jacobs, G. (1997). Flowering Synchronization of 'Sensation' Mango Trees by Winter Pruning. *Acta Horticulturae*, 455, 422–30.

Schaffer, B., Urhan, L., Miyara, I. and Whiley, W. (2009). 6 Ecophysiology. In Litz, R. E. (Ed.), *The Mango, Botany, Production and Uses*, 2nd edn. CAB International. Wallingford and Oxon, 170–209.

Singh Gill, M., Navprem, S., Singh, N. and Gill, P. P. S. (2015). Performance of Mango Cultivars under Sub-Mountane Zone of Sub-Tropics of India. *Acta Horticulturae*, 1066, 27–33. https://doi.org/10.17660/actahortic.2015.1066.2.

Tomer, E. and Charny, Y. (1983). Inhibition of Flowering in Mango by Gibberellic Acid. *Hassadeh*, 290–3 (in Hebrew).

Tongumpai, P., Chantakulchan, K., Subhadrabandhu, S. and Ogata, R. (1997). Foliar Application of Paclobutrazol on Flowering of Mango. *Acta Horticulturae*, 455, 175–9.

Torres, M. D., Guirado, E., Farré, J. M. and Hermoso, J. M. (2009). Pre-bloom Shoot Tipping of Mango Influence on Yield, Fruit Size and Quality. *Acta Horticulturae*, 820, 391–4.

Vázquez-Valdivia, V., Pérez- Barraza, M. H. and Osuna-García, J. A. (2009). Effect of Gibberellic Acid on Flowering of 'Ataulfo' Mango. *Acta Horticulturae*, 820, 413–17. https://doi.org/10.17660/actahortic.2009.820.50.

Verheij, E. W. M. (1986). Towards a Classification of Tropical Tree Fruits. *Acta Horticulturae*, 175, 137–50.

Watson, B. J. and Moncur, M. (1985). *Criteria for Determining Survival: Commercial and Best Minimum July Temperatures for Various Tropical Fruits in Australia (S. hemisphere)*. Wet Tropical Region Publication. Queensland, Australia.

Whiley, A. W., Rasmusen, T. S. Saranah, J. B. and Wolstenholme, B. N. (1989). Effect of temperature on Growth, Dry Matter Production and Starch Accumulation in Ten Mango (*Mangifera indica* L.) Cultivars. *Journal Horticultural Science*, 64, 753–65.

Chapter 7

Mango cultivation in greenhouses

John Y. Yonemoto, Japan Tropical Fruit Association, Japan

1 Introduction

Mango production in Japan has increased from 440 tons in 1992 to 3327 tons in 2013. The entire mango crop is grown in greenhouses because of low winter temperatures and rainfall during the flowering period. Most production takes place in the southern part of Japan, including Okinawa, Kyushu and Honshu Islands. However, some growers have begun to cultivate mangoes in Hokkaido Island, the most northern part of Japan, where winter temperatures fall to –20°C.

The cultivar 'Irwin' (Fig. 1) is preferred due to its red colour, mild sweet flavour and lack of fibre which makes it popular with Japanese consumers; other cultivars such as 'Osteen', 'Kent' and 'Keitt' are also cultivated under greenhouse conditions in Spain (Galán Saúco, 2015). Although over 7000 tons of cultivars including 'Carabao', 'Tommy Atkins', 'Nam Doc Mai', 'Kensington' and 'Kent' are imported to Japan, they are sold at much lower prices (around 500 yen/kg) compared to the domestically produced Irwin (2000–4000 yen/kg). Irwin is harvested at full maturity when the fruit drops in a protective bag. Fruit are packed and shipped soon after harvesting, and due to door-to-door delivery systems, they reach the customers within two days throughout the country. Brand names are important for maximising prices. Irwin fruit must show 100% colour development, be without insect damage, have a perfect shape and measure more than 15.5 Brix if they are to command high prices. Growers use near-infrared spectrum sugar detectors (Fig. 2) to measure the sugar content of each fruit before packing. The proximity of Spain to the rest of the European Union market and good logistics systems also facilitates the rapid delivery of mangoes to European customers. Delivering mangoes in a fresh 'ready-to-eat' condition allows growers to charge higher prices.

http://dx.doi.org/10.19103/AS.2017.0026.10

Figure 1 Cultivar Irwin.

Figure 2 Measuring the sugar content of single fruit by near-infrared spectrum.

Special techniques (described below) are used in Japan while harvesting perfect greenhouse-grown fruit. These include training and pruning, floral initiation, flower panicle hanging, fruit thinning and hanging, bagging, root limitation and cultivation practices such as fertilisation, irrigation, and pest and disease control. Cultural techniques for mangoes cultivated under greenhouse conditions in Spain do not differ much from those used in open-air cultivation and, in consequence, will not be described here. Readers are referred to sources such as Galán Saúco (2015). The main advantages of greenhouse cultivation in Japan are:

1 Shortening of the juvenile period;
2 Protection against adverse climatic conditions and reduction of 'sunburn' damage;
3 Increase of flowering and fruit set;

4 Increase of foliar surface and consequent increase of photosynthetic capacity;
5 Better control of irrigation allowing better plant growth and out-of-season production;
6 Earlier harvesting season due to better temperature control;
7 Possibility of obtaining good late harvesting crops;
8 Better fruit appearance;
9 Better control of pests and diseases; and
10 Better possibilities for organic cultivation due to lower incidence of pests.

2 Training and pruning

2.1 Training

Young grafted trees are grown in 20-litre pots for two years. Holes are prepared for more than a month before planting and a mixture consisting of 20 kg of decomposed manure, 100 g of phosphorus with some boron and manganese, and 250 g of magnesium calcium fertiliser is applied to the hole. This is mixed with soil which is then made into the planting mound. After planting, the trees are watered, organic mulching is placed around the trunk and the tree is stabilised by a stake. Shoots grow after planting and these new shoots will produce flowers and set fruit. However, only three or five fruit are harvested, and the rest left for thinning to develop the tree canopy for the following year.

Trees are trained in an open vase form (Fig. 3 and 4). The main stem is cut at a height of 30–40 cm above the ground and three to four shoots growing from the cut part are used as scaffold branches. These scaffold branches are also cut at a length of 30–40 cm, and two to three shoots from each of the scaffold branches are used as sub-scaffold branches. Lateral shoots grow from the sub-scaffold branches to become the fruiting branches. The scaffold and sub-scaffold branches are trained low to promote spreading, and the tree height is kept to 1.5 m or less. Stakes and pipes are used on the ground to tie down the limbs and branches (Fig. 5) and water-permeable non-woven cloths are used

Figure 3 Open vase form of training: young trees.

Figure 4 Open vase form of training: mature trees.

Figure 5 Using stakes and ground pipes to tie down limbs and branches.

to limit root growth. A hole or trench with a depth of 50 cm is dug and lined with cloth prior to filling with soil and planting (Fig. 6). Some growers use plastic or non-woven cloth pots to control tree growth (Fig. 7 and 8). Tree spacing is 2.5 m × 2.5 m and the trees are thinned out when the canopies become crowded, generally five to seven years after planting (Fig. 9).

The yield in the second year will be 10–15 fruit per tree; in the third year, 25–35; and in the fourth, 40–50. If the tree spacing is 2.5 m × 2.5 m, there will be at least 150 trees per 1000 m^2, that is, 40 fruit per tree × 150 trees = 6000 fruit per 1000 m^2. At an average fruit weight of 380 g, the yield will be 0.38 kg × 6000 fruit = 2280 kg per 1000 m^2.

Figure 6 Hole lined with cloth and filled with soil before planting.

Figure 7 Planting in a non-woven cloth pot to control tree vigour.

Figure 8 Non-woven cloth pots are partially buried in soil to prevent water loss.

2.2 Pruning

Pruning is carried out soon after harvesting. Heavy pruning is the norm in Kyushu and the northern part of Japan (Fig. 10a,b). However, gentler pruning has been practised in Okinawa because heavy pruning had been found to diminish flower initiation in the warmer climate. The practice of limiting pruning to cutting the shoot just below the node directly under the peduncle (Fig. 11) has increased recently. The shoots appearing after pruning (Fig. 12a) are thinned at an early stage, leaving two shoots to promote fruit yield. In general, where mangoes put out shoots by mid-August, flower buds may be expected to appear in the following year, but their appearance is unstable if shoots come out in late August or later. The fruiting shoots which appear after pruning are trained horizontally to increase the flowering rate (Fig. 12b).

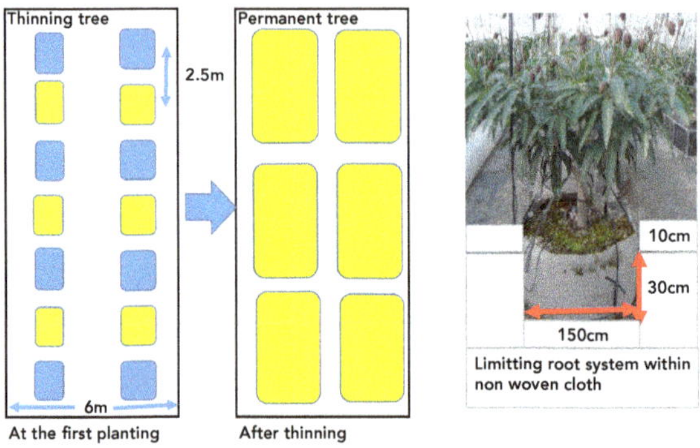

Figure 9 Illustration of a high-density planting and tree thinning.

Figure 10 Before pruning (top left); after pruning (top right); close-up view showing cut-back pruning (bottom).

Figure 11 Cutting the shoot just below the node directly under the peduncle.

Figure 12 Left: new shoots appearing after pruning; right: new shoots maturing. These new shoots will produce flowers in the next season. They will be trained horizontally to avoid additional shoot germination and promote floral initiation.

3 Control of flowering

3.1 Floral initiation

Floral initiation occurs during the winter season. In Okinawa, the protective plastic film is removed in May when the temperature rises, and replaced in January before the flower buds sprout. In Kyushu and other northern areas, plastic covers remain in place all year round or are put in place before December. Maturing the fruiting branch is important for good floral initiation. If the new shoots fail, floral initiation will not occur. Naphthalene acetic acid (NAA) was previously utilised to prevent new shoot emission; however, its use is now prohibited. Paclobutrazol is used in many mango-producing countries although it is no longer permitted in Japan. The current practices for promoting floral initiation are to put trees under water stress, to bend fruit-producing branches horizontally and to reduce greenhouse temperatures.

Growers try to maintain day/night temperatures of 25/10°C and to keep the soil pF2.7–2.9 (0.5–0.8 Bars) until the tip of fruiting shoots swell (Fig. 13), after which the day/night temperatures are raised to 28/23°C and the soil pF1.9–2.2 (0.08–0.15 Bars) decreased for flowering, fruit set and growth.

3.2 Flower panicle hanging

Each panicle is hung upward (Fig. 14) to increase available light, decrease anthracnose incidence and enable optimum pollination and colour development. Some growers will thin the basal flower panicle to create space between fruit and leaves (Fig. 15) which allows more light to reach the fruit and prevents scars caused by friction with the leaves.

3.3 Pollination

Honeybees and flies are used for pollination. The price of a beehive (Fig. 16) containing a queen and around 4000 honeybees is a little over 20 000 yen. However, honeybees

Figure 13 Swelling of tip of fruiting shoot.

Figure 14 The space between the leaves and fruit can be improved by thinning the basal peduncle; each floral panicle is hung upwards to increase the available light.

Figure 15 Plant showing space between leaves and fruit.

Figure 16 Honeybee hive containing 4000 bees.

are not always effective pollinators in greenhouses due to the use of insecticide spray for controlling thrips before the flowering season. Consequently, many seedless, small and un-pollinated fruit may develop in 'Irwin' mangoes. Flies are used as pollinators in Okinawa. The cost of using flies is low as growers use fish waste to rear them. The disadvantage of this practice is the odour arising during the pollination period.

4 Care of fruit

4.1 Minimising resin exuded from the fruit skin

If the minimum temperature during fruit development is 19–20°C, resin will exude from the rind (Fig. 17). The formation of dew on the rind is thought to be the cause of this problem and it reduces the market value of the fruit. Growers, therefore, raise the minimum temperature to 23–24°C to minimise the problem. Irrigation is also important, and it is decreased during the harvesting period.

4.2 Fruit thinning and hanging

The first thinning takes place a month after full blooms have appeared (Fig. 18a). Three to five of the largest fruit without scars or malformation are selected for each panicle, and the rest of them are thinned off. The second thinning will be done when the fruit have become egg-sized, and it is carried out by leaving one or two fruit per shoot (Fig. 18b). The leaf to fruit ratio should be more than 70 leaves to one fruit for optimum sugar content, colour development and stability of production in the following year. Efficient growing will yield two tons of fruit per 1000 m^2 every year. It is possible to produce 2.5 tons per 1000 m^2 by applying 1000 ppm CO_2 to maximise photosynthesis in a closed greenhouse (Fig. 19). However, it seems that a yield of around 2 tons per 1000 m^2 is the maximum achievable in the open vase form of training system which has a smaller canopy area when compared to the central leader or hedge low training system.

Figure 17 Resin exude on rind.

Figure 18 Left: fruit before thinning; right: the second fruit thinning is undertaken when fruits become the size of an egg: one or two fruits are left per shoot.

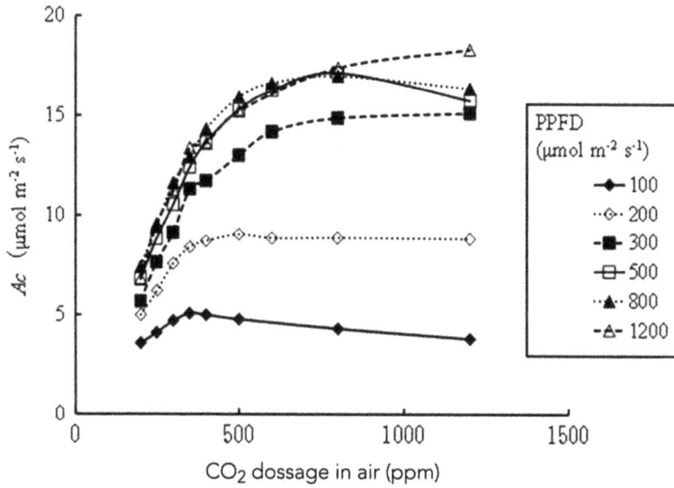

Figure 19 Effect of PPFD and CO_2 on photosynthesis in Irwin mango leaves; PPFD (Light Intensity for effective photosysthesis), *AC* (rate of photosynthesis).

4.3 Bagging of fruit

'Irwin' mangoes are harvested at the fully mature stage. This variety has the characteristic of dropping at maturity, and so the fruit is bagged to avoid damage which might be caused while dropping (Fig. 20). Growers in Okinawa place each fruit in a white paper bag to protect them against sun damage (Fig. 21a). In Kyushu and other northern parts of Japan where the light is less intense, clear net bags are used (Fig. 20) to permit the passage of the ultraviolet (UV) rays necessary for the development of the red colour which commands the highest prices. Some growers place a piece of white paper-like cloth directly under the fruit apex to reflect the light (Fig. 20). Figure 22 shows the effects of reflecting UV 250 nm in this way. Some growers change the position of the hanging fruit to enable uniform sunlight cover which will produce the perfect fruit colour.

4.4 Harvesting and shipping

Because the fruit is collected after it has dropped, no specialist knowledge is required to avoid harvesting immature fruit. Growers, therefore, hire labourers for harvesting fruit in bags, while the grower will sort and pack the harvested fruit. Usually, two persons can manage 5000 m^2 of the greenhouse. At the highest yield of 10 tons per 5000 m^2 and when sold at 2500 yen per kg, the gross income will be 25 million yen per grower. Figure 23 shows prices per kg at the Tokyo central wholesale market in 2011. The high prices in March, April and May are for mango from heated greenhouses. The heating oil consumed was 2.0–2.5 kl for a 1000-m^2 greenhouse. Most domestically produced mangoes will be sold as gifts in Japan (Fig. 21b). There are two gift exchange seasons: one in summer and one at the end of the year. Mangoes are considered suitable gift items for the summer season. However, it is possible to harvest mangoes at the end of the year by initiating

Figure 20 Bagging with clear net bag; fruit drops into the bag when it is fully matured. The white paper beneath the fruit reflects ultra violet light to develop a red color on the fruit apex.

Figure 21 Left: bagging with white paper bags; right: high-quality mangoes (Brand name: Egg of Sun) are sold in boxes of three at premium prices.

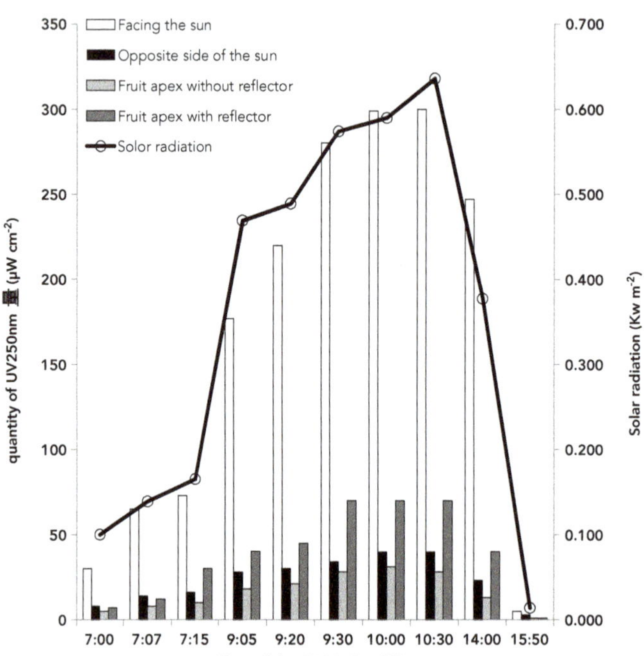

Figure 22 Changes in ultraviolet light from daytime solar radiation in greenhouse mangoes (data from 31 Oct, 2010).

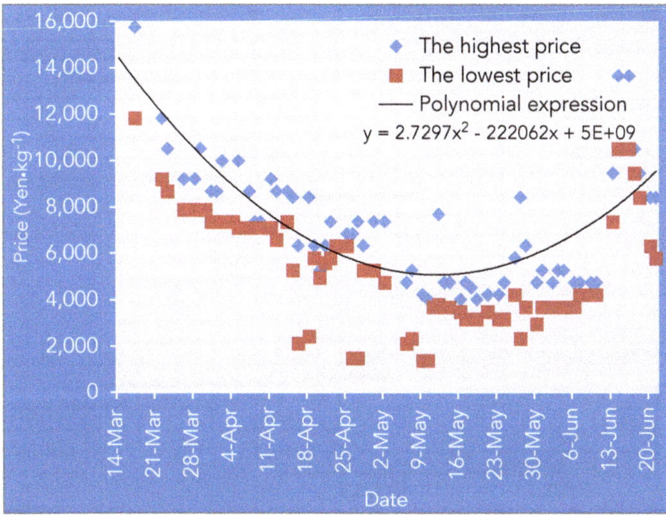

Figure 23 Mango prices at Tokyo central wholesale market in 2011.

flowering during the cool summer in Hokkaido. The present author has successfully followed this practice during the last six years.

5 Cultivation practices

5.1 Summer and winter harvesting

As was the case before the development of forced heating cultivation, most mangoes in Japan are shipped in June. However, in the Miyazaki and the Kagoshima Prefectures, the earliest shipping starts in March and continues up to May. This is followed by fruit from non-heated greenhouse culture in the Okinawa Prefecture from June to August. Kinki University in Wakayama Prefecture has developed a later cultivation of mangoes and harvests the fruit in September by delaying flowering. All year round harvesting is now possible in Hokkaido as flowering can be initiated at any time of year. However, the price of heating oil is very high in Hokkaido and hot spring water is now being used to minimise heating costs.

Examples of plant growth habitats for summer and winter harvesting are shown in Fig. 24 and 25. A heat pump was used to lower the night temperature in October and November to initiate flowering for shipping fruit in March and April. Under these conditions, two shoot growths (in May and July) occur after pruning, which is done soon after the harvest. As a third growth is not wanted, the techniques described in Section 3.1 are utilised.

Late-raising cultivation is practised by pinching flower clusters at the base when they have reached a length of 10 cm. Under cool conditions, this will produce further flower clusters, thus delaying the flowering season by a month. Since the harvesting season for late-raising cultivation is in September, pruning after harvest is not practised because of the difficulty of producing new fruiting shoots at that time of year unless the greenhouse temperature is raised. The yield per 1000 m^2 is generally low in this type of cultivation

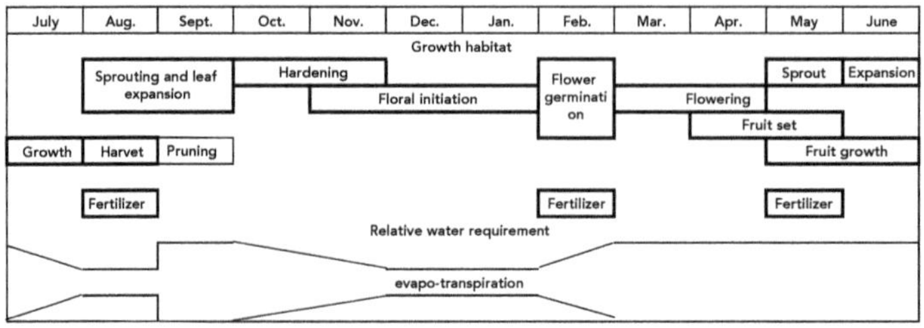

Figure 24 Plant growth habitat and water requirement in summer harvesting mango.

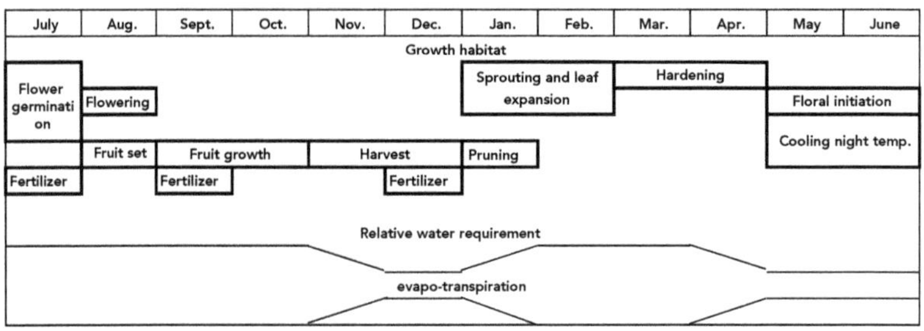

Figure 25 Plant growth habitat and water requirement in winter harvesting mango.

because the accumulation of photosynthate in the tree is small due to lower sunlight levels and low temperatures in the autumn and winter season.

5.2 Fertiliser application

Fertiliser is applied soon after harvest, at flowering time and at fruit development time (Fig. 19 and 22). Compound (N, P and K) fertilisers are used on each occasion: 50% is applied soon after harvest, and 25% at flowering and again at the fruit development stage. Ca and Mg are applied after harvesting. Calcium is applied every year to keep soil pH between 7.0 and 5.5 (the optimum range for mango tree growth). Mixed micronutrients are added to irrigation water (Fig. 26). Some growers use a fertigation system. Diluted N, P, K and all required micronutrients are always applied by irrigation. This is particularly important where pot culture is used since soil nutrients can leach out.

A 500–1000 ppm boron solution is applied at full bloom and again after blooming by foliar (flower cluster) spray to improve the fruit set. Fruit produced under conditions of low boron may become malformed (Fig. 27). The total amount of N per 1000 m^2 is between 10 kg and 20 kg according to soil fertility. Organic mulching (Fig. 28a) is preferred for growing feeder roots and increasing yield. Growing mango trees in a non-woven cloth pot may also increase the feeder roots instead of developing a taproot (Fig. 28b).

5.3 Irrigation

Irrigation systems are utilised in greenhouses (Fig. 29). The cost is around one million yen, but it delivers a saving on the labour costs associated with irrigation and fertiliser

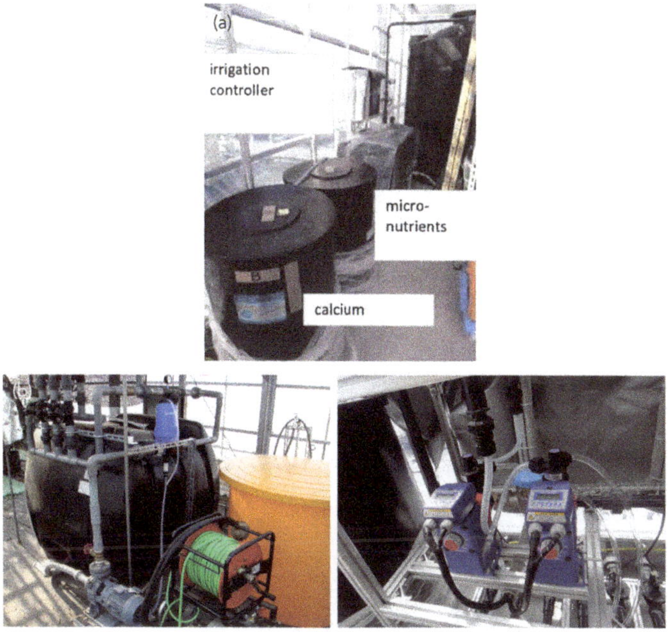

Figure 26 Micronutrients added into irrigation water.

Figure 27 Malformation from boron deficiency.

application. Figure 30 shows the daily evapo-transpiration rate on a single Irwin tree with 1000 leaves, grown in a 60-litre pot. The peak water requirement was 10 litres per day in the summer months, decreasing to one litre per day during the winter. Using an irrigation system maintains soil moisture by spreading irrigation through the course of a day.

A spray pen emitter is used for irrigating the whole soil surface under a canopy or pot. Spraying water evenly under the canopy is important for the effective use of fertiliser

Figure 28 Growing in a non-woven cloth pot encourages feeder (rather than tap) roots.

Figure 29 Greenhouse irrigation system using a pen-type spray.

compounds and for preventing the fruit from splitting. The amount of water supplied is adjusted according to the season and climatic conditions: 120% will be applied in conditions of good sunlight and 50% on wet days.

5.4 Tree replacement

Growers will plant larger trees (Fig. 31) in order to reduce the period until the first fruit production. If three-year-old grafted trees are planted in spring, fruit may be expected in the following summer. Tree spacing and thinning have been already mentioned in Section 2.1.

Although some growers manage to produce fruit under plastic housing from the same trees for over 20 years, problems arise in achieving adequate levels of production in

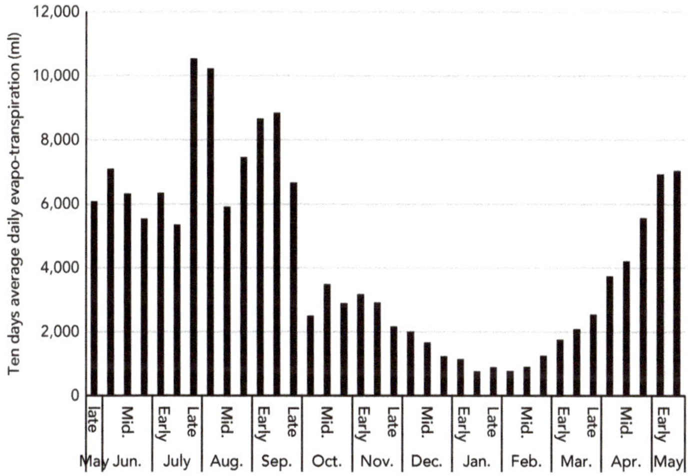

Figure 30 Changes of daily evapo-transpiration rate on Irwin tree with 1000 leaves grown in a 60-litre pot. The tree was grown in a greenhouse in Hokkaido, Japan (2011, May-2012, May).

Figure 31 Re-planting larger trees to produce fruit in the next season.

trees over ten years old. It becomes difficult to control the tree habit for optimum fruit production, especially in achieving even floral differentiation which is easier to control on younger trees. Trees are, therefore, generally replaced when they are around ten years old.

5.5 Key tasks from planting through to harvest

- **April:** planting and watering.
- **May:** watering and thinning new shoots, leaving two vigorous new shoots from each point.

- **June:** spraying fungicide for controlling anthracnose and pesticide for thrips.
- **July:** fertilising, watering and training branches.
- **August:** Thinning new shoots. Spraying pesticide for thrips.
- **September:** fertilising and training branches.
- **October:** horizontal bending of fruit-producing branches to prevent autumn germination.
- **November:** reducing water supply.
- **December:** protecting from frost by fitting a covering film on greenhouses.
- **January:** ventilation to lower the day temperature and prevent dew forming on the plant.
- **February:** spraying fungicide and pesticide before flower bud germination.
- **March:** fertilising and watering for flower bud development. Hanging flower stalks. Caring for honeybee hives.
- **April:** ventilation for lowering humidity to prevent fungal diseases. Spraying boron solution on the flower stalks at full bloom.
- **May**: fruit thinning and spraying of fungicide and pesticide. Fertilising for fruit development.
- **June:** fruit hanging.
- **July:** fruit bag or net covering for catching mature fruit. Decreasing water supply to improve fruit taste.
- **August:** harvesting fruit as they drop in bags or nets. Cut-back pruning and fertilising soon after the harvest.

6 Disease and pest control

Controlling anthracnose (Fig. 32) and thrips (Fig. 33) are of primary importance in Japanese greenhouse culture. Anthracnose is prevented by plastic covering, good ventilation and spraying with fungicide. Even where there are no anthracnose symptoms on fruit at packing, these may appear a few days later after sale.

Thrips are the most difficult insect pests to control because there is no effective pesticide to control them. Some growers use natural predators; however, this is insufficient for the production of perfect fruit. There are many registered pesticides for thrips in Japan (http://lib.ruralnet.or.jp/cgibin/ruralnouyakulist.php?ARG1=534b3d837d26534e3d837d83938353815b265246383d313032).

Figure 32 Symptoms of anthracnose on the rind.

Figure 33 Rind showing thrip damage.

6.1 Guidelines for the efficient use of pesticides

Using a variety of pesticides such as organophosphorus, carbamate, pyrethroid, nicotine or neo-nicotinoid is more effective than using the same pesticide continuously. A second spraying is recommended 10–14 days after the first, as surviving eggs may be hatching. Weed growth in a greenhouse should be avoided as this will encourage the development of thrips. Greenhouses should be covered with a fine mesh net to prevent the entry of thrips. Clothes should be changed after working in an infested greenhouse.

7 Future trends and conclusion

High production costs are preventing further increases in mango production. Only high-quality mangoes can be sold at a high price. Growers who produce such high-quality fruit are able to continue mango production. There are so many growers who give up growing mango as they lack the ability to produce high-quality fruit. In Japan, only the growers who have strong will and high-end techniques to produce high-quality mango can survive.

8 Where to look for further information

CABI 2009. *The Mango: Botany, Production and Uses*. 2nd edn. Edited by Richard E. Litz. CABI International, Cambridge, USA, pp. 680.

Fairchild Tropical Garden 1992. *Mango A Guide to Mangos in Florida*. Florida, USA, pp. 200.

9 References

Galán Saúco, V. 2015. Ventajas y desventajas del cultivo del mango (*Mangifera indica* L.) en zonas subtropicales y potencial del cultivo bajo invernadero. *Acta Horticulturae* 1075: 167–78.

Yonemoto, Y. 2006. Tropical fruit in Japan. *Horticulture in Japan* 63: 709–25.

Yonemoto, Y. 2008. Series of new rare fruit trees Mango–How to grow full maturity mango-. Nousangyoson-bunka-kyoukai, Tokyo, Japan, pp. 190 (in Japanese).

Yonemoto, Y. 2009. Cultivation of tropical fruit trees –Grow and enjoy full maturity 28 tropical fruit trees-. Nousangyoson-bunka-kyoukai, Tokyo, Japan, pp. 184 (in Japanese).
Yonemoto, Y. 2014. Backyard cultivation of tropical fruit. -Growing 34 testy tropical fruit trees as small as possible-. Nousangyoson-bunka-kyoukai, Tokyo, Japan, pp. 124 (in Japanese).

Chapter 8

Management of an ultra-high-density mango orchard and benefits of the small-tree system

Steven A. Oosthuyse, HortResearch SA, South Africa

1 Introduction

Majumder et al. (1982) reported increased yields during the initial years in planting mango trees at a density of 1600 trees per ha as opposed to at lesser tree densities. Ram and Sirohi (1988) communicated their findings after planting mango trees at the spacing of 3 by 2.5 m. In relation to trees planted at wider spacings, yield per ha was significantly enhanced during the initial years after planting due to greater land use efficiency during these years. It was recognized that pruning to contain canopy size was required to avoid orchard overcrowding at a stage after planting. The question of whether regular pruning will adversely affect cropping once containment pruning was required was posed. Majumder and Sharma (1988) stated that it was important to maintain the high level of production attained in planting trees at high density. They indicated that success in this regard might be achieved in using regular bearing, dwarf varieties or varieties grown on dwarfing rootstocks. Accomplishment in growing mangoes at the density of 1600 trees per ha was reported. Judicious pruning was required, however, and the system was

http://dx.doi.org/10.19103/AS.2017.0026.11

reported to only suit certain varieties, such as Amrapali, which are slow-growing and bear regularly. Charnivichit and Tongumpai (1991) were able to reduce canopy spread and effect flowering in treating Nam Doc Mai trees in a 2.5 by 2.5 m orchard, which was hard-pruned, with paclobutrazol. The treated trees were noted to flower more profusely and considerably earlier than the untreated trees.

Oosthuyse (1993) proposed *ultra-high-density* planting of mango, and discussed the probable benefits of such a system. Reference was made to '*the ideal tree spacing*', the requirement for regular canopy size maintenance, and branching encouragement by terminal shoot tipping of newly planted trees to increase growth rate and tree sturdiness during initial growth. Branching encouragement, in previous research (Oosthuyse and Jacobs, 1995), was shown to markedly increase tree-cropping capacity when trees are permitted to bear for the first time one to three years after planting. Thus, in addition to increased orchard cropping capacity due to the high number of trees planted per ha in a high-density orchard, tipping further increases yield during the initial years after planting.

In 1996, a Tommy Atkins ultra-high-density orchard, where the trees were spaced 3 by 1 m apart (between row-within-row spacing), was established on HortResearch SA's research farm by the author. It is noteworthy that the farm is in a cool subtropical zone of South Africa. This was done to ascertain the merits of the ultra-high-density system in mango, and to determine the key management actions required for success of the system. It is noteworthy that neither a dwarfing rootstock nor a growth retardant was used. Success in warmer regions may require the use of a growth retardant, for example, paclobutrazol. Dwarfing rootstocks may or may not impart the necessary physiological state eliciting propensity to flower.

The present chapter details this orchard at various stages from the time of its establishment to the stage when the trees filled their allotted space in the orchard row and were bearing fully. Benefits, and management actions required for sustained production are disclosed. Furthermore, success criteria are indicated for different growing conditions and various cultivars.

2 Orchard establishment

A two-hectare Tommy Atkins orchard, comprising trees spaced 1 m apart *in the row* and 3 m apart *between rows*, was established in 1996 at the HortResearch SA research farm located on the outskirts of Tzaneen, South Africa. The 3 × 1 m spacing was considered to be ideal in the sense of it rendering trees that are 'hand-manageable', and which are of a size bestowing optimal light utilization by the orchard. By 'hand manageable' the intended meaning is that of the trees always being of a size such that workers are able to reach all parts of the canopy without having to resort to the use of elevation equipment, for example, ladders.

Prior to planting, the area of land was cleared and ripped to a depth of 0.8 m. The soil utilized is deep (>2 m), fertile and well drained, and has a clay content of 40–50%. Prior to ripping, single super phosphate was broadcast evenly at a rate recommended from the results of the analysis of the soil samples taken after land-clearing. After site levelling, a drip irrigation system was installed with spacing matching the tree spacing adopted. Four-litre-per-hour button drippers were inserted into 15 mm diameter polyethylene irrigation pipe, the distance between drippers being 1 m. At the site of each dripper, Sabre rootstock seeds (poly-embryonic) were planted 1 to 2 cm below the soil surface. Two or three de-husked seeds were inserted in wetted soil adjacent to each dripper in January

1996. After seedling emergence, the strongest tree at each planting point was retained. The remaining seedlings were cut-off at ground level. The retained trees were made to develop as vertical single stems until exceeding 80 cm in height. Prior to this time, any lateral shoots were removed by pinching them out at an early stage of their development. A number of months after planting, the trees were top grafted to Tommy Atkins at a height of less than 80 cm above ground level. Once the trees were approximately 1 m in height they were headed back to 80 cm to initiate the phase of forced branching (Fig. 1).

After first heading, the apical bud on each new, vigorous terminal shoot was removed (pinched off) once the shoot had fully matured (tipping) (Fig. 2).

Figure 1 Tree headed at 80 cm to initiate branching.

Figure 2 Tipping of only the new vigorous terminal shoots to encourage prolific branching.

This was done after every flush to effect continued branching. Approximately 60% of the new shoots were treated in this way after each growth flush. This action markedly hastened tree development, giving rise to compact canopies with an abundance of vigorous terminal shoots (Fig. 3).

Branching encouragement was carried out for one season, after which the trees were allowed to bear for the first time. All pruning actions, tipping or otherwise, were carried out during the warm months of the year, that is, from September and before April of each year. The trees were permitted to grow to become hedgerows (Fig. 4).

Hedge width was maintained at 1 m and hedge height at 1.8 m in hedging to 1.6 m height (20 cm of growth occurring after hedging).

Figure 3 Compact, sturdy trees developing as a result of terminal shoot tipping after each new flush. Note canopy closure due to the high number of terminal shoots present.

Figure 4 Trees growing to become hedgerows. Alleyway width = 2 m. Hedge width = 1 m.

3 Basis for increased productivity

Table 1 shows the optimal hedgerow surface area per ha for a number of between-row tree spacings. It is taken that the optimal hedgerow height is a function of alleyway width, the relationship relating alleyway width to height being *Optimal Height = 0.8 × Required Alleyway Width + Canopy Initiation Height* (Jackson, 1980; Jackson and Middleton, 1988; Jackson, J. E. pers. comm., 1992, Horticultural Research Station, Marondera, Zimbabwe).

Alleyway width is the between-row distance separating the hedge sides. Comparisons are drawn between hedgerows, having vertical and horizontal sides, of differing sizes. The dimensions of the hedgerows closely optimize light utilization for the between-row spacings presented, viz., 3, 5, 7 or 9 m. Figure 5 shows the differences in hedgerow surface area per hectare for these spacings.

It is noteworthy that the differences in surface area are small in relative terms. In fact, the surface area per hectare of the 9 m between-row tree spacing was slightly greater than that of the 7 m between-row tree spacing. In assuming that bearing potential and hedgerow surface area per ha are directly related, the conclusion can be drawn that once the optimal hedgerow dimensions are attained for various between-row tree spacings, orchard cropping potential is similar, irrespective of the planting spacing. In considering that the rate of increase in surface area per tree with time after planting has no direct bearing on the planting spacing, the difference regarding orchard performance can be noted to pertain to the time taken for the optimal hedgerow

Table 1 Determination of hedgerow surface area per hectare for a number of tree spacings (3 x 1; 5 x 1.5; 7 x 3; 9 x 4 m)

Surface area per ha of vertically sided hedges having ideal dimensions, for a number of tree spacings				
Between-row spacing (m)	3	5	7	9
Hedge surface area determination - vertically sided hedges				
Alleyway width (m)	2	3	4	5
Hedge length (m)	100	100	100	100
Hedge width (m)	1	2	3	4
Hedge height (m)	1.6	2.4	3.2	4
Sun exposed areas (square m)				
Hedge top	100	200	300	400
Hedge side 1	160	240	320	400
Hedge side 2	160	240	320	400
Hedge end 1	1.6	4.8	9.6	16
Hedge end 2	1.6	4.8	9.6	16
Total hedge area	423.2	689.6	959.2	1232
Number of hedges per ha	33	20	14	11
Total surface area (square m)	13966	13792	13429	13552

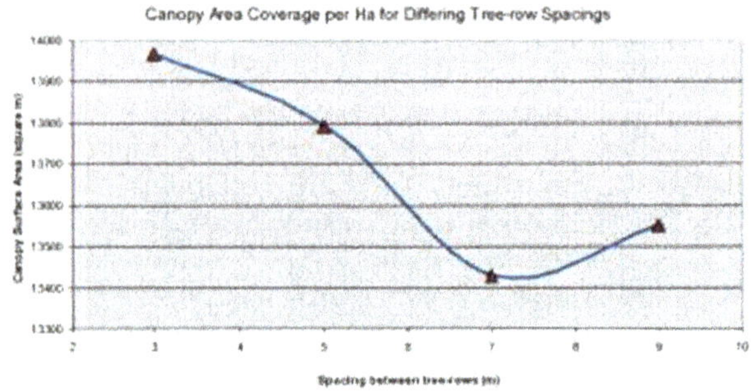

Figure 5 Hedgerow surface area in relation to between-row spacing; once the trees attain their ideal dimensions.

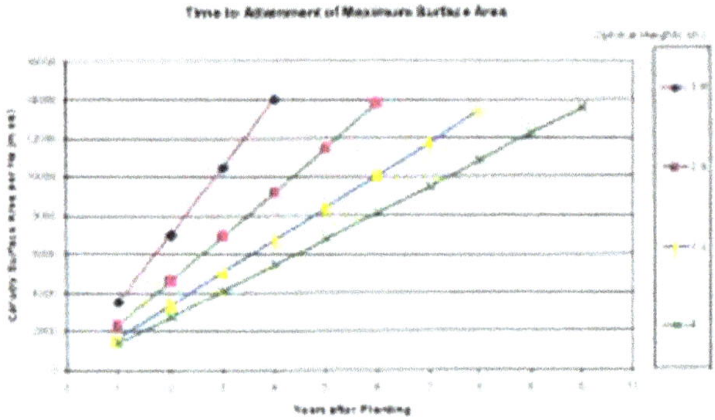

Figure 6 Estimated time to the attainment of optimal canopy dimensions of orchards with trees at a number of spacings (3 x 1; 5 x 1.5; 7 x 3; 9 x 4 m).

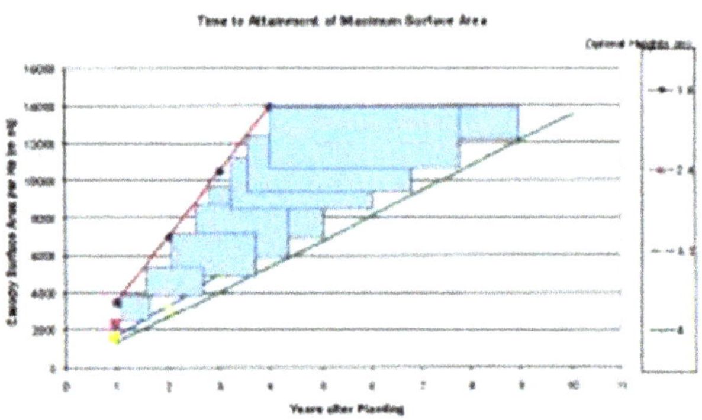

Figure 7 Orchard productivity differences relating to tree spacing depicted by the area between hedgerow-surface-area increase lines with time.

dimensions to be attained for differing spacings. Optimal hedgerow dimensions are attained in a shorter time for narrower as opposed to wider tree spacings (Fig. 6).

The difference in orchard productivity over a fixed period, 'say' 10 years, between the 3 and 9 m between-row spacings, can be depicted by the area between the surface-area-increase lines for these spacings (Fig. 7).

Table 2 Probable yields of trees spaced at 3 by 1 m for a number of cultivars

Number of trees per ha	Number of fruits per tree	Average fruit weight (kg)	Yield (kg)
Tommy Atkins			
3333	10	0.4	13332
	15	0.35	17498
	20	0.35	23331
	25	0.3	24998
	30	0.3	29997
	35	0.25	29164
Kent			
3333	10	0.4	13332
	15	0.35	17498
	20	0.35	23331
	25	0.3	24998
	30	0.3	29997
	35	0.25	29164
Keitt			
3333	10	0.4	13332
	15	0.4	19998
	20	0.4	26664
	25	0.35	29164
	30	0.35	34997
	35	0.3	34997
Sensation			
3333	10	0.35	11666
	15	0.3	14999
	20	0.3	19998
	25	0.28	23331
	30	0.25	24998
	35	0.23	26831

It might be concluded that ultra-high-density orchards attain maximum production potential in a markedly shorter period than conventionally spaced orchards. However, final production potential is similar. Maintaining the maximum level of productivity attained for and indefinite period is, however, the challenge regarding success of the ultra-high-density system. In the experience of the author, maximum productivity can be maintained if hedgerow size-maintenance (pruning) actions are taken at the correct times during the yearly production cycle. Variation in this regard has been noted, and relates to growing region, soil condition and the cultivar grown. Probable orchard yields for a number of cultivars are shown in Table 2. Maximum yield can be attained in three to five years after planting. The time taken depends on how vegetative the trees are during the initial years after planting. Canopy expansion is far less rapid in 'heavy' as opposed to 'light' bearing trees.

4 Observations relating to flushing and pruning practices

Figure 8 shows the orchard in question when flowering, four years after first heading (12 September 2003).

At this stage, the hedgerows had almost attained their optimum dimensions. Once bearing was permitted, canopy expansion rate slowed markedly. During the seasons when set was poor, due either to reduced flowering or to excessive inflorescence disease colonization, canopy expansion was pronounced, this resulting from more frequent flushing during the season.

Figure 9 shows the orchard shortly after fruit set in October 2003. If set was moderate, terminal shoots not bearing fruits tended to remain dormant until after harvest. However, if set was poor, new shoot development occurred from the non-bearing terminal shoots during the period of fruit development. Once the optimal hedgerow dimensions were attained, the trees were pruned back twice each season; in early November (after natural post-flowering fruit drop had occurred) when the terminal shoots not bearing fruits were headed back, and again after harvest when the terminal shoots having borne fruits were headed back. Figure 10 shows trees after fruit set in November during a poor cropping year.

The second pruning was delayed after harvest to ensure that the new shoots developing after pruning were at the ideal maturation stage for flowering when winter conditions (air temperatures, soil dryness) were maximally inductive. Only the terminal shoots extending past or having extended to the point of the hedgerow width line (0.5 m into the inter-row space) were headed back. These branches were pruned such that the cut ends were 20–30 cm inside the 0.5 m width line. At the same time, canopy height was reduced to 1.6 m, where this height was exceeded.

Figures 11 and 12 show the trees shortly after pruning in February 2004 and 2005, respectively (harvest was in late December or early January).

Figure 13 shows the trees when post-harvest flushing occurred during February–March 2005.

Figures 14 and 15 show the trees after terminal shoot extension had occurred and when terminal shoot maturation was occurring, respectively (February–March 2005).

The trees were noted to generally flush only once after post-harvest pruning. In April of each year, the entire outer canopy comprised non-diseased new shoots. It is noteworthy

Figure 8 The orchard flowering in September 2003, four years after first heading.

Figure 9 The orchard shortly after fruit set in October 2003.

Figure 10 Trees after fruit set in November during a poor cropping year. Non-bearing terminal shoots have been headed back.

Figure 11 Trees shortly after post-harvest pruning in February 2004.

Figure 12 Trees shortly after post-harvest pruning in February 2005.

Figure 13 Trees when post-harvest flushing occurred in February–March 2005.

Figure 14 Terminal shoot extension having occurred after post-harvest pruning. Photo taken in March, 2005..

Figure 15 Trees when terminal shoot maturation was occurring (February–March 2005).

Figure 16 Trees in May 2005, when the new terminal shoots had matured.

Figure 17 Trees in flower during July 2005.

Figure 18 Trees bearing fruits in December 2005.

that these new shoots often developed after the cessation of summer rains each year. Figure 16 shows the trees in May 2005, when the new terminal shoots had matured. The terminal shoots generally matured during March–April, remaining dormant until flowering in July–August.

Figure 17 shows the trees in flower during July 2005, and Figure 18 shows trees bearing fruits in December 2005.

5 Observations concerning flowering

During the period of hedgerow expansion, yearly production increased roughly in relation to the increase in yearly hedgerow growth. Then and subsequently, poor flowering occurred during certain years due to terminal bud development occurring shortly before or after the period when environmental conditions were inductive (June–July–August). Late bud development was deemed to be due to terminal shoots not being sufficiently mature during the inductive period; this resulted from pruning not having been carried out early enough after harvest. The extent to which the terminal shoots had matured during the flowering period bore a strong relationship with post-harvest pruning date. Pruning directly after harvest did on occasions give rise to terminal bud development occurring prior to the time when

Figure 19 At stages, a number of trees were not pruned to assess the effect of later heavy pruning, to get the trees back to size, on subsequent cropping. It was noted that these trees flowered earlier than the pruned trees; this apparently relating to both time-of-last flushing prior to flowering, and reduced terminal shoot vigour due to the absence of pruning.

conditions were ideally inductive. Here, shoots as opposed to inflorescences developed. In spite of hand removal of these new shoots when they were developing during the month prior to flowering, inflorescence development did not occur from the distal lateral buds thereafter. Bud development was delayed until after the inductive period. These observations indicate that post-harvest pruning date, as it relates to terminal shoot maturation state when conditions became inductive, subsequently, has the greatest bearing on flowering intensity. It might alternatively be stated that the extent to which the terminal shoots had matured by early to mid-winter was the determinant of whether terminal bud inflorescence development would occur or not when conditions were inductive. Issarakraisila et al. (1997) noted that the probability of a terminal shoot flowering was related to whether that shoot had undergone a period of vegetative growth and maturation during the period before flowering.

In certain years, a number of trees were not pruned to assess the effect of heavy pruning, to get the trees back to size, on subsequent cropping. Generally, these trees, before they were pruned, flowered earlier than the pruned trees (Fig. 19); this apparently relating to the time of last flushing prior to flowering, and reduced terminal shoot vigour due to the absence of pruning.

These trees were allowed to become large and overcrowd one another. Subsequently, they were brought back to size by heavy pruning shortly after harvest. New shoot growth thereafter was vigorous, and the trees tended to flush during the flowering period or just before. Consequently, flowering intensity in these trees was markedly reduced during the spring succeeding pruning. However, in pruning these trees as the others thereafter to maintain canopy size (November and February pruning), their flowering and growth patterns became synchronized with the remaining trees. Increased vigour after pruning was probably due to enhanced root system development resulting from the more extensively grown canopy.

It was noted that vigorous terminal shoots (thick stemmed and having large leaves), which were mature when conditions were inductive, often produced new shoots as opposed to inflorescences, when inflorescence development generally occurred. However, this did not occur when the winter temperatures were lower than usual or when conditions became dry earlier during autumn. It might be concluded that the inductive requirement of the highly vigorous terminal shoots, developing due to excess vigour, possibly imparted by a disproportionately large root system, is greater than that of moderately vigorous or non-vigorous terminal shoots.

Flowering intensity was adequately intense when canopy expansion was maintained properly in the sense of pruning being carried out at the right times, that is, in early November, and subsequently during late January or early February three to six weeks after harvest. In pruning at this time, any new shoots having already started to develop were removed, this in itself being de-vigourizing. Reasonably vigorous shoot development after pruning was noted to be desirable, ensuring adequate canopy recovery prior to the onset of winter when conditions became cool (night minimum temperatures generally <12°C). It is noteworthy that Tommy Atkins trees in the Tzaneen region generally set relatively poorly, often growing substantially during the fruit-bearing period. Poor set in Tommy Atkins trees grown around Tzaneen is ascribed to conditions not being sufficiently warm during the inflorescence development period for this variety.

Removal of unexpected late flush by hand or by chemical means shortly after its materialization may be a manipulation enabling flowering if such action takes place at least 6–8 weeks prior to the flowering period.

6 Cultivar and environment attributes suiting ultra-high-density planting

The following is stated in view of observations made of ultra-high-density mango orchards in Egypt and Mexico.

6.1 Cultivar

In cultivars that generally crop heavily, such as Keitt or Sensation, the annual need for canopy containment pruning is less. During years of heavy cropping, pruning shortly after fruit set is not required, post-harvest pruning being all that is required. In these varieties, flushing after post-harvest pruning may be delayed, thus negatively impacting on flowering intensity the succeeding season. Reduced flowering is apparently due to the terminal shoots being insufficiently mature when conditions are inductive. In off-year trees, general heading both shortly after the time of set as well as after the time of harvest is required; the non-bearing branch ends being headed twice in the season. After heavy cropping, it is important that influences hindering re-growth, such as reduced soil water availability, reduced soil fertility or applications of saline/high-pH water, are minimized after post-harvest pruning. The trees usually bear regularly and adequately if flushing occurs shortly after pruning, and is intense.

In cultivars such as Tommy Atkins or Kent, appreciable shoot development usually occurs during the fruit growth and development period, due to non-excessive fruit set and retention. Pruning non-bearing branches after set and bearing branches after harvest often suffices in these cultivars.

6.2 Growing environment

The growing environment may be such that growth is not vigorous and the flushing frequency is low. Such situations particularly suit ultra-high-density planting. This may be

Figure 20 A tree suffering from excess sodium uptake – Egypt.

explained as follows: The time taken for trees to fill their allotted space in the orchard row depends on canopy expansion rate. Irrespective of planting density, this time is prolonged if conditions favouring growth are poor. In a high-density planting, the allotted space is markedly less than that of a conventionally spaced planting, and hence, the time until the allotted space is filled and maximum productivity potential attained is substantially less in the ultra-high-density situation.

Conditions reducing tree growth rate include environmental coolness, tree reliance on limited rainfall for water (rain-fed situation), or poor soil due to shallowness, a high proportion of rock in the upper layers, high pH (>7.3), limited nutrient content, or a high concentration of salts. Currently, truly dwarfing rootstocks have not been declared in mango. Mango trees are particularly susceptible to high soil and irrigation water salinity and pH, which is typical of soils in arid regions. Figure 20 shows a stunted tree; stunting resulting from excess sodium uptake, and Fig. 21 shows leaf damage due to excess chloride uptake. Figure 22 shows a tree ailing from trace element deficiency due to the soil pH being high (>7.8).

Figure 21 A tree showing leaf damage resulting from excess chloride uptake – Egypt.

Figure 22 A tree suffering from general trace element deficiency resulting from high soil pH (>7.8) – Egypt.

7 Additional benefits and their significance

Many of the benefits described below relate directly to the canopies (hedgerows) of ultra-high-density trees being entirely accessible to farm workers on foot.

- Harvesting is carried out with ease, and for this reason is relatively quick. The need for ladders, hoist harvesters (cherry pickers) or picking sticks (picking bags attached to sticks) is obviated. The loss of fruits due to bruising or cracking resulting from impact damage caused by accidental fruit drop is reduced.
- Unwanted new shoots are easily broken-off by hand (Fig. 23). Terminal shoots arising during the flowering period can be removed at this time to reduce canopy expansion and fruit shading.
- General or selective branch heading, which the system requires, is easily accomplished by workers.
- Malformed shoots, branch sections or inflorescences, resulting from *Fusarium mangiferae* infection, are easily *broken out* (removal by cutting is not desirable as it

Figure 23 Hand removal of unwanted new shoots is easy to achieve due to the entire canopy being accessible. Unwanted flush occurring in early winter can be removed in this way.

Figure 24 Branches and inflorescences seriously affected by inflorescence malformation caused by *Fusarium mangiferae* – Egypt. If malformation is not managed by affected-tissue break-out during or shortly after the inflorescence development period each year, complete tree colonization may occur. Malformed inflorescences do not set fruits in most commercially grown mango cultivars.

promotes *malformation* spread). Inflorescence malformation is a serious problem in most mango growing regions. If not managed by break-out during or shortly after the inflorescence development period each year, complete tree colonization can occur (Fig. 24), thus markedly reducing cropping ability. Malformed inflorescences do not set fruits in most commercially grown cultivars. There are exceptions, however. Chokanan, a Thai variety, may set fruits on malformed inflorescences.

- Due to close proximity between spray outlet and target (Fig. 25), the effectiveness of fruit and tree spraying with chemicals preventing or limiting disease colonization or pest infestation is enhanced. Moreover, the fruits can be specifically targeted by using hand-held spray guns, thus reducing the quantity of chemical required and the amount of environmental damage caused. Of concern in mango culture is the over-use of the protectant fungicides, copper hydroxide, copper oxychloride and mancozeb. Close access also allows for ease in totally covering the fruits with chemical products, or specifically covering the fruit's stem-end where much of the fungal infection occurs (Fig. 26).

Figure 25 Close proximity spraying. Due to the close proximity between spray outlet and target, fruit and tree spraying with chemicals to prevent or limit disease colonization or pest infestation is more effective. Moreover, particular tissues such as the inflorescences, for example, can be specifically targeted.

Figure 26 Close fruit access regarding spraying enables ease in totally covering the fruits with protectants, or specifically applying cover to the fruit's stem-end where much of the fungal infection occurs.

Figure 27 Specific fruit treatments, like that of the application of sleeves to prevent disease infection, wind abrasion and solar injury, can easily be carried out.

Figure 28 Due to the 'smallness' of the individual canopies, canopy renewal occurs yearly as a result of the pruning actions required.

Figure 29 Mulch occurs due to the accumulation of pruned off material on the orchard floor. Prevention of grass and weed growth and nutrient recycling are benefits.

- Specific fruit treatments such as the application of caps to limit solar injury or sleeves to prevent disease infection, wind abrasion and solar injury can easily be carried out (Fig. 27).
- Due to the 'smallness' of the individual canopies, almost complete canopy renewal occurs yearly as a result of the pruning actions required (Fig. 28). This may not be the case where problems relating to tree recovery exist. Excessive cropping during an on-year may, for example, give rise to the inability of trees to respond in initiating strong re-growth after post-harvest pruning. In general, since the majority of outer leaves and shoots are renewed yearly, the capacity for

diseases and pests to build-up in the canopy is greatly impaired. Consequently, fruit diseases resulting from pre-harvest infections are more easily controlled. In considering pests, the ability of gallfly, for example, to increase in population is largely reduced.

- Mulch accumulation occurs due to the build-up of prunings on the orchard floor (Fig. 29). As a consequence, weed and grass growth between the trees is inhibited to a large extent, thus resulting in a reduction in the need for herbicide use or other weed and grass control measures. Mulch accumulation also gives rise to nutrient recycling from tree to soil, an increased abundance of soil flora and microfauna, a thicker surface organic soil-layer and an increased soil concentration of organic-matter breakdown products (humic acid, fulvic acid and humin).
- The time taken from reworking of the trees to a new variety to canopy re-filling is short in relative terms, due to the high number of trees per ha and the final requirement of relatively small canopies.
- The 'shortness' of the trees renders both the trees and fruits less susceptible to wind damage. Fruit damage due to solar injury is markedly reduced due to the 'closeness' of the hedgerows.

8 System adoption to date

Ultra-high-density plantings have been established in Mexico (Escuinapa) (Fig. 30), Egypt (Nimos Farm and others) (Fig. 31), India, Peru, Northern Brazil, Malawi, and South Africa with success.

Cultivars grown using the system include Tommy Atkins, Kent, Irwin, Keitt, Heidi (South African), Eweisse (Egyptian), Saddik (Egyptian) and Hindi Bi Sinnara (Egyptian). Of these varieties, Keitt has proven to be particularly suited to the system. Keitt is often dwarfed due to exceptionally heavy cropping reducing the ability of the trees to grow seasonally. In Egypt, trellising wires are erected above high-density Keitt trees to force branch growth at upright angles of 45 degrees or more from the horizontal (Fig. 32).

Figure 30 An ultra-high-density orchard in Escuinapa, Mexico.

Figure 31 3 by 1 m Ewiess orchard at Nimos Estate, Egypt.

Figure 32 Keitt grown at high density in Egypt is often trellised to keep the branches upright (angles of 45° or more from the horizontal).

9 Conclusion and future trends

The booklet titled, *Greatly Enhanced Ease in Growing Mangoes on Small Trees*, has been published by SQM Europe N.V. (Sint Pietersvliet 7, bus 8, 2000, Antwerpen, Belgium) and is available on the website, http://sqm.com. In the opinion of the author, the ultra-high-density system of producing mangoes is the ultimate cultivation system for this tree crop, both from the point of view of convenience and productivity. It is noteworthy, however, that a system's success depends primarily on the execution of pruning actions at specific times during the yearly growth cycle. The times when branch heading is required depend on the environment in which the trees are grown, the specific season to an extent, the heaviness of bearing and the cultivar grown. Key is the understanding of how to obtain suitably mature terminal shoots when conditions are maximally inductive during the year. If suitably mature shoots remain vegetative, soil-drench treatment with paclobutrazol is recommended. Also, the trees should never be allowed to overcrowd. A great benefit of the system is the ability to carry out novel manipulations, which are largely impractical in large-tree, conventional systems. An example is the application of sleeves over each fruit after the period of post-set drop. The ability to easily remove malformed inflorescences is a highly significant benefit.

10 Where to look for further information

The following are recommended sources of additional information:

Iyer, C. P. A. and Kurian, R. M. 1992. Tree size control in mango (*Mangifera indica* L.) – some considerations. *Acta Horticulturae*, n. 321, pp. 425–36.
Kulkarni, V. and Hamilton, D. 1997. An integrated approach towards improving mango productivity. *Acta Horticulturae*, n. 455, pp. 84–91.
Nath, V., Das, B. and Rai, M. 2007. Standardization of high-density planting in mango (*Mangifera indica*) under sub-humid Alfisols of Eastern India. *Indian Journal of Agricultural Sciences*, v. 77, pp. 3–7.
Oosthuyse, S. A. Tree spacing trends and options for yield improvement in mango. S.A. Mango Growers' Association Yearbook v. 13, pp. 34–9, Merensky Technological Services, P.O. Box 14, Duivelskloof 0835
Oosthuyse, S. A. 2009. Management of a 'Tommy Atkins', ultra-high density orchard and recognized benefits associated with small tree mango orchards. *Acta Horticulturae*, n. 820, pp. 335–8.
Oosthuyse, S. A. 1994. Pruning of Sensation mango trees to maintain their size and effect uniform and later flowering. *South African Mango Grower Association Yearbook*, v. 14, pp. 1–5.
Ram, S. and Sirohi, S. C. 1991. Feasibility of high density orcharding in Dashehari mango. *Acta Horticulturae*, n. 291, pp. 207–12.
Yeshitela, T., Robbertse, P. J. and Stassen, P. J. C. 2003. The impact of panicle and shoot pruning on inflorescence and yield related developments in some mango cultivars. *Journal of Applied Horticulture*, v. 5, pp. 69–75.

11 Acknowledgements

Nadia Niazi (Egypt) and Carlos Menchaca (Mexico) are thanked for adopting the systems for assessment.

12 References

Charnivichit, S. and Tongumpai, P. 1991. Effect of paclobutrazol on canopy size control and flowering of mango cv. Nam Doc Mai Twai No. 4, after hard pruning. *Acta Hort.* 291:60–6.
Issarakraisila, M., Considine, J. A. and Turner, D. W. 1997. Vegetative and reproductive growth aspects of mango growing in a Mediterranean climate in Western Australia. *Acta Hort.* 455:56–63.
Jackson, J. E. 1980. Light interception and utilization by orchard systems. In Janick, J. (Ed.), *Horticultural Reviews*, Volume 2. John Wiley & Sons, Inc.: Hoboken, NJ, USA, Chapter 5.

Jackson, J. E. and Middleton, S. G. 1988. Modeling of orchards for maximum productivity and fruit quality. In: *Coltura del melo verso gli anni '90, Conference*, Cordenons, Italy, 18–20 December 1986, pp. 309–20.

Majumder, P. K., Shara, D. K. and Singh, R. N. 1982. A study on 'High Density' orcharding in mango (*Mangifera indica* L.) var. Amrapali. *Punjab Hort. J.* 22:123–7.

Majumder, P. K. and Sharma, D. K. 1988. A new concept of orcharding in mango. *Acta Hort.* 231: 335–8.

Oosthuyse, S. A. 1993. Mango tree spacing trends and options for yield improvement with special reference to South Africa. *J. South. Afr. Soc. Hortic.-Sci.* 3:92–6.

Oosthuyse, S. A. and Jacobs, G. 1995. Relationship between branching frequency, and growth, cropping and structural strength of 2-year-old mango trees. *Sci. Hort.* 64:85–93.

Ram, S. and Sirohi, S. C. 1988. Studies on high density orcharding in mango cv. Dashehari. *Acta Hort.* 231: 339–44.

Chapter 9

Organic mango production: a review

Víctor Manuel Medina-Urrutia, Jaime Eduardo Reyes-Hernández, Gil Virgen-Calleros and Enrique Pimienta-Barrios, Universidad de Guadalajara, Mexico; and Marciano Manuel Robles-González, Campo Experimental Tecomán, Mexico

1 Introduction

Consumer demand for organic tree fruit products in Europe and North America, the dominant organic food markets, has spurred increases in organic agricultural land area and production globally. From 2008 to 2013, the area of production increased 109%, 42% and 53% for organic temperate tree fruits, citrus and tropical/subtropical fruits, respectively, with much lower growth over the total area (non-organic plus organic) in these categories. Most organic tree fruits represent approximately 1–2% of total production area for a specific fruit; avocado is the exception, with 8% of avocado-producing area worldwide under organic management. Mexico, Italy and China are the top three countries in terms of organic tree fruit area. Published data on the relative yield of organic to conventional tree fruits are scarce. Comparisons of yields from organic cultivation of tree fruits range from 42% to 126% of yields from conventional cultivation. Reported production costs for organic tree fruits are generally higher than those for conventional tree fruits, with higher cost for fertilizers

http://dx.doi.org/10.19103/AS.2017.0026.12

Table 1 Mango-growing area and production in the world

Country	Area (ha)	Production (ton)	Exports (%)
India	2 515 970	18 431 330	5.2
China	536 169	4 522 019	nd
Thailand	410 707	3 597 589	nd
Indonesia	251 000	2 431 329	nd
Philippines	196 412	899 014	7.8
Mexico	196 216	1 754 609	10.3
Pakistan	170 714	1 716 882	5.2
Nigeria	130 200	875 000	nd
Brazil	70 317	1 132 463	6.0
Bangladesh	56 296	992 296	nd

Source: FAOSTAT, 2015.

being a common reason. Higher selling prices for organic tree fruits compensate for lower yields and higher costs in most cases, leading to a higher net return under certified organic management (Granatstein et al., 2016). However, organic mango production around the world is still at a low level (Willer and Lernoud 2015).

Details of conventional mango production areas are presented in Table 1 (FAOSTAT, 2015). The main mango-producing countries are India, China, Thailand, Indonesia, Pakistan, Mexico and Brazil. Although there is a lack of precise information, it is known that a number of producing countries have active organic mango research and production programmes. However, organic mango production around the world is still at a low level. Mexico, Thailand, Brazil, Peru and Pakistan are the main mango-exporting countries. The biggest markets for export are the United States and Europe. Australia is one of the most important producers of organic mangoes. Other areas undergoing expansion are India, several countries in Latin America and South Africa. Mexico reported that 12 462 ha were undergoing organic conversion (Gómez-Cruz et al., 2010). Other countries include Peru, Ecuador, Guatemala and Colombia (García, 2003; Granatstein et al., 2016).

The demand for organic products along with the health concerns of consumers has increased. Few technologies are currently available to support organic mango production systems in the main mango-growing regions. A comprehensive mango research programme aimed at developing successful organic production methods is needed to facilitate organic and sustainable production and enhance demand. In this review, information regarding current technologies for field sustainable and organic mango production and postharvest is presented and discussed.

2 Climate and soil selection

Mango trees are cultivated under tropical and subtropical conditions (Litz, 2009; Ramírez and Davenport, 2010; Laxman et al., 2016). Diverse soil, rainfall and temperatures prevailing in these regions play an important role in realizing the potential yields. Under these variant conditions mango trees are subjected to many biotic and abiotic stress. Though the existing knowledge identifies the extent of damage caused by various abiotic stresses, further studies are needed to quantify the likely impacts and also to develop adaptation strategies for sustainable yields (Laxman et al., 2016). The site used for producing organic mangoes must have soil, climate and water availability that favour good growth, high yields and good fruit quality (McCoy, 2004). Soil texture is a special concern in organic agriculture. Loamy soils are likely to require relatively low nutrient inputs and have lower water demands than sandy soil types. In addition, the clay content of loamy soils can accommodate organic matter, the development of good soil biological activity and retain humus suitable for organic production (McCoy, 2004, 2007a).

Growers may be able to improve some soil traits over time but they cannot alter the subsoil layers or modify climate conditions to any significant extent. Conventional growers often rely on chemical fertilizers and pesticides to compensate for poor site conditions, whereas organic growers cannot. In tropical dry climates, orchards are exposed to fewer diseases and pests than those in tropical wet regions. Good drainage and air circulation are essential for disease control. The presence of certain weeds and forage species is a concern for organic growers. For example, Bermuda grass, Johnson grass and several other aggressive species are difficult to control using organic methods.

3 Cultivars and rootstocks

3.1 Cultivars

Breeding programmes are currently in progress in several mango-growing regions of the world (Krishna and Singh, 2007; de Queiroz-Pinto, 2009; Dinesh et al., 2011; Byrne, 2012; Hiwale, 2015). Mango breeding objectives for conventional and organic production are quite similar. The following factors need to be taken into account while selecting mango cultivars for organic production: 1) harvest season; 2) adaptability to the environment; 3) water requirements; 4) tree size; 5) resistance to pests and diseases; 6) fruit quality and marketability and 7) proximity to appropriate markets (McCoy, 2007a). Information on mango cultivars that can assist growers in the selection of cultivars for organic production is available in the literature (Campbell, 1992; Iyer, 1997; Zheng and Ploetz, 2002; Human et al., 2009; Knight et al., 2009; Krishna and Singh, 2007; de Queiroz-Pinto, 2009; Dinesh et al., 2011; Byrne, 2012; Hiwale, 2015).

Mango cultivars that are tolerant to pests and diseases are appropriate for organic production, although few cultivars have tolerance or resistance characteristics. Cultivars should be tolerant to anthracnose, postharvest diseases and physiological fruit disorders. Some mango cultivars are moderately tolerant to anthracnose, but none are tolerant enough to be cultivated without fungicides in humid areas (Litz, 1997, cited by Ploetz, 2004). Resistance to bacterial black spot varies among cultivars, and those with moderate tolerance should be chosen where disease prevalence is high (Ploetz et al., 1994). In Brazil, 'Bhadauran' is resistant to the artificial inoculation of a causal agent of malformation, whereas 'Palmer', 'Parvin', 'Sensation', 'Van Dyke' and 'Zill' have a low incidence of the same agent (Zaccaro et al., 2004). Cultivars resistant to malformation can help reduce dispersion and the costs of control by pruning. Of 113 cultivars screened for shoot gall caused by *Apsylla cistellata* Buckton, 20 had no gall formation. These mango cultivars exhibit field resistance to shoot gall by escaping attack or by resisting egg laying (Singh, 2000b).

Suitable mango cultivars have been identified for organic systems, but growers need to select those that combine desirable characteristics with good market value. Currently there are a range of varieties that possess desirable qualities such as dwarfing, acceptable tolerance to major tree and fruit diseases as well as good fruit quality. As an example, 'Amrapalli' of short stature, regular and prolific bearing is the most suitable variety for high-density planting in India (Hiwale, 2015). 'Tommy Atkins', 'Ataulfo' and 'Kent' are preferred for organic systems in Mexico because 'Tommy Atkins' is more tolerant to anthracnose and well adapted to calcareous soils, 'Ataulfo' is an early season crop with a longer shelf life and 'Kent' is a late-season crop with a good market. Additionally, 'Tommy Atkins' showed better nutritional response to saline solutions than other varieties (de Lucena et al., 2012).

3.2 Rootstocks

Dinesh et al. (2011) reported that the main desirable rootstock features are: polyembryony, dwarfing, tolerance to the adverse soil conditions (high pH, calcareous soil, salinity) and good scion compatibility. In addition, for organic production, rootstocks need to show tolerance to soil diseases. Dwarfing rootstocks tolerant to adverse soil conditions are preferred because more trees can be planted per unit area. In addition, tree size facilitates pest and disease control as well as pruning and harvesting. In India, 'Vellalkulamban' and 'Olour' were reported as promising rootstocks for the mango variety 'Alphonso' compared to others six rootstocks. 'Vellalkulamban' was found to be an appropriate rootstock for potential intensive cultivation of 'Alphonso' in India (Reddy and Singh, 1993; Kurian et al., 1996; Reddy et al., 2003). In Australia the 'Brodie' and 'MYP' rootstocks have been grafted with 'Kensington' to combine high yield with small tree size (Smith et al., 2003, 2008). Recently, Dayal et al. (2016) reported that the rootstock 'K-5' restricted the vigour of cvs. 'Pusa Arunima', 'Pusa Surya' and 'Dushehari'. Also, the 'Olour' rootstock caused a similar effect on 'Amrapalli' and 'Mallika' cultivars in India. In the Canary Islands, Zuazo and Tarifa (2006) reported that 'Gomera-3' was an efficient mango rootstock for 'Keitt' cultivar as this combination showed an effective leaf-nutrient content correlated with fruit yield. However the cultivar, 'Osteen' has slightly more tolerance to salinity on Gomera-1 rootstock than on Gomera-3. Sensitivity of Gomera-3 was reflected in smaller fruits and lower yield in the most saline treatments (Zuazo et al., 2004).

Tolerance of rootstocks to salt and calcium carbonate is also an important factor for organic production. A rootstock from Israel (13-1) was reported as being tolerant to calcareous and saline soil conditions (Gazit and Kadman, 1980; Lavi et al., 1993; Crane et al., 1997). 'Olour' and 'Kurakkan' rootstocks could withstand salinity level of upto 2.15 d/Sm with mild necrosis and scorch on leaves (Dubey et al., 2007). Pandey et al. (2014) found that 'Olour', 'Turpentine' and 'Nekkare' rootstocks can exclude Cl^- ions; however, 'Kurukkan', 'Bappakai' and 'Moovandan' rootstocks were also found to exclude Na+ ions at a lower level of salt concentrations. Most commercial rootstocks used for mangoes belong to *Mangifera indica*, but disease tolerance has been reported in *Mangifera casturi, Mangifera odorata* and *Mangifera pentandra* species. The graft compatibility and horticultural traits of these species must be evaluated in a wide range of conditions (Campbell, 2004). Galli et al. (2011) observed that after 17 years of the planting, the rootstock 'Manila' was resistant to *Ceratocystis* spp.; on the contrary 'Coquinho' was not resistant. In another work the same authors found that the polyembryonic rootstocks

'Vitória', 'IAC 112', 'Dura' and 'Bocado' were resistant to the isolates of *Ceratocystis mangicola* M. van Wyk and M. J. Wingf., under greenhouse conditions. The variety 'Juliana' had the same degree of susceptibility as the variety 'Coquinho'; both these varieties should be avoided in areas where the disease is present. Rootstocks can also influence the susceptibility of trees to pests. As an example, a local polyembryonic rootstock promoted firmness, 3-carene levels and main flavonoid content in 'Manila' trees, resulting in a greater resistance to infestation by *Anastrepha obliqua*. The obtained results support the use of this rootstock because this was found to improve the resistance of mango fruit to infestation by this fruit fly (Vazquez-Luna et al., 2011).

Performance of mango rootstocks can be improved by using soil microorganism isolated from the rhizosphere of mango plants. *Gigaspora margarita* was found to be the most efficient vesicular arbuscular mycorrhizal fungi for improving growth and nutrition of 'Nekkare' mango rootstock, resulting in greater plant height, stem diameter, shoot biomass and shoot phosphorus content (Reddy and Bagyaraj, 1994).

3.3 Interstocks

Mango interstocks are integrated by the combination of three different cultivars in the same tree, that is, a rootstock for the development of a root system that is well adapted to the soil conditions, an intermediate grafted onto the rootstock to control tree size and a desirable cultivar grafted onto the interstock.

Dwarf cultivar/rootstock and cultivar/interstock/rootstock combinations can also reduce mango tree size, which facilitates management and pest control (Vargas-Ramos et al., 2004; Smith et al., 2008). 'Esmeralda', a polyembryonic selection, has been used as an interstock between native polyembryonic rootstocks and the cultivar 'Manila' (Ávila-Reséndiz et al., 1993; García-Pérez et al., 1993, 1996; Mosqueda-Vázquez et al., 1996). Similar results were found with the cultivar 'Ataulfo' on 'Esmeralda' interstock and native polyembryonic rootstock (Vázquez-Valdivia et al., 2000, 2006), which resulted in much smaller tree size. Using the combination integrated by 'Ataulfo'/'Esmeralda'/'native polyembryonic' seedling rootstock, it may be possible to have 400 trees/ha and a cumulative 3-yr yield of 39 t/ha, compared to 11 t for traditional orchards. This strategy of tree size control facilitates orchard management practices and increases the rate of investment return (Vazquez-Valdivia et al., 2000).

4 Soil preparation and planting

Soil selection and preparation is one of the main components of organic production. General information on soil preparation, planting and management for conventional orchards is available in publications on mango (Galán Saúco, 2008; Litz, 1997). Few reports providing specific information about preparation and soil enrichment for an organic production are available in the literature.

Important considerations during site preparation include alleviating soil compaction, enhancing fertility, adjusting pH and managing weeds, pests and diseases. Many growers rip the soil to loosen layers of compaction before they plant a new orchard, because deep tillage can disrupt the soil structure of established orchards (Ames and Kuepper, 2004).

In general, sandy to clay loamy soil of 5.5–7.0 pH at 1.2–1.5 m depth is considered appropriate for mango production (Galán Saúco, 2008). The pH can be adjusted by the addition of lime to raise it or by addition of sulphur to lower it. Soil analysis helps guide the application of compost, lime gypsum or any other rock powders that create good soil conditions, which satisfy the nutritional needs of the orchard (Ames and Kuepper, 2004).

Nutritionally balanced soils, proper soil pH and adequate levels of organic matter are fundamental components of fertility management plans. The soil improvement of organic fruit plantings usually involves pre-planting with other crops and the application of organic fertilizer in compost, natural or mineral form (Ames and Kuepper, 2004).

4.1 Cover crops and soil incorporation

Cover crops produce a thick green layer that can shade or choke out weeds. Combined with a well-planned sequence of tillage, planting cover crops is an effective pre-planting weed suppression strategy that contributes to soil fertility and stabilizes humus. The strategy begins with the ploughing or disking of existing soil and weeds, followed by ripping to loosen the soil and then planting a cover crop. Finally, the mature cover crop is mowed and tilled to prepare the soil for tree planting. Several cover crops and tillage sequences may be necessary before planting trees. Characteristics of some cover crops are presented in Table 2. The cover crops selected for site preparation before planting an orchard may be entirely different from those required once the orchard is established (Ames and Kuepper, 2004).

The application of heavy straw mulch during the first few years of establishment may be beneficial for new tree plantings.

4.2 Planting certified trees

It is important to use disease- and pest-free trees from reputable nurseries that are certified as organic. Planting clean trees prevents the spread of diseases and pests during the early years. In addition, the cost of controlling diseases is reduced, and organic certification is facilitated.

Some microbial agents can help to develop healthy and rapid growth of mango plants in the nursery and also help to adapt well to the soil conditions. Mohandas (2012) reported that *Glomus fasciculatum* and *Glomus mosseae* mycorrhizal species improved the plant height and growth and nutrient content of mango rootstocks compared to non-mycorrhizal species in pot culture. Also, under field conditions 'Totupari' rootstock inoculated with these vesicular arbuscular mycorrhizal fungi yearly promoted the number of branches, available soil P, leaf P, Zn and Cu compared to uninoculated plants. Similarly, inoculation of mango plants with either fungal species *Glomus fasciculatum*

Table 2 Cover crop characteristics related to weed suppression and nitrogen (N) fixation in the soil for fruit trees (adapted from Ames and Kuepper, 2004)

Cover crop	Biomass production	N fixation	Time to decomposition	Weed suppression
Cowpea	+	++	+++	++
Crotalaria	++	+++	++	++
Sesbania	+++	++	++	+++
Sudan grass	+++	NI	NI	++++

NI indicates no information.

and *Glomus magnicaulis* and/or *Azotobacter chroococ* (AZ1 and AZ2) bacterial strains led to a significant increase in plant height, stem diameter, leaf area and total root length in comparison to non-inoculated control, and it was also demonstrated that its inoculation with *Glomus fasciculatum* and AZ1 stimulated maximum growth of the seedlings with reduced application of N and P inorganic fertilizer sources. The authors conclude that the selected species and strains should be preferred for use as a beneficial growth activator, especially for a viable and sustainable nursery and field production (Sharma et al., 2014).

5 High density, pruning and shading

Orchard layout influences the long-term health of trees and field operations such as pruning, irrigation, fertilization and management of weeds and pests. Tree size determines spacing, the number of trees per hectare and the time required to obtain an economic return. Standard trees produce more fruit at the adult stage, which minimizes the initial purchasing and planting costs. In contrast, smaller trees have higher planting costs because more plants are needed. However, dwarf and semi-dwarf trees are expected to come into production sooner (Koepke and Dhingra, 2013). Smaller trees simplify many field operations including pruning, pest management and harvesting. Efficiency and safety are greater when the majority of operations can be accomplished from the ground rather than on ladders.

Most mango varieties are vigorous and poorly adapted to high-density plantings which is in turn an appropriate strategy for organic production. Hiwale, 2015, reported that 'Amrapalli' mango trees planted at 2.5 × 2.5 (1600 plants/ha) in a triangular method may produce two times the yield of the more open traditional planting systems. A plant spacing of 2.5 × 3.0 m (1333 plants/ha) has been recommended for better production of 'Dashehari', a more vigorous cultivar than 'Amrapalli'. Tipping of shoots is recommended during the first years to avoid excessive tree growth and stimulate the early flowering. According to Hiwale, 2015, plants at high density are given normal dosages of manures and fertilizers. 'Amrapalli' mango orchards usually decline in yield after the 12th year of planting due to overcrowding and intermingling of the tree canopies. This is the same scenario with a 10-year-old 'Dashehari'. Therefore, severe pruning is recommended. In the case of the more vigorous 'Dashehari' cultivar, 50% of the branches should be dehorned in the 11th year and 25% in the 12th year. Pruning of trees should be done after harvesting to produce fruit regularly.

A diversity of dwarfing variety and rootstock combinations, together with a better knowledge of tree physiology and root activity in high-density conditions, will be required in the future to improve the management of high-density orchard production. Zaharah and Razi (2009) showed that restriction of tree roots and leaf expansion led to reduced stomatal conductance and leaf water potential as well as an increase of leaf proline and ABA. These results suggest that reduction of soil volume and water stress could effectively control tree size through physiological and morphological changes, thereby bringing higher sustainable returns per hectare and greater effectiveness in orchard management.

In many regions, high-density mango trees are exposed to excessive radiation. Long periods of exposure to high radiant energy can lead to temperature stress which can limit photosynthesis and production. Therefore, Jutamanee and Onnom (2016) investigated the use of a transparent plastic roof as shading in brining diurnal changes in photosynthetic gas exchange, chlorophyll fluorescence, fruit set and quality of mango (*Mangifera indica*

L. cv. 'Nam Dok Mai') growth in the field conditions in Thailand. Their results indicated that shading could maintain the high photosynthetic activity by reducing stomatal limitations for carbon supply and was effective in alleviating the photoinhibitory damage to PSII during bright and clear days with excessive radiation. Finally, shading could increase the number of fruits and improve mango peel colour, which can help to improve mango production in regions with similar problems.

6 Management of established mango orchards

Organic mango growers can also switch from conventionally established orchard into an organic orchard. The transition to organic cultivation can be risky and difficult to implement, given the strict standards imposed by quality regulations. Such standards and regulations may not be clear to farmers in developing countries. The following procedures are recommended to guide the transition from conventional to organic systems.

6.1 Basic considerations during conversion to organic systems

Setting up an organic mango production system is time-consuming. A minimum of three years is required before certifying a farm as organic. This reflects the significant changes that must take place for an organic system to begin to function properly. There may be some unfavourable results during the change process because organic crops usually perform differently than conventional crops. For example, they may have different growth patterns or different vulnerabilities to pests and diseases. Risks may be reduced if a small trial area is dedicated to organic methods so that growers can gain experience, knowledge and confidence. If possible, it is better to move towards a more biological approach several years prior to implementing organic production. This will ensure a successful transition to a fully organic certified system without dramatic changes in crop quality and yield (McCoy, 2007a).

Growers should consider the following points when changing to organic cultivation: a) diseased trees or trees in poor condition are not appropriate to be considered for conversion; b) invasive perennial weeds should be minimized before establishing an organic system; c) trees that are less prone to pests and diseases and with a low tendency towards biennial bearing should be selected; and d) orchards containing very large trees should be avoided because they require more time and labour to convert. High-density plantings may be subject to less airflow, which makes them more susceptible to fungal disease, although they can facilitate some non-pesticide methods for controlling fruit flies, like bagging individual fruits. Fungal diseases can be avoided with a careful layout and pruning in order to facilitate air flow (McCoy, 2007a).

6.2 Selecting the site

When converting to an organic system, it is desirable, but not always possible, to select a location that is isolated from potential sources of pests, diseases or weeds to reduce the risks of infestation. Soil and water contamination by chemical sprays from adjacent orchards or by animal waste products also represents potential health risks. Orchards with good soil traits should be selected. Chemical or heavy metal residues in the soil must not exceed the limits set by organic standards.

6.3 Production strategies

To meet organic certification requirements, the conversion from conventional to organic systems is likely to involve changes in practices and the adoption of new strategies. Changes in management go beyond stopping the use of synthetic chemicals and fertilizers. Naturland (2001) provides a technical guide to producing and processing ecological mangoes. Specific techniques and strategies adopted by individual organic growers will vary according to specific circumstances, location of the property and the type of enterprise (McCoy, 2007a). Good organic managers rely on close observation, anticipation and prevention to develop a robust and productive system in a specific situation. The overall management strategy needs to reflect the following key organic farming principles: a) soil health largely determines plant health; b) organic systems are biological systems; c) organic farms should operate as closed systems as far as possible and d) a holistic approach ensures good integration.

6.4 Soil health and nutrient management

The requirements to produce 1 t/ha of mangos are as follows: 0.9–1.3, 100–200 g, 1.5–6.2 kg, 1.5–1.6 kg and 120–200 g of the macronutrients N, P, K, Ca and Mg, respectively, and 28, 24, 12.4 and 8.3 g of the macronutrients Fe, Zn, Cu and Mn, respectively (Iyer, 2004; Huett and Dirou, 2000). The following recommendations for organic fertilization have been published in India (Chadha et al., 1980, cited by Iyer, 2004): 1) application of 225 g of liquid manure per plant every two weeks before planting for three-month-old mango seedlings grown in nurseries; 2) application of 12.5–50 kg per hole of farm yard manure, 0.75–2.5 kg of bone meal and 1.25–4.0 kg of wood ash; 3) application of farm yard manure (4.5–15.0 kg per tree) during the juvenile period. More research is needed to formulate general recommendations for the fertilization of organic mango.

Reports of chemical fertilizer soil application on mango production have shown mixed results. Many studies found no differences in yield between chemical fertilizer treatments and the controls (Baluyut et al., 1983; Ireta-Ojeda and Guzmán-Estrada, 2002). However, chemicals improved tree growth and yield in some areas (Avilán, 1983; Sergent et al., 1995, 1997; Morales and Rivas, 2004). The mineral content of mango leaf petioles from trees fertilized using farm yard manure did not differ from the content of those fertilized using synthetic agents or those that were not fertilized (Chaudhary et al., 1985). However, a combination of chemical and organic fertilizers was found to improve the flowering and yield of mango trees (Feungchan et al., 1989, cited by Litz, 1997; Das et al., 2009).

Mango orchards have specific nutritional requirements that must be maintained at optimal levels. The mango is sensitive to excess of N during fruit development and ripening. This can result in poor fruit quality and green fruit with poor storage traits. The Ca and K levels are also thought to influence fruit quality and storage. Important trace elements include Zn, Cu and Mn, while the role of B in flowering and internal fruit quality is particularly crucial (McCoy, 2007a). Shaaban (2012) proved that a small portion of the soil-added fertilizers is taken by the plant's roots and higher amount (more than 60%) is lost by leaching, volatilization and fixation. Growth of the injection-fertilized mango trees was 20–25% higher than soil-fertilized plants.

The use of chemical fertilizers should be avoided completely. This means that the animal and plant organic residues used as fertilizers must fuel the biological activity that enhances soil fertility and plant nutrient availability. Mango trees continuously remove nutrients from the superficial soil layers. Thus, organic fruit growers have to increase the organic matter content of the soil through the incorporation of plant residues, livestock manure and compost, which will increase microbial activity and improve the soil structure. Nutrient release from organic matter is a slow but steady process that provides macronutrients and micronutrients without losses from the soil. This allows nutrient absorption during several seasons (Nampoothiri, 2001, cited by Iyer, 2004).

6.4.1 Compost and green manure

The use of plant residues as a source of mulch in mango orchards should increase the nutrient availability and improve the soil water content. In Zimbabwe it was found that miombo woodland releases nutrients and lignin more rapidly than mango litter (Musvotoa et al., 2000). Mulch should not come from fields that have been treated with agrochemicals, or include leaf litter, weeds or any other plant residues from the orchard or nearby fields. If these residues are used, soil organic reserves that promote microbial activity will decay gradually (Iyer, 2004).

The production of compost biomass and its incorporation into the soil is imperative for maintaining soil fertility and the productivity of the entire cropping system. The biomass for composting could include fallen mango leaves, weeds and biomass from intercrops. It has been estimated that 5 kg of dry weight mango leaf litter contains 57 g of N, 8 g of P and 48 g of K, (El-Naggar and Gaffar, 1973, cited by Iyer, 2004). Mango litter also could be used to provide S and polyphenols and to maintain or enhance soil carbon (Musvotoa et al., 2000). Leaf litter can meet the nutritional requirements of mango orchards. However, if increased yields are desired, additional manure or compost should be incorporated into the soil (McCoy, 2007a). Green manure can be derived from certain plant species that provide a rich source of nutrients including *Pongamia*, neem, *Gliricidia* and *Leucaena*. They are typically planted at the edge of orchards, ponds and country roads. Soils can be enriched further using phosphate minerals (Iyer, 2004).

Due to variability in soil and climate of mango plantations, boric acid + FYM compost is ideal for maintaining enough B content in the high-rainfall region of Konkan, whereas for the low-rainfall region of Bangalore, Borax + FYM is the best combination for maintaining enough B content in soil. Application of B in organic matrix is needed for the humid tropical soils that are low in B content (Reuveni et al., 1998).

6.4.2 Vermicompost

The earthworm (*Eisenia fetida*) has been widely used as an agent to hasten the decay of organic matter. The compost produced by earthworm activity is known as vermicompost. Earthworms enhance soil productivity through their effects on soil pH, physical decomposition, humus formation, soil structure and soil enrichment (Iyer, 2004). The range of nutrients produced in vermicompost from plant and animal residues should be determined in each local region to ensure the efficient application of nutrients. A wide range of nutrients and a high content of hormones and humus produced by vermicompost have been found in the soils of India. Clearly, information on vermicompost is important in the assessment of environmental conditions for growth and development of mango trees (Kale, 1996, cited by Iyer, 2004).

Compost can be a valuable input in an integrated soil fertility management programme. However, the availability, quality, purpose and cost of compost are important characteristics that depend on the location. Mango production and quality are sensitive to excess N, so careful attention must be given to the timing of organic matter decomposition and subsequent N release to ensure that this occurs at the appropriate stage in the annual growth cycle of mango trees (McCoy, 2007a).

6.4.3 Bio-fertilizers

Whenever organic manure is not easy to collect, use and store, bio-fertilizers may be utilized to fix a sizeable portion of N from the atmosphere and also to increase P availability in the soil. Experimental studies have validated the utility of bio-fertilizers in organic mango cultivation. Soil application of *Azospirillum* (300 g/plant) with manure (20 kg/tree) in a high fruit-bearing year was found to increase the mango yield by up to 134 kg/tree; the second best treatment was the application of neem cake, which produced an increased yield of 106 kg/tree; the use of *Azospirillum* in a low fruit-bearing year produced an increase of only 33 kg/tree, which was low when compared with the yield produced with the use of manure and other bio-fertilizers (Anonymous, 2002, cited by Iyer, 2004). Guzmán-Estrada (2004c) found that application of the bio-fertilizer Maya-Magic during full bloom (containing *Azotobacter vinelandii*, *Clostridium pasteurianum*, *Klebsiella pneumoniae* and *Citrobacter freundii*) improved the weight and total soluble solids content of fruit from the cultivar 'Kent'. Soil inoculations using mycorrhizae from *Glomus fasciculatum* and *Glomus mosseae* increased the availability of N, P, Zn, Cu, Mn and Fe compared with the control. It was suggested that inoculation with VAM fungi can reduce the need for P fertilizer by more than 500 g of P_2O_5/tree (Anonymous, 2002, cited by Iyer, 2004).

Cultural practices and soil amendments can also improve the rhizosphere microflora, which favours the growth of soil microorganisms and promotes better mango growth and yield. For example, the addition of cow dung increases VAM colonization and the population of P-solubilizing bacteria in the root zone. Other organic sources include neem cake and green manure, which can be combined with *Beijerinckia indica* to improve N fixation by indigenous diazotrophs (Nampoothiri, 2001, cited by Iyer, 2004). These microorganisms have more positive and pronounced effects in intercropping and mixed farming situations because these crops continuously add plant residues that undergo organic recycling (Nampoothiri, 2001, cited by Iyer, 2004). Soil and leaf tissue analyses are important for identifying nutrient deficiencies. The general approach for correcting any deficiency is via the soil, rather than by direct application to the plant, for example, by foliar spraying. However, some foliar applications may be required while soil imbalances are being corrected during the early years of conversion (McCoy, 2007a).

Warschefsky et al. (2016) reported that also bacteria species like *Burkholderia phytofirmans* strain PsJN can colonize the root system and protect the shoot vine from pathogenic *Botrytis* and *Pseudomonas* while also modulating carbohydrate metabolism and increasing freezing tolerance.

7 Weed control

Weed control is a major problem in organic mango systems. Many annual and perennial weeds are present in most mango orchards, especially in tropical wet conditions. Strategies for controlling weeds include elimination by hand or by plugging, cultivation with leguminous plants as green manure, cultivation with cover crops, plant intercropping, biological control and the use of natural herbicides. Most of these methods require substantial time and intensive labour. Australian organic mango growers use leguminous crops, because they suppress weed growth and also add N to the soil (McCoy, 2007a). All methods of weed control, such as hand weeding, mechanical cultivation, paper and hay mulching and cover crop cultivation, have their own advantages and disadvantages, where each method performs well in at least one area and does not in at least one area. Weed cover between rows in mango orchards decreases the N content and the yield of mango trees after two years (Torres et al., 2009). However, no general recommendations for weed control are available and more work is needed to develop effective weed control options for organic orchards on a farm-by-farm basis (McCoy, 2007a).

In the future, sustainable cropping systems will avoid the use of agrichemicals. As a result, natural herbicides must be developed by screening a range of crop residues for allelopathic effects on weed germination and growth. Allelopathy is the beneficial or harmful effect of a plant on another plant via the excretion of toxic substances. For example, annual rye releases a chemical from its roots that prevents certain seeds from germinating (Ware, 1996). Rotations could be employed where allelopathy might reduce weed problems. Extracts of medicinal plants have been subjected to bioassay tests for the control of weeds in the Philippines. *Cinnamomum mercadoi, Coccinia grandis* and *Tinospora rumphii* contain active compounds that are effective in inhibiting the shoot and root growth of plants, which are comparable to 2, 4-D (Medina-G. et al., 1995). Studies in India showed that the incorporation of 2.5 kg m–2 of dry *Emblica officinalis* into the soil significantly reduced the abundance of weeds, that is, *Pluchea wallichiana, Saccharum spontaneum, Ageratum conyzoides, Launaea pinnatifida, Crotalaria medicaginea and Imperata cylindrica*. In contrast, the mulching of mango leaves in rose gardens caused allelopathic suppression of weeds such a *Cyperus, Cynodon, Euphorbia* and *Blumea*. Subsequently, it was found that the compound responsible for the allelopathic activity of mango leaves was mangiferin (Venkateshwarnlu et al., 2001, cited by Iyer, 2004). However, more works are needed to confirm this. Seasonal hand weeding of creepers that climb up tree trunks may be required in tropical regions.

7.1 Cover crops

When choosing the planting density, the grower should consider the possibility of including multiple cover crops between a specific number of mango rows to reduce weed problems. It is highly recommended that cover crops are grown to maintain and improve the soil health in organic mango orchards. They prevent soil erosion and increase organic matter, thereby improving microbial activity, soil structure and water infiltration. Cover crops also aid nutrient cycling, reduce soil temperature fluctuations, provide a habitat for beneficial insects, attract bees and other pollinators and suppress weed populations (Ware, 1996). Several cover crops can suppress the growth of weeds by producing thick canopies that prevent light from reaching the soil surface. This weakens competing weed species and suppresses weed germination. Leguminous cover crops such as cowpeas, *Crotalaria* and *Sesbania* are good weed suppressors that also supply biomass and N to the soil (Ware, 1996).

7.2 Intercrops and rotations

In addition to weed control and organic matter supply, plant diversity in the form of intercrops between tree rows helps to provide green manure, compost and beneficial insects, which is of capital importance for organic mango production. In India, recommended intercrops include green and black gram, wheat, groundnut, cabbage, cauliflower, peas, tomato, potato and capsicum. Forage peanut, kudzu, perennial soybean and velvet beans are recommended for organic mango production in Mexico (Ireta-Ojeda and Guzmán-Estrada, 2002). Other crop rotation pairs include cowpea–potato, green gram–wheat, black gram–wheat, green gram–gram, cowpea–wheat and cowpea–gram (Anonymous, 1998, cited by Iyer, 2004). In Florida, weed suppression and biomass production were accomplished in citrus orchards using hairy indigo, cowpea and Alyce clover (Linares et al., 2008).

Chandra et al. (2013) recommended to intercrop preferably leguminous and oilseed crops with mango economically up to seven years on degraded lands and thereafter shade-tolerant crops for better use of resources in India. The fruit-based models also improve livelihood, food and nutritional security of rural people through a wiser integration of leguminous and oilseed intercrops. The fruit-based models improve not only the fertility status of degraded soils but also productivity of poor lands by continuous addition of manures, fertilizers, crop residues and leaf litter, which ultimately improve soil organic carbon levels and their moisture-holding capacity. The mango-based agri-horticulture model with intercropping of 15 years can add more monetary value to the farming business as compared to those having ten years of intercropping or no intercropping with reduced payback period and increase economic benefits.

Crop rotation does not entail changing the economic crop. It involves diversifying the vegetation that grows around the orchard. A diversity of orchard cover crops along the farm perimeter and irrigation line paths provides shelter and food for a variety of beneficial species (Ames and Kuepper, 2004). Guidelines for the selection and management of ground cover crops include: a) specifying the objectives in order of priority, for example, weed suppression, N addition or organic matter incorporation; b) taking into account rainfall patterns and other climate factors; c) considering the soil type and potential soil erosion; d) identifying the desired growth patterns and characteristics of the cover crop; e) considering planting techniques and timing and f) choosing a single cover crop or a series of different cover crops (Ames and Kuepper, 2004).

7.3 Mowing and mulching

A well-established plant cover of a preferred species provides the basis for outcompeting and controlling weeds without application of chemical control. Managing the orchard floor cover requires periodic mowing and mulching, which can be planned and timed to optimize the impact on target weeds. A tractor-mounted mower is commonly used with a spring-loaded retractable outrigger that moves around tree trunks. The height and timing of cutting can influence the growth and flowering of various orchard floor species (McCoy, 2007a).

8 Irrigation

Mango trees need water for most of the year to achieve better growth and yield. The source of water needs to be of good quality in order for salinity and soil contamination to be avoided. Water application with excessive salt not only affects tree growth and yield but also results slowly in soil contamination. Clean water is also required for fruit production free of microbial contamination which when present represents a risk for mango consumers. Irrigation water needs to be carefully chosen to avoid salt and microbial contamination.

Water requirements of mango trees in Nayarit, Mexico, are met by rainfall during summer and autumn. In fact around 1000 ha of organic mango on the hills of Nayarit, México, are supplied with 1000–1200 mm of rainfall annually (Pérez-Barraza et al., 2007). However, most mango-growing regions require good sources of available water for irrigation. The main period of water application is from middle autumn to spring in most mango-growing regions.

Water use efficiency (WUE) is an important parameter in mango trees for helping them avoid excessive water loss. According to Teixeira et al., 2008, the evapotranspiration (ETo) of mango tree cultivar 'Tommy Atkins' is less than that for grass by the relatively large bulk surface resistance due to the presence of older leaves, shadow of the canopy and the absence of a ground cover crop. The annual water percolation flow of a wet light-textured soil was estimated as 800 mm which increases the threat of groundwater contamination and soil salinity build-up in Brazil. Schulze et al. (2013) used a micro-sprinkler irrigation in 9-year-old mango tree cultivar 'Nam Dok Mai Seethong' on sandy loam soil, to reduce or even eradicate the run-off and high infiltration problems and to avoid dry intervals between irrigation events. The use of a micro-sprinkler system showed increased WUE as the price of fruits per unit was higher and gave more profitability to farmers in Thailand. In Brazil, using a micro-sprinkler system in 'Tommy Atkins', Rodrigues da Silva et al. (2009) found that the WUE can be improved by irrigation scheduling. Schulze et al. (2013) found that a deficit irrigation, of 50% ETo using micro-sprinkler, increases the crop water productivity substantially and stabilizes yield during drought, and the profit can be increased by 55% under full irrigation (100% ETo) in 'Nam Dok Mai' in Thailand.

Water availability is expected to be limited in the near future. Therefore, strategies to minimize water use and application in critical phenological phases also need to be incorporated to improve mango production and fruit quality. Fukuda et al. (2013), based on random forest models in the estimation of mango fruit yield under different irrigation regimes in cv. 'Chok Anan' in Thailand, found that water supply early in the mango season is important for the total yield, while water supply towards the end of the mango-growing season is important for improving both total yield and the marketable mango fruit. However, the relevance of irrigation during fruit set and shortly before harvest needs to be clarified using specifically designed experiments. Spreer et al. (2007) reported that yields of mango cv. 'Chok Anan' were reduced when trees were subjected to a regulated deficit irrigation and partial rootzone drying (PRD) treatments as compared to fully irrigated control. However, in areas where water is a limiting factor for production, PRD may be the key for a sustainable increase in mango production. Recently, Spreer et al. (2009) have reported that in an experiment conducted over four years 'Chok Anan', the average yield in the different irrigation treatments was found to be 83.35 kg/tree (100% ETo), 80.16 kg/tree (50% ETo), 80.85 kg/tree (50% ETo in alternating sides of the root system) and 66.1 kg/tree (no irrigation). It was concluded that deficit irrigation strategies can save considerable amounts of water without affecting the yield to a large extent, possibly increasing the average fruit weight, apparently without negative long-term effects.

Casadesús et al. (2011) reported an irrigation system for mango based on the strong relationship between transpiration and the amount of radiation intercepted (RI) by canopy and its direct use for automated irrigation. It is almost as good as that of an expert using a water balance approach but without any rescheduling of labour.

In general, water application before flowering is not recommended. González et al. (2004) determined that the water application 30–36 days before flowering increased the overall photosynthetic activity of mango trees at the time of flowering compared to water irrigation during the period of intensive flowering. From this report, caution is recommended during prolonged water stress to induce flowering as a penalty on fruit set and retention is likely to occur on trees where carbon reserves are depleted and the tree photosynthetic capacity is reduced. Water stress is an important tool to flowering and sustainable production; however, the time and duration of water application need to be determined for each region and variety.

9 Flowering habit and induction

Mangoes grown in the low-latitude tropics rely less on temperature for floral induction than trees grown in the upper-latitude tropics and subtropics. The age of the last flush, the key factor for mango flowering, is influenced by the duration of water stress (Ramírez and Davenport, 2010). Mango cultivars, in low latitudes, initiate floral morphogenesis following exposure to an extended period of mild water stress (Singh, 1960; Bally et al., 2009). However, water stress does not induce flowering. It is the age of the last flush impacted by the stress duration that drives flowering (Ramírez and Davenport, 2010). Mango flowers are produced from the terminal buds of the most recently maturing shoots. In the subtropics, most mango cultivars flower once a year during the winter or spring, following a dormant period. Flower initiation is usually triggered by cool nights and dry conditions. Some mango cultivars, such as 'Kensington Pride' in Australia and 'Haden' in Mexico, are prone to inconsistent flowering and irregular bearing, especially in climates with a short dormant period (Ireta-Ojeda and Guzmán-Estrada, 2002; McCoy, 2007a). Synthetic chemicals, that is, paclobutrazol, ethephon, potassium nitrate and ammonium nitrate, are widely used to promote flowering in conventional systems (Tongumpai et al., 1991; Burondkar and Gunjate, 1993; Medina-Urrutia, 1994, 1995a,b; Medina-Urrutia and Alavez-Ramírez, 1994; Sergent et al., 1995, 1997; Medina-Urrutia and Núñez-Elisea, 1996a; Salazar-García and Vázquez-Valdivia, 1997; Davenport and Núñez-Elisea, 1997; Tongumpai et al., 1997; Salazar-García et al., 2000; Crane et al., 2009; Pérez-Barraza et al., 2007). These chemicals are not permitted to be used in organic systems.

The following issues should be considered during organic production to minimize any losses from inconsistent flowering and irregular bearing (Anderson et al., 1982; Medina-Urrutia and Núñez-Elisea, 1996a,b; Pérez-Barraza et al., 2000; Rebolledo-Martínez and Del Ángel-Pérez, 2004; Guzmán-Estrada, 2005; McCoy, 2007a).

9.1 Before flowering

a) Nutrition is important after harvest to ensure good growth and carbohydrate accumulation prior to flowering. Excessive N can result in excessive vegetative growth, which leads to a decrease in flowering. Urban et al. (2004) reported that leaf nitrogen content per unit mass was low in leaves close to inflorescences and even lower in leaves close to panicles bearing set fruits. b) Early pruning after harvest can improve the uniformity of shoot growth and subsequent uniformity of flowering, especially after a heavy crop. In contrast, late or heavy pruning can be detrimental to yield. c) Withholding water for 2–3 months can help to ensure dormancy and improve floral induction. d) Tip pruning prior to flowering improves flowering in tropical areas (Ramírez and Davenport, 2010). This technique involves light tree pruning of the mature wood immediately before flowering. The new emerged shoot, reach maturity and remain ready to the floral stimulus. According to Ramírez and Davenport (2010), cool temperatures determined the flowering response in the upper latitude tropics and subtropics. But, the age of the last flush is the key factor for mango flowering influenced by the duration of water stress in the low-latitude tropics.

9.2 After flowering

a) Wind and insects facilitate mango pollination. Wasps, bees and some flies must be present in the orchard because they are efficient pollinators. During the period of flowering it is important to avoid the use of chemicals in order to protect the activity of entomologic agents. b) Boron is important for pollination and fruit development and must be either readily available from the soil or applied as a foliar spray prior to bud break. c) Water stress during flowering can disrupt flower development and reduce fruit set. Any stress after flowering can increase the number of fruit dropped. Pollen germination and viability are highly dependent on temperature. Temperatures lower than 15°C can affect pollen viability. In the tropics, cool temperatures could have a negative impact on pollen germination, tube growth, zygote formation and eventual fruit set (Ramírez and Davenport, 2016). A second axillary flowering can be achieved under organic cultivation in cooler, subtropical climates by manually removing the flowering wave occurring normally after winter time (Galán Saúco, 2008)

9.3 Flowering induction

Mango trees can be maintained in good condition with adequate manure, irrigation, plant protection and other cultural practices; however, external inductive stimulation may be required to ensure high annual yields (Iyer, 2004). The old practice of smudging used to induce flowering in the Philippines can be applied in organic orchards. However, the experience of growers in many countries has shown that they cannot always rely on artificial flowering induction.

Girdling and cincturing produce very thin wounds around the trunk or main branches, which interrupts photosynthetic product transport from the leaves to the roots. Thus, photosynthetic products are redistributed to the foliage, which promotes flowering, fruit set or increased fruit size. These practices were evaluated for their ability to serve as alternatives for promoting early flowering in 'Tommy Atkins', 'Kensington Pride' and 'Cogshall' cultivars (Reboucas, 1997; Blaikie et al., 1999; Leonardi et al., 1999; Urban and Léchaudel, 2005; Urban et al., 2009). Girdling alone promoted flowering 15 days earlier compared with untreated trees (Urban et al., 2009). The application of girdling 60–75 days before nitrate sprays promoted a 23-day earlier harvest compared with ungirdled sprayed control (Reboucas, 1997). Cincturing has been repeatedly shown to lead to early intense flowering and high fruit production in 'Kensington Pride' in Australia (Blaikie et al., 1999; Leonardi et al., 1999). Presumably, the presence of twine causes long-term disruption of the distribution of hormones or assimilates, which improves fruit production over the long term. However, more work including other cultivars is required to determine the duration of positive effects after treatment with twine. More deep work needs to be done on girdling branch nutritional and accumulated reserve conditions and alternative inductors to substitute nitrate sprays under organic cultivation to increase flowering response and avoid resistance to use of fertilizers.

10 Pest and disease management

Successful organic production requires an integrated approach to the management of pests and diseases (Heath et al., 1996; Peña, 2004; Peña et al., 2009). This approach involves a number of basic preventive strategies that can minimize the likelihood and severity of problems. Preventive measures are suitable for new orchards as well as established ones. Conventional growers who adopt integrated pest management (IPM) in their orchard practices will find the transition to organic systems less daunting than those without IPM knowledge and experience. All of the principles of IPM can be applied to organic systems, although some of the substances used for specific pests or diseases may need to be changed. Building biodiversity into an organic system by establishing and managing an orchard floor that attracts and harbours beneficial predators can increase the effectiveness of IPM (McCoy, 2007a). Accurate identification, regular monitoring and timely intervention are essential for successful pest and disease management. In the event of an outbreak, several biorational compounds are permitted for the control of pests and diseases during organic production.

10.1 Pest monitoring

Sampling techniques and economic threshold availability are very useful tools for minimizing pest damage. Unfortunately, useful reports on these subjects are scarce or incomplete. Information about sampling techniques suitable for pest monitoring is presented in Table 3. This information can help identify practices that should be adopted for controlling pests, for example, pruning, soil and water management and the use of traps and pheromones.

10.1.1 Cultural practices

Several strategies can be applied to facilitate pest control in organic mango farming. Damage caused by hoppers, *Idioscopus clypealis*, can be reduced by avoiding very high-density planting. The soil in mango orchards should be plugged during November to expose mealy bug (*Drosicha mangiferae*) eggs, diapause fruit fly (*Bactrocera dorsalis* or other fruit flies) pupae and midge (*Erosomyia indica*) larvae, which are later destroyed by solar heat. Orchard flooding is also recommended during October to kill mealy bugs (Iyer, 2004). This practice avoids the emergence of nymphs that damage tender shoots (Peña, 2004). Three species of mango stone weevil have been reported: *Sternochetus mangiferae*, *Sternochetus frigidus* and *Sternochetus olivieri*. They can be controlled by maintaining good field sanitation, destroying fallen fruit, raking the soil under the tree canopy to expose insects to light and avoiding the close planting of mango trees or through appropriate pruning (Pinese and Holmes, 2005; Peña et al., 2009). The elimination of infested and fallen fruit also helps to reduce fruit fly (*Bactrocera*. dorsalis) infestation. Pruning can eliminate shoots and branches affected by the leaf webber (*Orthaga cuadrusalis*), shoot borer (*Chlumetia transversa*), and stem borer (*Batocera rufomaculata*).

Table 3 Recommended techniques for monitoring pests of mango

Pest	Sampling recommendation
Seed weevil *Sternochetus mangiferae* and *Sternochetus frigidus*	Fruits from lower parts of the tree (up to 2 m high) (eggs) (Shukla et al., 1988). Nearly ripe and fallen fruits prior to harvest; rejected or fallen fruits during harvest (eggs).
Blossom midge *Erosomyia mangiferae, Erosomyia indica, Asynapta mangiferae, Dasineura mangiferae*	Midge: Affected tissue (newly emerged panicles and very young fruits); sticky traps on tree canopy (Peña, 2004). *Erosomyia mangiferae*: Sampling could be done biweekly during flowering and fruit formation by checking the presence of the pest or symptoms in buds from eight branches (two for each quadrant); in the inflorescence, by checking four panicles per plant, one panicle per quadrant; and in fruits, by checking one fruit per quadrant at fruit formation; the action level is 5% or more of infected branches (2% for inflorescences and/or fruits) (Haji et al., 2004).
Hoppers *Idioscopus clypealis, Idioscopus niveosparsus, Idioscopus magpurensis, Amritodus atkinsoni*	Hoppers: Direct visual examination of tender leaves, flowers, or shoots (Peña, 2004).
Bud mite *Aceria Eriophyes mangiferae* Acarines *Tetranychidae Oligonychus mangiferae, Oligonychus yothersi, Oligonychus punicae* and spider mites *Tetranychus cinnabarinus Frankliniella schultzei*	Bud mites: Number of mites per bud on 25–220 apical bud samples (Peña, 2004). Aceria: Eight branches per tree, allocating two branches per quadrant; action level is one branch showing malformation; action level is 40% or more branches infested (Peña, 2004; Haji et al., 2004). Acarines: Little or no information available on sampling techniques. *Frankliniella*: From the beginning of flowering until fruitlet appearance, randomly shaking each of four new panicles per plant five times over a white plastic tray, one panicle for each quadrant; action level is 10% or more of the inflorescence with ten or more thrips; in racemes, ten or more fruit infected by thrips (Haji et al., 2004).
Leaf scales *Aulacaspis mangiferae, Coccus viridis, Philephedra tuberculosa*	Scales: Crawler movement is detected using double-sided, sticky, dark-coloured bands close to infested leaves, and microscopic examination (Peña, 2004).
Leaf mealy bugs *Drosicha stebbingi, Rastrococcus invadens, Rastrococcus iceryoides, Rastrococcus mangiferae*	Mealy bugs: Inspection of external canopy 2 m above the ground (Narashimhan and Chacko, 1988).
Trunk and twigs: Coleoptera and scales	No information on monitoring is available.
Lepidoptera: Geometridae	Sampling is done once a week by randomly shaking four panicles per plant from the beginning of flowering until fruit formation; action level is 10% or more inflorescences with caterpillars (Haji et al., 2004).

Citrus thrips (*Scirtothrips aurantii*), mango weevil (*Sternochetus mangiferae*) and coconut bug (*Pseudotheraptus wayi*) have been effectively controlled using Surround WP (wettable powder), which is a non-toxic commercial kaolin product, and mixtures of Surround WP with sulphur and lime sulphur in a 'Sensation' orchard in South Africa. The increased consumer resistance to the use of toxic chemicals means that Surround WP should be registered and used in an IPM programme or under organic cultivation for the mango either in South Africa (Joubert et al., 2000, 2004) or elsewhere.

10.1.2 Natural insecticides

Several natural insecticides have been shown to control different pests. They include sprays of neem (active ingredient azadirachtin) products at concentrations of 3000 or 1000 ppm to control the mango hopper (*Idioscopus niveosparsus*). The efficiency of azadirachtin extracts seemed to be dependent on the hopper density, but at low densities (< four per panicle) it was as effective as synthetic chemicals (Verghese, 2000). Neem seed extract (4%) and 0.2% nimbicide were effective in minimizing mealy bug damage (Verghese, 2000). Neem oil mixed with garlic extract has proved useful in controlling mites (*Aceria mangiferae*) on mango trees (Iyer, 2004). Aliakbarpour et al. (2011) showed promising results using 2% neem oil against adult thrips. Although neem oil was less effective (59.8% thrips reduction) than imidacloprid (68.7%), it caused less mortality (24.9%) to pollinators than imidacloprid (92.7%). Future improvement of neem oil formulation should be aimed at improving its efficacy against thrips and further reducing toxicity to pollinators.

Neem seed kernel extract (4%) prepared fresh in the laboratory and sprayed on trees was effective for controlling hopper populations in mango panicles. However, nimbecidine at 0.2% (azadirachtin 1500 ppm plus neem oil) was more potent. These neem products were comparable to the chemical insecticides lindane (0.05%) and imidacloprid (0.02%). Multineem applications of 0.4% (similar to nimbecidine) were quite effective against the leaf webber (*Orthaga cuadrusalis*) for ten days after spraying (Singh, 2000a). Most scale species can be suppressed to economic levels by the application of oils (Peña, 2004).

10.1.3 Biological control

Several natural enemies of mango pests can be useful in organic cultivation. Natural control of fruit flies can be achieved using the predators *Xenophygus analis*, *Belonuchus rufipennis* and *Solenopsis geminata*, which feed on larvae and pupae, as well as the parasitoid wasp *Diachasmimorpha longicaudata*. Biological control was successful in Hawaii where *Diachasmimorpha tryoni* was released to control the Mediterranean fruit fly (*Ceratitis capitata*), while the parasitoid *Fopius arisanus* was used to control the oriental fruit fly (*Bactrocera dorsalis*). In Florida, *Diachasmimorpha longicaudata* was released to control *Anastrepha suspensa*, while *Fopius arisanus* and *Diachasmimorpha longicaudata* were utilized in Fiji to control *Bactrocera passiflorae* and *Bactrocera xanthodes* (Cabrera-Mireles and Ortega-Zaleta, 2004). Biological control of *Anastrepha* species by the release of *Diachasmimorpha longicaudata* has been practised in the mango orchards in the South and Pacific West regions of Mexico, although the usefulness of this strategy is limited in fruit fly control (Ireta-Ojeda and Guzman-Estrada, 2002; Cabrera-Mireles and Ortega-Zaleta, 2004). Sterile insect release has been used to eradicate Mediterranean and Mexican fruit flies in south Mexico. Eradication was achieved by the release of 52 sterile fruit flies for each wild fruit fly captured in traps (Cabrera-Mireles and Ortega-Zaleta, 2004).

The biological control of mango scale (*Aulacaspis tubercularis*) in South Africa was achieved using a parasitoid *Aphytis chionaspis* and a predator *Cybocephalus binotatus* from Thailand. Up to 46.3% scale parasitism occurred, while *Cybocephalus binotatus* successfully controlled the remaining scale populations, reducing them to 2%. The release of 500–1000 Cybo*cephalus binotatus* beetles per hectare was recommended for effective scale control (Joubert et al., 2000; Le Lagadec, 2004; Daneel and Joubert, 2009). Good control of mango scale can also be achieved using the predator *Aulerodothrips fasciapennis* and the parasitoid *Aspidiotiphagus citrinus* (Kfir and Rosen, 1980, cited by Iyer, 2004). Mango scale (*Milviscutulus mangiferae*) can be controlled in Israel using parasitoids such as *Coccophagus lycimnia* (Iyer, 2004). The scales *Chrysomphalus aonidum* and *Aonidiella aurantii*, which are found on Manila trees, have the following wasps as natural enemies: *Aphytis lingnanensis, Aphytis holoxanthus, Aphytis chrysompali (Hymenoptera: Aphelinidae)* and *Encarsia* spp., which are well distributed in the state of Veracruz, Mexico (Cabrera-Mireles and Ortega-Zaleta, 2004). A wasp, *Anicetus beneficus*, has been identified in Australia for the control of the pink wax scale (*Ceroplastes rubens*) (Cunningham, 1984). Similarly, the parasitoid *Goethana parvipinnis* has been known to attack mango thrips (*Scirtothrips mangiferae*) (Bartlet, 1938, cited by Iyer, 2004).

The predators *Mallada boninensis* and *Chrysoperla lacciperda* destroy leaf hopper nymphs, while the fungus *Verticillium lecanii* infects adults (Fasish and Srivastava, 1990, cited by Iyer, 2004). The coccinellid beetles, *Menochilus sexmaculatus, Sumnius renardi* and *Rodolia fumida*, are important predators of the mealy bug on mango trees. Leafhoppers can be successfully controlled using parasitoids such as *Anagrus* spp. and *Gonotecerus* spp. in Mexico (Cabrera-Mireles and Ortega-Zaleta, 2004).

Ants are effective in controlling the major mango insect pests in Australia and their use in non-spray or organic mango orchards leads to the production of similar or more fruit per tree, with 20.3% more first class fruit and a 55.5% higher profit per tree per year compared with orchards lacking ants (Peng and Christian, 2006).

10.1.4 Other methods of control

McPhail traps (glass and plastic versions) are the most widely used traps for monitoring and, in some cases, for controlling different species of fruit flies. They are baited with a mixture of protein (occasionally hydrolysed cotton seed plus borax, molasses or fermented juices) and water, although non-liquid trapping substances are better options. The monitoring of fruit flies that attack the mango and alternate hosts also informs control strategies (Balock and Lopez, 1969). Complementary control methods include avoiding alternative intercrop hosts of mango flies, the collection and destruction of infested fruit, soil raking to expose pupae to light irradiation, sanitary pruning and the establishment of crops as traps (Ireta-Ojeda and Guzmán-Estrada, 2002).

Mango fruit flies can be controlled by using methyl eugenol (0.1%) bait traps, hot water treatments (48°C for 60 min), heat vapour (47°C) to raise the temperature in the centre of the fruit and salt water (5%) as postharvest treatments (Gaffney et al., 1990; Mangan and Ingle, 1992; Verghese et al., 2002, cited by Iyer, 2004).

Pheromone traps and adhesive bands are used for pest control in other fruit crops. Tape coated with Vaseline is used to trap scale crawlers (Peña, 2004). Banding tree trunks with various materials to prevent mealy bug nymphs from climbing or dusting hydrocarbons on the soil have been used with little or no success (Srivastava, 1981, cited by Peña, 2004). Fastening a 25-cm wide band of 400-gauge polyethylene 30 cm above ground level on the trunk is now a commercial practice for managing mealy bug crawlers in mango orchards (Iyer, 2004). Organic mango orchards should be free of weeds to prevent damage from thrips. The use of wild hosts (weeds) and frequent irrigation are common practices in mango orchards for controlling thrips (Cabrera-Mireles and Ortega-Zaleta, 2004). Damage caused by stem borers (*Xiloborus* spp.) can be controlled by pruning dead branches 50–60 cm below the point of entrance where damage is initiated (Ireta-Ojeda and Guzmán-Estrada, 2002).

10.2 Disease management

Mango trees are hosts to many diseases. Leaves, twigs, branches, inflorescence panicles and fruit are all affected by several diseases. Tropical wet conditions are more favourable for the development of diseases than tropical dry conditions. Anthracnose, powdery mildew, mango malformation, branch die back, fruit scab and trunk canker are very common (Ploetz, 2004; Rodríguez-Escobar and Rodríguez-Acosta, 2004. Bacterial black spot, stem-end rot (SR) and blossom blight occur in hot wet climates, whereas *Alternaria* rot, powdery mildew and decline syndrome are important problems in hot, dry climates (Ploetz, 2004). Al Adawi et al. (2006) presented evidence that the vascular wilt pathogen *Ceratocystis fimbriata* causes mango sudden decline disease in Oman, possibly in concert with *Lasiodiplodia theobromae* and the recently described *Ceratocystis omanensis*. Most of these diseases are also present in the subtropical conditions (Galán Saúco, 2008). Prasad et al. (2006) developed an expert system for the diagnosis of pest and diseases and disorders of mango named AMRAPALIKA in India.

10.2.1 Resistant cultivars

Anthracnose is one of the most important diseases found in mango. Weir et al. (2012), reported 22 species plus one subspecies within the *Colletotrichum gloeosporioides* complex. Mango breeding programmes to develop resistant varieties to face biotic problems are under way in several regions. However, resistant varieties to principal diseases, required for organic systems are not available, yet. According to Kamle and Kumar, 2016, in the case of anthracnose this is because of variability in the presence and occurrence of virulent isolates and cultivars to the disease from one to another location. Although mostly all cultivars are susceptible to mango anthracnose, some less susceptible and resistant cultivars to infection caused by *Colletotrichum gloeosporioides* include 'Tommy Atkins', 'Van Dyke', 'Alphonso', 'Baramasi', 'Carabao', 'Carrie', 'Early gold', 'Kent', 'Kishan Bhog', 'Rad' and 'Saigon'. The Egyptian cultivar Zebda has also been reported as highly resistant to anthracnosis infection (Ploetz, 2015).

Ebrahim et al., (2011) revealed that the activity of chitinase and -1,3-glucanase in the leaves was significantly high in mango cultivars resistant to malformation ($r = -0.90$ and $r = -0.91$, respectively) during the flowering period, whereas lignin content did not show a significant correlation with malformation. The highest activity of chitinase (1.977–2.011 units) and -1,3-glucanase (80.54–82.06 units) was recorded in resistant mango cultivars 'Bhadauran' and' Elaichi'. In contrast, these activities were less than 1.010 and 25.21, respectively, in highly susceptible mango cultivars such as 'Amrapali', 'Eldon' and 'Neelum'. Ebrahim et al. (2012), also found that physical parameters like leaf wax content, small leaf size, etc with fewer numbers of stomata per unit leaf area can be successfully used for identification of mango varieties resistant or tolerant to malformation in mango. Chitinase activity and physical parameters could be used to evaluate a wide number of mango cultivars on collection blocks in other regions to select cultivars with better characteristics for organic production.

Ceratocystis wilt is one of the most destructive diseases in Brazil and other areas producing mango. Oliveira et al, 2016, tested the response of cultivars to the most aggressive isolates of *Ceratocystis fimbriata* from highly resistant to highly susceptible. The cultivar 'Ubá' was found to be the most resistant regardless of the isolate tested. These results demonstrate that there are significant differences in aggressiveness among *Ceratocystis fimbriata* isolates from mango in Brazil, as well as cultivar × isolate interaction and also validate the better suitability of Uba, the cultivar used in the processing industry for organic cultivation of mango. Efforts for finding fresh fruit cultivars resistant to *Ceratocystis* wilt must continue.

10.2.2 Cultural practices

Several cultural practices can help to avoid or mitigate the impact of mango diseases. Sanitary pruning is recommended to eliminate and destroy branches and leaves that are affected by anthracnose (*Colletotrichum gloeosporioides*), powdery mildew (*Oidium mangiferae*), mango malformation (*Fusarium subglutinans*) and bacterial black spot (*Xanthomonas campestris*). Removed foliage must be destroyed (Guzmán-Estrada, 2000; Ploetz, 2004; Iyer, 2004). Kamle and Kumar, 2016, recommended pruning of trees on a yearly basis and recommended that dead and fallen leaves and debris be removed and burnt from the ground area near mango tree. Plant vigour plays an important role in keeping the plants free from twig and flower infection by anthracnose disease. Therefore, proper irrigation and fertilizer application are essential to maintain the tree vigour. Interplanting mango trees with other plants that are non-hosts of mango anthracnose will avoid chances of epidemics.

A marked reduction in the number of branches and flowering panicles that are affected by mango malformation can be obtained by pruning branches 60–100 cm below the affected point (Ireta-Ojeda and Guzmán-Estrada, 2002; Guzmán-Estrada, 2004a). The disease can be reduced to insignificant levels if these measures are implemented continuously for 2–3 years (Ploetz, 2004).

Die back disease is currently controlled in India by removing the affected portions of the canopy and treating the wounded areas with a 5:5:50 Bordeaux mix (Prakash and Raoof, 1989, cited by Ploetz, 2004). Commonly employed Bordeaux mix ingredients include calcium, copper and water as mineral ingredients which are authorized for organic production. Damping off disease (*Rhizoctonia solani*) can be prevented in nurseries by avoiding excess watering.

So far none of the horticultural practices have provided satisfactory control of powdery mildew. However, reduction of inoculum potential of the pathogen at early stages is likely to decrease disease incidence (Joubert, 1998). Regular inspection of mango orchards and removal/pruning of infected leaves and malformed panicles reduce the load of primary inoculum and improve the use of alternative products to control (Prakash and Misra, 1992, 1993a,b; Prakash and Raoof, 1994, cited by Ploetz, 2004). Raisinghani reported that losses were reduced by 50–75% if the infected branches on the mango trees were shaken after each shower.

Mango fruit can be protected from anthracnose by covering individual fruit with paper bags during fruit development. This process is labour intensive and costly, but higher fruit prices can compensate for such costs. Covering mango fruits with paper bags from fruit set to maturity is also recommended in Mexico to reduce fruit fly infestation and for an increased firmness, and resistance to postharvest management (Guzmán-Estrada, 2000; Cabrera-Mireles and Ortega-Zaleta, 2004).

Some advantages of fruit bagging have been demonstrated for some cultivars. Bagging of 'Nam Dok Mai #4' fruit increased the fruit weight and peel colour development from green to yellow, as a result of lower chlorophyll a and chlorophyll b levels. Furthermore, fruit bagging did not affect the internal development of fruit quality (Watanawan et al., 2008) or the skin or flesh calcium concentration (Hofman et al., 1999). Bagging using different paper materials resulted in mango fruit with lower insect and disease damage, and fewer fruit quality defects compared with unbagged fruit. Thick, waxy paper prevents fruit fly damage, whereas brown paper bags and local newspapers are only recommended during the dry season because these materials are easily destroyed by rain. However, fruit bagging does not improve fruit colour for 'Haden', 'Tommy Atkins', 'Kent', 'Irwin' and 'Sensation' cultivars (Guzmán-Estrada, 2000). Moreover, the use of paper bags during the rainy season decreased their effectiveness in controlling *Colletotrichum* sp., *Capnodium sp.* and *Coccus mangiferae* (Guzmán-Estrada, 2004b).

Bagging of mango fruit before harvest and postharvest treatment for 10 min in hot water treatment (HWT at 52–55°C) was reported to reduce anthracnose infection successfully by 83%. The temperature and duration of the treatment depend on the size or weight of the fruit, stage of maturity, cultivar type, growing conditions and intensity of disease severity. It is recommended that fruit be subjected to HWT within 24 h after harvest (Arauz, 2000). HWT widely preferred due to increased profit, lower damage caused and high market value of fruit. This technology is consumer friendly, easy accessible and adopted frequently by farmers also due to its environment friendly approach (Arauz, 2000).

Texas root rot (*Phymatotrichum omnivorum*) can be controlled by the cultivation and incorporation of green manure (mainly sorghum and corn), organic manure, organic matter and sulphur to reduce the soil pH. The construction of ditches (100 cm deep) around the orchard filled with organic manure and sulphur is also recommended (Ireta-Ojeda and Guzman-Estrada, 2002; Rodríguez-Escobar and Rodríguez-Acosta, 2004).

10.2.3 Biocontrol of diseases in the field

Trichoderma harzianum and *Trichoderma viride* are reported as antagonists against: *Lasiodiplodia theobromae; Diplodia natalensis; Botryodiplodia theobromae; Fusarium moniliforme* var *subglutinans*; *Penicillium oxalicum* Currie; *Penicillium sclerotienum* Yamamoto; *Aspergillus niger* van Tiegh; *Aspergillus tamarii* Kita and *Rhizoctonia sp.* (Verma et al., 2007).

According to Verma et al. (2007), *Trichoderma* spp. has been widely used as antagonistic fungal agents against several pests as well as plant growth enhancers. Faster metabolic rates, antimicrobial metabolites and physiological conformation are key factors chiefly contributing to antagonism of these fungi. Mycoparasitism, spatial and nutrient competition, antibiosis by enzymes and secondary metabolites, and induction of plant defence system are typical biocontrol actions of these fungi.

De los Santos-Villalobos et al. (2013b) isolated 20 *Trichoderma* species from mango orchards located in south Mexico. One member, identified as *Trichoderma asperellum* T8a, was able to control *Colletotrichum gloeosporioides* ATCC MYA 456 *in vitro* and *in vivo*, as well as five *Colletotrichum gloeosporioides* isolates obtained from mango orchards from the State of Oaxaca. Assay of the lytic enzymes involved suggest that cellulases of *Trichoderma asperellum* T8a play a role in biological control against *Colletotrichum gloeosporioides*

ATCC MYA 456 more than chitinase or glucanase. Thus, *Trichoderma* strain T8a, should be evaluated against *Colletotrichum gloeosporioides* under different growth conditions (*in vitro*, in the greenhouse, and in the field) and their whole potential as biological control agents suppressing anthracnose to enhance mangoes production should be evaluated.

Bautista-Rosales et al. (2013) reported that the yeast *Meyerozyma caribbica* L6A2 performed as an efficient biocontrol for the phytopathogenic *Colletotrichum gloeosporioides*, in the mango cv. 'Ataulfo' presenting different antagonistic mechanisms of action such as competition for space and nutrients, production of hydrolytic enzymes, parasitism and biofilm formation through quorum sensing. Patiño-Vera et al. (2005) reported that the yeast *Rhodotorula minuta* has also been used as a biocontrol agent for an effective control of mango anthracnose. The authors are currently involved in the development of a stable liquid formulation that preserves high viability.

Also, preharvest applications of *Pseudomonas fluorescens* (FP7) with chitin formulation at monthly intervals showed long-lasting protection against mango anthracnose. The pathogen suppression and fruit yield recovery indicated that this bioformulation may become an attractive and alternative source for the management of reduced anthracnose incidence of *Mangifera* in future (Vivekananthan et al., 2004a, b).

Bacillus subtilis was found to be relatively effective by controlling 50% of the anthracnose disease on fruits at postharvest. Anthracnose disease of mango is biologically controlled by *Bacillus subtilis* LB5 through inhibiting conidial germination. *Bacillus subtilis* isolated from the avocado phylloplane has been successfully exploited for control of preharvest and postharvest diseases of mango (De Villiers and Korsten, 1996; Korsten et al., 1993; Ruangwong et al., 2012 cited by Reddy 2014). Koomen and Jeffries (1993) found that among 121 isolates antagonistic to the anthracnose pathogen, only *Pseudomonas fluorescens* gave significant reduction of mango anthracnose. Preharvest foliar application of talc-based fluorescent pseudomonad strain FP7 supplemented with chitin at fortnightly intervals (5 g/L; spray volume 20 L/tree) on mango trees from pre-flowering to fruit maturity stage induced flowering to the maximum extent and reduced the latent infection by *Colletotrichum gloeosporioides* besides increasing the fruit yield and quality (Vivekananthan et al., 2004a, b). Vivekananthan et al. (2004a,b) found that the plant growth–promoting rhizobacteria *Pseudomonas fluorescens* (FP7) amended with chitin sprayed at fortnightly intervals gave the maximum induction of flowering, a yield attribute in the preharvest stage, and consequently reduced latent anthracnose symptoms at the postharvest stage.

Silimela and Korsten (2007) reported that preharvest *Bacillus licheniformis* applications alone and alternated with copper sprays applied at 3-week intervals from flowering until harvest controlled moderate levels of anthracnose, bacterial black spot and soft rot. The stickers and spreaders did not improve the ability of *Bacillus licheniformis* to attach to and colonize the leaf surface. Another advantage of using a biological product to control plant diseases is its general acceptance in organic farming. Once product registration has been attained, commercialization of *Bacillus licheniformis* may provide an effective alternative to mango fruit producers

Bautista-Rosales et al. (2014) observed that the antifungal yeast *Cryptococcus laurentii* ((Kuff.) C.E. Skinner) strain L5D showed a high antagonistic potential *in vivo*, with significant inhibition of anthracnose *Colletotrichum gloeosporioides* (Penz) Penz. & Sacc.) in mango *Cryptococcus laurentii* showed competition for nutrients, specifically for sucrose. Scanning electron microscopy (SEM) showed that the yeast biofilm adheres to the fruit and to *Colletotrichum gloeosporioides* hyphae showing competition for space; *Cryptococcus laurentii* was not washed off from treated *Colletotrichum* hyphae as observed with SEM.

Activity of all three hydrolytic enzymes (nagase, glucanase activity and suppressed chitinase) was detected *in vitro*, but only nagase was induced by addition of autoclaved pathogen mycelium. Parasitic activity of yeast on pathogen was not detected.

Recently, Siddiqui and Ali (2014) reported *Pseudomonas putida, Trichoderma viride, Streptomyces hygroscopicus, Streptomyces noursei, Streptomyces natalensis* and *Cryptococcus magnus* as potential biological control agents (BCAs) as they helped in effectively controlling anthracnose disease in fruits.

According to Nasir et al. (2014), mildew of mango is incited by the fungus *Pseudoidium anacardii* (F. Noack) U. Braun and R.T.A. Cook 2012 (formerly known as *Oidium mangiferae* Berthet), which causes the most severe losses at flowering and young shoots flushes during cool and dry conditions. A minimum temperature range of 11–14°C and maximum of 27–31°C along with 64–72% RH are the most conducive for disease development. In spite of a large number of fungicides in use, substantial incidences of powdery mildew are still encountered and the inoculum potential of the pathogen is substantially building up.

The decreasing efficacy of many of the fungicides used to control plant diseases, as well as risks associated with fungicide residues on the leaves and fruit, has highlighted the need for more effective and safer alternative control measures, particularly in the case of organic cultivation. Dag et al. (2000) found that fungicides also reduced pollen germination and fertility in mango. One of the potential methods of reducing the severity of powdery mildew in an environmentally safe manner is the use of BCAs and/or natural compounds. Sztejnberg et al. (1989) reported that an isolate of *Ampelomyces quisqualis* parasitized powdery mildew of mango and reduced the disease in field trials. He also found that *Ampelomyces quisqualis* was tolerant to many fungicides currently used to control this disease.

The *in vitro* application of BCAs as *V. lecanii, Bacillus subtilis* and *Tilletiopsis minor* to leaf disks before inoculation with *Oidium mangiferae* markedly decreased conidial germination and leaf infection. The BCAs applied at 15-day intervals effectively controlled *Oidium mangiferae* on blossom clusters and fruit set on naturally powdery mildew-infected mango cvs. 'Alphonso' and 'Seddek'. Mixing kaolin and MKP with BCAs increased the efficacy of BCAs. The mechanisms implicated in antagonism towards biological control of powdery mildew fungus include mycoparasitism, antibiosis, competition and induced resistance (Nofal and Haggag, 2006). The biofungicide Biomix (*Thrichoderma* sp.) has a similar effect on the control of powdery mildew caused by *Erysiphe polygoni* as that of seven different chemical products when applied to 'Tommy Atkins' trees in Brazil.

Chen et al. found that the active compound D-pinitol obtained from vegetable soybean stalks, which are usually treated as agricultural waste, is effective in controlling cucumber powdery mildew disease. Future studies to explore their use as a potential, economical phytochemical fungicide against powdery mildew of mango are needed.

Cazorla et al. (2006) carried out a field study to evaluate control treatments against bacterial apical necrosis caused by *Pseudomonas syringae* pv. *syringae* and necrotic lesions produced on mango buds and leaves, causing a decrease of flowering and fruit set in cv. Tommy Atkins in Spain. This disease, of particular relevance under windy and cold subtropical conditions, can be controlled by appropriate pruning and use of Bordeaux mixture (Galán Saúco, 2008). Kishun 2003, cited by Reddy 2014, described that *Bacillus coagulans* is a potential bioagent which reduced the disease incidence to the extent of 75%. According to Reddy, 2014 also antagonistic bacteria like *B. coagulans, B. subtilis* and *B. amyloliquefaciens* and fluorescent pseudomonads have been reported to be effective biocontrol agents against the disease *Xanthomonas campestris* pv. Mangiferae indicae.

10.2.4 Safety of fungicides and biofungicides

Very few chemical fungicides are permitted for use in organic production. Copper oxychloride sprays at 0.3% are used extensively to control branch die back, phoma blight, scab, SR and bacterial black spot (Arauz, 2000; Guzmán-Estrada, 2000; Rodríguez-Escobar and Rodríguez-Acosta, 2004). Powdery mildew (*Oidium mangiferae*) in 'Manila' is effectively controlled with three 2% wettable sulphur applications (Ireta-Ojeda and Guzmán-Estrada, 2002; Rodríguez-Escobar and Rodríguez-Acosta, 2004). Powdery mildew is readily controlled using sulphur, although sulphur may burn the flowers and young fruits in warm, sunny conditions (Ploetz et al., 1994).

Newman et al. (2012) mentioned that as mango malformation disease is caused by various fungi of *Fusarium*, it is difficult to diagnose and control this disease. Therefore, in this study the authors developed a polymerase chain reaction (PCR) diagnostic tool for detection of *Fusarium mangiferae* by generating primers which flank fungus-specific sequences. Sprays with monopotassium phosphate (MKP) concentration of 0.5–1.0% were effective in controlling the development of powdery mildew (*Oidium mangiferae*) in 'Kent','Keitt' and 'Tommy Atkins' (Reuveni et al., 1998; Oosthuyse, 2000) but cannot be used in organic cultivation. Biological products formulated from *Chaetomium cupreum*, *Chaetomium globosum*, *Trichoderma harzianum* and *Trichoderma hamatum* have been applied to mango trees in the field and were more effective for reducing pathogen inocula and the incidence of anthracnose (*Colletotrichum gloeosporioides*) than chemical products. The incidence of anthracnose on fruit from trees treated with biofungicides or chemical products was similar, although biological treatment gave a better yield than chemical fungicide treatment (Noiaium and Soytong, 2000).

Palou et al. (2015) discussed the use of low-toxicity chemical extracts of natural origin as essential oils, antifungal peptides and small proteins, and coatings based on chitosan or plant gels like those from Aloe spp as alternative fungicides to control postharvest diseases. The authors discussed the overall performance, advantages, disadvantages, limitations and potential combinations of these alternatives for postharvest decay.

10.3 Control of postharvest diseases

Anthracnose, SR and bacterial black spot are the main postharvest diseases affecting mango fruits at postharvest. The strategy to control these postharvest diseases depends on the specific requirements of the markets. Several practices were implemented to decrease the damage impact on the fruit.

10.3.1 Hot water

Anthracnose control is one the main challenges in postharvest treatments. The major strategy for effective management of postharvest mango anthracnose disease is regular fungicide sprays and HWT after harvest. Anthracnose could be controlled by dip treatment of fruits in (0.1%) carbendazim in hot water at 52°C for 15 min, which is recommended for export quality mangoes. Cultivars also vary in their tolerance to the hot water and temperature treatments should never exceed 55°C for 5 min. HWT as a decay control treatment is applied commercially in few countries due to its efficacy. However, as carbendazim is not permitted in organic production, more research is needed to evaluate other alternative or biological products for better disease prevention in postharvest ((Mitcham and Yahia, 2009)).

10.3.2 Irradiation

Irradiation is recommended as a quarantine or phytosanitary treatment but according to US regulation, foods that carry the USDA Organic Seal cannot be irradiated and in consequence cannot be used in certified organic mangoes.

10.3.3 Biocontrol

Efforts to control postharvest diseases using biological strategies have been developed in several regions. But few results are available to use in organic production. Korsten (2004) reported that *Bacillus* spp. effectively controlled *Cercospora* spot of avocado, anthracnose of mango and avocado, *Dothiorella/Colletotrichum* fruit rot complex and SR of avocado and mango, soft brown rot on mango and postharvest decay and secondary infections on litchi and citrus. Control was achieved through semi-commercial preharvest sprays or postharvest packhouse dip and spray applications. Innovative alternatives to apply the antagonists in the field included the use of foraging bees to disseminate the antagonists to flowers and woolly based plastic caps to provide a slow release effect for the antagonist under field conditions. *Bacillus subtilis* is also available against *Cercospora* spot on avocado. Sharma et al. (2009) reported the following microbial species controlling postharvest diseases in mango: *Bacillus licheniformis* (Weigmann) Verhoeven versus anthracnose (*Colletotrichum gloeosporioides*) and SR (*Dothiorella gregaria* Sacc.) (Govender et al., 2005); *Brevundimonas diminuta* (Leifson & Hugh) Segers and *Candida membranifaciens* Hansen against anthracnose (Kefialew and Ayalew, 2008). Also *Brevundimonas diminuta* and the yeast B-65-23 were as effective as HWT in controlling postharvest diseases at 55°C for 5 min (Sharma et al., 2009).

Govender and Korsten (2006) evaluated different formulations of *Bacillus licheniformis* alone and in combination with prochloraz and stroburilin for their ability to reduce mango postharvest fruit diseases (anthracnose and SR when applied as a dip treatment in a mango pack house. The antagonist provided an effective alternative to fungicides. Furthermore, the powder formulation of the antagonist can be successfully incorporated into the existing pack line. Fruits of cultivar 'Keitt', subjected to the prochloraz-biocontrol (*Bacillus licheniformis*)-hot water combination, showed reduced anthracnose and SR incidence under simulated market conditions (storage temperature, 10°C, Relative humidity 90%, and at room temperature, 20°C at 75% RH for seven days). This integrated treatment also retained most effectively the fruit colour and firmness with high marketability than other treatments (Govender et al., 2005).

Zheng et al. (2013) screened four bacterial strains of *Bacillus* spp. antagonistic to the mango anthracnose. *Bacillus pumilus* and *Bacillus thuringiensis* showed 88.87% and 80.07% of *in vitro* mycelia growth inhibitions, and 94.28% and 87.06%, of *in vivo* inhibition, respectively. These *Bacillus* species produced volatiles and artificial mixtures that could be promising methods for control of anthracnose in harvested mango fruit.

Cell suspensions and culture filtrates of microorganisms isolated from mango trees in Ethiopia inhibited spore germination and hyphal growth of *Colletotrichum gloeosporioides in vitro* and reduced significantly the severity of anthracnose in artificially inoculated mango fruit (Kefialew and Ayalew, 2008). Undescribed species *Brevundimonas diminuta* (B-62-13), *Stenotrophomonas maltophilia* (L-16-12) *Enterobacteriaceae* (L-19-13), *Candida membranifaciens* (F-58-22) and the yeast isolate B-65-23 were effective in controlling the severity of anthracnose in naturally infected fruit. They kept anthracnose severity (lesion development) below 5% during 12 days, while severity on untreated fruit reached 29%. *Brevundimonas diminuta* and the yeast B-65-23 were as effective as HWT at 55°C for

5 min. Only a single application of the isolates showed a potential for the control of mango anthracnose on harvested fruit.

According to Sharma et al. (2009), it appears that the biocontrol system strategy is still in its initial steps compared to fungicidal treatment, but progress made in this area during the last two decades has been remarkable. However, if the same pace continues, the use of microbial antagonists for the control of postharvest diseases of fruits and vegetables will be greatly expanded in the future and will definitely become an internationally adopted practice and will be of great utility for organic mango production.

10.4 Mycorrhizal and plant growth-promoting rhizobacteria (PGPR)

Several mycorrhizal and bacterial microorganisms have been reported by their favourable effects on mango plants. According to Mohandas (2012), the mycorrhizal species of *Glomus* (*Glomus fasciculatum and Glomus mosseae*) and *Acaulospora* were found colonizing the rhizosphere of three-year-old mango rootstocks in India. The colonization of the root was higher in 'Vellakullamban' and 'Totapuri' rootstocks. Rootstock seedlings with mycorrhizal inoculations in pot culture showed improved plant height, growth and nutrient content compared to non-mycorrhizal inoculations. 'Totapuri' rootstocks grafted with 'Arka Aruna' and 'Arka Puneeth' cultivars inoculated with Arbuscular Mycorrhizal fungi improved plant growth and available soil P, leaf P, Zn and Cu compared to uninoculated plants, and this effect was observed during eight years. The inoculum can be applied to the rootstock at nursery and also in field every year. In organic production of mango and in integrated nutrient management, mango mycorrhizal inoculums can be safely used for reaping maximum benefit.

Sharma et al. (2014) determined the effectiveness of fungal species *Glomus fasciculatum* (Thaxter sensu Gerdemann), *Glomus magnicaulis* (Hall) and *Azotobacter* strains on growth promotion of mango seedlings under limited nitrogen (N) and phosphorus (P) soil fertilization for sustainable nursery management in a rainfed ecosystem. *Glomus fasciculatum* and *Azotobacter* (AZ1) had a greater effect on all vegetative growth parameters under reduced N and P fertilizer rates; these authors conclude that the selected species and strains should be preferred for use as a beneficial growth activator, especially for a viable and sustainable nursery production. The antagonist when used in mango pack house treatments could provide an effective alternative to fungicides. Furthermore, the powder formulation of the antagonist can be successfully incorporated into the existing pack line. Recently, Kumar et al. (2016) isolated the strain *Arthrobacter kerguelensis* by genomic analysis and designated it as VL-RK 09; the strain under *in vitro* conditions exhibited high antimicrobial activity against 23 bacteria species, one yeast and five moulds.

De los Santos-Villalobos et al. (2013a) evaluated a managed microbial inoculation treatment using plant growth–promoting rhizobacteria (PGPR) isolated from the mango rhizosphere as an alternative to induce mango growth promotion and flowering cv. 'Ataulfo' by integrating endogenous and exogenous signals. Data indicated that inoculation of mango trees with PGPR (*Burkholderia caribensis* and *Rhizobium sp.*, associated with this crop) is a potential alternative way to promote growth and induce flowering in mango, greatly reducing the high economical costs and environmental contamination associated with traditional agricultural practices and particularly the use of paclobutrazol for inducing flowering in the hot tropics.

10.4.1 Interplanting shrubs

Hernandez et al. (2015) showed that mango-shrub interplanting significantly lowered pH and increased arbuscular mycorrhizal fungi (AMF) colonization of mango roots, enzyme activities and microbial biomass compared to mango alone. It was concluded that *Piliostigma reticulatum* enhances soil biological functioning and that there is a synergistic effect of intercropping mango with the native shrub, *Piliostigma reticulatum*, in soil quality with a more diverse community, greater AMF infection rates and greater potential to perform decomposition and mineralize nutrients, which can be of great utility for organic mango cultivation. The results also suggest that this intercrop system would be valuable for increasing resilience to climate change, but this needs further research. The research raises other questions that are yet to be answered (e.g. if the presence of *Piliostigma reticulatum* increases mango crop yields once mangoes reach the fruiting stage). Soil microbial community structure is different but higher resolution data such as next-generation sequencing technology are needed to enable phylogenetic characterization of the communities in shrub + mango intercrops and to determine how seasonal dynamics affect soil biochemical and nutrient properties.

11 Comparing organic and conventional systems

Research on organic fruit production is generally designed to evaluate the separate effects of organic fertilizers, biofungicides, weed control, pest biocontrol, etc. Research is needed in real field situations to test the interactions of several components of agroecosystems. Very few reports that compare organic systems with conventional systems for the mango and other fruit trees are available. The expansion of organic production will depend on the development of systems that integrate pest-resistant cultivars, ground cover management, biological control and the maintenance of biodiversity.

McCoy (2007b) reported a very interesting comparison of organic, organic plus compost and conventional mango orchards, where the tree health, soil health, yield, fruit quality and economic returns were evaluated for 'Kensington Pride' in Australia. Compost and straw mulch produced higher soil microbial activity and soil organic matter levels compared with unmulched conventional plots. This elevated biological activity may contribute to nutrient availability. Mineralization of organic matter may increase soil N during fruit growth and contribute to greater N accumulation. Over time, organic management enhanced and stabilized the soil biology while decreasing the levels of organic carbon. These results confirmed that it is not easy to sustain elevated organic matter levels for long periods in extremely dry tropical conditions. Growing ground covers is one strategy for maintaining organic matter cycles and protecting the soil surface from harsh climatic conditions. Similar experiments with apples showed that the basal respiration of the soil, its CO_2 efflux and the abundance of earthworms were significantly higher in organic orchards than in conventional orchards (Jamar et al., 2008). Trees in organic systems tend to produce similar or higher yields than conventional orchards two or more years after initiating the organic system (McCoy, 2007b). Long-term research comparing organic and conventional systems is required to confirm these results in terms of tree performance, fruit quality, shelf storage and the relationship between soil health microbiology and nutrition in order to determine the sustainability of soils.

In the second year of evaluation of the Australian experiment, the yields from mango trees with organic treatments were similar to those of conventional systems (McCoy, 2007b). Organic apple orchards produced two-thirds less than conventional orchards during the first year, but one-third more in the second year when compared with conventional or integrated systems (Peck et al., 2006). The costs were slightly higher with organic mango production, but the gross margin returns increased because of the average price premium of organic mangoes (McCoy, 2007b).

Organic apples showed improvements in quality when compared with those under conventional or IPM (Peck et al., 2006). Compared with conventional fruit, organic grapefruit had an increased level of naringenin, which is a desirable compound associated with a lowered incidence of heart disease (Lester, 2006). Organic grapefruit also had a lower level of bergamottin, which interferes with the absorption of certain drugs including lipid-lowering medications (Fukuda et al., 2000). Similar to these mentioned crops, mangoes produced with organic systems may or may not be superior in taste and nutrition. To ascertain any possible quality differences, comparative studies of organic and conventional produce should consider the following guidelines: 1) appropriate study approaches, for example, retail market, farm, or research-centred studies and 2) standardized preharvest production sites, harvest procedures, postharvest handling and analytical methodology (Lester, 2006). To increase the confidence of producers, packers, retailers and consumers, additional research on fruit quality and nutrition, sensorial evaluation, harvest and postharvest handling is required with organic mangoes. Results of these studies may increase the price and consumption of organic products in the near future.

In summary, the mango organic orchard is a relatively new production system with advantages and disadvantages. Some noteworthy benefits of MOO include improvement of soil health, reduction in the use of chemical products, a low risk to human health and the high commercial value of the fruit. Constraints on organic systems include generally low production, little available technology and risks to consumers via the ingestion of bacterially contaminated fruit if the trees are treated with uncontrolled animal organic residues. Thus, growers need to carefully select and apply the most appropriate technology.

12 Conclusion

It is expected that the global market for organic mangoes will continue to increase because people prefer to eat healthier foods. The area of land dedicated to organic mango farming represents a very small proportion of the total area of global mango cultivation. The adoption of organic mango systems is expected to occur more rapidly in dry areas than the wet tropics because there are fewer pest and disease problems. Organic mango production is regulated by International Federation of Organic Agriculture Movements (IFOAM) standards. Organic production is based on drastic reductions in traditional pesticides and chemical fertilizers, and is focused on the conservation of natural resources for the production of high-quality mangoes. As a result, major changes are likely to be required to ensure adequate soil health and nutrient management, pest and disease control, weed management, flowering habit and postharvest treatment. It is feasible to change from conventional to organic systems. Early reports comparing organic and conventional systems have shown that soil microorganisms, organic matter and N content can be higher in organic orchards. Yields of organic mango orchards tend to be similar to those of conventional ones. Future research should focus on weed control, nutrition, pest and disease management and postharvest treatments. Research into fruit quality and nutritional value is also needed to increase the confidence of consumers in organic mangoes.

13 References

Al Adawi, A. O., Deadman, M. L., Al Rawahi, A. K., Al Maqbali, Y. M., Al Jahwari, A. A., AlSaadi, B. A., Al Amri, I. S. and Wingfield, M. J. 2006. Aetiology and causal agents of mango sudden decline disease in the Sultanate of Oman. *European Journal of Plant Pathology* 116:247–54.

Aliakbarpour, H., Che Salmah, M. R. and Dzolkhifli, O. 2011. Efficacy of neem oil against thrips (Thysanoptera) on mango panicles and its compatibility with mango pollinators. *Journal of Pest Science* 84:503–12.

Ames, G. K. and Kuepper, G. 2004. *Tree Fruits: Organic Production Overview.* ATTRA Horticulture System Guide. National Center for Appropriate Technology, USDA, 32p.

Anderson, D. L., Sedgley, M., Short, J. R. T. and Allwood, A. J. 1982. Insect pollination of mango in northern Australia. *Australian Journal of Experimental Agriculture* 33:541–8.

Arauz, L. F. 2000. Mango anthracnose: Economic impact and current options for integrated management. *Plant Disease* 84:600–11.

Ávila-Resendiz, C., Pérez-García, E., Matheis-Toledano, E. and Mosqueda-Vázquez, R. 1993. Production efficiency of compact 'Manila' mangos grafted onto different interstock/rootstocks combinations. *Acta Horticulturae* 341:281–7.

Avilán, R. L. 1983. La fertilización del mango *Mangifera indica* L en Venezuela. *Fruits* 38:183–8.

Bally, I. S. E., Lu, P. and Johnson, P. R. 2009. Mango breeding. In: S. M. Jain and P.M. Priyadarshan (Eds), *Breeding Plantation Tree Crops: Tropical Species*, pp. 51–82. Springer Science+Business Media, LLC.

Balock, J. and López, D. F. 1969. Trapping to control Mexican fruit fly in mango and citrus groves. *Journal of Economic Entomology* 62:53–6.

Baluyut, N. M., Baclig, E. V., Rubio, P. P. and Ancheta, C. 1983. Mango Research. Crop Series # 3. PCARRD, Los Baños, Laguna, Philippines.

Bautista-Rosales, P. U., Calderon-Santoyo, M., Servín-Villegas, R., Angélica Ochoa-Álvarez, N. and Ragazzo-Sánchez, J. A. 2013. Action mechanisms of the yeast Meyerozyma caribbica for the control of the phytopathogen *Colletotrichum gloeosporioides* in mangoes. *Biological Control* 65(3):293–301.

Bautista-Rosales, P. U., Calderon-Santoyo, M., Servín-Villegas, R., Angelica Ochoa-Alvarez, N., Vazquez-Juarez, R. and Ragazzo-Sanchez, J. A. 2014. Biocontrol action mechanisms of Cryptococcus laurentii on *Colletotrichum gloeosporioides* of mango. *Crop Protection* 65:194–201.

Blaikie, S. J., Leonardi, J., Chacko, E. C., Muller, W. J. and Steele Scott, N. 1999. Effect of cincturing and chemical treatments on growth, flowering and yield of mango *Mangifera indica* L. cv 'Kensington Pride'. *Australian Journal of Experimental Agriculture* 39:761–70.

Burondkar, M. M. and Gunjate, R. T. 1993. Control of vegetative growth and induction of regular and early cropping in 'Alphonso' mango with paclobutrazol. *Acta Horticulturae* 341:206–15.

Byrne, D. H. 2012. Trends in fruit breeding. In: D. H. Byrne and M. L. Badenes (Eds), *Fruit Breeding, Handbook of Plant Breeding* 8:3–36. Springer Science+Business Media, LLC.

Cabrera-Mireles, H. and Ortega-Zaleta, D. A. 2004. Principales plagas del mango, pp. 13–31. In: Tecnología para la producción forzada de mango 'Manila' con calidad fitosanitaria en Veracruz. INIFAP-CP-Veracruz (Eds), INIFAP-CIRGOC-Campo Experimental Cotaxtla. Memoria Técnica # 15. Division Agricola, Veracruz, Mexico

Campbell, R. J. 1992. *A Guide to Mangos in Florida*. Fairchild Tropical Garden, Miami, Florida, p. 227.

Campbell, R. J. 2004. Graft compatibility between *Mangifera* species and *Mangifera indica* Turpentine rootstocks and their subsequent horticultural traits. *Acta Horticulturae* 645:341–3.

Casadesus, J., Mata, M., Marsal, J. and Girona, J. 2011. Automated irrigation of apple trees based on measurements of light interception by the canopy. *Biosystems Engineering* 108:220–6

Cazorla, F. M., Arrebola, E., Olea, F., Velasco, L., Hermoso, J. M., Pérez-García, A., Tores, J. A., Farre, J. M. and de Vicente, A. 2006. Evaluation of treatments for the control of the bacterial apical necrosis of mango (*Mangifera indica*) caused by Pseudomonas syringae pv. Syringae. *European Journal of Plant Pathology* 116:279–88.

Chandra, R. A., Saroj, P. L., Lal, H., Sharma, N. K., Jayaprakash, J., Chaturvedi, O. P., Raizada, A., Tomar, J. M. S. and Dogra, P. 2013. Performance of mango based agri-horticultural models under rainfed situation of Western Himalaya, India. *Agroforestry Systems* 87:1389–404.

Chaudhary, S. K., Nijjar, G. S. and Rehalia, A. S. 1985. Nutritional status of mango orchards of Gurudaspur in Punjab. *Acta Horticulturae* 231:282–5.

Crane, I. H., Bally, I. S. E., Mosqueda-Vázquez, R. V. and Tommer, E. 1997. Crop production, pp. 203–56. In: R. E. Litz (Ed.), *The Mango Botany, Production and Uses*. CAB International, Wallingford, Oxon.

Crane, I. H., Salazar-García, S., Lin, T. S, de Queiroz-Pinto, A. C. and Shu, Z. H 2009. Crop production, pp. 432–83. In: R. E. Litz (Ed.), *The Mango: Botany, Production and Uses*. Second Edition. CAB International, Wallingford, Oxon.

Cunningham, C. 1984. Mango insect pest, In: *Australian Mango Research Workshop*, pp. 211–24. 26–30 November 1984. CSIRO, Melbourne.

Daneel, M. S. and Joubert, P. H. 2009. Biological control of the mango scale Aulacaspis tubercularis Newstead (Coccidae: diaspididae) by a parasitoid Aphytis chionaspis Ren (*Hymenoptera: Aphelinidae*). *Acta Horticulturae* 820:567–74.

Dag, A., Eisenstein, D. and Gazit, S. 2000. Effect of temperature regime on pollen and the effective pollination of 'Kent' mango in Israel. *Scientia Horticulturae* 86:1–11.

Das, A., Mandal, K. K., Hasan, M. A., Bhattacharya, B., Majumder, D. and Bandopadhyay, B. 2009. Effect of organic and inorganic nutrients on improving flowering. *Acta Horticulturae* 820:371–80.

Davenport, T. and Núñez-Elisea, R. 1997. Reproductive physiology in mango, pp. 69–146. In: Litz, R. E. (Ed.), *The Mango: Botany, Production and Uses*. CAB International, Wallingford, Oxon.

Dayal, V., Dubey, A. K., Singh, S. K., Sharma, R. M., Dahuja, A. and Kaurc, C. 2016. Growth, yield and physiology of mango (*Mangifera indica* L.) cultivars as affected by polyembryonic rootstocks. *Scientia Horticulturae* 199:186–97.

de los Santos-Villalobos, S., de Folter, S., Delano-Frier, J P., Gómez-Lim, M A., Guzman-Ortiz, D. A. and Pena-Cabriales, J. J. 2013a. Growth promotion and flowering induction in mango (*Mangifera indica* L. cv 'Ataulfo') trees by burkholderia and rhizobium inoculation: Morphometric, biochemical, and molecular events. *Journal of Plant Growth Regulators* 32:615–27.

de los Santos-Villalobos, S., Guzmán-Ortiz, D. A., Gómez-Lim, M. A., Délano-Frier, J. P., de-Folter, S., Sánchez-García, P. and Peña-Cabriales, J. J. 2013b. Potential use of Trichoderma asperellum (Samuels, Liechfeldt et Nirenberg) T8a as a biological control agent against anthracnose in mango (*Mangifera indica* L.). *Biological Control* 64:37–44.

De Lucena, C. C., de Siqueira, D. L., Martinez, H. E. P. and Cecon, P. R. 2012. Salt stress effect on nutrient absorption in mango tree. *Revista Brasileira de Fruticultura* 34(1):297–308.

De Queiroz-Pinto, C. A., Faleiro, F. G., Ramos, V. H. V., Cordeiro, M. C. R., de Andrade, S. R. M., Junqueira, N. T. V. and Dias, J. N. 2009. Performance of seven new mango (*Mangifera indica* L.) hybrid selections at central region of Brazil. *Acta Horticulturae* 820:137–46.

De Villiers, E. A. and Korsten, L. 1996. Alternative strategies for control of mango fruit diseases. *South African Mango Growers Association Yearbook* 16:61–4.

Dinesh, M. R., Vasanthaiah Hemanth, K. N., Ravishankar, K. V., Thangadurai, D., Narayanaswamy, P., Ali, Q., Kambiranda, D. and Basha, S. M. 2011. Chapter 4: Mangiferae. In: C. Kole (Ed.), *Wild Crop Relatives: Genomic and Breeding Resources, Tropical and Subtropical Fruits*, pp. 61–74. Springer-Verlag, Berlin Heidelberg.

Dubey, A. K., M. Srivastav, Sharma, Y. K., Pandey, R. N. and Deshmukh, P. S. 2007. Dry mass production and distribution of nutrients in two mango rootstocks as affected by salinity. *Indian Journal of Horticulture* 64(4):385–90.

Durigan, J. F., de Almeida Teixeira, G. E., Castanharo, N. M. and Domarco, R. E. 2004. Postharvest conservation of 'Tommy Atkins' mango fruit influenced by gamma radiation, wax, hot water and refrigeration. *Acta Horticulturae* 645:601–4.

Ebrahim, S., Usha, K. and Singh, B. 2011. Pathogenesis-related (PR)-proteins: Chitinase and -1,3-glucanase in defense mechanism against malformation in mango (*Mangifera indica* L.). *Scientia Horticulturae* 130:847–52

Ebrahim, S., Usha, K. and Singh, B. 2012. Plant architectural traits and their role in defense mechanism against malformation in mango (*Mangifera indica* L.). *Scientia Horticulturae* 139:25–31

Food and Agriculture Organization (FAO), 2015. FAOSTAT. Rome, Italy. http://faostat3fao.org/download/Q/QC/E.

Fukuda, K., Guo, L., Ohashi, N. and Yoshikawa, M. 2000. Amounts and variation in grapefruit juice of the main components causing grapefruit-drug interaction. *Journal of Chromatography B* 741:195–203.

Fukuda, S., Spreer, W., Yasunaga, E., Yuge, K., Sardsud, V. and Müller, J. 2013. Random Forests modelling for the estimation of mango (*Mangifera indica* L. cv. Chok Anan) fruit yields under different irrigation regimes. *Agricultural Water Management* 116:142–50

Gaffney, J. J., Hallman, G. J. and Sharp, J. L. 1990. Vapour heat research unit for insect quarantine treatments. *Journal of Economic Entomology* 83:1965–71.

Galán Saúco, V. 2008. *El cultivo del mango*. 2nd Edition. Mundi-Prensa, Madrid, 340p.

Galli, J. A., Mello Martins, A. L., Fumiko Ito, M., Toma Braghini, M., N Narita and Rossetto, C. J. 2011. Seca–da–mangueira XXII sobrevivência de variedades poliembriônicas. *Revista Brasileira de Fruticultura*, 33(4):1119–26.

García, G. J. E. 2003. Situación y perspectivas de la agricultura orgánica con énfasis en Latino-América, pp. 3–4. In: C. Gómez (Ed.), *Producción, Comercialización y Certificación de la Agricultura Orgánica en América Latina*. Universidad Autónoma de Chapingo, Chapingo, México.

García-Pérez, E., Becerril-Román, A. E., Ávila-Resendiz, C., Matheis-Toledano, L. and Mosqueda-Vázquez, R. 1993. Vegetative growth analysis of mango 'Manila' trees grafted onto several interstocks/rootstocks combinations. *Acta Horticulturae* 341:256–63.

García-Pérez, E., Mosqueda-Vázquez, R. and Ávila-Reséndiz, R. 1996. Interstock and pruning to control 'Manila' mango tree size. In: *5th International Mango Symposium*. Tel-Aviv, Israel, 1–6 September, p. 50 (Abstract).

Gazit, S. and Kadman, A.. 1980. Mango rootstock selection. *HortScience* 15(5):69.

Gómez-Cruz, M. A., Swentesius-Rinderman, R., Ortigoza-Rufino, J. and Gómez-Tovar, L. 2010. *Situación y desafíos del sector orgánico de México*. Revista Mexicana de Ciencias Agrícolas, México, 1(4): 593–608.

González, A., Lu, P. and Müller, W. 2004. Effect of pre-flowering irrigation on leaf photosynthesis, whole-tree water use and fruit yield of mango trees receiving two flowering treatments. *Scientia Horticulturae* 102(2):189–211.

Govender, V. and Korsten, L. 2006. Evaluation of different formulations of *Bacillus licheniformis* in mango pack house trials. *Biological Control* 37:237–42.

Govender, V., Korsten, L. and Sivakumar, D. 2005. Semi-commercial evaluation of *Bacillus licheniformis* to control mango postharvest diseases in South Africa. *Postharvest Biology and Technology* 38:57–65.

Granatstein, D., Kirby, E., Ostenson, H. and Willer, H. 2016. Global situation for organic tree fruits *Scientia Horticulturae* 208:3–12.

Guzmán-Estrada, C. 2000. Guía para la prevención y control de la antracnosis del mango en Sinaloa. Folleto para Productores # 12. INIFAP-CIRNO-Campo Experimental Sur de Sinaloa, p. 22. Centro de Investigaciones Regional del Noroeste, Culiacán, Sinaloa, México.

Guzmán-Estrada, C. 2004a. Malformación o escoba de bruja o flor macho del mango. Folleto Técnico # 5. INIFAP-CIRNO-Campo Experimental Sur de Sinaloa, p. 39. Centro de Investigaciones Regional del Noroeste, Culiacán, Sinaloa, México.

Guzmán-Estrada, C. 2004b. Effect of fruit bagging on sanitation and pigmentation of six mango cultivars. *Acta Horticulturae* 645:95–199.

Guzman-Estrada, C. 2004c. Evaluation of a biofertilizer, clearing and fruit bagging in mango Kent. *Acta Horticulturae* 645:217–21.

Guzmán-Estrada, C. 2005. Programa para la floración forzada y producción extemporánea de mango en el trópico mexicano. Tecnología al Servicio de los Productores. Fundación Produce Guerrero A.C., México, 101p.

Haji, F. N. P., Barbosa, F. R., Lopes, P. C. R., Moreira, A. N., de Alencar, J. A. and Ferreira, R. C. F. 2004. Monitoring mango pests within an integrated production program in Brazil. *Acta Horticulturae* 645:163–5.

Heath, R., Epsky, N., Dueben, B. and Meyer, W. 1996. Systems to monitor and suppress Mediterranean fruit fly (Diptera: Tephritidae) populations. *Florida Entomologist* 79:144–53.

Hernandez, R. R., Debenport, S. J., Leewis, M-C. C. E., Ndoye, F., Nkenmogne, K. I. E., Soumare, A., Thuita, M., Gueye, M., Miambi, E., Chapuis-Lardy, L., Diedhiou, I. and Dick, R. P. 2015. The native shrub, Piliostigma reticulatum, as an ecological 'resource island' for mango trees in the Sahel. *Agriculture, Ecosystems and Environment* 204 :51–61.

Hiwale S. 2015. Chapter 7: Mango (*Mangifera indica* L.). In: Shrikant, H. (Ed.), *Sustainable Horticulture in Semiarid Dry Lands*, Springer India. pp. 97–114.

Hofman, P. J., Beasley, D. R. and Joyce, D. J. 1999. Effect of bagging and embryo abortion on calcium levels in 'Kensington pride' mango fruit. *Australian Journal of Experimental Agriculture* 39:345–9.

Huett, D. O. and Dirou, J. F. 2000. An evaluation of the rationale for fertilizer management of tropical fruits. *Australian Journal of Experimental Agriculture* 40:1137–43.

Human, C. F., Rheeder, R. and Sippel, A. D. 2009. New cultivars and hybrid selections from the mango breeding program of the Agricultural Research Council in South Africa. *Acta Horticulturae* 820:119–26.

Ireta-Ojeda, A. and Guzmán-Estrada, C. 2002. Guía técnica para la producción de mango en Sinaloa. Folleto Técnico # 22. INIFAP-CIRNO-Campo Experimental Valle de Culiacán, 150p.

Iyer, C. P. A. 1997. Classical breeding and genetics, pp. 49–68. In: Litz, R. E. (Ed.), *The Mango: Botany, Production and Uses*. CAB International, Wallingford, Oxon.

Iyer, C. P. A. 2004. Growing mango under organic system. *Acta Horticulturae* 645:71–84.

Jamar, L., Aubinet, M., Fievez, T., Magein, H. and Lateur, M. 2008. Soil biological properties in conventional and organic orchards. In: *16th IFOAM Organic World Congress*, Modena, Italy, 18–20 June 2008 (Abstract). http://orgprints.org/view/projects/conference.html.

Joubert, P. H. 1998. Pests: Mango weevil. In: E. A. de Villiers (Ed.), *The Cultivation of Mangoes*. Institute for Tropical and Subtropical Crops, ARC LNR, Nelspruit, South Africa, pp. 142–5.

Joubert, P. H., Daneel, M. S. and Grove, T. 2000. Progress towards integrated pest management (IPM) on mangoes in South Africa. *Acta Horticulturae* 509:811–17.

Joubert, P. H., Grove, T. de Beer, M. S. and Steyn, W. P. 2004. Evaluation of kaolin (Surround WP) in an IPM program on mangoes. *Acta Horticulturae* 645:493–9.

Jutamanee, K. and Onnom, S. 2016. Improving photosynthetic performance and some fruit quality traits in mango trees by shading. *Photosynthetica* 54:542–50.

Kamle, M. and Kumar, P. 2016. *Colletotrichum gloeosporioides*: Pathogen of Anthracnose Disease in Mango (*Mangifera indica* L. In: P. Kumar, V. K. Gupta, A. K. Tiwari and M. Kamle (Eds), *Current Trends in Plant Disease Diagnostics and Management Practices, Fungal Biology*, pp. 207–19. Springer International Publishing, Switzerland.

Kefialew, Y. and Ayalew, A.. 2008. Postharvest biological control of anthracnose (*Colletotrichum gloeosporioides*) on mango (*Mangifera indica*). *Postharvest Biology and Technology* 50:8–11.

Koepke, T. and Dhingra, A. 2013. Rootstock scion somatogenetic interactions in perennial composite plants. *Plant Cell Reports* 32:1321–37.

Korsten, L. 2004. Biological control in Africa: Can it provide a sustainable solution for control of fruit diseases?. *South African Journal of Botany* 70(1): 128–39.

Korsten, l., De Villiers, E. E. and Lonsdale, J. H. 1993. Biological control of mango postharvest disease in packhouse. *South African Mango Growers Association Yearbook* 13:117–21.

Knight Jr., R. J., Campbell, R. J. and Maguire, I. 2009. Important mango cultivars and their descriptors, pp. 42–6. In: Litz, R. E. (Ed.), *The Mango: Botany, Production and Uses*. Second Edition. CAB International, Wallingford Oxon.

Koomen, I. and Jeffries, P., 1993. Effects of antagonistic microorganisms on the postharvest development of *Colletotrichum gloeosporioides* on mango. *Plant Pathology* 42, 230–7.
Krishna, H. and Singh, S. K. 2007. Biotechnological advances in mango (*Mangifera indica* L.) and their future implication in crop improvement — A review. *Biotechnology Advances* 25:223–43.
Kumar, M. R., V. Muvva, S. Konda, K. Naragani, U. K. Mangamuri, K. Reddy Dorigondla and D. M. Akkewar. 2016. Antimicrobial profile of Arthrobacter kerguelensis VL-RK 09 isolated from Mango orchards. *Brazilian Journal of Microbiology* 47(4):1030–1038.
Kurian, R. M.; V. V. P. Reddy and Y. T. N. Reddy. 1996. Growth, yield, fruit quality and leaf nutrient status of thirteen year old 'Alphonso' mango trees on eight rootstocks. *Journal of Horticultural Science* 71(2):181–6
Lavi, U. S., D. Tomer, E. Adato and S. Gazit. 1993. Conventional and modern breeding of mango cultivars and rootstocks. *Acta Horticulturae* 341:146–51
Laxman, R. H., C. J. Annapoornamma and G. Biradar.2016. Mango. In: N. K. S. Rao, K. S. Shivashankara and R. H. Laxman (Eds), *Abiotic Stress Physiology of Horticultural Crops*, pp. 169–81. Springer, India.
Le Lagadec, M. D. 2004. The control of mango scale in commercial orchards through the use of the predatory beetle Cybocephalus binotatus. *Acta Horticulturae* 645:509–15.
Leonardi, J., Blaikie, S. J., Mueller, W. J., Steele Scott, N. and Chacko, E. C. 1999. Effect of cincturing and chemical treatments on growth, flowering and yield on mango cv. 'Kensington Pride'. *Australian Journal of Experimental Agriculture* 39:761–70.
Lester, G. E. 2006. Organic versus conventionally grown produce: Quality differences, guidelines for comparison studies. *HortScience* 41:296–300.
Linares, J., Scholberg, J. M. S., Chan, C, McSorley, R. and Ferguson, J. 2008. Evaluation of annual warm-season cover crops for weed management in organic citrus. In: *16th IFOAM Organic World Congress*, Modena, Italy, 18–20 June 2008 (Abstract). http://orgprints.org/view/projects/conference.html.
Litz, R. E. 1997. *The Mango: Botany, Production and Uses.* CAB International, Wallingford, Oxon, 587p.
Litz, R. E. 2009. *The Mango: Botany, Production and Uses.* Second Edition. CAB International, Wallingford, Oxon, 680p.
Mangan, R. L. and Ingle, S. J.. 1992. Forced hot air quarantine treatments for mangoes infested with West Indian fruit fly. *Journal of Economic Entomology* 85:1859–64.
McCoy, S. 2004. *Organic Horticulture: Strategic Opportunities for Western Australia.* Bulletin 4622. Department of Agriculture Western Australia, Australia, p. 23.
McCoy, S. 2007a. *Organic Mangoes: A Production Guide.* Bulletin 4712. Department of Agriculture and Food Western Australia, Australia, p. 42.
McCoy, S. 2007b. *Organic Conversion Demonstration at Kununurra.* Australian Government, Rural Industries Research and Development Corporation, Department of Agriculture Western Australia, Australia, p. 27.
Medina, M. J. B., Calumpang, S. M. F., Magsino, G. L. and Medina, J. R. 1995. Inhibition of cucumber, corn and rice shoot and roots by selected extracts of plants. *The Philippine Agriculturist* 78(2):129–36.
Medina-Urrutia, V. M. 1994. Poda y paclobutrazol afectan el crecimiento y producción de árboles jóvenes de mango 'Tommy Atkins'. *Proceedings of the Interamerican Society of Tropical Horticulture* 38:50–5.
Medina-Urrutia, V. M. 1995a. Effect of paclobutrazol and severe pruning on mango trees cv. 'Tommy Atkins' at high densities. *Proceedings of the Florida State Horticultural Society* 108:364–8.
Medina-Urrutia, V. M. 1995b. Poda mecanizada lateral y de copete en árboles adultos de mango 'Tommy Atkins'. *Horticultura Mexicana* 3:124–30.
Medina-Urrutia, V. M. and Alavéz-Ramírez, J. 1994. Adelanto de floración en mango 'Tommy Atkins' con aplicaciones de paclobutrazol. *Proceedings of the Interamerican Society of Tropical Horticulture* 38:56–61.
Medina-Urrutia, V. M. and Nuñez-Elisea, R. 1996a. Mechanical pruning to control tree size, flowering and yield of mature 'Tommy Atkins' mango trees. *Acta Horticulturae* 455:305–14.

Medina-Urrutia, V. M. and Nuñez-Elisea, R. 1996b. Summer promotion of vegetative shoots to increase early flowering response of mango trees to ammonium nitrate sprays. *Acta Horticulturae* 455:188–201.

Mitcham, E. and Yahia, E. (2009). Alternative treatments to hot water immersion for mango fruit report to the national mango board. http://www.mango.org/media/73313/alternatives to hot water treatment-Final report.pdf (Accessed 10 August 2010).

Mohandas, S. 2012. Arbuscular mycorrhizal fungi benefit mango (*Mangifera indica* L.) plant growth in the field. *Scientia Horticulturae* 143:43–8.

Morales, V. D. and Rivas, S. J. 2004. Uso eficiente de la fertilización y su efecto sobre el rendimiento del mango *Mangifera indica*, L en el municipio Mara del estado de Zula. *Revista de la Facultad Agronomia-LUZ* 1:237–43.

Mosqueda-Vázquez, R., C. Ávila-Reséndiz, E. García-Pérez, F. De los Santos-De la Rosa and A. Ireta-Ojeda 1996. 'Esmeralda' un clon para utilizarse como interinjerto y reducir el tamaño del árbol de mango cultivar 'Manila'. Folleto Técnico # 12. INIFAP-Colegio de Postgraduados, Chapingo, Edo de México, México, p. 21.

Musvotoa, C., B. M. Campbell and H. Kirchmann. 2000. Decomposition and nutrient release from mango and miombo woodland litter in Zimbabwe. *Soil Biology & Biochemistry* 32:1111–19.

Narashimhan, U. A. and Chacko, M. J. 1988. The distribution of some Rastrococcus spp. (Homóptera: Pseudococcidae) on mango in India. *Bulletin of Entomology Research* 81:445–8.

Nasir, M., Muhammad Mughal, S., Mukhtar, T. and Zaman Awan, M. 2014. Powdery mildew of mango: A review of ecology, biology, epidemiology and management. *Crop Protection* 64:19–26.

Naturland E. V. 2001. *Organic Farming in the Tropics and Subtropics: Mango*. Second Edition. Grafelfing, Germany, p. 37

Newman, Z.; Freeman, S., Biton, I., Saada, D., Paz, T., Maymon, M. and Lavi, U. 2012. Molecular diagnosis of mango malformation disease and phylogeny of Fusarium mangiferae. *Phytoparasitica* 40:287–97.

Nofal, M. A. and Haggag, W. M.. 2006. Integrated management of powdery mildew of mango in Egypt. *Crop Protection* 25:480–6.

Noiaium, S. and Soytong, K.. 2000. Integrated biological control of mango var. Choke Anan. *Acta Horticulturae* 509:769–78.

Oliveira, L. S. S., Damacena, M. B., Guimarães, L. M. S., Siqueira, D. L. and Alfenas, A. C. 2016. *Ceratocystis fimbriata* isolates on *Mangifera indica* have different levels of aggressiveness. *European Journal of Plant Pathology* 145(4):847–56. Published online February 2016.

Oosthuyse, S. A. 2000. Cost reduction of powdery mildew control in mango with mono potassium phosphate. *Acta Horticulturae* 509:719–23.

Palou, L., Ali, A., Fallik, E. and Romanazzi, G. 2015. GRAS, plant- and animal-derived compounds as alternatives to conventional fungicides for the control of postharvest diseases of fresh horticultural produce. *Postharvest Biology and Technology* (122:41–52.

Pandey, P., Singh, A. K., Singh, A. K. , Dubey, A. K., Dubey, A. K. and Awasthi, O. P.. 2014. Effect of salinity stress on growth and nutrient uptake in polyembryonic mango rootstocks. *Indian Journal of Horticulture* 71(1):28–34

Patiño-Vera, M., Jiménez, B., Balderas, K., Ortiz, M., Allende, R., Carrillo, A. and Galindo, E. 2005. Pilot-scale production and liquid formulation of *Rhodotorula minuta*, a potential biocontrol agent of mango anthracnose. *Journal of Applied Microbiology* 99(3):540–50.

Peck, G. M., Andrews, P.K., Reganold, J.P. and Fellman, J.K. 2006. Apple orchard productivity and fruit quality under organic, conventional, and integrated management. *HortScience* 41:99–107.

Peña, J. E. 2004. Integrated pest management and monitoring techniques for mango pests. *Acta Horticulturae* 645:151–61.

Peña, J. E., Aluja, M. and Wysoki, W. 2009. Pests, pp. 317–66. In: R. E. Litz (Ed.), *The Mango: Botany, Production and Uses.* Second Edition. CAB International, Wallingford, Oxon.

Peng, R. K. and Christian, K. 2006. Potential for organic mango production in the north territory of Australia using weaver ant, Oecophylla smaragdina (*Hymenoptera: Formicidae*). *Acta Horticulturae* 767:81–8.

Pérez-Barraza, M. H., Salazar-García, S. and Vázquez-Valdivia, V. 2000. Delayed inflorescence bud initiation, a clue for the lack of response of the 'Tommy Atkins' mango to promoters of flowering. *Acta Horticulturae* 509:567–72.

Pérez-Barraza, M. H., Vázquez-Valdivia, V., Osuna-García, J. A., Ríos-Torres, A. and López-Arriaga, G. 2007. Diagnóstico del cultivo del mango en Nayarit. Folleto Técnico No. 7. Centro de Investigación Regional del Pacífico Centro, Campo Experimental Santiago Ixcuintla, 47p.

Pinese, B. and Holmes, R. 2005. *Managing Mango Seed Weevil*. Department of Primary industries and Fisheries Queensland, Queensland Government. http://www2. dpi. qld. gov. au/horticulture/18269. 2005.

Ploetz, R. C. 2004. The major diseases of mango: Strategies and potential sustainable. *Acta Horticulturae* 645:137–50.

Ploetz, R. C. 2015. The mango disease trilogy: antracnos, malformation and sudden decline. *Invited lecture at the XIth International Mango Symposium*, 28th September – 2 October 2015, Darwin, Australia.

Ploetz, R. C., Zentmyer, G. A., Nishijima, W. T., Rohrbach, K. G. and Ohr, H. D. 1994. *Compendium of Tropical Fruit Diseases*. APS Press, St. Paul.

Prakash, O. and Misra, A. K., 1992. Important diseases of mango and their effect on production. *Biological Memoirs* 18:39–55.

Prakash, O. and Misra, A. K.. 1993a. Fungal diseases of sub-tropical horticultural fruit crops. In: K. L. Chadha (Ed.), *Advances in Horticulture*. Malhotra Publishing House, New Delhi (India), pp. 1275–372.

Prakash, O. and Misra, A. K. 1993b. Integrated approach in the management of mango diseases (Abstract). In: *National Conference on Eco-friendly Approaches in the Management of Pests/ Diseases and Industrial Effluents*, 20–22 December at CSAAU & T, Kanpur, pp. 70–1.

Prasad, R., Rajeev Ranjan, K. and Sinha, A. K. 2006. AMRAPALIKA: An expert system for the diagnosis of pests, diseases, and disorders in Indian mango. *Knowledge-Based Systems* 19:9–21.

Ramírez, F. and Davenport, T. L. 2010. Mango (*Mangifera indica* L.) flowering physiology. *Scientia Horticulturae*, Amsterdam, 126:65–72.

Ramírez, F. and Davenport, T. L. 2016. Mango (*Mangifera indica* L.) pollination: A review *Scientia Horticulturae*, Amsterdam, 203:158–68.

Rebolledo-Martínez, A. and Del Ángel-Pérez, A. L. 2004. Producción forzada: Técnicas para el control de la floración en mango manila, pp. 1–11. In: Rebolledo-Martinez, A. (Ed.), *Tecnología para la Producción Forzada de Mango 'Manila' con Calidad Fitosanitaria en Veracruz*. INIFAP-CIRGOC-Campo Experimental Cotaxtla, Memoria Técnica # 15. División Agrícola, Veracruz, Mexico.

Reboucas, S. J. A. 1997. Effect of girdling treatment on flowering and production of mango trees cv. 'Tommy Atkins'. *Acta Horticulturae* 455:132–4.

Reddy, P. P. 2014. Subtropical and temperate fruit crops. In *Plant Growth Promoting Rhizobacteria for Horticultural Crop Protection*. Springer India, pp. 107–27. doi:10.1007/978-81-322-1973-6_7.

Reddy, Y. T. N. and Singh, G. 1993. Effect of rootstock on growth and yield of alphonso mango (*Mangifera indica*). *Indian Journal of Agricultural Sciences* 63(4):208–10

Reddy, B. and Bagyaraj, D. J. 1994. Selection of efficient vesicular-arbuscular mycorrhizal fungi for inoculating the mango rootstock cultivar Nekkare. *Scientia Horticulturae* 59(1):69–73

Reddy, Y. T. N., Kurian, R. M., Ramachander, P. P., Gorakh, S. and Kohli, R. R. 2003. Long term effects of rootstock on growth and fruit yielding patterns of 'Alphonso' mango *Mangifera indica* L. *Scientia Horticulturae* 97:95–108.

Reuveni, M, Harpaz, M. and Reuveni, R. 1998. Integrated control of powdery mildew on field-grown mango trees by foliar sprays of mono-potassium phosphate fertilizer, sterol inhibitor fungicides and the strobilurin Kresoxym-methyl. *European Journal of Plant Pathology* 104:853–60.

Rodríguez-Escobar, G. and Rodríguez-Acosta, F. 2004. Principales enfermedades del mango, pp. 33–44. In: INIFAP-CP-Veracruz (eds), Tecnología para la Producción Forzada de Mango 'Manila' con Calidad Fitosanitaria en Veracruz. INIFAP-CIRGOC-Campo Experimental Cotaxtla, Memoria Técnica # 15. División Agrícola. Veracruz, Mexico.

Rodrigues da Silva V. de P., Baracuy da Cunha Campos, J. H. and de Azevedo, P. V. 2009. Water-use efficiency and evapotranspiration of mango orchard grown in northeastern region of Brazil. *Scientia Horticulturae* 120:467–72.

Salazar-García, S. and Vázquez-Valdivia, V. 1997. Physiological persistence of paclobutrazol on the 'Tommy Atkins' mango (*Mangifera indica* L.) under rainfed conditions. *Journal of Horticultural Sciences* 72:339–45.

Salazar-García, S., Pérez-Barraza, M. H. and Vázquez-Valdivia, V. 2000. Effect of ammonium nitrate sprays on flowering and harvest time of 'Manila', 'Ataulfo', and 'Tommy Atkins' mango in Nayarit, Mexico. *Acta Horticulturae* 509:573–9.

Schulze, K., Spreer, W., Keil, A., Ongprasert, S. and Müller, J. 2013. Mango (*Mangifera indica* L. cv. Nam Dokmai) production in Northern Thailand—Costs and returns under extreme weather conditions and different irrigation treatments. *Agricultural Water Management* 126 :46–55.

Sergent, E., Casanova, E. and Leal, F. 1995. Aplicación de nitrógeno y potasio en mango (*Mangifera indica* L.). *Agronomía Tropical* 45:293–312.

Sergent, E., Ferrari, D. and Leal, F. 1997. Effect of nitrate potassium and paclobutrazol on flowering induction and yield of mango *Mangifera indica* L. cv 'Haden'. *Acta Horticulturae* 455:180–7.

Shaaban, M. M. 2012. Trunk nutrition. In: A. K. Srivastava (Ed.), *Advances in Citrus Nutrition*, pp. 101–12. Springer Science Business Media B.V.

Sharma, R. R., Singh, D. and Singh, R. 2009. Biological control of postharvest diseases of fruits and vegetables by microbial antagonists: A review. *Biological Control* 50:205–21

Sharma, S. D., Kumar, P. and Kumar Yadav, S. 2014. Glomus–Azotobacter association affects phenology of mango seedlings under reduced soil nutrient supply. *Scientia Horticulturae* 173:86–91.

Shukla, R. P., Tandon, P. L. and Suman, C. L.1988. Intra-tree distribution of eggs and diapausing adults of the stone weevil. *Acta Horticulturae* 231:566–70.

Siddiqui, Y. and Ali, A. 2014. *Colletotrichum gloeosporioides* (Anthracnose). In: S. Bautista-Banos (Ed.), *Postharvest Decay*, Elsevier Inc. pp. 337–64. http://dx.doi.org/10.1016/B978-0-12-411552-1.00011-9.

Silimela, M. and Korsten, L. 2007. Evaluation of pre-harvest *Bacillus licheniformis* sprays to control mango fruit diseases. *Crop Protection* 26 :1474–81

Singh, L. B. 1960. *The Mango, Botany, Cultivation and Utilization.*Leonard Hill, London, 438p.

Singh, G. 2000a. Efficacy of neem formulations compared to chemical control insecticides against hoppers and leaf webber. *Acta Horticulturae* 509:745–9.

Singh, G. 2000b. Field resistance in mango varieties against shoot gall. *Acta Horticulturae* 509:759–63.

Smith, M. W., Hulme, M. D. and Bright, J. D. 2003. Rootstock affects yield, yield efficiency, and harvest rate of Kensington Pride mango. *HortScience* 38:273–6.

Smith, M. W., Bright, J. D., Hoult, M. D., Renfree, R. A. and Maddern, T. 2008. Field evaluation of 64 rootstocks for growth and yield of 'Kensington Pride' mango. *HortScience* 43:1720–5.

Spreer, W., Nagle, M., Neidhart, S., Carle, R., Ongprasert, S. and Muller, J. 2007. Effect of regulated deficit irrigation and partial rootzone drying on the quality of mango fruits (*Mangifera indica* L., cv. 'Chok Anan'). *Agricultural Water Management* 88:173–80.

Spreer, W., Ongprasert, S., Martin, H., Wunsche, J. N. and Muller, J. 2009. Yield and fruit development in mango (*Mangifera indica* L. cv. Chok Anan) under different irrigation regimes. *Agricultural Water Management* 96:57484

Sztejnberg, A., Galper, S., Mazar, Shlomit and Lisker, N., 1989. Ampelomyces quisqualis for biological and integrated control of powdery mildews in Israel. *Journal of Phytopathology* 124, 285–95.

Teixeira, A. H. de C., W. G. M. Bastiaanssen, M. S. B. Moura, J. M. Soares, M. D. Ahmad and M. G. Bos. 2008. Energy and water balance measurements for water productivity analysis in irrigated mango trees, Northeast Brazil. *Agricultural and Forest Meteorology* 148:1524–37

Tongumpai, P., Jutamanee, K. and Subhadrabandhu, S. 1991. Effect of paclobutrazol on flowering of mango cv Khiew Sawoey. *Acta Horticulturae* 291:67–70.

Tongumpai, P., K. Chantakulchan and S. Subhadrabandhu 1997. Foliar application of paclobutrazol on flowering in mango. *Acta Horticulturae* 455:175–9.

Torres M. O., Guirado, E., Farre, J. M. and Hermoso, J. M. 2009. Influence of weed cover on mango nutrition, productivity and fruit quality. *Acta Horticulturae* 820:351–6.
Urban L., and Léchaudel, M. 2005. Effect of leaf-to-fruit ratio on leaf nitrogen content and net photosynthesis in girdled branches of *Mangifera indica* L. *Trees* 19:564–71.
Urban, L., Lu, P. and Thibaud, R. 2004. Inhibitory effect of flowering and early fruit growth on leaf photosynthesis in mango. *Tree Physiology* 24:387–99.
Urban L., Lechaudel M. and Alphonsoult, L. 2009. The effect of flowering and leaf net photosynthesis in mango cv. Cogshall. *Acta Horticulturae* 820:251–8.
Vargas-Ramos, V. H., Pinto, A. C. Q., Junqueira, N. T. V., Gomes, A. C., Andrade, S. M. R. and Cordeiro, M. C. R. 2004. Effect of mono and polyembryonic rootstocks on growth, yield and fruit quality of four mango cultivars in the central region of Brazil. *Acta Horticulturae* 645:201–7.
Vazquez-Luna A., Rivera-Cabrera, F., Pérez-Flores, L. J. and Diaz-Sobac, R. 2011. Effect of rootstock on mango fruit susceptibility to infestation by Anastrepha obliqua (plagas). *Journal of Economic Entomology* 104(6):1991–8.
Vázquez-Valdivia, V., Salazar-García, S. and Pérez-Barraza, M. H. 2000. 'Esmeralda' interstock reduce 'Ataulfo' mango tree size with no reduction in yield: results of first five years. *Acta Horticulturae* 509:291–6.
Vázquez-Valdivia, V., Pérez-Barraza, M. H. and Salazar-García, S. 2006. Control del crecimiento de árboles de mango, pp. 51–86. In: V. Vázquez-Valdivia and M. H. Pérez-Barraza (Eds), *El Cultivo del Mango: Principios y Tecnología de Producción*. Libro Técnico # 1. INIFAP-CIRPAC-Campo Experimental, Santiago, Ixcuintla.
Verghese, A. 2000. Effect of imidacloprid, Lambdacyhalothrin and Azadirachtin on the mango hopper, Indioscopus niveosparsus (Leth.) (Homoptera: Cicadellidae). *Acta Horticulturae* 509:733–5.
Verma, M., Brar, S. K., Tyagi, R. D., Surampalli, R. Y. and Valero, J. R. Review: Antagonistic fungi, Trichoderma spp.: Panoply of biological control. *Biochemical Engineering Journal* 37:1–20.
Vivekananthan, R. R., Ravi, M., Saravanakumar, D., Kumar, N., Prakasam, V. and Samiyappan, R. 2004a. Microbially induced defense related proteins against postharvest anthracnose infection in mango. *Crop Protection* 23:1061–7.
Vivekananthan, R.; Ravi, M., Ramanathan, A. and Samiyappan, R. 2004b. Lytic enzymes induced by Pseudomonas fluorescens and other biocontrol organisms mediate defense against the anthracnose pathogen in mango. *World Journal of Microbiology & Biotechnology* 20: 235–44.
Ware, A. 1996. Cover crops for soil improvement in horticultural crops. Publication H. 1100. Kerr Center for Sustainable Agriculture, Inc., Poteau, 4p.
Warschefsky, E. J, Klein, L. L., Frank, M. H., Chitwood, D. H., Londo, J. P., von Wettberg, E. J. B. and Miller, A. J. 2016. Rootstocks: Diversity, domestication, and impacts on shoot phenotypes. *Trends in Plant Science*, 21(5):418–37.
Watanawan, A., Watanawan, C. and Jarunate, J. 2008. Bagging 'Nam Dok Mai # 4' mango during development affects color and fruit quality. *Acta Horticulturae* 787:325–8.
Weir, B. S., Johnston, P. R. and Damm, U. 2012. The *Colletotrichum gloeosporioides* species complex. *Studies in Mycology* 73:115–80. Published online: 21 August 2012. doi:10.3114/sim0011. Hard copy: September 2012.
Willer, H. and Lernoud, J. (Eds), 2015. *The World of Organic Agriculture: Statistics and Emerging Trends 2015*. Research Institute of Organic Agriculture (FiBL), Frick, Switzerland; International Federation of Organic Agriculture Movements(IFOAM), Bonn, Germany, 300p. https://www.fibl.org/fileadmin/documents/shop/1663-organic-world-2015.pdf.
Zaccaro, R. P., Donadio, L. C. and Lemos, E. G. M. 2004. Mango cultivar behavior in relation to Fusarium subglutinans. *Acta Horticulturae* 645:343–6.
Zaharah, S. S. and I. M. Razi.2009. Growth, stomata aperture, biochemical changes and branch anatomy in mango (*Mangifera indica*) cv. Chokanan in response to root restriction and water stress. *Scientia Horticulturae* 123: 58–67.
Zheng, Q. and Ploetz, R. 2002. Genetic diversity in mango, and development of a PCR assay for identifying the mango malformation pathogen. *Plant Pathology* 51:208–16.

Zheng, M., Shi, J., Shi, J., Wang, Q. and Li, Y. 2013. Antimicrobial effects of volatiles produced by two antagonistic Bacillus strains on the anthracnose pathogen in postharvest mangos. *Biological Control* 65:200–6

Zuazo, V. H. D. and Tarifa, D. F.. 2006. Rootstock influence on fruit yield, growth and mineral nutrition of mango (*Mangifera indica* L. cv. 'Keitt'). *European Journal of Horticultural Science* 71(3):102–8.

Zuazo, V. H. D., Raya, A. M. and Ruiz, J. A. 2004. Impact of salinity on the fruit yield of mango (*Mangifera indica* L. cv. 'Osteen'). *European Journal of Agronomy* 21(3):323–34.

Chapter 10

Improving fertilizer and water-use efficiency in mango cultivation

A. G. Levin, Supplant Ltd, Israel

1 Introduction

In all regions, with increasing land and water scarcity and the added pressures of a globalizing world, the future of agriculture is intrinsically tied to better stewardship of natural resources. The solution is not to slow agricultural development but to seek more sustainable production systems (World Bank 2008).

The future of the mango industry worldwide will depend on the efficient use of water and land resources.

'*Efficiencies*' in the use of critical resources (water, land and labour) are today the bottlenecks of the future of agriculture in general and of the mango industry, in particular. The mango growers must also adapt to potentially more extreme weather conditions associated with climate change whilst minimizing its environmental impact by the adoption of modern agricultural techniques. Modern technology in the agriculture sector must be accompanied by the development of precise agriculture strategies, technical support and proper finance systems supporting the development of the mango industry from the technological point of view.

Water is a scarce resource worldwide due to increased consumption, mismanagement and pollution. The agricultural sector is one of the main contributors to this unsustainable

http://dx.doi.org/10.19103/AS.2017.0026.13

situation. Irrigated agriculture is a major consumer of water and accounts for about two-thirds of the total freshwater diverted for human uses (Fereres and Evans 2006).

In vast areas of the world mango production, the main irrigation strategy during the dry season and/or at critical phenological stages is by gravitation or furrow irrigation (personal observations). The amount of available water needed for this irrigation strategy compared to drip irrigation, for example, is significantly higher (Tiwary et al. 1998; Singandhupe et al. 2006). In addition, the groundwater may be contaminated and salinized (Ishfaq 2002) by inefficient fertilizer and water management under the furrow irrigation system (Suarez 1989).

1.1 Production and water

The use of irrigation contributes to two important factors in a competitive global economy, increasing yields and improvement of fruit quality (Coelho and Borges 2004). Irrigation in fruit trees provides protection against drought and contributes to increased or stabilized production (Fereres and Evans 2006). Fruit trees, like mango, with high economic value can suffer from poor water and fertilizer management because of the carry-over effects into subsequent years.

For this purpose, a sustainable water use in agriculture in general and in mango in particular must be developed as crop-specific water and fertilizing-saving irrigation techniques that do not negatively affect crop productivity. However, to develop such strategies, crop-specific water and fertilizer needs under different growing conditions and yield levels must be understood.

The mango industry, unfortunately, is still too far from developing a specific crop coefficient and fertilization needs for this crop, as mango is cultivated around the world from tropical and subtropical areas of Latin America to semi-arid regions like those present in Spain, Israel, Brazil and Mexico (Galán Saúco 2011; Mirjat et al. 2011).

In addition, specific information regarding water and nutrient needs for mango at different phenological stages and under different production levels is still missing. Today, the average mango yields per hectare around the world range between 8 and 9 tons in places like Pakistan (Bakhsh et al. 2006) and 75 tons in Israel (Levin et al. 2015a,b). This data points to the wide range of crop strategies used by mango growers around the world, but mainly indicates the potential production level for newly planted mango areas around the world. We assume that the main factors causing such differences in mango production level are mostly related to irrigation and fertilization strategies adopted by the mango growers around the world.

In many of the main mango producing countries and/or areas where the average yield per hectare is below 10 tons/ha (Oosthuyse 1993), like China (Xiuchong et al. 2001; Gao 2011), India (Negi 2000), Pakistan (Bakhsh et al. 2006), Mexico, Peru and Dominican Republic (personal observations 2015), gravity or furrow irrigation is the main water and fertilizer strategy used by farmers to supply water and fertilizer at critical phenological periods of the trees. On the other hand, in places where the average yield per hectare can increase to over 15 tons/ha, like Brazil (Pinto et al. 2004; personal observations 2016), Spain, Israel (Levin 2015a,b), Peru (Rosario Benitez, Camposol, 2016, pers. comm.), Puerto Rico (Dr. Yair Aron, Martex, 2016, pers. comm.) and Dominican Republic among others (personal observation 2016), drip or sprinkle irrigation is the main way to apply water and fertilizer (fertigation: application of liquid fertilizer through the irrigation system) to the trees at critical phenological stages or dry periods.

1.2 Production and fertilizer

Fertilization is critical to obtain a satisfactory mango yield. One of the most important problems in commercial mango orchards is the nutritional disorders. However, assessment of the mineral needs of mango trees is a significant challenge. The interactions between environmental conditions, soil type and tree characteristic (i.e. cultivar, yield level, age and phenological stage) are among the main factors causing this challenge. In addition, fertilizer type (e.g. mineral source), technology (e.g. by hand, drip or furrow irrigation) and strategy (e.g. constantly, every other, quantitatively or proportionally) used for the fertilizer application make it even more difficult to reach an accepted fertilizer protocol for mango growers.

The importance of the application of fertilizer to mango trees has been already reported by Young and Koo (1974). The increased fruit yield due to frequent fertilizer application was reported as well by Feungchan et al. (1989), Sharma et al. (2000) and Nasreen et al. (2014) among others.

The study of the effects of minerals on mango production and fruit quality is problematical because of the difficulty to quantify the impact of the individual parameters (i.e. soil type, soil pH, cultivar, water quality and climate) interacting in the final result evaluated in the experiment. Eventually, many fertilizer guidelines may result from different studies dealing with the different challenges faced by the mango industry (e.g. fruit size, fruit colour, yield, control of vegetative growth, internal quality and post-harvest behaviour). In addition, timing and technology are among other factors also affecting growers' fertilizer strategies.

1.3 Fertilizer and water-use efficiency

Nutrient-use efficiency (NUE) is a critically important concept in the evaluation of crop production systems. It can be greatly impacted by fertilizer management as well as by soil- and plant water management (Fixen et al. 2015). The objective of nutrient use is to increase the overall performance of cropping systems by providing economically optimum nourishment to the crop while minimizing nutrient losses from the field (Fixen et al. 2015).

Water-use efficiency (WUE) is commonly defined as the ratio of crop yield to evapotranspiration (Simsek et al. 2005; Zhang et al. 2004), although Aujla et al. (2005) obtained WUE as the relation between crop yield and total water applied (water productivity).

An irrigation-deficit strategy can be implemented in various ways, differing mainly in how the water restriction is applied. Particularly, sustained-deficit irrigation (SDI) is based on a uniform water restriction, depending on the crop water requirements that can be applied at less susceptible phenological periods from the production point of view (Durán Zuazo et al. 2011a). This approach allows the crop to adapt to the stressful situation.

Complementary approaches in mango are still needed to increase water and fertilizer use efficiency (WFUE). Although mango is considered a drought-resistant crop and exhibits some adaptive features that give it drought tolerance, such as deep tap/sinker roots, long life, tough leaves with thick cuticles for nutrient retention and recycling, and resin ducts to reduce wilting (Bally 2006), the irrigation requirement of mango has been not well investigated (Spreer et al. 2007), unlike many other fruit trees.

Relatively small amount of research regarding deficit irrigation strategies has been carried out in mango. Early studies were focused on the development of irrigation strategies for the entire fruiting period as a whole (Pavel and Villiers 2004). However, the irrigation and fertilization requirements of mango trees at specific phenological stages and under specific crop yield conditions, soil types, climates and water quality are still missing.

Mango has complicated and variable phenological cycles that influence the trees' uptake and translocation of minerals (Bally 2009). In addition, the impact of different climate, soil types, yield level and evaluated cultivar in WFUE makes the field research a very difficult task.

In tropical areas, for example, where the water is not a limiting factor, irrigation systems are not generally used. However, the soils are generally poor in mineral and organic content, with low pH due to the strong leaching of calcium. As a result, plant nutrition became the limiting factor for a sustainable mango production (de Almeida 2012).

Sustainability of mango production in many producing countries will depend on the application of proper irrigation and fertilization strategies, based on modern techniques and proven strategies. In some of mango producing countries like Mexico, Thailand, Peru, Israel and China, freshwater available for agriculture use is rapidly declining because of climate change and water pollution (Anon 2006). In addition, further water price increases are to be expected. Therefore, the adoption of water- and fertilizer-efficient strategies by mango producers is becoming crucial.

Well-proven fertilizing strategies associated with water-saving strategies in mango based on scientific research are lacking.

1.4 Water quality

Considering the scarcity of water sources for agricultural use, there is an urgent need to search alternative water sources replacing the high-quality water required for human consumption (Marecos do Monte et al. 1996; Angelakis et al. 1999; Oron et al. 2001; Toze 2006). The reuse of municipal wastewater, or other water alternatives like saline water, for irrigation could be a realistic way of reducing water shortage, as has been demonstrated in many countries in the Mediterranean region such as Israel, Cyprus, Jordan and Tunisia (Angelakis et al. 1999).

In the preparation of this chapter, scarce data have been found in the bibliographical review regarding the impact of different water sources and quality (i.e. recycled water) on the performance of the mango tree. Therefore, a big research field is potentially open for this industry.

1.5 Production and technology

In areas with dry and hot climates, drip irrigation and protected soil cultivation have improved WUE, mainly by reducing run-off and evapotranspiration losses, but also by protecting the soil from the detrimental effects of heavy rain events characteristic in some mango production areas. As a consequence, improving mainly crop WFUE by the incorporation of modern technology into mango farms has been a matter of concern to researchers and agronomists in recent years.

The aim of this chapter is to *suggest* how research in the field of irrigation and fertilization can help to solve the challenges faced by the mango industry and be translated into practical outcomes for farmers by making mango production more sustainable. To achieve our goal, we first present an extensive and detailed review of the most relevant articles which survey existing research on these topics. Finally, we suggest potential applied research areas that can significantly contribute to more sustainable mango agriculture in small, medium and large mango farms in developed and developing countries. *High-quality research is critical to meeting these challenges.*

2 Assessing water requirements of mango trees

Although the mango tree is drought tolerant and may survive for many months without rain or irrigation, water deficits during the reproductive cycle can adversely affect fruit retention and early fruit growth (Schaffer et al. 2009; Whiley and Schaffer 1997) and consequently final yield.

Climatic and soil conditions, cultivar and yield are among some of the main factors affecting mango water needs. These needs also vary between the non-productive period (juvenile phase) that goes from planting until the beginning of production and the productive phase from flowering to harvest (Coelho et al. 2002).

Irrigation management is critical to the success of the mango production and sustainability. Even though 'when to irrigate and how much?' are the most frequent questions asked by mango growers, how to irrigate is also a critical question. Its answer will determine in a big part the irrigation strategy to be followed and therefore the potential to achieve maximum fruit yield in a sustainable production system. The central gaps in these areas of knowledge to develop an efficient irrigation strategy are related to a crop's water requirements in relation to:

1. impact of seasonal and environmental local conditions (however, most of the time these kinds of data are regional rather than local);
2. effects of variety and crop load level on mango water needs; and
3. identification of reliable plant and/or soil and/or weather parameters to describe seasonal water use and requirement.

To maintain an optimal water status to maximize yield in the short, medium and long term (in a sustainable way), or to prevent trees from a significant loss of productivity under limited water availability, water relation of the mango tree (water requirement – time and quantity, physiological response to water deficit, etc.) needs to be understood.

2.1 When to irrigate

For irrigation to be efficient there is a need to supply water to mango trees when it is needed and at the required quantity. To achieve such requirements, the utilization of modern irrigation technology is essential. These technologies include the use of suitable supportive irrigation technology to determine the irrigation requirement – time and quantity (e.g. soil tensiometers, tree xylem sap flow, trunk and fruit dendrometers, leaf temperature sensors, and evaporation pan), and the delivery system (e.g. pipes, dripline, microsprinklers and irrigation controller/programmes). These combined irrigation technologies will allow the mango grower to maximize WUE by optimizing water quantities applied and irrigation timing. These technologies are also crucial for fertigation (the application of fertilizer through the irrigation system).

Most of the research carried out in this field in mango for the last 20 years in order to establish or understand mango water needs have three main constraints:

1 they have been focused mainly in saving water rather than in maximizing crop;
2 the effects of the different irrigation strategies on crop have been mainly evaluated for the entire fruiting period as a whole, rather than differentiating the two natural phenologic periods during mango fruit development (as reported by Subramanyam et al. (1975) and Tharanathan et al. (2006), where the first period is characterized by fruit exponential growth by cell division and then the second period where the fruit growth is characterized by cell enlargement (Subramanyam et al. 1975; Tharanathan et al. 2006); and
3 a high percentage of the research was carried out only for two productive seasons, meaning that some of the irrigation accumulative effects (positive and/or negative) carried over by the trees, season after season, mainly under heavy soil conditions (like those present in many of the mango growing areas in the world) under different irrigation strategies, cannot be properly assessed.

Nevertheless, some studies on mango water use and irrigation covered a number of important areas, mainly in tropical production areas, including assessing pre-flowering irrigation treatments (Gonzalez and Muller 2004), pre-harvest irrigation cessation (Diczbalis et al. 1993; Anon 1988) and season-long reduced volume applications, on fruit quantity and/or quality (Pavel and Villiers 2004; Spreer et al. 2007, 2009a,b; Durán Zuazo et al. 2011a,b), effects of irrigation at different phenological periods (Levin 2015a,b; Santos et al. 2014) and the effects of pre-flowering irrigation on trees receiving flowering-promoting treatments (Gonzalez and Muller 2004; Bithell, 2010).

Two groups of methods were used to determine mango irrigation requirements:

1 those related to the use of plant- or soil-based parameters; and
2 those related to the determination of crop coefficient (Kc) based on local or regional evapotranspiration (ETP) or pan evaporation.

2.2 Irrigation scheduling based on plant- or soil-based parameters

Irrigation scheduling has conventionally aimed to achieve an optimum water supply for productivity, with soil water content being maintained close to field capacity. In recent years, there has been a wide range of proposed novel approaches to irrigation scheduling which have not yet been widely adopted. Many of these are based on sensing the plant response to water deficits rather than sensing the soil moisture status directly (Jones 1990a). The advent of precision irrigation methods such as drip irrigation has played a major role in reducing the water required in agricultural and horticultural crops, but has highlighted the need for new methods of accurate irrigation scheduling and control.

The choice of irrigation scheduling method depends to a large degree on the objectives of the irrigator and the irrigation system available. Effective operation of irrigation systems requires a sensing system that determines irrigation need in real time or at least at frequent intervals. This rules out large-scale manual monitoring programmes for such purposes and indicates a need for automated monitoring systems.

Irrigation scheduling is conventionally based either on 'soil water measurement', where the soil moisture status (whether in terms of water content or water potential) is measured directly to determine the need for irrigation, or on 'soil water balance calculations', where the soil moisture status is estimated by calculation using a water balance approach in which the change in soil moisture over a period is given by the difference between the inputs (irrigation plus precipitation) and the losses (run-off plus drainage plus evapotranspiration). A potential problem with all soil-water-based approaches is that many features of the plant's physiology respond directly to changes in water status in the plant tissues, whether in the roots or in other tissues, rather than to changes in the bulk soil water content (or potential).

Very few reports can be found in the scientific literature regarding the use of soil sensors to establish mango water needs. Mostert and Hoffman (1997) carried out in South Africa one of the longest irrigation experiments recorded in mango worldwide. Over a six-year period, they monitored the water use of 12-year-old (at beginning of the experiment) mango trees cv. Fascell planted 7X7 m using tensiometers placed at depths of 300, 600 and 900 mm. In this experiment, two water cycles were created, 1-short and 2-long, where the average readings at all three depths had reached −30 kPa and −60 kPa, respectively. The total annual water use averaged over the six years for the frequently irrigated (short cycle) treatment was 1200 mm (range 1050–1390 mm). Trees in the long cycle treatments used less water and produced lower yields than those in the short cycle treatments (Mostert and Hoffman 1997). The reported range of mango water use is in agreement with those found by the author of this chapter (see Section 7) for the best production levels.

The plant response to a given amount of soil moisture therefore varies as a complex function of evaporative demand. Thus, it has been suggested (Jones 1990) that greater precision in the application of irrigation can potentially be obtained by direct sensing of plant stress.

If soil-water-based measures are to be replaced by plant-based measures, it is important to consider what measures might be most appropriate for irrigation scheduling purposes. Possible measures include direct measurements of some aspects of plant water status as well as measurements of several plant processes that are known to respond sensitively to water deficits.

It appears that changes in stomatal conductance are particularly sensitive to developing water deficits in many plants and therefore potentially provide a good indicator of irrigation need in many species. At early years, most effort was concentrated on developing stomatal conductance as a practical, plant-based irrigation scheduling approach. Although stomatal conductance can be measured accurately using widely available diffusion parameters, to reach an overall view of the plant water status of a section of the mango orchard, or even a single tree canopy, measurements are extremely labour-intensive and not practical for farm use.

In fruit trees, stem water potential has been shown to be a reliable indicator of tree water status (Shackel et al. 1997). There were a limited number of reports of mango leaf water potential measurements (Pongsomboon 1991; Castro et al. 2004); however, it is generally found that excessive latex exudation in mango petioles makes measurements of leaf or stem xylem water potential unreliable (Lu 2005, 2013).

Sap flow was also found to be a good indicator for plant water status in mango. Xylem sap flow in the tree trunk, micro-variation (shrinking and expansion) of branch/trunk diameter (microdendrometry) and canopy temperature have been studied in mango in the seasonally wet–dry tropics in northern Australia (Lu and Chacko 1997, 1998, 2000; Lu et al. 2000; Lu 2002, 2005, 2013). Under experimental conditions, whole-tree sap flow measurement provides not only a measure of tree water use (a key part of the irrigation need) but also an indicator of plant water status. Lu (2013) showed a general correlation between whole-tree water use (sap flow) and tree size (trunk diameter) for mango trees (cvs. Kensignton Pride, Irwin, Nam Dok Mai) grown in the Darwin and Katherine region, Australia (Fig. 1). By improving the efficiency of water delivery, it is possible to match the irrigation input as closely as possible to the actual tree water use/need (estimated by sap flow).

On mature Kensington Pride mango trees grown on deep sandy-loam soil in the field in Northern Australia, the ratio of the daily sap flow (dry/control) repeatedly fell to about 0.85 when volumetric soil moisture fell to 15% (Fig. 2) and other indicators (microdendrometry, stomatal conductance) revealed signs of stress. This suggested that the sap flow may also reliably indicate the onset of the plant water stress for field-grown mature trees (Lu 2002, 2005).

High-resolution stem or trunk diameter variations are widely recognized as a useful drought stress indicator and have therefore been used in many irrigation scheduling studies in many fruit crops like lemon (Ortuño et al. 2006), almond (Goldhamer and Fereres 2001), peach (Goldhamer et al. 1999) and plum (Intrigliolo and Castel 2004) but to a lesser extend in mango.

To utilize the information of microdendrometers to control irrigation on mango, Lu (2005) developed a 'shrinkage index' to quantify the plant water status. The 'shrinkage index' was calculated as $100^*(1-DS_{actual}/DS_{ref})$. DS_{actual} is the actual daily shrinkage value and DS_{ref} is the reference daily shrinkage value which is the average shrinkage value observed during well-watered periods or wet season. This index was reliable over two growing seasons: the index of the control trees never rose above 25 (Fig. 3), while the microdendrometer-controlled trees ('Dendro') showed signs of early stress when the index

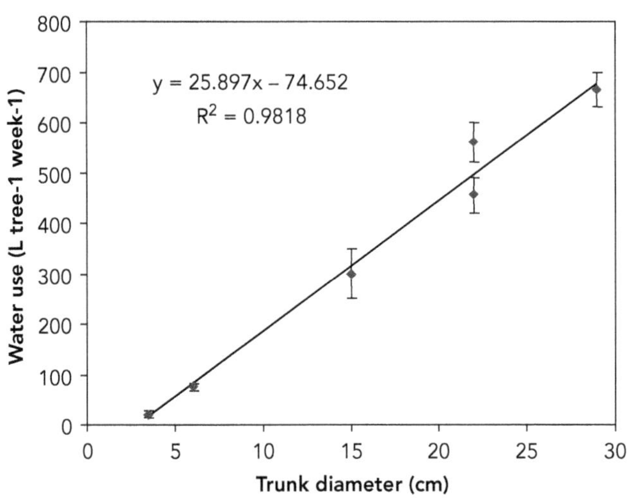

Figure 1 Relationship between tree trunk diameter and whole-tree water use (sap flow) on orchard-grown mango trees (cvs. Kensington Pride, Irwin, Nam Dok Mai) in Darwin and Katherine areas in northern Australia (Mean ± SD) (Lu 2013).

hit a threshold of 30. They purposely withheld irrigation when the index reached 30 for the first time (on 13 August) to determine that this index value of 30 or above remained until rewatering. Comparing to the changes in concurrent measurements of soil moisture and sap flow during the first drying cycle (in Fig. 2), change in Dendro's shrinkage index values appears to be a good indicator of the onset of water stress.

Lu (2005, 2013) compared the sensitivity of the indicators examined (sap flow, microdendrometry, stomatal conductance and soil moisture) and showed that sap flow was not as sensitive as the microdendrometry. From well-watered to stressed conditions, the range of changes in sap flow ratio was only 15%, while microdendrometers exhibited a change of 30%. Furthermore, sap flow was a less sensitive indicator of water stress under field conditions in soil with high water-holding capacity. This finding is consistent with Goldhamer et al. (1999) who evaluated the relative sensitivity of trunk diameter-derived parameters and conventional indicators of plant water status in peach trees, including

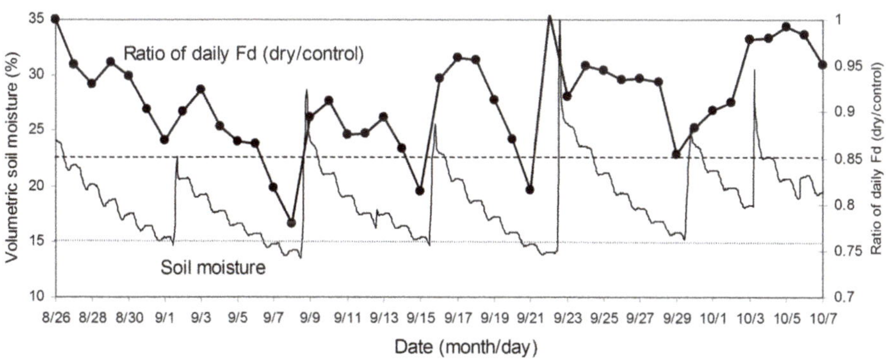

Figure 2 Ratio of the daily sap flow (Fd) measured on dry trees vs. well-watered control trees over several irrigation cycles as indicated by variation in volumetric soil moisture in the top 30 cm soil profile (7-year-old Kensington Pride trees on sandy-loam soil in an orchard near Darwin) (Lu 2002).

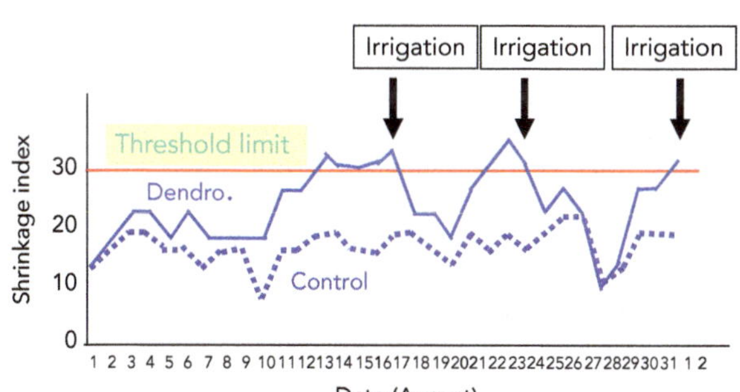

Figure 3 Daily changes in shrinkage index of a group of trees which were irrigated at local authority recommended rate (Control) and of a group of trees that received irrigation only when the shrinkage index hit the threshold limit of 30 (Dendro). Irrigation events for the 'Dendro' trees are indicated by arrows.

various measurements of tree water potential, stomatal conductance, fruit growth and photosynthesis, and they also concluded that parameters derived from stem diameter variation were more sensitive than the other indicators in that they detected water stress earlier in a drying cycle.

Results from the above two-year field trial on mango (Lu 2005) also showed that trees under irrigation controlled by the dendrometers using the shrinkage index as indicator had better crop WUE than fully irrigated trees at 80% evaporation replacement (Control) and farmer's own method.

Despite its notorious interest, so far none of the plant-based methods are yet well adapted for long-term automation of irrigation scheduling or control of a commercial farm because of the difficulties of long-term reliable measurement of any of their variables, especially at seasonal level. Lu (2005, 2013) concluded that plant-based measurements are better suited for research in improving irrigation methods, but not reliable and robust tools for field operation.

Nevertheless, with the rapid development of the sensor technology, pressures for enhanced WUE and for greater precision in irrigation systems are likely to provide a real impetus for the development of new precision irrigation scheduling systems based on plant/organs growth, and may well involve greater use of plant-based sensing systems combined with soil and weather sensors, called growth-based irrigation (GBI) (Ben Ner, Z. Agro Web Lab, 2016, pers. comm.).

2.3 Irrigation based on crop coefficient (Kc)

Several researchers have investigated yield responses in mangoes to irrigation below potential evapotranspiration (ETP) replacement, also termed reference evapotranspiration (ETo), using or evaluating different crop coefficients (ETcrop/ETP), where ETcrop is the actual crop evapotranspiration. Results indicated that the reduced water inputs can still successfully maintain yields (Spreer et al. 2007, 2009a,b; da Silva et al. 2009). Recommended water requirements for mango production in the Northern Territory, Australia, are based on pan evaporation levels (Epan) and a crop factor (water use/Epan) of 0.7 (Bithell et al. 2010). Durán-Zuazo et al. (2011b) reported the changes of the average crop coefficient (Kc) for mango over two monitoring seasons estimated by the water balance from experimental drainage lysimeters. The crop coefficients presented at three main growing stages (flowering, fruit set and fruit growth) were fitted by a polynomial function. The average crop coefficient values were 0.56, 0.71 and 0.61 for flowering, fruit set and fruit growth period, respectively. After fruit harvest, crop coefficient decreased quickly to 0.26.

de Acevedo et al. (2003) studied the water requirements of irrigated mango orchards in San Francisco River Valley in Petrolina-PE, Brazil. Field measurements were taken during the productive cycle of a seven-year-old mango variety 'Tommy Atkins' orchard. The mango trees evapotranspiration was obtained by two methods: Bowen ratio-energy balance (BREB) and soil water balance. Daily mango orchard evapotranspiration increased slowly from 3.1 mm per day at the beginning of the experimental period (mid-July) to 4.9 mm per day at the maximum growth period of the fruit. Then, it decreased to reach a 4.1 mm per day value, approximately at the full maturation fruit. The accumulated mango orchard water consumption for the whole productive cycle was 551.6 and 555.1 mm by the soil water and BREB methods, respectively.

Teixeira et al. (2008) measured actual evapotranspiration and its components in a commercial mango orchard in northeast Brazil where yields reached 41.5 t ha^{-1} in 2003/2004 and 48.4 t ha^{-1} in 2004/2005. The corresponding values for water productivity for each of the two years were 4.8 and 4.3 kg m^{-3}, when based on the volume of irrigation water applied; 2.8 and 3.6 kg m^{-3}, when based on evapotranspiration; and 3.6 and 5.4 kg m^{-3}, when based on transpiration, respectively.

Studies on mango water use and irrigation have covered several important areas; however, irrigation protocols for mango to maximize yields in a sustainable way have not been accomplished yet. The impacts of tree age, cultivar, rootstock, yield and soil characteristic, among other variables, have been less studied and, therefore, are less understood. Irrigation delivery technology (i.e. via drip, sprinklers or gravity irrigation) has a great influence on WUE. However, finding common soil, plant and/or environmental variables which provide the best information regarding mango trees' response to water and actual mango water needs under different growing conditions is paramount to develop precise irrigation strategy/ies for the different mango world growing areas for sustainable production.

de Souza et al. (2016) evaluated the water requirement estimate for the reproductive period of mango orchards in the northeast of the state of Pará, Brazil. The aim of this study was to estimate the water consumption in mango orchard during each phenological stage. For this purpose, a micrometeorological tower was installed and implemented in a mango orchard, cv. Tommy Atkins, with trees which were 22 years old. Data were collected during the crops of 2010/2011 and of 2011/2012. The actual crop evapotranspiration was estimated from the energy balance using the Bowen ratio technique. The phenological stages evaluated were flowering, start of fruit fall, fruit formation, fruit maturation and harvest. They concluded that the actual crop evapotranspiration during its reproductive period ranged between 402.9 and 420 mm with a mean daily water consumption of 3.8 mm at flowering, of 4.25 mm at fruit fall, of 3.56 mm at fruit formation, of 3.0 mm at fruit maturation and of 3.73 mm for the whole period.

3 Evaluation of main mango irrigation strategies

In the past 20 years, several strategies have been developed to improve WUE in agriculture in general (Anon 2002; Oster and Wichelns 2003; Costa et al. 2007; Geerts and Raes 2009) and in mango in particular (Spreer et al. 2007, 2009a,b; da Silva et al. 2009; Durán-Zuazo et al. 2011a,b; Levin et al. 2015a,b). One strategy is 'deficit' irrigation' (DI). This is described as irrigation maintenance at rates below that expected to meet water required for evapotranspiration (Costa et al. 2007; Geerts and Raes 2009). The DI approach can be achieved by two different methods:

1 by applying water at a constant deficit throughout the dry season, or on a supplemental basis if irrigation occurs for a short period only, for example, during the rainy season. This method is defined as 'sustained-deficit irrigation' (SDI) and has been used in mango studies, for example, replacing 70%, 80% or 90% evapotranspiration (ETo) along the irrigation season for crop coefficients of 0.75 or 0.8 (da Silva et al. 2009; Spreer et al. 2009a,b).
2 by varying the degree of DI according to critical phenology. This is described as regulated deficit irrigation (RDI) (Costa et al. 2007) or as 'drought stress differentiated by phenological stage' (Geerts and Raes 2009).

The latter method, although identified as appropriate for mango trees (de Azevedo et al. 2003), has not been well investigated under different growing conditions. This strategy requires comprehensive knowledge of the water requirements at each phenological stage under different growing conditions and yield (Goodwin and Boland 2002; Costa et al. 2007; Levin et al. 2015a,b).

In comparison, SDI is easier to implement and requires less information, but is not tailored to the crop's changing requirements through time. Another application method is that of partial root zone drying (PRD) which splits applications from one side of the tree to the other side, with alternate drying of root zones being the underlying principle. The volumes of application or scheduling are usually applied on a deficit irrigation basis. This method has received widespread attention, for example, in grapes in Spain (De la Hera et al. 2007), Australia (Dry et al. 1996; Dry and Loveys 1998) and China (Du et al. 2008) and in other high-value horticultural crops (Costa et al. 2007). However, recent studies comparing RDI and PRD from many crops indicated that PRD resulted in only small increases in yield (average ~5%), compared with RDI (Sadras 2009). A comparison of RDI and PRD in mango production demonstrated no clear advantage (Spreer et al. 2007). There is a need to evaluate RDI strategies once seasonal water use or crop coefficients become available.

Below, main studies carried out on mango using the three different irrigation approaches and their impact on the main crop production variables, yield, fruit number, fruit size and vegetative growth are considered. Table 1 summarizes some of the main studies evaluating different irrigation strategies on mango worldwide and the suggested best treatment according to the results.

3.1 RDI, PRD and SDI

Increased WUE in mango through DI has been reported by Pavel and de Villiers (2004) and Spreer et al. (2009a,b) among others.

Pavel and de Villiers (2004) reported no significant yield decrease under deficit irrigation and no difference in fruit size under conditions of light drought stress between irrigated and non-irrigated trees. Spreer et al. (2007) evaluated the 'effect of RDI and PRD on the quality of mango fruits in cv. "Chok Anan"'. This study evaluated four irrigation treatments with respect to mango yield and fruit quality, control (CO = 100% of ETc); RDI = 50% of ETc; PRD = 50% of ETc, applied to alternating sides of the root zone and no irrigation (NI). The author reported a significant yield reduction in the PRD treatment in both years with respect to CO. On the other hand, the fruit size distribution was most favourable in the former one, mainly under low productive conditions (7.7 and 8 t/ha, in PRD and CO treatment, respectively).

Spreer et al. (2009a) evaluated the 'effect of deficit irrigation on fruit growth yield and WUE of mango in Northern Thailand' for two consecutive seasons. The author reported a significant yield reduction in deficit irrigation with 50% of ETc (DI50) when compared to 100% irrigation per ETc (FI) under low productive conditions. The PRD treatment showed a significant WUE in both years.

Spreer et al. (2009b) reported a slight yield reduction over four years of experiment in RDI (50% of ETc) and PRD (50% of ETc) in mango (cv. Chok Anan) under Chiang Mai, Thailand growing conditions compared to a full irrigated treatment (CO = 100% of ETc). As in previous reports of the same author, however, the fruit size distribution was more favourable in the RDI treatment and the WUE was also significantly higher in the last treatment compared to the CO one.

Table 1 Evaluated irrigation strategies on mango worldwide in general and in Latin America in particular and the suggested best treatment according to the results

		Evaluated technique*									Evaluated phenological period**				
No.	Author (year)	SDI	RDI	PRD	Other	Number of treatments	Evaluated years	Cultivar	Orchard place	Outstanding treatment	F-FS	MFG	FFG	EFPP	(PH)
1	Chandel and Singh (1992)	X				4	1	Dashehari	Himachal Prades India	20% available soil moisture (ASM)	–	–	–	X	–
2	Pavel and de Villiers (2004)		X			5	1	Kent	Hoedspruit, S. Africa	Co (85% from farm control)	–	–	–	X	–
3	Durán Zuazo et al. (2004)				Salinity	4	4	Osteen	El Zahorí-Spain	Control = dSm−1	–	–	–	X	–
4	Spreer et al. (2006)	X		X		4–5	2	Chok Anan	Chiang Mai, Thailand	DI-75% and PRD 50% of Etc	–	–	–	X	–
5	Spreer et al. (2007)		X	X		4	2	Chok Anan	Chiang Mai, Thailand	partial root zone drying (PRD = 50% of ETc, applied to alternating sides of the root zone)	–	–	–	X	–
6	da Silva et al. (2009)		X			4	2	Tommy Atkins	Semi-Arid Brazil	T3 = 90% of reference evapotranspiration – Eto	–	–	–	X	–
7	Spreer et al. (2009)		X	X		4	2	Chok Anan	Chiang Mai, Thailand	c) PRD with 50% of ETc (PRD)	–	–	–	X	–
8	Coelho Filho et al. (2009)			X		5	1	Kent	Fazenda Boa Vista-Brazil	T-1 = Treatment without deficit	–	–	–	X	–
9	Cotrim et al. (2011)		X			10 and 8	2	Tommy Atkins	Semi-Arid Brazil	T1-full irrigation (100% ETc) in phases II and III and 40% of ETc in Phase I (NS)	X	X	X	–	–

(Continued)

Table 1 (*Continued*)

No.	Author (year)	Evaluated technique* SDI	RDI	PRD	Other	Number of treatments	Evaluated years	Cultivar	Orchard place	Outstanding treatment	Evaluated phenological period** F-FS	MFG	FFG	EFPP	(PH)
10	Durán Zuazo et al. (2011)	X				4	3	Osteen	El Zahorí-Spain	SDI-2 = 50% of Etc	–	–	–	X	–
11	Santos et al. (2014)		X			5	2	Tommy Atkins	Semi-Arid Brazil	T4 = 100% of ETc from early flowering to late fruit expansion and 50% of ETc during physiologic ripening	X	X	X	–	–
12	Levin et al. (2015a,b)		X			12	4	Keitt	Israel	T-4 in FFG and PH= 9.2 and 5.4 mm/day, respectively	–	X	X	–	X
13	Santos et al. (2015)			X		5	1	Tommy Atkins	Ceraíma, Guanambi, Bahia, Brazil	5) 40% of ETc with alt. irrig. side of 15 days	–	–	–	X	–
14	Faria et al. (2016)		X			5	2	Tommy Atkins	Vales do São Francisco-Brazil	T1 (0% of ETc without irrigation in flowering induction period (FI) and 100% in fruiting phase (FII)), T2 (25% of ETc in FI and 100 % in FII)	X	–	–	X	–
15	Santos et al. (2016)		X	X		12	2	Tommy Atkins	Ceraima, Bahia, Brazil	RDI50S3 – full irrigation, 100% of ETc in stages I and II and 50% of ETc in stage III-(RDI at fruit maturation only)	X	X	X	–	–

* SDI = sustained-deficit irrigation; RDI = regulated deficit irrigation; PRD = partial rootzone drying.
** F-FS = flowering-fruit set; MFG = main fruit growth; FFG = final fruit growth; EFPP = entire fruit production period; PH = post-harvest.

Durán Zuazo et al. (2011a) reported the results of a field irrigation experiment conducted near Granada over three seasons (2006–2008), where the highest yield and WUE were obtained with the sustain deficit irrigated treatment of 50% of ETcrop, and thus the greatest amounts of water did not result in the highest yield.

Coelho Filho et al. (2009) reported a significant yield and fruit number reduction in SDI (50% less water than control treatment) with respect to control treatment (CO = maintained at field capacity at 15 and 30 cm depth) in cv. Kent under semi-arid conditions of São Paulo state in Brazil. However, no significant difference was reported between the PRD treatments (50% applied alternating side every 7 days (T3); 14 days (T4); and 21 days (T5)) and the CO one.

Santos et al. (2015) reported no significant difference in production parameters (yield and number of fruit) between PRD treatments (100, 80, 60 and 40% of ETc) and full irrigated (100% of ETc) one in cv. Tommy Atkins under PRD irrigation system in Bahia, Brazil. Significant WUE was recorded only in the PRD 40% treatment.

According to the results presented above, it could be concluded that mainly PRD and RDI irrigation strategies can potentially be applied in mango trees without significantly affecting yield parameters but increasing, not always, significantly the WUE. Nevertheless, these results should be taken carefully, mainly because the clear majority of these studies were carried out only for two seasons, too short for irrigation studies to evaluate the medium- and long-term effects of these deficit irrigation treatments; secondly, also the yield on some of these studies were relatively low (under 15 T/ha), which means that the water demands may be relatively low compared with trees that produced more than 25 or 30 T/ha (personal observations); then it is more difficult to produce a real impact (significant difference) with different irrigation treatments, mainly those related to deficit irrigation and full irrigated one.

3.2 Mango response to RDI, PRD and SDI

To evaluate the effect of different irrigation strategies on different yield components, a clear definition of the phenological period during which the treatments were applied is crucial. For irrigation purposes the following phenological stages can be considered:

1 Post-harvest (PH) vegetative growth.
2 Pre-flowering to flowering.
3 Fruit set.
4 Final fruit growth period (FFG-pit hardening to harvest).

Yield increase due to irrigation normally results from a higher crop load (number of fruit) rather than greater fruit size (Pavel and de Villiers 2004; Spreer et al. 2009b).

3.2.1 Post-harvest vegetative growth

Several studies showed that mango vegetative growth is reduced by soil water deficit. Marloth (1947), Yan and Chen (1980), Tahir et al. (2003) and Levin et al. (2015a,b) reported that PH vegetative growth in mango cv. Keitt under Israeli growing conditions was significantly reduced when water application during the PH period was reduced by 50% compared to the standard farm water application (control), mainly after low production years. Also, reduced water application during the final fruit development/growth (FFG) period had a significant impact on PH vegetative growth, mainly under high crop loads, even though the most irrigated trees also produced significantly higher yields and all the trees received the same amount of water during PH (Levin et al. 2015b). Conversely, Larson et al. (1989) reported no difference in mean shoot growth in cv. Tommy Atkins under Florida growing conditions, among the three evaluated treatments (irrigation every 7, 14 days and NI).

3.2.2 Pre-flowering to flowering

Flowering and fruit set are the most critical stages for determining the mango tree crop. In nature, mango trees may produce large numbers of flowers while only a small proportion set fruit. Understanding mango flowering in the tropics and subtropics is essential for efficient utilization of crop management systems such as irrigation and fertilization management, which may extend both the flowering and crop production seasons (Chacko 1991; Whiley et al. 1991) and ensure a sustainable production system. Under subtropical conditions, temperatures of 15°C or below promote mango flower induction (Lu and Chacko 2000), while temperatures close to 20°C promote vegetative growth (Davenport and Nunez-Elisea 1997). In the tropics where temperatures may remain too high for flower induction by cool nights, a dry period preceding flowering may be necessary to achieve a commercial crop (Chacko 1986). Nevertheless, the effect of plant water stress on flowering response is still a matter of controversy (Lu and Chacko 2000).

Singh (1960) and Bally et al. (2000) reported floral morphogenesis initiation after mango cultivars were exposed to an extended period of mild water stress in the low-latitude tropics. However, according to Ramirez and Davenport (2010) water stress does not induce flowering. It is the age of the last flush impacted by the stress duration that drives flowering. Water stress prevents shoot initiation and maintains trees at rest until age accumulation in leaves takes place and trees flower due to the age-dependent reduction of the vegetative promoter (Ramirez and Davenport 2010). The accumulating age of stems is greater in water-stressed trees than in trees maintained under well-watered conditions, which can vegetatively flush more frequently (Davenport 1993; Schaffer et al. 1994).

Mango tree's response to water stress during the pre-flowering period may be cultivar-dependent, at least under tropical conditions. Lu and Chacko (2000) reported a different response of cvs. Nam Dok Mai, Irwin and Kensington Pride to water stress during the pre-flowering period. While cv. Nam Dok Mai did not need a strong floral stimulus such as that promoted by water stress, cv. Irwin appeared to have the strongest water stress requirement for flowering, even more than cv. Kensington Pride.

Núñez and Davenport (1994) reported on an experiment conducted in Homestead, Florida, to determine whether water stress induces floral morphogenesis in mango during July (mean minimum temperatures about 20°C, non-floral inductive), and October and November (mean minimum temperatures about 15°C, floral inductive) in cv. Tommy Atkins grown in containers. Water stress for 35 days during October advanced floral bud break by nearly two weeks in nearly 40% of the buds. In this experiment Núñez and Davenport (1994) conclude that low temperatures thus promoted floral induction of mango, whereas water stress promoted growth of florally induced buds. Levin et al. (2015a,b) reported that more severe water stress during the PH period results in earlier flowering. Trees under deficit irrigation treatments (mainly T-1 = 0.57 of ETc) flowered earlier than trees receiving T-3 or T-4 (0.92 and 1.09 of ETc, respectively) irrigation. However, the treatments did not affect flower intensity. However, no effect on flowering time was observed in the trees belonging to the second phenological period (pit hardening – harvest) due to different irrigation levels applied at that time of the season (Levin et al. 2015a,b).

3.2.3 Fruit set

Mango productivity throughout the tropics and subtropics is relatively low compared to its potential, due mainly to severe fruit drop, especially during the initial 3–4-week growth period following anthesis (Dahsham and Habib 1985; Chen et al. 1995; Searle et al. 1995). Yield depends on the number of fruits that progress through various growth and developmental stages from initial fruit set until maturity (Singh et al. 2005).

Pongsomboon (1991) reported that water stress (−1.2 MPa compared to 0.3 MPa) during the first two months after fruit set induced heavy fruit drop during the first 30 days of fruit development; however, final fruit retention was unaffected by this level of water stress, but fruit size was reduced. Similarly, Schaffer et al. (1994) reported that fruit abscission on trees was enhanced under severe drought stress. Larson et al. (1989) and Spreer et al. (2007) reported that fruit drop in mango at an early developmental stage was associated with low soil moisture. According to Larson et al. (1989) and Spreer et al. (2007) an excessive loss can be prevented by adequate irrigation, particularly during flowering and the first six weeks after fruit set.

On the other hand, Roemer (2011) reported that fruit drop under northern Vietnam growing conditions, before the beginning of the rainy season, was affected by cultivar rather than by irrigation treatments during full bloom and fruit set period.

3.2.4 FFG period

For simplicity purpose, the research evaluating the impact of different irrigation strategies on these two phenological stages will be presented together.

Lakshminarayana et al. (1970) reported that mango fruit growth pattern followed a sigmoid curve in which the main fruit growth (MFG) stage is characterized by rapid fruit expansive growth that is associated mainly with high rate of cell division and expansion (Subramanyam et al. 1975; Ram et al. 1983; Tharanathan et al. 2006). Approximately 80% of the final fruit size is achieved in the MFG stage while the FFG stage counts for approximately 20% of the final fruit size at harvesting (personal observations). The fruit at this last stage refers to growth by cell enlargement only (Subramanyam et al. 1975; Tharanathan et al. 2006).

Coelho Filho and Coelho (2005) reported no significant differences among different RDI treatments (50, 70 and 85% of ETc) in yield, number of fruits per plant and average fruit weight on cv. Tommy Atkins. The RDI treatments were applied at one out of the three phenologial stages of the fruit production.

Cotrim et al. (2011) reported no significant effects of RDI treatments (40, 60 and 80% of ETc) applied at different phenological stages during two consecutive production cycles on productivity, number of fruits per plant, average fruit weight and WUE on Tommy Atkins under Brazilian semi-arid conditions when compared to full irrigated treatment (100% of ETc).

Santos et al. (2014) evaluated the impact of RDI treatment (50% from ETc) during two consecutive production cycles at three different phenological stages on cv. Tommy Atkins at Southwest of Bahia, Brazil.

The author reported that the use of RDI with 50% of ETc in the fruit set phase caused yield reduction even though such reduction was not significant for the two-year study period.

Santos et al. (2016) reported a significant reduction in fruit number and yield in RDI treatments (50 and 75% of ETc) applied during the flowering-fruit set period and 50% of ETc at fruit development period in cv. Tommy Atkins in Southwestern State of Bahia, Brazil. On the other hand, when PRD treatments (80, 60 and 40% of ETc) were applied from flowering to harvest, the number of fruit and yield was significantly reduced by the PRD treatments 40 and 60% in the former parameter and 80% as well in the last parameter, when compared to the full irrigated treatments.

According to the results from the different studies presented above, including the results which will be later presented in the 'case study' in this chapter, the impact of irrigation at different phenological stages can be summarized as follows: I) Irrigation at flowering-fruit set period may impact primarily in the number of fruit and as a consequence potentially also in the final yield; II) Irrigation at the MFG period may impact on the number of fruit but not necessarily in the final yield due to a fruit size compensation growth under lower number of fruit later in the FFG period; III) Irrigation at the FFG period may affect mainly final fruit size, potentially PH vegetative growth, under semi-arid and/or subtropical conditions and as a consequence in the medium and long term (more than two seasons) may significantly reduce yield as well; IV) Irrigation during the PH period (after pruning) may affect vegetative growth and as a consequence final yield, either by affecting fruit number and/or final fruit size.

4 Impact of water quality on mango productivity

Considering the scarcity of conventional water sources for agricultural use, there is an urgent need for alternative water sources for agriculture to replace the high-quality water required for human consumption (Marecos do Monte et al. 1996; Angelakis et al. 1999; Oron et al. 2001; Toze 2006). The reuse of municipal wastewater or other water alternatives like saline water for irrigation could be a realistic way of reducing water shortage, as has been demonstrated in many countries in the Mediterranean region such as Israel, Cyprus, Jordan and Tunisia (Angelakis et al. 1999). In Israel, for example, treated sewage effluent is expected to be the main (70%) source of water for irrigation by 2040 (Haruvy et al. 1999). In many parts of the world, treated wastewater has been successfully used for irrigation, and many researchers have recognized its benefits (Asano and Levine 1991; Levine and Asano 2004). In Mediterranean countries, treated wastewater is increasingly used in areas with water scarcity and its application in agriculture is becoming an important addition to water supplies.

Several studies have shown the advantages and disadvantages of using wastewater for irrigation of various crops (Reboll et al. 2000). However, in the preparation of this chapter, scarcity data were found regarding the impact of different water sources and quality (i.e. recycled water, saline water, etc.) on mango production parameters. The main research found in scientific literature was carried out in Spain. Durán Zuazo et al. (2004a) evaluated the impact of four saline irrigation treatments (1.00, 1.50, 2.00 and 2.50 dS m^{-1} of NaCl) irrigated to the equivalent of crop evapotranspiration, estimated from a class-A evaporation pan, on the fruit yield of mango cv. Osteen grafted on two different rootstocks, Gomera 1 (G1) and Gomera 3 (G3). According to the reported results, the toxic effect of salinity on fruit yield was significant in all salt-stressed trees compared to the control. As salinity increased, yield diminished in both scion-rootstock combinations. Nevertheless, the intensity of the negative effect on yield was slightly greater in G3 than in G1, especially at the two highest saline treatments (2.00 and 2.50 dS m^{-1}). The number of fruit also was significantly reduced with increasing salt concentration.

Previous research by the same authors examined the effect of rootstocks on mineral nutrition of mango cv. Keitt (Durán Zuazo et al. 2002), salt tolerance of mango rootstock cv. Osteen (Durán Zuazo et al. 2003) and the impact of salinity on macro- and micronutrient uptake in mango cv. Osteen with different rootstocks (Durán Zuazo et al. 2004b).

Trials in the Canary Islands (Galán Saúco et al. 1988) demonstrated that the polyembryonic rootstock 'Gomera 1' (G-1) was found to be more tolerant to salinity than the rest of the polyembryonic types tested ('Gomera 3', 'Gomera 4', 'Peach', 'Turpentine' and 'Kensington'). 'G-1' could behave well in water containing 560 ppm of chlorine and 560 ppm of sodium, similar to levels tolerated by '13/1' in Israel. In addition, the lower concentration of Na and Cl on roots and leaves of 'G-1' compared with the other rootstocks may indicate that tolerance to salinity may be due to selective uptake of salts by this rootstock, which is of capital importance for any scion grafted on it. Although all Gomera types were locally selected from the traditional mangoes cultivated in the island of La Gomera (Galán Saúco and García Samarín 1979), 'Gomera 1' was later found to be not different from what Popenoe (1920) named as 'Manga blanca' (Grajal-Martín 2012), still used as rootstock in Cuba. Both the tolerance of 'Gomera 1' to salinity and the association of the tolerance with the capacity of this rootstock to restrict the uptake and transport of Cl− and Na+ ions from the root system to the above-ground parts were later confirmed by Durán Zuazo et al. (2003, 2004b) in a trial with plants of the cultivar Osteen grafted on 'Gomera 1' and 'Gomera 3' exposed to salinized irrigation waters measured by electrical conductivity (1.02, 1.50, 2.00 and 2.50 dS m^{-1}). It is of interest to note that 'G-1' as rootstock was also found to produce higher yield (on kg/tree) of 'Osteen' than 'Gomera 3', although the reverse occurred with 'Keitt' that yielded higher on 'Gomera 3' than on 'G-1'. 'Gomera 1' also produces smaller trees on both 'Keitt' and 'Osteen' (Durán Zuazo et al. 2005) which may be an advantage for high-density plantings. Similar observations regarding smaller size of mango trees grafted on 'Gomera1' have been observed by the author of this report for most cultivars in the Canary Islands where until recently only 'Gomera 1' was used commercially as rootstocks.

This lack of significant information, regarding the response of mango trees to varying water quality, opens up a significant research field that may be very relevant for the mango industry in the near future.

5 Technologies for more efficient water management

This section will be focused mainly on the impact of different irrigation strategies using different technologies like drip irrigation, sprinkles and/or furrow or gravitational irrigation in mango production parameters. The second field 'software' is beyond the scope of this chapter, and the third field 'supportive irrigation technology' has been previously evaluated in 'Irrigation based on plant parameters and/or plant/soil sensors' in section 2.2 of this chapter.

The type of irrigation used for mangoes is an important consideration for production and sustainability. At present, in many mango production areas in the tropics (e.g. Mexico, Pakistan, India, Thailand and Dominican Republic) irrigation supplies water to trees through the typical six-month annual dry season using furrow irrigation along the tree rows (Davenport 2007; Mirjat et al. 2011 and personal observation). In-row furrow irrigation has the disadvantage of providing water and fertilizer quantities inaccurately periodically around the base of trees. In addition, in the medium and long terms, furrow irrigation may significantly increase soil salinity, depending on soil texture (Rajak et al. 2006).

Unlike furrow/flood irrigation method, drip irrigation method supplies water directly to the root zone of the crop; therefore, the water losses occurring through evaporation and deep drainage can be significantly lower (Narayanamoorthy 1996, 1997; Dhawan 1988). The on-farm irrigation efficiency of properly designed and managed drip irrigation system is estimated to be about 90%, while the same is only about 35–40% for surface method of irrigation (INCID 1994).

Mango in Brazil is grown on a wide range of soil types, from those latossols with high percentage of sand in Northeast to loamy latossols in Southeast Region (Pinto et al. 2004). Commercial mango orchards in Southeast Region, mainly in São Paulo state, do not use irrigation. However, in Northeast Region, under semi-arid tropical conditions, irrigation is a common practice used throughout the hot and dry season in most of the commercial orchards. According to Gomes et al. (2002) there is a wide range of irrigation systems being used in the Northeast mango orchards. Around 41% of the orchards are irrigated with micro-sprinkler system, 21% with other irrigating systems (furrows, drip, basin, etc.) and 33% of the orchards do not use any type of irrigation.

Mattar (2007) evaluated the irrigation system effect on growth and productivity in mango orchard in El-Sharkya Governorate, Egypt, during the period from 2002 to 2006 seasons. The experiment involved three different irrigation systems as follows: furrow (FI), drip (DI) and subsurface irrigation (SSI).

According to Mattar (2007) using SSI increased growth parameters for all age of trees more than drip irrigation and furrow irrigation by 15.2% and 21% for plant height, 8.75% and 19.6% for stem girth, respectively. Also, the average yield of mango under SSI increased by 13.88% and 26.25% over DI and FI, respectively.

The manual irrigation of mangoes using water hoses is a common practice in northern Thailand (Schulze et al. 2013). Schulze et al. (2013) evaluated mango (cv. Nam Dokmai) production in Northern Thailand – costs and returns under extreme weather conditions during the productive period (flowering to harvest) and different irrigation treatments: (1) FI, 100% of ETC; (2) DI, 50% of ETC; (3) conventional irrigation with micro-sprinkler (CIm); and (4) conventional irrigation with traditional equipment (CIt).

According to Schultze et al. (2013), cost–benefit analyses showed that an investment in a micro-sprinkler system can be recommended. As the marketable yield of fruits can be increased substantially (31% increase in class I fruits larger 300 g) with improved irrigation, especially during a drought year, it is worthwhile to change traditional irrigation into modern, water-efficient and flexible systems. DI increases the crop water productivity substantially and stabilizes yield during drought. The profit can be increased by 55% under FI with micro-sprinklers.

Important saving of water and fertilizer can be achieved by drip irrigation when compared to furrow irrigation methods. Also, fertilizer may be lost by run-off or percolation in undesired soil areas. On the other hand, drip irrigation applies water and fertilizer to the trees in slow mode which may prevent trees from waterlogged and significantly improve mineral absorption by the root system. Also, drip irrigation can keep a much better soil water-oxygen balance which allows the trees to maximize water and mineral absorption by the root system.

According to Lu (2013), drip irrigation is not commonly used for mango trees in Australia. In tropical regions of Australia, almost all mango growers use under-tree micro-sprinkler systems. There were several growers who trialled drip irrigation, but sooner or later they all changed to a micro-sprinkler system. The driving forces for the change have been the ease of field management (spotting the failures of the sprinklers from a motor-bike) and reduced need for water filtration and treatment when using sprinklers.

Even though drip irrigation may provide the most efficient way to irrigate from the resources (water and fertilizer) and from the water and fertilizer management point of view, it does not necessarily mean that drip irrigation is the only way to go and/or is the best option under all different cultivation conditions. Most of the published studies on drip irrigation on mango were made in India in the last decade (Lu 2013). Desai et al. (1999), for example, highlighted economic constraints like the high cost of spare parts and heavy initial expenses for installation of a drip irrigation system as a limiting factor in the development of drip irrigation projects in India. Parmar and Thorat (2016) identified time-to-time attention for minor repairs, inadequate and uncertainly in power supply, frequent clogging of drippers and microtubes, heavy initial expenses and lack of technical know-how as well as lack of proper training as major constraints faced by farmers in adopting drip irrigation system in the Gujarat, India.

Drip irrigation demands higher management skills to achieve its maximum efficiency in water and fertilizer use. Poor water quality (e.g. high level of organic material dissolved in the water), lack of proper maintenance and/or design among other factors may promote frequent clogging of the drippers and therefore it is more likely that drip-irrigated plants may suffer from more severe water stress than plants under other systems when a failure of the irrigation system occurs due to a lower buffer capacity of the soil with respect to water content (drippers wet a much smaller volume of soil) than other irrigation systems, such as micro-sprinkles or farrow irrigation. When mango growers are planning to adopt new irrigation systems, they are advised to be fully aware of the advantages and disadvantages of different systems and should make careful decisions according to their particular growing conditions and financial situations (Lu 2013).

6 Fertilization

6.1 Assessing nutrient requirements of mango trees/fruit

Best management practices for improving fertilizer use efficiency include applying nutrients according to plant needs, placed correctly to maximize uptake, at an amount to optimize growth and using the most appropriate source. These principles are reflected in nutrient stewardship programmes (e.g. 4R or the 'four rights', viz. right source, at the right rate, at the right time, in the right place; Drechsel et al. 2015). This concept can be potentially only satisfied by using fertigation (fertilizer applied through the irrigation system) in a 100% automatized irrigation system.

When not properly managed, up to 70–80% of the added N can be lost in rain-fed conditions and 60–70% in irrigated fields (Ladha et al. 2005; Roberts 2008). In contrast, NUE levels close to those observed in research plots can be achieved by farmers when using precision farming techniques under temperate conditions in the absence of other limiting factors (Drechsel et al. 2015).

The subject of macro- and micronutrient deficiency symptoms in mango will not be presented here; however, information about specific deficiency symptoms in mango will be highlighted where necessary. For characteristic mineral deficiency symptoms in mango it is recommended to consult the chapter 'Crop production, mineral nutrition' by Bally (2009) in *The Mango, Botany, Production and Uses* and Prado et al.'s (2012) chapter 'Macronutrient and micronutrient deficiency symptoms in mango'.

The nutrient demands of mango, expressed as the accumulated amounts of the elements found in different plant organs, vary according to factors such as genotype, soil, climate, use of irrigation, water quality, plant health, phenological stage and expected and past crop load among the main factors. Understanding and visualizing nutrient deficiency symptoms would enable improvement of fertilization programmes and increases in yield. In mango, visual examination of the plants, as well as soil and leaf analysis, is an important additional tool since it permits modifications in the fertilization programme during the same cropping year (Prado 2004). Thus, it becomes possible to intervene in a situation of nutritional disorders within a short period of time and therefore to guarantee more fruits with better quality (Prado 2004). The symptoms exhibited by a plant have a direct relationship with the functions the mineral plays in plant metabolism (Bally 2009). So, after the occurrence of the biological events, the symptoms will be related to the mineral element causing the nutritional disorder and this is linked to its functions and mobility in the plant (Prado 2004).

The importance of fertilizing mango for commercial production was emphasized previously by Young and Koo (1974) who reported a significant yield increase in cvs. Parvin and Kent grown in the sandy soils of Lakewood, Florida, when nitrogen (N) fertilization was increased threefold for the four-year average. Increasing potassium (K) fertilization threefold increased yield of 'Parvin' significantly in the second and fourth years, and for the four-year average. There was NI or cultivation. Highest yields in 'Parvin' were obtained with high rates of both N and K fertilization. Potassium rates had no significant effect on yield of 'Kent'. The same authors also reported a generally good correlation between treatment and leaf concentration of N and K. They observed that a heavy crop tended to decrease the level of N and K in leaves.

Recommendations for N supply for example indicate that 400 g N per tree per year are needed for acceptable commercial yields (Chia et al. 1988; Wanitprapha et al. 1991; Xiuchong et al. 2001). Crane and Campbell (1994) suggested that N amounts could be increased per tree size and site conditions. In sandy soils, fertilization practices may raise environmental concerns about rapid N leaching to groundwaters. The recommended fertilizer levels (N, P, K) in Brazil for the mango crop vary according to the expected productivity (from <10 to >50 t fruit ha^{-1}), the nutrient content of the leaf, the element itself and whether or not the trees are irrigated (Pinto et al. 2007). The timing and proportions of the total annual application also vary according to whether the crop is rain-fed or irrigated. Raij et al. (1996) recommended maximum nutrient inputs of up to 50 kg N ha^{-1}, 80 kg P ha^{-1} and 80 kg K ha^{-1}, for a rain-fed crop with high expected yield (>20 T-ha^{-1}) and low leaf mineral concentration and/or low nutrient availability from the soil (mainly N-P-K). For an irrigated crop these figures are increased to 120 kg N ha^{-1}, 150 kg P ha^{-1} and 250 kg K ha^{-1} (Silva et al. 2002). 'Fertigation' is encouraged with drip or micro-sprinklers. In Brazil, yields of up to 40 t ha^{-1} are possible with irrigation, but average yields under rain-fed conditions are in the range 8–12 t ha^{-1} (Carr 2014).

Nguyen et al. (2004) evaluated the effect of different nitrogen soil applications or by foliar spray on the skin colour and other quality attributes of ripe fruit of mango cv. Kensington Pride in high green (HG) skin and low green (LG) fruit skin orchards. The authors reported a significantly ($P < 0.05$) increased in the proportion of green colour on the ripe fruit in soil applications of 150 g N or more per tree. Foliar sprays resulted in a higher proportion of green colour than the highest soil treatment in the HG orchard, but not in the LG orchards. Anthracnose disease severity was significantly ($P < 0.05$) higher with 300 g of N per tree or foliar treatment in the HG orchard, compared with no additional N. The authors concluded that N can reduce mango fruit quality by increasing green colour and anthracnose disease in ripe fruit.

Morales and Rivas (2004) evaluated the efficiency use of fertilization and its effects on mango yields in the Zulia state, Venezuela (rain-fed orchard is assumed, four-year-old trees of cv. Haden). The authors report highly significant differences in yield and efficiency for dose (D), frequency of application (E) and the interaction E × D. The efficiency was calculated as the kg of fruit produced for kg of fertilizer applied. The authors concluded that higher fertilizer doses combined with higher frequency application achieved the best yield results. However, the annual fertilizer doses are significantly more efficient when the number of application was increased.

Durán-Zuazo et al. (2011b) reported that according to balance of nutrients from lysimeter data, only 13.3, 3.3 and 29.1% of N, P and K, respectively, were taken up by mango trees annually. Nutrient losses in the leachates of the mango lysimeters represented 6.6% for inorganic N (NO_3^-+NH_4^+), 0.23% for P and 16% for K. Soil N, P and K residual accounted for 48.0, 47.0 and 46.6%, respectively. Thus, conventional application rates of fertilizers (N, P and K per tree 638, 274 and 221 gr, respectively) in the coast of Granada (SE Spain) were excessive in mango orchards. In addition, the authors reported that in mango trees with an average yield of 24.3 kg per tree, N, P and K removed by pruning per year represented 5.8, 1.4 and 9.1%, respectively, while 7.5, 1.9 and 20.0%, respectively, by the fruit.

Tropical soils, characteristic of many mango production areas worldwide, are usually highly acidic and this may impede suitable mango tree nutrition, and therefore, commercial production. de Almeida et al. (2012) reported the impact of lime applied during the production phase in an orchard of mango cv. Haden. The lime treatments were determined with respect to the calculated reference lime dose (3.1 t/ha^{-1}) required to achieve 80% base saturation at a depth of 0–20 cm. The treatments were as follows: T1, no lime; T2, 1.55; T3, 3.10; T4, 4.65; T5, 6.20 t lime per hectare. Soil liming improved soil reaction chemistry, leading to higher pH, and lower H and Al. Concentrations of Ca and Mg also increased, leading to increases in both sum of bases and base saturation. The liming procedure did not affect fruit number (F = 0.37, ns) or production (F = 0.54, ns) in the first year of the study. However, in the second year, soil liming promoted an increment in fruit number and yield with a quadratic adjustment curve with respect to the applied lime doses. The highest fruit yield was achieved with a lime dose of 4.6 t ha^{-1} at which point soil base saturation was 72%.

Prakash et al. (2015) reported a significant increase in fruit number and fruit weight in mango cv. Alphonso when combined application of 24 L day^{-1} plant^{-1} and 120.0/75.0/100.0 g NPK/tree/year was compared with the combination of similar or less amount of water combined with less amount of fertilizer yield.

Salazar-García et al. (2016) found that fertilization treatments (from zero dose to two levels of balanced fertilizer (N, P, K, Ca, Mg, Fe, Mn, Zn and B)) did not affect the proportion of parthenocarpic mango (cv. Ataulfo) that reached harvest maturity, in Mexico.

6.2 Organic fertilization

Organic agriculture has been recently defined as a production system that sustains the health of soil, ecosystems and people. According to 'The International Federation of Organic Agricultural Movements (IFOAM)' it relies on ecological processes, biodiversity and cycles adapted to local conditions rather than on the use of artificial inputs with adverse effects on the environment and potentially on human health (IFOAM 2009). Eight per cent of the total planted mango area in Mexico is in the process of conversion from conventional to organic systems, mainly in the Pacific Coast under tropical dry and wet conditions. Mango cultivars in organic systems are Tommy Atkins, Kent and Ataulfo, which are demanded by market for their early or late harvest season (Medina-Urrutia et al. 2011).

According to Medina-Urrutia et al. (2011) mango cultivars for organic production in Mexico are selected according to a range of characteristics: 1) harvest season (early or late); 2) adaptability to the environment; 3) water requirements; 4) tree size (dwarf to semi-dwarf); 5) better tree and fruit tolerance to pests and diseases; 6) fruit quality and marketability; and 7) proximity to markets.

Fertilization studies in Mexico to determine optimum nutrient dosages on mature mango trees did not show yield differences between treated and control trees (Ireta-Ojeda, pers. comm. In, Medina-Urrutia et al. 2011). Conversely, Avilan (1983) reported increased yield and vegetative growth because of organic fertilizer application. Also, Das et al. (2009) reported improved flowering and yield when chemical fertilizer was combined with organic fertilizer. To date, fertilizer use in Mexican mangoes is based on technical reports adapted from studies in other countries (Vazquez-Valdivia et al. 2006; Ireta-Ojeda and Estrada-Guzman 2002). Diversity of physical and chemical soil conditions, rootstocks and water availability are more important factors interacting with tree nutrition. Under these conditions, organic nutrition is chosen according to a diversity of local formulations prepared by growers using local resources as a result of their own accumulated experience (Medina-Urrutia et al. 2011).

Silva et al. (2013) evaluated the impact of three different organic composts (A, B and C) which were prepared mixing different proportions of animal (e.g. goat manure) and vegetable (e.g. sugarcane, banana leaves, leaves and branches of mango, etc.) materials at three different concentrations (0, 5 and 10 t ha^{-1}) each, on soil chemical characteristics, nutrient contents in leaves and production level of mango crops grown in the organic system under the semi-arid conditions of northeastern Brazil. The organic composts increased the levels of soil organic matter (SOM), especially compost C, whose soil analysis showed higher levels of total P, K, Ca, Mg, B, Cu, Mn and Zn. The SOM increased linearly with increasing compost levels. Production and fruit number per plant were higher when using the B and C composts, which showed higher total nutrient contents. Nitrogen leaf content increased linearly with the compost level. The production of fruit (kg ha^{-1}) and fruit number per plant was defined by a quadratic function with respect to the compost concentration, without reaching a maximum.

Peralta-Antonio et al. (2014) reported that chicken manure at 10 t ha^{-1} had a similar effect to the mineral doses of 230-0-300 and 230-0-0 g NPK tree^{-1}/year recommended by Mosqueda et al. (1996) on number of fruit and final yield in cultivars Manila, Tommy Atkins and Ataulfo and on soil contents of N, K, Cu and Zn.

The above results may indicate that the organic production system can be relevant for mango growers interested in producing either naturally or target-specific niche with potentially higher price. Even though in the last few years a huge growth in the options available for organic production system have been observed, in both pest control and fertilizer products, this production system still have certain limitation in both fields, mainly when compared to conventional systems. In addition, weed control in organic system is in general more expensive and problematic than in conventional ones. Consequently, not all potential mango growing areas will be suitable enough for organic production system. For example, areas with sandy soil and/or with heavy precipitations will be more challenged to produce mango in commercial quantity and quality level under organic system than in dry climate areas and heavier soil.

Despite the wider options of organic fertilizer present today, some limitations can be highlighted. In general, the number of organic liquid fertilizer and the different combination of minerals is very limited compared to conventional ones. In addition, their mineral concentration is in general significantly lower (e.g. N around 5%) than conventional ones; therefore, the amount of fertilizer to be applied per irrigation (fertigation) is significantly higher. Therefore, the fertigation system can become a constraint for the application of the total fertilizer need under organic system. On the other hand, solid organic fertilizers may be very labour-intensive and nutrient ratios may be also often unknown. However, the following advantage of organic fertilizer should be highlighted:

- In addition to releasing nutrients, as organic fertilizers break down, they improve the structure of the soil and increase its ability to hold water and nutrients. Over time, organic fertilizers may make your soil and plants healthy and strong.
- Since they are the ultimate slow-release fertilizers, it is very difficult to over fertilize (and harm) your plants.
- There is little to no risk of toxic build-ups of chemicals and salts that can be deadly to plants.
- Organic fertilizers are renewable, biodegradable, sustainable and environmentally friendly.
- Although rather expensive in packages, you can make your own organic fertilizer by composting or find inexpensive sources – such as local dairy farms – that may sell composted manure. However, electrical conductivity in all these materials should be checked to avoid salinity problems and potential plant injury by low quality compost.

7 Case study

The objective of our case study is to present of how research has been used to improve mango cultivation in practice. In this occasion the study to be presented is 'The response of field-grown Mango (cv. Keitt) trees to RDI at three phenological stages' under semi-arid conditions and drip irrigation.

Differential irrigation levels were applied at each phenological stage where commercial irrigation levels were applied in the rest of the season. The evaluated phenological stages were the following:

MFG – From fruit set to pit hardening (the beginning of May to the beginning of July).

FFG – From early July up to harvest, from the beginning of July to the beginning of September (~7/7–10/9).

PH – After harvest until the first meaningful rain (>20 mm), from the beginning of September to the middle of November (~11/9–15/11).

Average daily irrigation (mm), crop coefficient Penman–Monteith (Kc) (Allen et al. 1988) and total seasonal irrigation (mm) at the different irrigation treatments at the different phenological stages for the experimental period are presented in Table 2.

Table 2 Average daily irrigation (mm), crop coefficient Penman–Monteith (Kc) and total seasonal irrigation (mm) at the different irrigation treatments at the different phenological stages for the entire experimental period (2010–14). The phenological stages are main fruit growth (MFG) stage, final fruit growth (FFG) stage and post-harvest (PH) stage

Phenological period	mm/day				Kc				mm/year			
	T-1	T-2	T-3	T-4	T-1	T-2	T-3	T-4	T-1	T-2	T-3	T-4
MFG (1)	2.8	3.7	4.6	5.9	0.38	0.49	0.63	0.80	939.6	991.4	1066.8	1144.6
FFG (2)	3.8	5.6	7.5	9.2	0.48	0.72	0.96	1.17	874.7	928.3	1034.0	1132.8
PH (3)	2.8	3.7	4.5	5.4	0.57	0.75	0.92	1.09	949.7	951.5	1026.8	1067.4

7.1 Main outcomes of the research

7.1.1 MFG stage

The MFG stage is characterized by rapid fruit expansive growth that is associated mainly with high rate of cell division and expansion (Subramanyam et al. 1975; Ram et al. 1983; Tharanathan et al. 2006). Approximately 80% of the final fruit size is achieved in the MFG stage (data not shown). In our study the MFG stage lasted between 57 and 61 days. The following treatments were evaluated in this period: T-1 = 0.38; T-2 = 0.49; T-3 = 0.63 and T-4 = 0.80 Kc. The increasing irrigation level in this period, during high productive years (2011, 2013 and 2014), did not affect yield, increased the number of fruit per tree in all years, even though such increase was not significant (Fig. 4). Since the lower number of fruits in the lower treatments was already apparent in the first season, it may indicate that water stress tended to increase fruit drop. It seems that T3 (daily irrigation rate of 4.6 mm and Kc of 0.63) may avoid potential risk of increased fruit drop and there is no need for further increase in irrigation level in the MFG stage. The average fruit size in the T4 treatment for the three 'ON' seasons was significantly smaller than the T2 treatment despite the former one receiving higher irrigation rates (~46%) (Fig. 4). The yields presented at the MFG stage were fitted by a polynomial quadratic function (between yield and Kc), as reflected in Fig. 5a. Durán-Zuazo et al. (2011a) reported similar results under similar conditions in Granada, Spain, in cv. Osteen. Average fruit size in 'ON' seasons decreased with increasing number of fruit per tree (Fig. 6) indicating the number of fruit rather than water quantities (Fig. 7) as the main determinant of the final fruit size for the MFG stage.

7.1.2 FFG period

In the FFG period the following treatments were evaluated: T-1 = 0.48; T-2 = 0.72; T-3 = 0.96 and T-4 = 1.17. The highest irrigation treatment (T4) had significantly higher crop yield (Fig. 8), higher number of fruit per tree and higher fruit weight. Even though the major determinant of crop yield was the number of fruit per tree (Fig. 8), fruit size was also positive affected by increasing irrigation despite the increased number of fruit in the higher water treatments. This is contrary to what is assumed by the mango growers in general and in Israel in particular, where higher number of fruit means smaller fruit. These results also pointed out the significant meaning of irrigation in years of high productivity

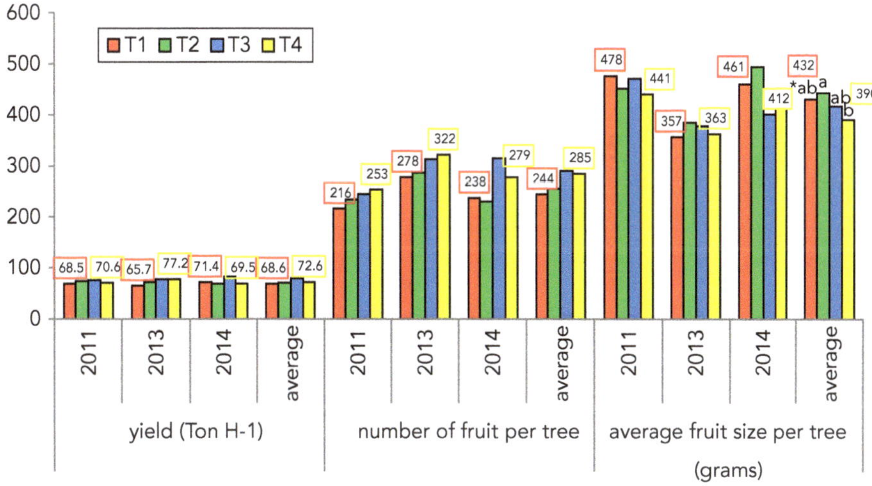

Figure 4 Total crop yield, number of fruit per tree and average fruit weight in response to different irrigation treatments (increasing from T1 to T4) at the main fruit growth stage (MFG). *Different letters within a year represent significant differences according to Tukey's test ($P < 0.05$). The average refers to the 'ON' seasons.

and the impact on critical productive parameters like number of fruit and fruit size. The yields presented at the FFG stage were fitted by a linear function (between yield and Kc), as reflected in Fig. 5b.

In addition, it seems that a trend of increasing number of fruit per tree with increasing irrigation rate was developed over the years (accumulative effect). It seems that long-term water stress in FFG stage may affect the long-term productivity. It might be supported by the lower PH vegetative growth in the lowest irrigation treatments during 'ON' productive years (Fig. 9) despite similar irrigation rates at the PH stage. Lower vegetative growth may end up with confined canopy size that may restrict the tree productivity and assimilate supply to the maturing fruit.

The effect of the irrigation treatments on fruit weight was significant only in 2013 (Fig. 8) where the number of fruit per tree was the highest and it was associated by low fruit weight compared with the other 'ON' seasons. It may indicate that the response to irrigation increases with increasing crop load, similar to apple (Naor et al. 2008) and olive (Naor et al. 2013). A quantitative evaluation of the interaction of irrigation and crop load may provide tools for the adjustment of irrigation levels to the actual crop load.

Our results are contrary to those reported by Santos et al. (2014) in cv. Tommy Atkins in northern part of Brazil where the highest crop yield was recorded in the deficit irrigation treatment (50% of ETc) applied at the same phenological stage (FFG) (15.4% higher than the control full irrigated treatment, not significant). The recorded differences between our and dos Santos et al. (2014) results could be attributed to the significantly higher yield recorded in our experiment (88 and 24.95 $t\text{-}ha^{-1}$, respectively) and/or may be influenced by remarkable difference in weather conditions (tropical vs. subtropical in our case). Cotrim et al. (2011) reported no significant difference in number of fruit, fruit size distribution and total yield among different irrigation treatments (between 0 and a 100% ETc) applied at the same phenological stage. They attributed these results to the rise of groundwater in the evaluated period.

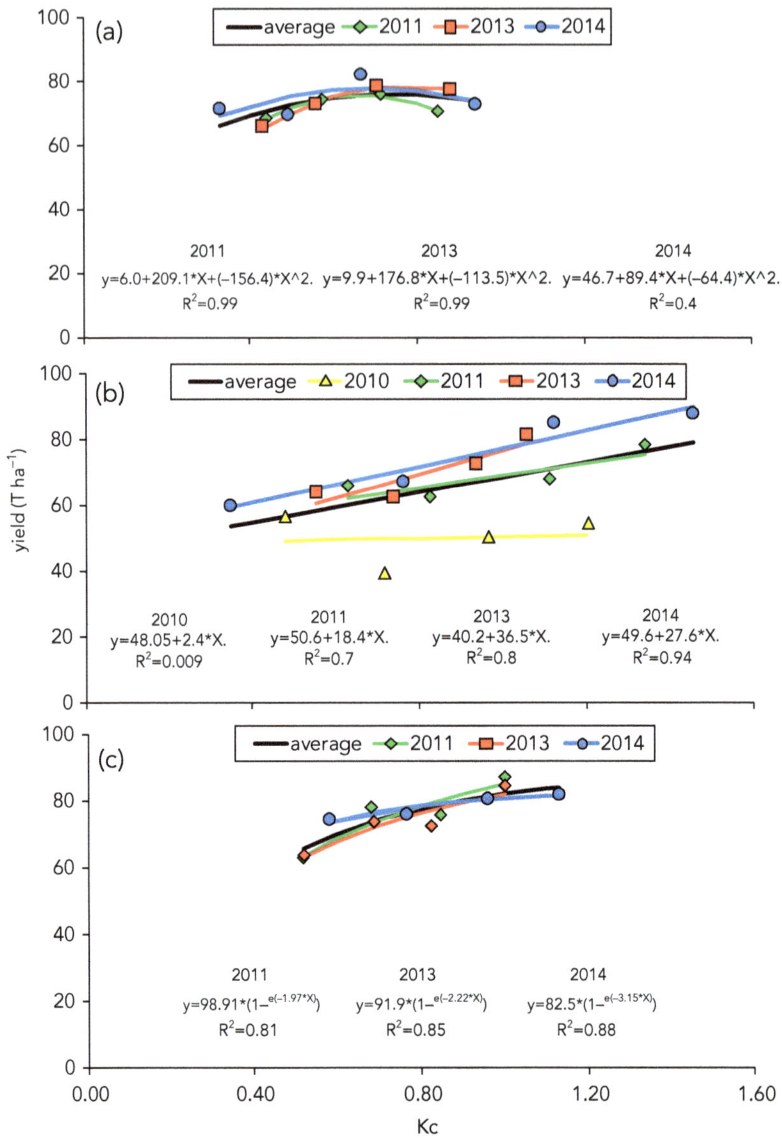

Figure 5 Relationship between the total crop yield (Y) and crop coefficients (Kc) from Penman–Monteith evapotranspiration in the three experiments, MFG – the main fruit growth stage (a), FFG – final fruit growth stage (b), PH – post-harvest stage (c). Data are from 'ON' seasons. The lines are polynomial regression. Polynomial regression values for the different phenological periods are a) *MFG*, $Y = 44.29 + 85.39 \cdot Kc - 57.46 \cdot Kc^2$; $r^2 = 0.41$. b) *FFG*, $Y = 48.57 + 25.06 \cdot Kc$; $r^2 = 0.72$. c) *PH*, $Y = 27.76 + 93.29 \cdot Kc - 38.74 \cdot Kc^2$; $r^2 = 0.76$.

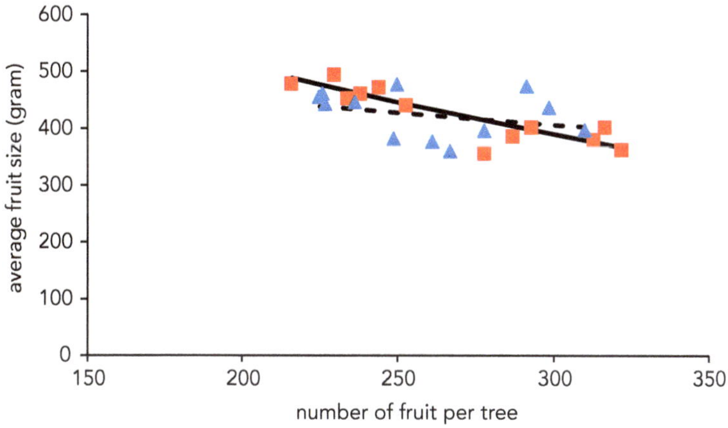

Figure 6 The relationship between the number of fruit per tree (N) and average fruit size (FW) in the main fruit growth stage (MFG, ■) and final fruit growth stage (FFG, ---, ▲) in 2011, 2013, 2014 (the 2012 which was an 'OFF' season was not included). The lines are linear regression for the three seasons in each phenological stage. MFG-FW = −1.126 X + 726.64 (r^2 = 0.77); FFG – FW = −0.403 X + 529.54 (r^2 = 0.09).

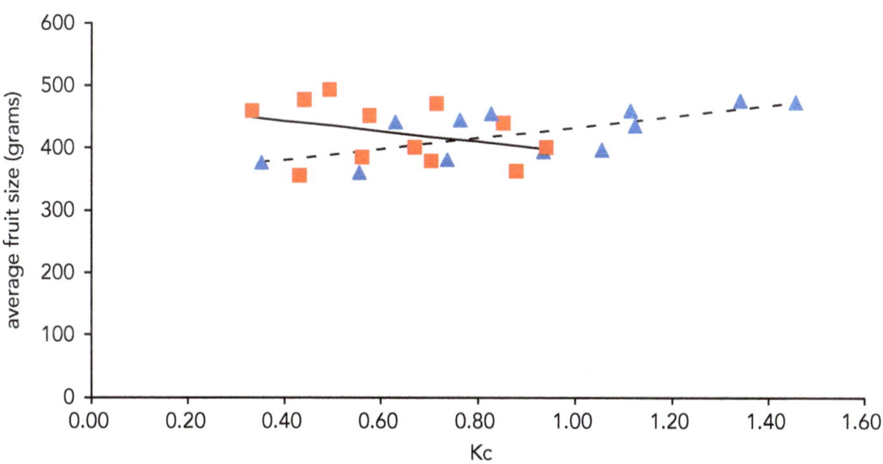

Figure 7 The relationship between water quantities (Kc) and average fruit size (FW) in the main fruit growth stage (MFG, ■) and final fruit growth stage (FFG, ---, ▲) in 2011, 2013, 2014 (the 2012 which was an 'OFF' season was not included). The lines are linear regression for the three seasons in each phenological stage. MFG-FW = −83.895 X + 476.89 (r^2 = 0.11); FFG – FW = 86.211 X + 346.72 (r^2 = 0.48).

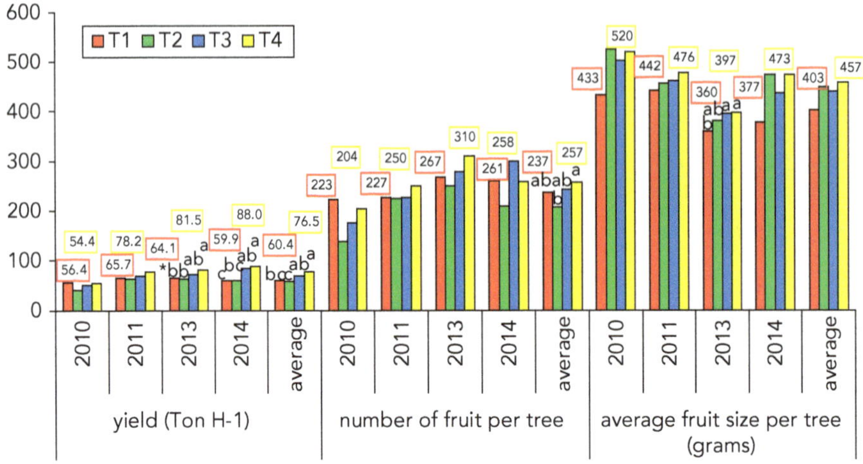

Figure 8 Total crop yield, number of fruit per tree and average fruit weight in response to different irrigation treatments (increasing from T1 to T4) at the final fruit growth (FFG) stage. *Different letters within a year represent significant differences according to Tukey's test ($P < 0.05$). The average refers to the 'ON' seasons.

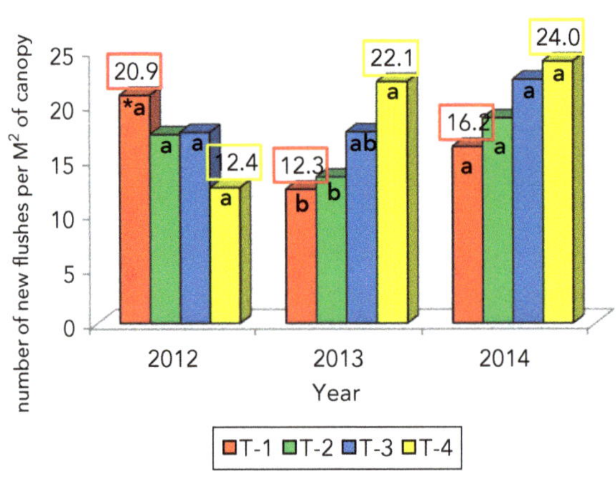

Figure 9 The relationship between the number of new post-harvest shoots per square meter of canopy in the final fruit growth (FFG) stage trees from 2012 to 2013. The crop yields in the FFG phenological period for 2012, 2013 and 2014 were 11, 69 and 75 T/ha^{-1}, respectively. The counting of the new shoots was carried out at the top third section of the canopy. *Different letters within a year represent significant differences according to Tukey's test ($P < 0.05$).

7.1.3 PH period

During the PH period the following treatments were evaluated: T-1 = 0.57; T-2 = 0.75; T-3 = 0.92; T-4 = 1.09. The PH irrigation levels increased crop yield (Fig. 10) but unaffected the number of fruit per tree, in spite of a clear positive response of vegetative growth to Kc (Fig. 11). The increased yield was due to increased fruit weight with increasing irrigation levels. It should be noted that all treatments received the same irrigation levels from bloom to harvest. It may indicate that the increased vegetative growth improved either potential fruit size or the photosynthetic capacity. The direct effect of PH stage irrigation on subsequent season fruit weight is unique and to the best of our knowledge was not reported in other fruit trees as well. Further research is required to explore the mechanism of this effect. The response of vegetative growth to Kc was similar in all seasons, except for 2011 (Fig. 11). It could be explained by the fact that the PH stage in 2011 was colder than other seasons and rains began earlier in 2011. In addition, 2012 was an extreme 'OFF' season, probably because of former seasons high crop load (2010 and 2011, 43.6 and 75.9 t/ha, respectively), and lower vegetative growth could be an early indicator of the fact that the trees were exhausted.

Crop yield responds to increasing Kc in the PH stage in the former season (Fig. 5c) and it reached a maximum at ~ Kc = 0.8. It indicates that irrigation level should not fall below that of T3 (daily irrigation rate of 4.5 mm and Kc = 0.92).

In summary, we found considerable evidence showing that the different evaluated phenological periods respond differently to increased water quantities under Israeli semi-arid conditions.

Much of the published literature evaluating the effect of different water quantities on mango production considered the fruit growth period as a single event. In our experiment, mango fruit responded differently to different water quantities at the two evaluated fruit growing periods, MFG and FFG. While during the MFG period the

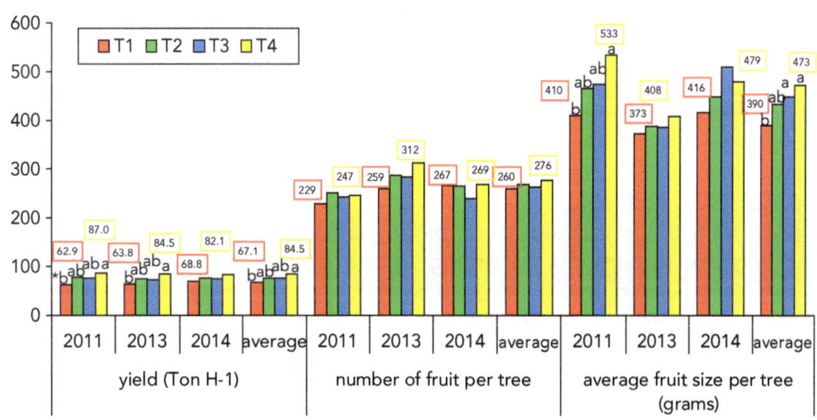

Figure 10 Total crop yield, number of fruit per tree and average fruit weight in response to different irrigation treatments (increasing from T1 to T4) at the post-harvest (PH) stage. *Different letters within a year represent significant differences according to Tukey's test ($P < 0.05$). The average refers to the 'ON' seasons.

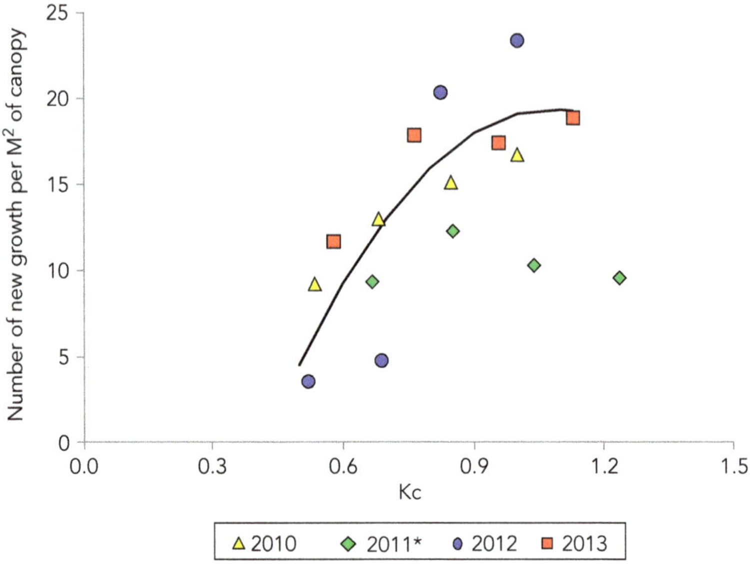

Figure 11 The relationship between the number of new shoots per square meter of canopy in the post-harvest (PH) stage in 2010 and 2012 and the crop yield in the subsequent season. The counting of the new shoots was carried out at the top third section of the canopy. The lines are polynomial regression. Polynomial regression values for the different seasons are a) 2010/2011, $y = 37.4 + 2.85X$, $R^2 = 0.86$; b) 2012/2013, $y = 65.6 + 0.61X$, $R^2 = 0.55$; c) 2013/2014, $y = 64.3 + 0.85X$, $R^2 = 0.58$; d) average, $y = -32.3 + 95.9X + (-44.5) X^2$, $R^2 = 0.67$. *2011 data was not included in the regression line.

irrigation mainly affected the number of fruit reaching harvest rather than fruit size, during the FFG period irrigation mainly affected final yield and fruit size and PH vegetative growth as well.

It is also assumed among mango and other fruit crops growers in Israel that water saving, if necessary, should be done primarily after harvesting, when there is no fruit on the trees. Our results clearly identify PH as the most sensitive phenological period to irrigation under Israeli growing conditions, having the largest impact on fruit production in terms of both quantity and quality (fruit size).

Our study also demonstrated a completely different response of mango trees to different irrigation treatments according to the production level. Important water saving can be achieved under conditions of low productivity. Water saving in the MFG period will primarily impact the final number of fruit, but not necessarily the yield, while water saving during the FFG period will primarily affect final fruit size and potentially also yield. Therefore, Kc optimal values for mango will depend not only on the phenological period but also in the production level mainly in relation to number of fruit per tree.

The results obtained in our study at different phenological stages, MFG, FFG and PH, indicated the existence of two phenological periods that are particularly sensitive or critical to irrigation in 'ON' or towards 'ON' productive years. The first critical period corresponds to the FFG period and the second to the PH period under Israeli growing conditions.

7.2 Practical outcome from the research

As the result of this research, irrigation strategies among many of the mango growers in Israel have been changed, mainly in relation to the PH and FFG phenological periods. In years of low production, the growers have significantly reduced the amount of applied water, from 850 to 1000 mm per season down to 650 mm (≈25–35%, less water) (Michael Noy, Extension Services, Ministry of Agriculture, Israel, pers. comm.). Contrarily, in years of high production the amount of water applied during the FFG period is significantly higher. There is almost no more water-saving strategy during the PH period. This water strategy, apparently, also starts to reduce high alternate bearing conditions in some of the cultivars and growers (Cleef Love and Michael Noy, Extension Services, Ministry of Agriculture, Israel, pers. comm.).

8 Conclusion

Even though significant progress in the field of irrigation, in both, technology and strategy, has been achieved in the mango industry, mainly in the last 20 years, several gaps in the main studies done in this field still exist. The clear majority of the irrigation studies were carried out for two consecutive seasons, meaning that the impact of the irrigation treatment in the medium and long term could not be evaluated, mainly in those where different irrigation strategies were evaluated under heavy soil conditions. In addition, the focus of many of these researches was related to the development of water-saving strategies rather than finding the optimal watering quantities to maximize production.

Also, this progress was significantly more important in semi-arid and subtropical region (e.g. Israel, Spain, semi-arid region of Brazil and Thailand) rather than in tropical ones (e.g. India, China, South Mexico and Vietnam) except for Australia where significant studies in the irrigation field have been carried out in their tropical region.

In the case of fertilization, the gaps in the knowledge of quantities of mineral needs for mango to produce high quantity and quality yield in a constant way is even bigger than in irrigation. This is mainly due to the difficulty to carry out fertilization studies under field conditions, mainly under heavy soil conditions, where the buffer capacity for nutrient content is big; then, many years could be needed to completely deplete a good formed heavy soil from N, for example, or other essential minerals generally evaluated. Then, the difficulty to establish a proper evaluation system either by soil or tissue analysis. Guzman-Estrada (2000), for example, reported low correlation values between foliar nutrients concentration and fruit production in cv. Manila under the Gulf of Mexico conditions. Prado (2010) reported phosphorous fertilization increased P concentration in the soil and influenced mango plant performance only after the second year of cultivation. However, phosphorous fertilization did not affect yield during the evaluated period (2005–2006). According to Obreza (2000) the best results in citrus are obtained when soil and leaf chemical analyses are made and compared. Also, in the mango industry, a fertilization strategy based on soil and tissue analysis will be probably the best way to go, considering the impact of different cultivars in the development of such strategy as well.

In the modern mango industry, fertilization should be considered together with irrigation, meaning fertigation, at least in medium (>30 ha) and big groves. The concept of fertigation should be further studied and developed for semi-arid or subtropical and tropical areas. While in semi-arid regions or subtropical areas the irrigation system provides both water and nutrient to the plant, under tropical conditions (>1200 mm rain) the irrigation system may provide mainly the nutrients for the plant principally, during high nutrient demanding periods (e.g. fruit set and fruit development). Under tropical conditions, in general, the soils can be poor because of the leaching effect of the intensive and long rain periods; then the plants may suffer more frequent deficiency of nutrients like Ca and N among others. Therefore, this fertigation system can significantly reduce the mineral leaching impact of the tropical rain and constantly provides the plant with the mineral needs at all phenological stages.

In case of small farms (<30 ha), the irrigation or the water needs for the plant can be achieved by simple portable irrigation systems rather than furrow irrigation, for example. The WUE in the former system will be significantly higher and the salinization impact on the soil significantly lower. Regarding the fertilization in small farms, many of this portable irrigation system allow the application of fertilizer through the system, even though its accuracy may be lower than a proper fertigation system. In addition, a strategy which combines foliar and granular application to the soil can be developed in rain-feeding farms. In this case, it is recommended to divide the fertilizer annual quantities in several applications, mainly to the soil, to diminish fertilizer leaching (mainly in tropical areas) on one hand and excessive fertilizer concentration on the other hand.

The aim of the recommended research fields in mango is related to the improvement of understanding of mango water and fertilizer needs and its sustainability from the economic and environmental point of view. This review of research conducted on past and present irrigation strategies in mango demonstrates a need to continue working in several fields, to determine:

1 Mango water requirements under lysimeter conditions (quantitative research).
2 Impact of different crop loads on water requirements under lysimeter conditions (quantitative research).
3 Mango mineral consumption under different crop loads under lysimeter conditions (quantitative research).
4 Impact of different crop loads on mango water requirements under particular growing conditions (qualitative research).
5 Mango water requirements for maximizing production in the short, medium and long term, with respect to quantity and quality, under different climate and soil conditions (qualitative and quantitative research).
6 Long-term response of mango (at least four seasons) to different water quantities, including deficit irrigation, on mango vegetative growth and associated yield at each phenological stage in an environment with no effective rainfall during the irrigation season and its impact on fruit PH behaviour in the main cultivars (qualitative research).
7 Impact of irrigation with different water qualities on production parameters such as number of fruit, fruit size distribution, fruit quality (chemical), total yield, vegetative growth, alternate bearing and fruit PH behaviour, under different soil and climate conditions (qualitative research).
8 Short-, medium- and long-term impacts of different irrigation methods (gravity, sprinkles and drip irrigation), under different soil conditions, on fruit PH behaviour of different cultivars (qualitative research).
9 Development of a fully automated one (e.g. GBI, based on plant, soil and weather sensors) which has long-term reliability and robust and low cost.
10 Usefulness of different foliar spray fertilizer strategies, with an emphasis on small- and medium-sized mango producers without proper irrigation systems in place, as alternative or complementary to manual soil fertilization (qualitative research).

The above-proposed studies are recommended to be conducted for a minimum of three to five production seasons in order to properly evaluate the impact of different irrigation and/or fertilization strategies on relevant production variables (including alternate bearing) pertinent for the mango growers. Also, it is recommended to conduct these evaluations or at least part of them, at the farms of growers who are interested in taking part in such research projects. This can facilitate subsequent dissemination of the information among the mango growers, locally and internationally, in a more effective way.

9 Where to look for further information

Further information regarding irrigation and mineral nutrition in mango can be found in Bally, I. S. E. (2009). *'Chapter 12: Crop production, mineral nutrition'*. In Litz , R. E. (Ed.). The Mango, Botany, Production and Uses. 2nd ed. CAB International, pp. 404–31. A comprehensive approach in nutritional physiology of crops can be found in Rengel, Z. (1999). *Mineral Nutrition of Crops: Fundamental Mechanisms and Implications.* Haworth Press, New York. p. 400, and in Marschner, H. (2011). Marschner's Mineral Nutrition of Higher Plants, https://books.google.com/?hl=en

10 References

Allen, R. G., Pereira, L. S., Raes. D. and Smith, M., (1988). *Crop Evapotranspiration: Guidelines for Computing Crop Water Requirements.* FAO Irrigation and Drainage, Paper No. 56, Rome, Italy, FAO.

Angelakis, A. N., Marecos do Monte, M. H. F., Bontoux, L. and Asano, T., (1999). 'The status of wastewater reuse practice in the Mediterranean basin, need for guidelines'. *Water Res.*, 10, 2201–17.

Anon., (1988). *Timing of Irrigation for Mango. Technical Annual Report 1987–1988.* Technical Bulletin No. 150, Northern Territory of Australia, Department of Primary Industry and Fisheries, 59–60p.

Anon., (2002). *Deficit Irrigation Practices.* Water report 22, Rome, Food and Agriculture Organisation of the United Nations, 101p.

Anon., (2006). World business council for sustainable development brochure. *Water facts and trends.* www.unwater.org/downloads/Water_facts_and_trends.pdf.

Asano, T. and Levine, D., (1991). Wastewater reclamation, recycling and reuse, past, present and future. In *Second International Symposium on Water Quality.* Iraklio, Greece, Wastewater reclamation and Reuse, pp. 5–17.

Aujla, M. S., Thind, H. S. and Buttar, G. S., (2005). 'Cotton yield and water use efficiency at various levels of water and N through drip irrigation under two methods of planting'. *Agric. Water Manage.*, 71, 167–79.

Avilan, R. L., (1983). 'La fertilización del mango Mangifera indica L en Venezuela'. *Fruits*, 38, 183–8.

Bakhsh, B., Hassan I. and Akhter, S., (2006). 'Profitability and cost in growing mango orchards'. *J. Agric. Soc. Sci.*, 2, 46–50.

Bally, I. S. E., Harris, M. and Whiley, A. W., (2000). 'Effect of water stress on flowering and yield of Kensington Pride mango (Mangifera indica L.)'. *Acta Hort.*, 509, 277–82.

Bally, I. S. E., (2009). 'Chapter 12: Crop production, mineral nutrition'. In Litz, R. E. (Ed.) *The Mango, Botany, Production and Uses.* 2nd ed. CAB International, pp. 404–31. ISBN 9781845934897. DOI: 10.1079/9781845934897.0404.

Bithell, S. L., Diczbalis, Y. and Moore, C. (2010). Review of Mango Irrigation Research in the Northern Territory. Technical Bulletin No. 334, Australia, Northern Territory Government.

Carr, M. K. V., (2014). 'The water relations and irrigation requirements of mango (Mangifera indica L.): A review'. *Experimental Agriculture*, Cambridge, 50(1), 1–23.

Castro Neto, M. T., Reinhardt, D. H. and Ledo, C. A. da S., (2004). Determination of water potential on mango trees by pressure chamber. In Pinto, A. C. Q. and et al. (Eds) *Proc. VIIth IS on Mango, Acta Horticulture 645, ISHS.*

Chacko, E. K., (1986). Physiology of vegetative and reproductive growth in mango (*Mangifera indica* L.) trees. In *Proc. 1st Australian Mango Wksp*, pp. 54–70.

Chacko, E. K., (1991). 'Mango flowering- still an enigma!'. *Acta Hort.*, 291, 12–21.

Chen, H., Huang, M., Chen, H. B. and Huang, M. Y., (1995). 'Fruit growth and abscission of the mango (Mangifera indica L.) Zihua'. *J. South China Agr. Univ.*, 16, 73–7.

Chia, C. L., Hamilton, R. A. and Evans, D. O., (1988). *Mango Commodity Fact Sheet MAN-3(A)*. Hawaii Coop. Ext. Ser., Univ. of Hawaii Press, p. 4.

Coelho, E. F., Oliveira, A. S., Neto, A. O. A., Teixeira, A. E. C., Araújo, E. C. E. and Bassoi, L. H., (2002). Irrigação. In Genu, P. J. C. and Pinto, A. C. Q. (Eds) *A cultura da mangueira. cap. 9.* Brasilia, pp. 167–89.

Coelho, E. F. and Borges, A. L., (2004). Irrigation and fertirrigation of mango. In Pinto, A. C. Q. and et al. (Eds) *Proc. VIIth IS on Mango, Acta Horticulture 645, ISHS.*

Coelho Filho, M. A. and Coelho, E. F., (2005). Uso da regulação do déficit de irrigação e produtividade de mangueira Tommy Akins em condições semi-áridas. In *CONGRESSO NACIONAL DE IRRIGAÇÃO E DRENAGEM, 15., 2005, Teresina.* Anais. Teresina, ABID. CD-ROM.

Coelho, E. F., Alves, A. A. C., Oliveira, V. V. M. and Velame, L. M., (2009). Manejo de irrigação com déficit (PRD) em pomar de manga 'Kent' no semi- árido baiano. In *JORNADA CIENTÍFICA, 3., 2009, Cruz das Almas.* Anais. Cruz das Almas, EMBRAPA MANDIOCA E FRUTICULTURA. 1 CD-ROM.

Costa, J. M., Ortuno, M. F. and Chaves, M. M., (2007). 'Deficit irrigation as a strategy to save water, Physiology and potential application to horticulture'. *J. Integr. Plant Biol.*, 49(10), 1421–34.

Cotrim, C. E., Coelho Filho, M. A., Coelho, E. F., Ramos, M. M. and Cecon, P. R. (2011). 'Regulated deficit irrigation and Tommy Atkins mango orchard productivity under microsprinkling in Brazilian semiarid'. *Engenharia Agrícola*, Jaboticabal, 31(6), 1052–3.

Cull, B. W., (1991). 'Mango crop management'. *Acta Hort.*, 291, 154–73.

da Silva, V. D. R, Campos, J. and de Azevedo, P. V., (2009). 'Water-use efficiency and evapotranspiration of mango orchard grown in northeastern region of Brazil'. *Sci. Hort.*, 120(4), 467–72. http://dx.doi.org/10.1016/j.scienta.2008.12.005.

Dahsham, D. I. and Habib, S., (1985). 'Seasonal changes in endogenous auxin like substances in relation to fruit drop in mango'. *Suez Canal Univ.*, Ismailevah, 2, 769–80.

Das, A., Mandal, K. K., Hasan, M. A., Bhattacharya, B., Majumder, D. and Bandopadhyay, B., (2009). 'Effect of organic and inorganic nutrients on improving flowering'. *Acta Hort.*, 820, 371–80.

Davenport, T. L., (1993). Floral manipulation in mangos. In Chia, L. E. and Evans, D. O. (Eds) *Proceedings of the Conference on Mango in Hawaii.* Cooperative Extension Service, Honolulu, USA, University of Hawaii, pp. 54–60.

Davenport, T. L. and Nunez-Elisea R., (1997). Reproductive physiology. In Litz, R. E. (Ed.) *The Mango, Botany, Production and Uses.* Wallingford, CAB International, pp. 69–146.

Davenport, T. L., (2007). 'Reproductive physiology of mango'. *Braz. J. Plant Physiol.*, 19(4), 363–76.

de Almeida, E. V., Fernandes, F. M., de Mello Prado, R., Boliani, A. C. and de Souza Corrêa, L., (2012). 'Mango tree response to lime applied during the production phase'. *Open J. Soil Sci.*, 2(2), 155–61. http://dx.doi.org/10.4236/ojss.2012.22021. Published Online June 2012 (http://www.SciRP.org/journal/ojss).

de Azevedo, P. V., da Silva B. B. and da Silva, V. P. R., (2003). 'Water requirements of irrigated mango orchards in northeast Brazil'. *Agric. Water Manage.*, 58, 241–54.

De la Hera, M. L., Romero, P., Gómez-Plaza, E. and Martinez, A., (2007). 'Is partial rootzone drying an effective irrigation technique to improve water use efficiency and fruit quality in field-grown wine grapes under semiarid conditions?' *Agric. Water Manage.*, 87, 261–74.

Desai, C. P., Patel, A. A. and Patel, M. R. (1999). 'Constraints faced by mango growers in adoption of drip irrigation system'. *Gujarat Agric. Univ. Res. J.*, 24(2), 52–7.

de Souza, P. J. O. P., Rodrigues, J. C., de Sousa, A. M. L., de Lima, R. T., da Rocha, E. J. P. and Ortega-Farias, S., (2016). 'Water requirement estimate for the reproductive period of mango orchards in Northeast of the state of Pará, Brazil'. *Rev. Bras. Frutic.*, Jaboticabal – SP, 38(3), e-311. http://dx.doi.org/10.1590/0100-29452016311.

Dhawan, B. D., (1988). 'Role of irrigation in raising intensity of cropping'. *J. Indian School Political Econ.*, 3(4), 632–71.

Diczbalis, Y., Kulkarni, V. and Toohill, B., (1993). *Mango Irrigation and Fruit Dry Matter Accumulation. Horticulture Branch Technical Annual Report 1992–1993.* Technical Bulletin No. 207, Northern Territory of Australia, Department of Primary Industry and Fisheries, 103–10p.

Drechsel, P., Heffer, P., Magen, H., Mikkelsen, R. and Wichelns, D. (Eds) (2015). *Managing Water and Fertilizer for Sustainable Agricultural Intensification.* First edition. Paris, France, International Fertilizer Industry Association (IFA), International Water Management Institute (IWMI), International Plant Nutrition Institute (IPNI), and International Potash Institute (IPI). Copyright 2015 IFA, IWMI, IPNI and IPI. All rights reserved ISBN 979–10-92366-02-0.

Dry, P. R., Loveys B. R., Düring, H. and Botting, D. G., (1996). 'Effects of partial rootzone drying on grapevine vigour, yield composition of fruit and use of water'. In Stockley, C. S., Sas, A. N., Johnstone, R. S. and Lee, T. H. (Eds) *Proceedings of 9th Australian Wine Industry Technical Conference*, Adelaide, Australia, pp. 128–31.

Dry, P. R. and Loveys, B. R., (1998). 'Factors influencing grapevine vigour and the potential for control with partial rootzone drying'. *Aust. J. Grape Wine Res.*, 4, 140–8.

Du, T., Kang, S., Zhang, J., Li, F. and Yan, B., (2008). 'Water use efficiency and fruit quality of table grape under alternate partial root-zone drip irrigation'. *Agric. Water Manage.*, 95, 659–68.

Durán-Zuazo, V. H., Martínez, R. A. and Aguilar, R. J., (2002). Effect of rootstocks on mineral nutrition of mango trees (Mangifera indica L. cv. 'Keitt'). In Villalobos, F. J. and Testi, L. (Ed.) *Proc. of the VII Congress of the European Society for Agronomy*, Cordoba, Spain, July 15–18, pp. 357–8.

Durán-Zuazo, V. H., Martínez, R. A. and Aguilar, R. J., (2003). 'Salt tolerance of mango rootstock (Mangifera indica L. cv. Osteen)'. *Span J. Agric. Res.*, 1(1), 67–78.

Durán-Zuazo, V. H., Martínez, R. A. and Aguilar, R. J., (2004a). Impact of salinity on the fruit yield of mango (Mangifera indica L. cv. Osteen). *Eur. J. Agron.*, 21(3), 323–34. Available in, www.sciencedirect.com [16 February 2004].

Durán-Zuazo, V. H., Martínez R. A., Aguilar R. J. and Franco T. D., (2004b). 'Impact of salinity on macro- and micronutrient uptake in mango (Mangifera indica L. cv. Osteen) with different rootstocks'. *Span. J. Agric. Res.*, 2, 121–33.

Durán-Zuazo, V. H., Aguilar-Ruiz, J. and Martínez-Raya, A. (2005). 'Fruit yield, plant growth and nutrient status in mango: Effect of rootstocks'. *Int. J. Fruit Sci.*, 5(4), 3–22.

Durán-Zuazo, V. H., Rodriguez, P. C. R. and Tarifa, D. F., (2011a). 'Impact of sustained-deficit irrigation on tree growth, mineral nutrition, fruit yield and quality of mango in Spain'. *Fruits* (Paris), 66, 257–68.

Durán-Zuazo, V. H., Rodríguez P. C. R. and Tarifa D. F., (2011b). 'Monitoring the pollution risk and water use in orchard terraces with mango and cherimoya trees by drainage lysimeters'. *Irrig. Drain. Syst.*, 25, 61–79.

Fereres, E. and Evans, R. G., (2006). 'Irrigation of fruit trees and vines'. *Irrig. Sci.*, 24, 55–7.

Feungchan, S., Yimsawat, T., Chindaprasert, S., Hongsbhanich, N. and Daito, H., (1989). 'The effect of the fertilizer application interval on the mango'. *Kaen Kaset Khon Kaen Agril. J.*, 17(2), 100–5.

Fixen, P., Brentrup, F., Bruulsema, T. W., Garcia, F., Norton, R. and Zingore, S. (2015). Nutrient/fertilizer use efficiency, measurement, current situation and trends. In Drechsel, P., Heffer, P., Magen, H., Mikkelsen, R. and Wichelns, D. (Eds) *Managing Water and Fertilizer for Sustainable Agricultural Intensification.* First edition. Paris, France, International Fertilizer Industry Association (IFA), International Water Management Institute (IWMI), International Plant Nutrition Institute (IPNI), and International Potash Institute (IPI). Copyright 2015 IFA, IWMI, IPNI and IPI. All rights reserved ISBN: 979-10-92366-02-0.

Galán Saúco, V. and García Samarín. J. (1979). *Pasado, presente y futuro del mango en Canarias.* Cuadernos INIA 9, 39p.

Galán Saúco, V., Acuña Barreto, J. F., Fernández Galván, D., Socorro Monzón, A. R. and Herrera Rodríguez, L. (1988). 'Tolerancia de distintos patrones de mango (Mangifera indica L.) a la salinidad'. *Acta Hort.*, 1, 83–9.

Galán Saúco, V. (2011). 'Worldwide mango production and market: Current situation and future prospects'. *Acta Hort.*, 992, 37–48. https://doi.org/10.17660/actahortic.2013.992.2

Gao, A., Chen, Y., Crane, J. H., Zhu, M., Huang, J. and Luo, H., (2011). 'Status and analysis on Mango production in China'. *Adv. Biomed. Eng.*, 12, 472–6.

Geerts, S. and Raes, D. (2009). 'Deficit irrigation as an on-farm strategy to maximize crop water productivity in dry areas'. *Agric. Water Manage.*, 96, 1275–84. http://dx.doi.org/10.1016/j.agwat.2009.04.009.

Goldhamer, D. A., Fereres, E., Mata, M., Girona, J. and Cohen, M., (1999). 'Sensitivity of continuous and discrete plant and soil water status monitoring in peach trees subjected to deficit irrigation'. *J. Am. Soc. Hort. Sci.*, 124(4), 437–44.

Goldhamer, D. A. and Fereres, E., (2001) 'Irrigation scheduling protocolsusing continuously recorded trunk diameter measurements'. *Irrig. Sci.*, 20, 115–25.

Gomes, U., Tozetto, L. J. and Pinto, A. C. de Q. (2002). *Censo Frutícola do Nordeste Brazileiro: Situação da cultura de Manga.* Boletín Técnico. Brasilia, Brazil, CODEVASF, p. 6.

Gonzalez, A., Lu, P. and Muller, W., (2004). 'Effect of pre-flowering irrigation on leaf photosynthesis, whole-tree water use and fruit yield of mango trees receiving two flowering treatments'. *Sci. Hort.*, 102(2), 189–211.

Goodwin, I. and Boland, A. M., (2002). Scheduling deficit irrigation of fruit trees for optimizing water use efficiency. In Anon. Ed.) *Deficit Irrigation Practices.* Rome, Food and Agriculture Organization of the United Nations, pp. 67–78.

Grajal-Martín, M. J. (2012). 'Selecciones de mango en Canarias'. *Acta Hort.*, 62, 235–6.

Guzman-Estrada, C. (2000). 'Relationship of macro and micronutrients content with some growth, production, dendrometric and climatic variables at the vegetative and reproductive phases of manila mango'. *Acta Hort.*, 509, 329–34.

Haruvy, N., Offer, R., Hadas, A. and Ravina, I., (1999). 'Wastewater irrigation-economic concerns regarding beneficiary and hazardous effects of nutrients'. *Water Resour. Manage.*, 13, 303–14.

Hernández Delgado, P.M., Aranguren, M., Reig, C.P, Fernández Galván, D., Mesejo, C., Martínez Fuentes, A., Galán Saúco, V. and Agusti, M. (2011). Phenological growth stages of mango (Mangifera indica L.) according to the BBCH scale. *Sci. Hort.*, 130, 536–40.

IFOAM. (2009). *The World of Organic Agriculture, Statistics and Emerging Trends 2009.* International Federation of Organic Agriculture Movements-Research Institute of Organic Agriculture (IFOAM- FiBL).

INCID. (1994). *Drip Irrigation in India.* New Delhi, Indian National Committee on Irrigation and Drainage.

Intrigliolo, D. S. and Castel, J. R., (2004). 'Continuous measurement of plant and soil water status for irrigation scheduling in plum'. *Irrig. Sci.*, 23, 93–102.

Ishfaq, M., (2002). *Water New Technology.* Lahore, Pakistan, Global Water Institute.

Ireta-Ojeda, A. and Guzmán-Estrada, C., (2002). *Guía técnica para la producción de mango en Sinaloa.* Folleto Técnico # 22. INIFAP-CIRNO-Campo Experimental Valle de Culiacán.

Jones, H. G. (1990). 'Plant water relations and implications for irrigation scheduling'. *Acta Hort.*, 278, 67–76.

Ladha, J. K., Pathak, H., Krupnick, T. J., Six, J. and van Kessel, C., (2005). 'Efficiency of fertilizer nitrogen in cereal production: Retrospects and prospects'. *Adv. Agron.*, 87, 85–156.

Lakshminarayana, S., Subhadra, N. V. and Subramanyam, H., (1970). 'Some aspects of developmental physiology of mango fruit'. *J. Hort. Sci.*, 45, 133–42.

Larson, K. D., Schaffer, B. and Davies, F. S., (1989). 'Effect of irrigation on leaf water potential, growth and yield of mango trees'. *Proc. Fla. State Hort. Soc.*, 102, 226–8.

Levin, A. G., Naor, A., Noy, M., Love, C., Gal, Y. and Peres, M., (2015a). 'Mango response to deficit irrigation at different phenological periods'. *Acta Hort.*, 1075, 103–13.

Levin, A. G., Naor, A., Noy, M., Love, C., Gal, Y., Peres, M. and Bernstein, N. (2015b). The response of field-grown Mango (cv Keitt) trees to regulated deficit irrigation at different phenological stages. In *XI International Mango Symposium*. Darwin, Northern Territory (Australia), 28 September – 2 October 2015.

Levine, A. and Asano, T. (2004). 'Recovering sustainable water from wastewater'. *J. Environ. Sci. Technol.*, 1, 201A–8A.

Lu, P. and Chacko, E. K., (1997). 'Xylem sap flow measurements in mango trees'. *Acta Hort.*, 455, 339–50.

Lu, P. and Chacko, E. K., (1998). 'Evaluation of Granier's sap flow meter in mango (Mangifera indica L.) trees'. *Agronomie*, 18, 461–71.

Lu, P. and Chacko, E. K. (2000). 'Effect of water stress on mango flowering in low latitude tropics of Northen Australia'. *Acta Hort.*, 509, 283–9.

Lu, P., Müller, W. J. and Chacko, E. K., (2000). 'Spatial variations in xylem sap flux density in the trunk of orchard-grown, mature mango trees under changing soil water conditions'. *Tree Physiol.*, 20, 683–92.

Lu, P. (2002). 'Measurement of whole-tree water use of some tropical and subtropical tree crops and its application in irrigation management'. *Acta Hort.*, 575, 781–9.

Lu, P., (2005). Mango water relations and irrigation scheduling. In Malik, A. U. (Ed.) *Mango and Date Palm, Culture and Export*. Faisalabad, University of Agriculture.

Lu P. (2013). 'Whole-tree water relations and irrigation scheduling for mango'. *Acta Hort. (ISHS)*, 992, 115–22.

Mai, Q., Jiang, J., Wu, N., Yu, X., Wu, Y., Jia, X. and Liu, Z., (2013). 'Fertilizer recommendations for off-season mango plantation based on yield response'. *Acta Hort. (ISHS)*, 992, 207–10. http://www.actahort.org/books/992/992_26.htm.

Marecos do Monte, M. H. F., Angelakis, A. N. and Asano, T., (1996). 'Necessity and basis for establishment of European guidelines for reclaimed wastewater in the Mediterranean region'. *Water Sci. Technol.*, 33, 303–16.

Marloth, R. H., (1947). 'The mango in South Africa'. *Farm. S. Afr.*, 22(25), 457–63.

Mattar, M. A., (2007). 'Irrigation system effect on growth and productivity in mango orchard'. *Misr. J. Agr. Eng.*, 24(1), 103–21.

Medina-Urrutia, V. M., Vázquez-García, M. and Virgen-Calleros, G., (2011). Organic mango production in Mexico: Status of orchard management. In Benkeblia, N. (Ed.) *Proc. Ist IS on Trop. Hort. Acta Hort. 894, ISHS 2011.*

Mirjat, M. S., Jiskani, M. M., Siyal, A. A. and Mirjat, M. U, (2011). 'Mango production and fruit quality under properly managed drip irrigation system'. *Pak. J. Agri., Agril. Engg., Vet. Sci.*, 27(1), 1–12.

Morales, D. V. and Rivas, J. S., (2004). 'Uso eficiente de la fertilizaron y su efecto sobre el rendimiento del mango, (Mangifera indica L.) en el municipio Mara del estado Zulia'. *Rev. Fac. Agron. (LUZ)*, Suppl. 1, 237–43.

Mosqueda, V. R., De los Santos, de la R. F., Becerra, L. E. N., Cabrera, M. H., Ortega, Z. D. A. and del Angel, P. A. L., (1996). *Manual para cultivar mango en la planicie costera del Golfo de México*. Folleto Num. 15, INIFAP, Centro de Investigación Regional Golfo Centro, Campo Experimental Cotaxtla, 130p.

Mostert, P. G. and Hoffman, J. E., (1997). Water requirements and irrigation of mature mango trees. In *V International Mango Symposium. Acta Horticulture 455*, pp. 331–7.

Naor, A., Naschitz, S., Peres, M. and Gal, Y., (2008). 'Responses of apple fruit size to tree water status and crop load'. *Tree Physiol.*, 28, 1255–61.

Naor, A., Schneider, D., Ben-Gal, A., Zipori, I., Dag, A., Birger, R. and et al. (2013). 'The effect of crop load and irrigation rate in the oil accumulation stage on oil yield and water relations of "Koroneiki" olives'. *Irrig. Sci.*, 31, 781–91.

Narayanamoorthy, A., (1996). *Evaluation of Drip Irrigation System in Maharashtra*. Mimeograph Series No. 42, Pune, Maharashtra, Agro-Economic Research Centre, Gokhale Institute of Politics and Economics.

Narayanamoorthy, A., (1997). 'Economic viability of drip irrigation: An empirical analysis from Maharashtra'. *Indian J. Agric. Econ.*, 52(4), 728–39.

Nasreen, S., Kamal, A. M., Siddiky, M. A., Rannu, R. P. and Islam, M. S., (2014). 'Effect of nitrogen, phosphorus, potassium and sulphur on the yield of mango'. *Bangladesh J. Agril. Res.*, 39(4), 631–40.

Negi, S. S., (2000). 'Mango production in India'. *Acta Hort.*, 509, 69–78. DOI:10.17660/ActaHortic.2000.509.4.

Nguyen, H., Hofman, P., Holmes, R., Bally, I., Stubbings, B. and Mcconchie, R., (2004). 'Effect of nitrogen on the skin colour and other quality attributes of ripe 'Kensington Pride' mango (Mangifera indica L.) fruit'. *J. Hort. Sci. Biotechnol.*, 79(2), 204–10. http://dx.doi.org/10.1080/14620316.2004.11511749.

Núñez-Eliséa, R. and Davenport, T. L., (1994). 'Flowering of mango trees in containers as influenced by seasonal temperature and waterstress'. *Sci.Hort.*, 58, 57–66.

Obreza, T. A. (2000). Fertilization of citrus in Florida – A general view. In Yamada, Y. and Abdalla, S. R. S. (Eds) *Simposio Sobre Fisiologia, Nutrição, Adubaçao E Manejo Para Produção Sustent ável De Citros [Symposium on Physiology, Nutrition and Fertilization Management for Sustainable Citrus]*. Piracicaba, Potafos [CD-ROM].

Oosthuyse, S. A. (1993). 'Research on mango aimed at increasing orchard productivity and export fruit quality'. *S. Afr. Mango Growers' Assoc. Yearb.*, 13, 40–4.

Oron, G., Armon, R., Mandelbaum, R., Manor, Y., Campos, C., Gillerman, L., Salgot, M., Gerba, C., Klein, I. and Enriquez, C., (2001). 'Secondary wastewater disposal for crop irrigation with minimal risks'. *Water Sci. Technol.*, 10, 139–46.

Ortuño, M. F., Garcia-Orellana, Y., Conejero, W., Ruiz-Sanchez, M. C., Alarcon, J. J. and Torrecillas, A., (2006). 'Stem and leaf water potentials, gas exchange, sap flow and trunk diameter fluctuations for detecting water stress in lemon trees'. *Trees*, 20, 1–8.

Oster, J. D. and Wichelns, D., (2003). 'Economic and agronomic strategies to achieve sustainable irrigation'. *Irrig. Sci.*, 22, 107–20.

Parmar, S. D. and Thorat, G. N. (2016). 'Constraints faced by farmers in drip irrigation system'. *Agric. Update*, 11(3), 229–33. DOI:10.15740/HAS/AU/11.3/229-233.

Pavel, E. W. and Villiers, A. J. D., (2004). 'Responses of mango trees to reduced irrigation regimes'. *Acta Hort.*, 646, 63–8.

Peralta-Antonio, N., Rebolledo-Martínez, A., Becerril-Román, A. E., Jaén-Contreras, D. and del Angel-Pérez, A. L., (2014). 'Response to organic fertilization in mango cultivars, Manila, Tommy Atkins and Ataulfo'. *J. Soil Sci. Plant Nutr.*, 14, 688–700.

Pinto, A. C. Q., Andrade, S. R. M., Amaro, A. A. and Gomes, U., (2004). 'Mango industry in Brazil'. *Acta Hort.*, 645, 37–50. https://doi.org/10.17660/ActaHortic.2004.645.1.

Pinto, A. C. de Q., Silva, D. J. and Pinto, P. A. da C. (2007). Mango. In Johnston, A. E. (Ed.) *Fertilizing for High Yield and Quality, Tropical Fruits of the Tropics*, IPI Bulletin 18. Horgen, Switzerland, International Potash Institute, pp. 123–41.

Pongsomboon, W., (1991). Effects of temperature and water stress on tree growth, flowering, fruit growth and retention of mango (Mangifera indica L.). Ph.D. Thesis, Bangkok, Thailand, Kasetsart University.

Prado, R. M., (2004). 'Nutrição e desordens fisiológicas na cultura da manga'. In Rozane, D. E., Darezzo, R. J., Aguiar, R. L., Aguilera, G. H. A., Zambolim, L. (Eds) *Manga, produção integrada, industrialização e comercialização*. Viçosa, Universidade Federal de Viçosa, pp. 199–232.

Prado, R. M., (2010). 'Phosphorus effects in the nutrition and growth of dedeveloping mango plants'. *J. Plant Nutr.*, 33, 2041–9.

Prado, R. M., Caione, G., Silva, D. J. (2012). Deficiency symptoms in mango due to macro and micro nutrients. In Valavi, S. G., Rajmohan, Govil, J. N., Peter, K. V. and Thottappilly, G. (Org.). *Mango*. Houston/EUA, Studium Press LLC.

Prakash, K., Vijayakumar, R. M., Balamohan, T. N. and Sundhar Singh, S. D., (2015). 'Effect of drip irrigation regimes and fertigation levels on yield and quality of mango cultivar "Alphonso" under ultra high density planting'. *Acta Hort.*, 1066, 147–50.

Raij, B. V., Cantarella, H., Quaggio, J. A. and Furlani, A. M. C. (1996). *Fertilizer and Lime Recommendations for the State Of Sao Paulõ*. Campinas, Brazil, Instituto Agronomico & Fundação IAC.

Rajak, D., Manjunatha, M. V., Rajkumar, G. R., Hebbara, M. and Minhas, P. S., (2006). 'Comparative effects of drip and furrow irrigation on the yield and water productivity of cotton (Gossypium hirsutum L.) in a saline and waterlogged vertisol'. *Agric. Water Manage.*, 83, 30–6.

Ram, S., Sirohi, S. C. and Rathore, V. S., (1983). 'Naturally occurring Cytokinins in mango (Mangifera indica L.) fruits'. *Aust. J. Plant Physiol.*, 10, 65–73.

Ramírez, F. and Davenport, T. L., (2010). 'Mango (Mangifera indica L.) flowering physiology (Review)'. *Sci. Hort.*, 126, 65–72.

Reboll, V., Cerezo, M., Roig, A., Flors, V., Lapeña, L. and García-Agustín, P., (2000). 'Influence of wastewater vs groundwater on young citrus trees'. *J. Sci. Food Agric.*, 80, 1441–6.

Roberts, T. L. (2008). 'Improving nutrient use efficiency'. *Turk. J. Agric For.*, 32, 177–82.

Roemer, M. G., (2011). Premature Fruit Drop in Mango (Mangifera indica L.) in Northern Vietnam. University Hohenheim (Ph.D. thesis) http://opus.ub.uni-hohenheim.de/volltexte/2011/657/pdf/Roemer2011.

Salazar-García, S., Álvarez-Bravo, A., Ibarra-Estrada, M. I., González-Valdivia, J. and Medina-Torres, R., (2016). 'Presence of parthenocarpic fruit in 'Ataulfo' mango and its relation with environmental temperature and fertilization treatments'. *Revista Mexicana de Ciencias Agrícolas*, Pub. Esp. Núm. 13, 2615–26.

Santos, M. R., Martinez, M. A., Donato, S. L. R. and Coelho, E. F., (2014). 'Fruit yield and root system distribution of "Tommy Atkins" mango under different irrigation regimes'. *Revista Brasileira de Engenharia Agrícola e Ambiental*, Campina Grande, 18, 362–9.

Santos, M. R., Neves, B. R., da Silva, D. L. and Rodrigues Donato, S. L., (2015). 'Yield, water use efficiency and physiological characteristic of "Tommy Atkins" mango under partial rootzone drying irrigation system'. *J. Water Res. Prot.*, 7, 1029–37. Published Online September 2015 in SciRes. http://www.scirp.org/journal/jwarp. http://dx.doi.org/10.4236/jwarp.2015.713084.

Santos, M. R., Rodrigues Donato, S. L., Coelho, E. F., Cotrim Jr., P. R. F. and de Castro, I. N., (2016). 'Irrigation deficit strategies on physiological and productive parameters of Tommy Atkins mango'. *Rev. Caatinga, Mossoró*, 29(1), 173–82.

Schaffer, B., Whiley, A. W. and Crane, J. H., (1994). Mango. In Schaffer, B. and Andersen, P. C. (Eds) *CRC Handbook of Environmental Physiology of Fruit Crops, Subtropical and Tropical Crops*. Boca Raton, CRC Press, pp. 165–97.

Schaffer, B., Urban, L., Lu, P. and Whiley, A. W. (2009). Ecophysiology. In Litz, R. E. (Ed.) *The Mango – Botany, Production and Uses*, 2nd ed. Wallingford, UK, CABI, pp. 170–209.

Schulze, K., Spreera, W., Keil, A., Ongprasert, S. and Müllera, J., (2013). 'Mango (Mangifera indica L. cv. Nam Dokmai) production in Northern Thailand–Costs and returns under extreme weather conditions and different irrigation treatments'. *Agric. Water Manage.*, Amsterdam, 126, 46–55.

Searle, C., Whiley, A. W., Simpson, D. R. and Saranah, J. B. (1995). A preliminary phenophysiological model for 'Kensington' mango in subtropical Australia, pp. 127–35. In *Mango 2000 – Marketing Seminar and Production Workshop*.

Shackel, K. A., Ahmadi, H. and Biasi, W., (1997). 'Plant water status as an index of irrigation need in deciduous fruit trees'. *Hortechnology*, 7, 23–9.

Sharma, R. C., Mahajan, B. V. C., Dhillon, B. S. and Azad, A. S., (2000). 'Studies on the fertilizer requirements of mango cv. Dashehari in sub-montaneous region of Punjab'. *Indian J. Agric. Res.*, 34(3), 209–10.

Silva, D. J., Quaggio, J. A. da C., Pinto, P. A. de Q., Pinto, A. C. and Magalhães, A. F. de J., (2002). Nutrição e Adubação, pp. 192–221. In Genu, P. J. de C. and Pinto, A. C. de Q. (Eds) *A Cultura da Mangueira*. Brasília, Embrapa Informação Tecnológica.

Silva, D. J., do Carmo Mouco, M. A., Tuão Gava, C. A., Giongo, V. and Pinto, J. M., (2013). 'Organic compost on mango trees (Mangifera indica L.) cultivated in the semi-arid of northeast region of Brazil'. *Rev. Bras. Frutic.*, Jaboticabal – SP, 35(3), 875–82.

Simsek, M., Tonkaz, T., Kaçira, M., Çomlekçoglu, N., Dogan, Z., (2005). 'The effects of different irrigation regimes on cucumber (Cucubis sativa L.) yield and yield characteristics under open field conditions'. *Agric. Water Manage.*, 73, 173–91.

Singandhupe, R. B., James, B. K., Srivastava, R. C. and Kumar, A., (2006). Prospect of adoption of micro-irrigation in eastern region of India. In *7th International Micro Irrigation Congress*, Kuala Lumpur, Malaysia, 10–15 September.

Singh, L. B., (1960). *The Mango, Botany, Cultivation and Utilization*. London, Leonard Hill.

Singh, Z., Malik, A. U. and Davenport, T. L., (2005). 'Fruit drop in mango'. *Hortic. Rev.*, 31, 111–53.

Spreer, W., Nagle, M., Neidhart, S., Carle, R., Ongprasert, S. and Müller, J., (2007). 'Effect of regulated deficit irrigation and partial rootzone drying on the quality of mango fruits (Mangifera indica L., cv. 'Chok Anan')'. *Agric. Water Manage.*, 88, 173–80.

Spreer, W., Hegele, M., Müller, J. and Ongprasert, S., (2009a). Effect of deficit irrigation on fruit growth and yield of mango (Mangifera indica L.) in Northern Thailand. In *Proceedings of the 8th International Mango Symposium, Acta Horticulture* 820, pp. 357–64.

Spreer, W., Ongprasert, S., Hegele, M., Wünsche, J. N. and Müller, J., (2009b). 'Yield and fruit development in mango (*Mangifera indica* L. cv. Chok Anan) under different irrigation regimes'. *Agric. Water Manag.*, 96, 574–84.

Suarez, D. L., (1989). 'Impact of agricultural practices on groundwater salinity'. *Agric. Ecosys. Environ.*, 26, 215–27.

Subramanyam, H., Krishnamurthy, S. and Parpia, H. A. B., (1975). 'Physiology and biochemistry of mango fruit'. *Adv. Food Res.*, 21, 223–305.

Tahir, F. M., Ibrahim, M. and Kamran, H., (2003). 'Effect of drought stress on vegetative and reproductive growth behavior of mango (*Mangifera indica* L.)'. *Asian J. Plant Sci.*, 2(1), 116–18.

Teixeira, A. H. D., Bastiaanssen, W. G. M., Moura, M. S. B., Soares, J. M., Ahmad, M. D. and Bos, M. G., (2008). 'Energy and water balance measurements for water productivity analysis in irrigated mango trees, Northeast Brazil'. *Agric. Forest Meteorol.*, 148(10), 1524–37.

Tharanathan, R. N., Yashoda, H. M. and Prabha, T. N., (2006). 'Mango (*Mangifera indica* L.), "The King of Fruits" – An Overview'. *Food Rev. Int.*, 22(2), 95–123. DOI:10.1080/87559120600574493.

Tiwary, K. N., Kannan, N. and Mal, P. K., (1998). Enhancing productivity of horticultural crops through drip irrigation. In *Proceeding of the National Seminar on Micro-Irrigation Research in India, status and Perspectives for the 21st Century*, Bhubaneswar, July 27–28.

Toze, S., (2006). 'Reuse of effluent water-benefits and risks'. *Agric. Water Manage.*, 80, 147–59.

Vázquez-Valdivia, V., Pérez-Barraza, M. H. and Salazar-García, S., (2006). Control del crecimiento de árboles de mango, pp. 51–86. In Vázquez-Valdivia, V. and Pérez-Barraza, M. H. (Eds) *El cultivo del mango, principios y tecnología de producción*. Libro Técnico # 1. México, INIFAP-CIRPAC-Campo Experimental Santiago Ixcuintla.

Wanitprapha, K., Yokoyama, K. M., Nakamoto, S. T. and Chia, C. L., (1991). *Mango Economic Fact Sheet No. 16*. University of Hawaii Press, p. 4.

Whiley, A. W., Rasmussen, T. S., Wolstenholme, B. N., Saranah, J. B. and Cull, B. W., (1991). 'Interpretation of growth responses of some mango cultivars grown under controlled temperatures'. *Acta Hort.*, 291, 22–31.

Whiley, A. W. and Schaffer, B., (1997). Stress physiology. In Litz, R. E. (Ed.) *The Mango, Botany, Production and Uses*. London, CAB International Press, pp. 147–73.

World Bank. (2008). *Agriculture for Development- World Development Report 2008*. World Bank, Washington, DC.

Xiuchong, Z., Guojian, L., Jianwu, Y., Shaoying, A. and Lixian, Y., (2001). 'Balanced fertilization on mango in Southern China'. *Better Crops Int.*, 15, 16–20.

Yan, S. S. and Chen, C., (1980). 'The effect of soil moisture on the photosynthesis, stomatal aperture and growth of mango (Mangifere indica L.)'. *J. Sci. Eng.* (Nat. Chung Hsing Univ., Taiwan), 17, 275–84.

Young, T. and Koo, R., (1974). 'Increasing yield of "Parvin" and "Kent" mangoes on Lakewood sandy by increased nitrogen and potassium fertilizer'. *Proc. Florida State Hort. Soc.*, 87, 380–4.

Zhang, Y., Kendy, E., Qiang, Y., Changming, L., Yanjun, S. and Hongyong, S., (2004). 'Effect of soil water deficit on evapotranspiration, crop yield, and water use efficiency in the North China Plain'. *Agric. Water Manage.*, 64, 107–22.

Part 3

Post-harvest management

Chapter 11

Monitoring fruit quality and quantity in mangoes

Kerry Walsh and Zhenglin Wang, Central Queensland University, Australia

1 Introduction

Management of the mango crop requires information on many parameters and processes. In this chapter, focus is directed to methods to assess fruit load (quantity) and quality (in its various forms), both infield and postharvest. Further, a decision support system (DSS) that utilises such information to guide management practices is described. This progression follows the dictum, 'Data is not Information, Information is not Knowledge, Knowledge is not Wisdom'. For example, fruit dry matter (DM) can be measured (data), interpreted in terms of fruit maturation (information) and used in guiding harvest timing (knowledge). Imperatives in this progression are knowledge of the biology of the issue and a robust measurement technique. Thus the first requirement is to define what attributes are useful to measure, the second to measure these attributes accurately and cost effectively and the third is to utilise such information within a management system.

Of course, reliable data are the foundation to a knowledge pyramid. The measurement of fruit quantity and quality can be done manually, but analogue measurement techniques that require human involvement are slow and at times unreliable. The dramatic march in the availability of measurement technologies that output a digitised signal and the expansion of telecommunication networks over the last decades offers opportunities in all fields, and mango production is no exception.

In this chapter, specifications for taste and maturity are explored, and then relevant measurement technologies are described.

http://dx.doi.org/10.19103/AS.2017.0026.14

1.1 A definition of quality

'Beauty is in the eye of the beholder', and so is quality. That is, quality has different meanings from different perspectives. Mango fruit quality, for example, can be seen from the perspective of a grower, a packer, a shipper, a retailer or a consumer, with overlapping sets of attributes of value (Walsh, 2015; Fig. 1). Specifications on quality will also be specific to the fruit maturity/ripening stage.

To the consumer of a 'dessert' mango, fruit quality is judged in terms of visual appearance in general (overall colour, presence of skin defects), size, stage of ripeness (judged on colour and firmness) and taste. Of course, taste involves destructive sampling, and so taste cannot be judged in store, except where taste samplers of fruit from the same consignment are provided. Consumers thus evaluate taste after purchase and will rarely return product to store on a bad taste experience, in contrast to a foreign object experience. However, marketing studies routinely report that a bad eating experience will result in loss of repeat purchases by the consumer for approximately six weeks (Diehl et al., 2013). Of course, there are also different consumers of mango, with different requirements. For example, some consumers prefer a redder external colour (e.g. the Chinese New Year market), while others prefer higher volatile levels (a stronger 'mango' aroma). There are also consumers purchasing mango for cooking purposes, for juice and for other purposes. Obviously a juice producer gives little attention to external appearance, except as it reflects internal problems.

The marketer/retailer is responsible for presenting the consumer with aesthetically pleasing dessert fruit that are free of significant defects and that will be ripe within a few days. Thus size, external colour, defects and shelf life are the primary quality attributes at this stage.

The packhouse is a point in the supply chain where sorting at an individual fruit level can occur, with technology available for grading fruit in-line. Preventative treatments can also be applied to fruit on the packline, for example, bactericidal and fungicidal treatments to prevent disorders further along the value chain. Of course, it is also possible to introduce 'downstream' disorders. For example, fruit harvested during rain events have turgid skin cells that can easily be damaged by brushes, etc. in the packline. Likewise, if the wash treatments are set at too high a temperature, skin scalding may result. This damage may not be noticeable in the packhouse but will become apparent some days later, after the fruit has been moved to market. The packer lives in compromise between the specifications of fruit taken from the field, the requirements of the marketer/retailer and consumer.

The grower is challenged to produce fruit that meets the concerns of the packer, the marketer/retailer and the consumer. Out-of-specification fruit can be rejected at the packhouse or indeed later in the value chain, but any rejection is lost profit. The primary tools available to the grower are choice of variety and agronomics, in particular disease control and timing of harvest.

1.2 Defining maturity and ripeness

It is important to have a common language for the sake of effective communication. Unfortunately, the terms 'maturity' and 'ripeness' are often used too loosely. There are four terms of importance – *physiological maturity*, *harvest maturity*, *commercial maturity* and *ripeness*.

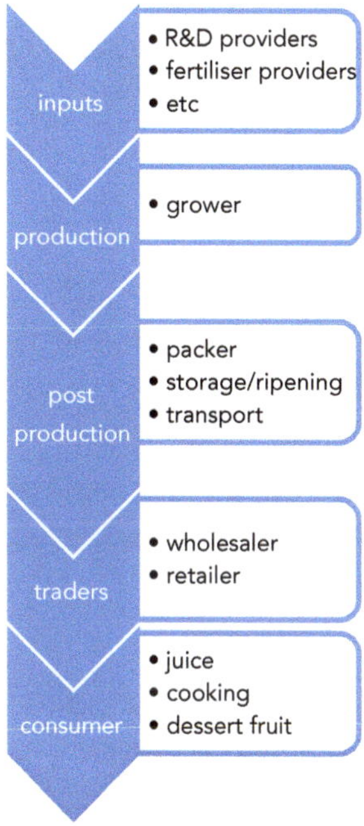

Figure 1 Steps in the mango value chain.

Mangoes are climacteric fruit, that is, after *physiological maturity* the fruit are able to undergo a *ripening* process triggered by ethylene, in which respiration rate rises, skin and flesh colours change, starch converts to soluble sugars, organic acid levels decrease, tissue softens and volatiles increase (Fig. 2). As an analogy, reproductive maturity in humans is reached (in early teen years) earlier than it is generally recommended to reproduce. Fruit physiological maturity occurs earlier than normal harvest or *harvest maturity*. Fruits reach *commercial maturity* when the fruit is at a stage acceptable to the consumer – for mango, when the fruit are partly ripened.

The decision on timing of harvest at some time after physiological maturity and before commercial maturity is a commercial decision involving compromise. As long as the fruit remains on the tree, there is opportunity to import photosynthate and increase reserves, thus increasing eating quality after ripening. However, the longer the fruit remains on the tree past physiological maturity, the shorter is its ripening time, and thus, the shorter the postharvest life. For some markets, it may be desired to have fruit ripen on tree, to have maximum carbohydrate accumulation (and taste) but no shelf life. The other extreme is a maximal shelf life and thus transport capability, at the expense of carbohydrate content and eating quality.

Ripening is a continuous process of senescence. The extent of ripening to achieve '*commercial maturity*' will vary by market, so generally fruit are sold in a partly ripened state, allowing the consumer some control over the extent of ripening at consumption. Generally, dessert fruit are consumed at a stage defined by flesh firmness, when starch has not quite fully converted to soluble sugar (Fig. 2).

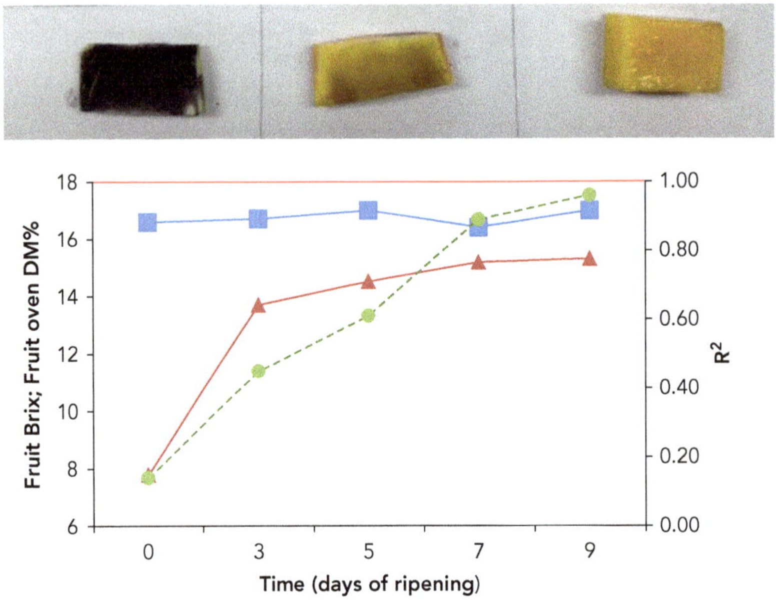

Figure 2 Top panel: IKI stained mango fruit at different stages of ripeness (left, hard green fruit, middle, fruit at eating stage; right, over ripened fruit). Bottom: change in DM (blue square), Brix (red triangle) and and R^2 of Brix-DM correlation (green circles) with time ripened (left axis). *Source* CQUniversity.

1.3 A taste specification

In general, fruit taste is indexed by soluble sugar content (SSC), organic acid content (TA) and flesh texture. In some fruit, with a low organic acid content and consistent texture (e.g. melon), SSC is the defining attribute. In other fruit, with low sugar content (e.g. lime), the organic acid content is of primary concern. For some fruit, for example, oranges, the soluble sugar:acid ratio is of importance. However, it is not only the ratio of sugars to acid that is relevant, but also the total level (creating a 'mouth feel'). Indeed, Florida oranges were graded to a matrix of SSC:TA ratios defined by the SSC level (Obenland et al., 2009). To address this complexity, Jordan et al. (2001) introduced the BrimA measure, calculated as Brix – k × total acid. This relationship was subsequently adapted slightly in the California standard for citrus, as [SSC – (4*TA)*16.5] (https://www.cacitrusmutual.com/marketing/the-california-standard/ Accessed date: 25 February 2017).

The taste of a mango is defined by its soluble sugar and organic acid content, flesh texture and the profile of volatile compounds. The levels of these attributes are determined by variety, growing and ripening conditions, fruit maturity and ripeness. In mango, SSC rises as starch and organic acid content and firmness decline during ripening. In optimally ripened fruit, fruit are of optimal firmness and low acid content, such that SSC becomes a dominant determinant of eating quality. Thus SSC of fully ripe fruit can be used as an index for determining eating quality (taste) (Fig. 3a). However, as noted earlier, mangoes are rarely eaten at a completely fully ripe (senescent) stage; rather, they are eaten slightly before full ripening, when they have slightly higher firmness and acidity. Thus the BrimA index also has relevance to mango (Fig. 4).

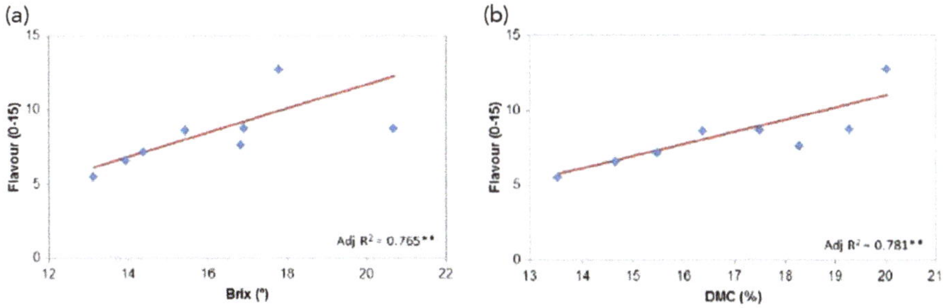

Figure 3 Plot of eating quality score as a function of (a) SSC (Brix) and (b) Dry matter content (DMC). *Source* Campbell (2015).

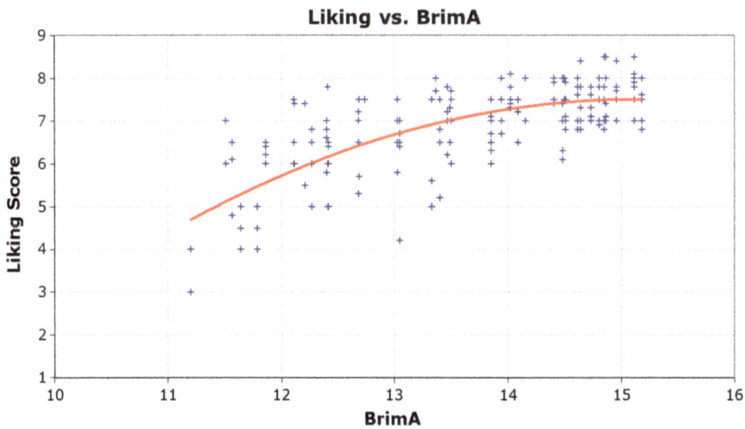

Figure 4 Plot of mango eating quality (scale of 1, extreme dislike, to 9, extreme liking) score vs Brim A. *Source* Mark Loeffen, Dyletics P/L.

However, at the time of consumption, there is a higher correlation of eating quality score to DM content than to SSC (Henroid et al., 2014; Fig. 3b). DM thus provides an index of the potential eating quality rather than the stage of ripeness, being a measure of both tissue-soluble and -insoluble solid content. DM is insensitive to a change between starch, organic acid and sugar pools, as occurs during ripening. Soluble solids will include sugars and organic acids. Insoluble solids will include starch, and also all cell structural material (e.g. cell membranes and wall). But as cell structural material remains fairly constant, change in mango tissue DM represents change in soluble sugars and starch levels, with a small contribution from organic acid levels. The DM content of mango tissue is thus well correlated with its SSC when fully ripe (Fig. 6). Therefore DM is a useful index for the determination of eating quality of a mango fruit.

Indeed, DM at any stage post physiological maturity should be proportional to SSC of fully ripened fruit. Specifically, DM at fruit harvest is well correlated with ripened fruit SSC and eating quality (Whiley et al., 2006; Fig. 5). A caveat applies if the total concentration of fruit starch and sugars changes during ripening - for example, fruit

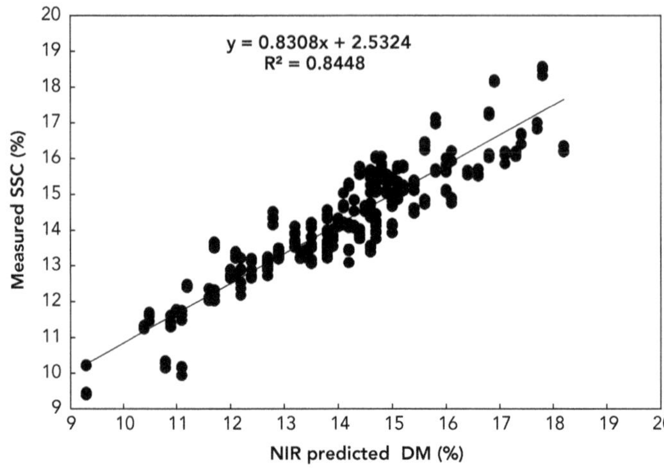

Figure 5 Plot of mango fruit SSC at fully ripe vs DM at fully ripe stage. SSC at fully ripe stage is approximately 1 unit lower than %DM. *Source* CQUniversity.

can lose water if ripened under low humidity conditions, leading to an increase in DM. In practice, industry is focused on retaining weight of fruit, which is achieved by maintaining high humidity. Conversely, fruit respiration will result in loss of carbohydrate and thus a tendency for lower DM. For example, if a respiration rate of 7 mL CO_2/kg/h (equivalent to 0.29 mmol/kg/h or 13 mg/kg/h) (Mitcham and McDonald, 1993) were maintained throughout a week of ripening (an overestimate as respiration rate decreases post climacteric), the fruit would lose 0.3 g/kg/day or 2.1 g/kg over seven days. This equates to a DM loss of 0.21%. Thus fruit loss due to respiration during fruit ripening may be measurable but is not a large amount, and likely offset by a small transpirational water loss.

On the basis of taste panel tests (Henroid et al., 2014), the Australian Mango Industry Association has recommended minimum DM levels by variety [15% DM for cultivars B74, Kensington Pride and Honey Gold, and 14% DM for R2E2; AMIA, 2016; http://www.industry.mangoes.net.au/my-mango/ Accessed date: 26 February 17] for ripened fruit to achieve an acceptable eating quality. The US Mango Importers Association has also commissioned taste tests through UC Davis, California, with similar recommendations (C. Crisosto, pers. comm., oral presentation at IHC2014, Brisbane, Australia). These specifications are for a DM value associated with an acceptable eating experience, not a good eating experience, which is associated with higher DM values. Applying this cut-off as a specification for the average of a population is, of course, generous, as 50% of any population is below average.

1.4 A harvest maturity specification

The decision to pick (harvest maturity) (Subedi et al., 2007; Walsh, 2015) is based on an assessment of the following:

- timing – calendar days from flowering
- timing – heat sums from flowering
- fruit shape (broadening of shoulders)
- skin colour (reddening or yellowing)
- internal flesh colour (white to yellow)
- DM

However, a number of varieties are well coloured and full in shape well before harvest maturity. Assessment of attributes such as flesh colour requires destructive sampling.

Several indices can be used together to improve decision-making. For example, the DM specification required for eating quality (e.g. 15% DM) can be used as a specification for harvest. This use of DM is separate from its use as an index for eating quality, although obviously the harvest maturity DM specification must have as a minimum the value of the eating quality DM specification. However, the level of DM in a fruit depends on photosynthetic conditions as well as the maturity of the fruit. High photosynthetic rates will favour higher DM levels, as will manipulations such as girdling. Thus the DM level associated with the desired harvest maturity as established by flesh colour, fruit shape or other attributes should be established for a given variety/ growing condition, in addition to the requirement to achieve the DM level required for eating quality.

1.5 A ripening specification

Ledger et al. (2012) have provided a useful treatise on mango ripening. In summary, fruit are typically harvested and transported in a hard green condition. Best practice for ripening involves use of a constant temperature, around 20°C. This may occur in transit (in container) for long-distance markets or in ripening centres close to markets. Ripening may be initiated using ethylene, although this may not be necessary in later season fruit or in fruit left longer on tree. Fruit are ripened until a desired firmness level is reached, typically gauged by hand feel and external colour, and then fruit are transported to stores in a condition with some days of shelf life remaining.

2 Monitoring harvest maturity: making the decision to pick

Relying on human expertise to judge harvest maturity is fraught, especially given the typically frantic pace of activity at harvest time. Further, assessing these attributes requires a level of skill that is often absent in a harvest crew. Quantifiable attributes that can be rapidly assessed are required, allowing sampling of a statistically significant number of fruit within an orchard.

2.1 Monitoring flowering

In a 'best practice' orchard, the date and extent of flowering are recorded to inform a heat unit-based model of time to harvest maturity and an estimate of the potential size of the crop. A record of early flowering trees could be used to enable variable rate spraying, with sprays turned off for non-flowering trees, and also to inform selective harvest, with earlier flowering trees having fruit that will mature earlier.

Different systems exist for estimating the date of flowering, from the swollen bud stage or later stages, for example, the 'Christmas tree stage', with two-thirds of flowers open on the inflorescence (Fig. 6a). Typically, assessing the extent of flowering involves a manual assessment of the percentage of terminals that have become reproductive. Flowering assessment can take many forms in commercial practice, for example, a drive through several rows of a block, with a mental estimation of the extent of flowering.

Attempts to introduce machine vision for the estimation of flowering have been made (Wang et al., 2016; Fig. 6a,b). While the colours of inflorescences are reasonably distinct from foliage, enabling segmentation of the image (Fig. 6a), the fact that the number of pixels associated with inflorescences increases due to increase in both the number of flowering terminals and the size of the inflorescences is a difficulty. This is not an issue if there is only a single flowering event; however, if flowering events overlap, identification of the individual events becomes difficult (Fig. 6c). Further progress with multiple flowering events will require the identification of flowering terminals, rather than simple counting of flower-associated pixel number.

Of course there is a large and variable flower and young fruit drop in mango, with only 8% or less of inflorescences setting a marketable fruit. Poor pollination and rain during flowering, with consequent disease issues, are associated with poor fruit set. Thus, in addition to flowering estimation, fruit set estimation is required (see Section 3.1).

2.2 Monitoring time and field temperature

Calendar time from flowering is a base guide to maturation time, based on the experience in previous years. However, development is temperature dependent, so an estimate of maturation time based on heat units will be more accurate. A heat unit is typically based on the summation of the difference between the average of daily maximum and minimum temperatures and a base temperature, for example, 12°C, from the date of flowering. The target value is set by variety and local experience, for example, 1600 heat units is required for maturation of the variety Kensington Pride in NT, Australia, counting from the date that the flowering bud is visible, or to 1300 units from the date when inflorescences have about two-thirds of open flowers (https://nt.gov.au/industry/agriculture/food-crops-plants-and-quarantine/fruit-crops/mango) (Fig. 7a).

Online heat sum calculators are available in some regions that utilise temperature data from a local government weather station, although better results will be obtained from use of an on-farm temperature record. Temperature is typically monitored using either a thermistor or a thermocouple, housed in a mini-Stevenson screen (i.e. a ventilated, shaded structure) (Fig. 7b,c). A thermistor is a temperature-sensitive resistor, while a thermocouple generates a voltage proportional to the temperature. Thermistors are typically used in horticultural applications, for which drift with temperature variation is greater than for thermocouples, but simpler electronic circuitry supports lower cost. A large number of commercially available on-farm weather stations or simple outdoor temperature loggers exist, including those able to broadcast wirelessly (radio, Bluetooth, Wi-Fi or phone networks) to a central station, linking via an Internet gateway to a cloud-based data centre for ease of data viewing (e.g. www.monnit.com). Utilising such data in decision support is considered in a later section.

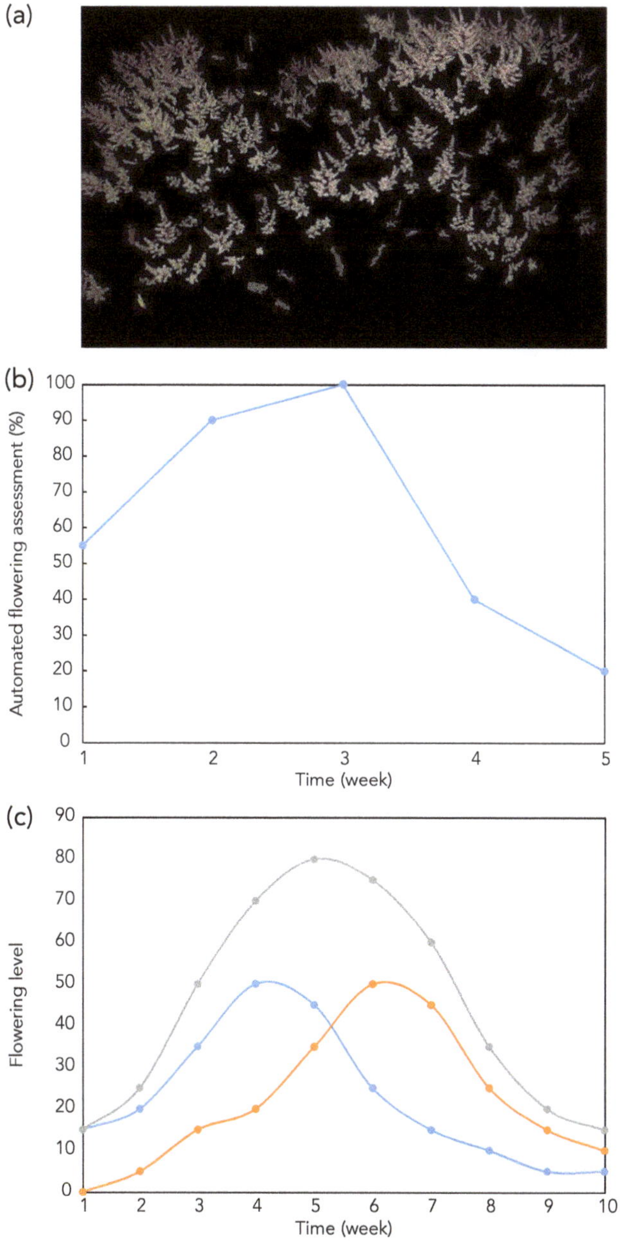

Figure 6 (a) Mango flowering at 'Christmas tree' stage – image segmented for inflorescences, (b) time course of inflorescence-associated pixels in canopy images, (c) artificial data: flower pixel count of two sequential flower events and associated total of inflorescence-associated pixels in canopy image. *Source* CQUniversity.

(a) (b)

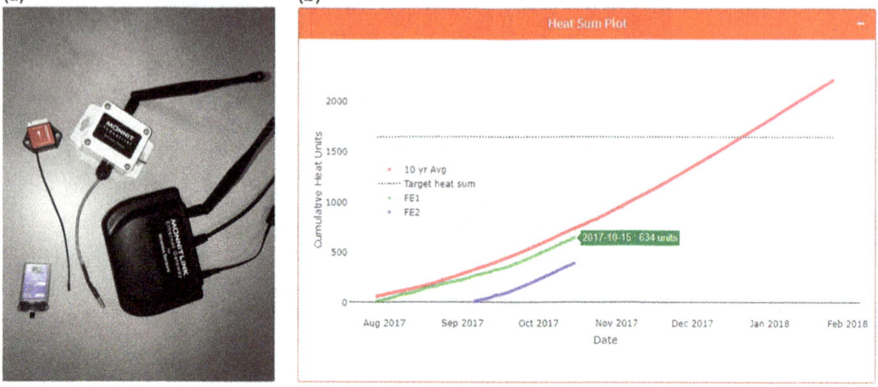

Figure 7 Example of (a) wireless temperature loggers and a receiving base station and (b) heat sums for a season (calculated for two flowering events, FE1 and FE2, for the current season temperature record, and for FE1 using the 10 year average temperature data). *Source* CQUniversity.

2.3 Monitoring size and external appearance

A change in fruit shape occurs as fruit mature for many, but not all, varieties, typically seen as a 'filling' of the fruit shoulders (i.e. an increase in fruit thickness rather than width or length). This visual clue can be used in gauging crop harvest maturity and in selective harvest of more mature fruit.

The size distribution of fruit on tree, pre-harvest, can be useful to support tray insert size purchase and marketing decisions. For some fruits (e.g. cherries, citrus, apples, kiwifruit; Li et al., 2015; Green et al., 1990), the rate of change of fruit size on tree is used as an index of harvest maturity. Fruit dimensions (length, width and thickness) can be related to fruit weight (e.g. for 'Chok Anan' mango, R^2 = 0.96) (Spreer and Müller, 2011; Anderson et al., 2017; see also Fig. 9a). The rate of increase of mango fruit weight is decreasing but not plateaued by the time of harvest maturity (Fig. 8b).

Mango fruit size can be estimated manually using callipers. This measurement requires a level of operator attention, particularly when manual transcription of results is required. Because of this requirement, infield fruit size measurements are rarely made in commercial mango production, and when undertaken, the labour requirement results in a small number of measurements, decreasing the statistical value of the information. Some digital versions of these measurement devices allow for transfer of data to a connected tablet, for example, Green et al. (1990); Gus P/L (http://www.gusstoday.com/datalogger.html). Other systems that allow continuous estimation of fruit size are available (e.g. Morandi et al., 2007; Phytek. http://www.phytech.com/); however, these systems require installation on a single piece of fruit, with equipment costs limiting multiple installations.

Vision-based systems are used for the estimation of fruit size in grading systems, under controlled lighting and background, typically estimating lineal dimensions to 1.0 mm, for example, Sadegaonkar and Wagh (2013) and Dhameliya et al. (2016); also see review by Moreda et al. (2009). A field machine vision-based fruit sizing (RMSE 3

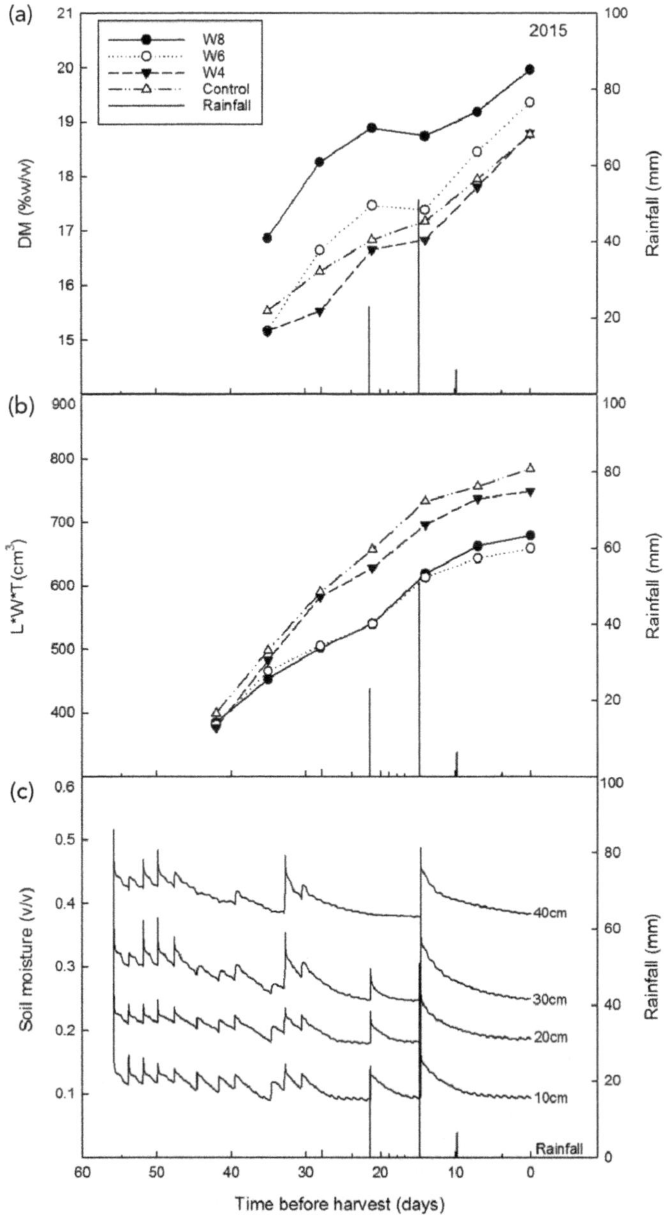

Figure 8 Water denial treatments involved trees denied water 0, 4, 6 and 8 weeks before harvest. (a) Weekly NIR DM of 30 fruits per treatment measured from five weeks before harvest, (b) L*W*T (cm^3) collected on the same fruit from six weeks before harvest, (c) soil moisture (v/v) at four soil depths from a probe in the 4-week denial treatment installed eight weeks before harvest and rainfall (mm) measured daily. Soil moisture values for different depths have been offset for ease of viewing. *Source* Anderson et al., 2017.

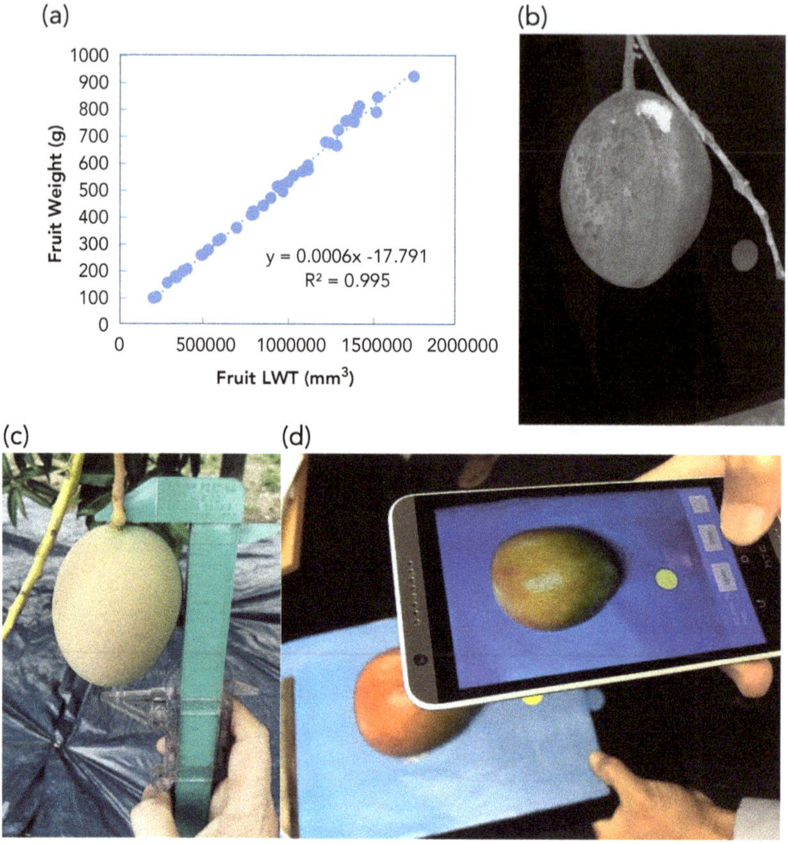

Figure 9 (a) Allometric relationship of fruit lineal dimensions with fruit weight and (b) CIE Lab b channel image of infield fruit against a blue background, (c) measurement of fruit dimensions infield using callipers, (d) measurement of fruit dimensions using phone app. *Source* CQUniversity.

mm) was described by Koirala et al. (2017), based on manual collection of images of fruit against an A4-sized blue background with scale. Images were collected using a mobile phone, with calculation of fruit weight based on the allometric relation with fruit minor and major axis dimensions, with information relayed with geolocation information to a cloud-based information system (Fig. 9).

Alternatively, infield fruit sizing information can be extracted from high-resolution images of whole canopies, from a tractor-mounted imaging system which might also do duty in flowering and fruit load assessment. In this approach only a representative number of fruit needs to be assessed, not all fruit, and thus analysis can be restricted to non-occluded and well-positioned fruit. Fruit size calculation requires depth information, from stereo vision or time-of-flight (TOF) laser imaging systems (e.g. Moreda et al., 2009; Wang et al., 2016). The TOF imaging systems typically use a wavelength of around 830 nm, so infield use is restricted to night time, with LED flood lighting.

2.4 Monitoring internal colour

Colour cards, for the estimation of flesh colour, are a useful aid to standardisation of human assessment – although these cards do fade and appear different with variation in lighting (Fig. 10a). Consistent measurement of colour can be achieved with commercial colorimeters (Fig. 10b); however, the cost of such units and the need for destructive sampling of fruit place a barrier to their use in farm production. There is a market need

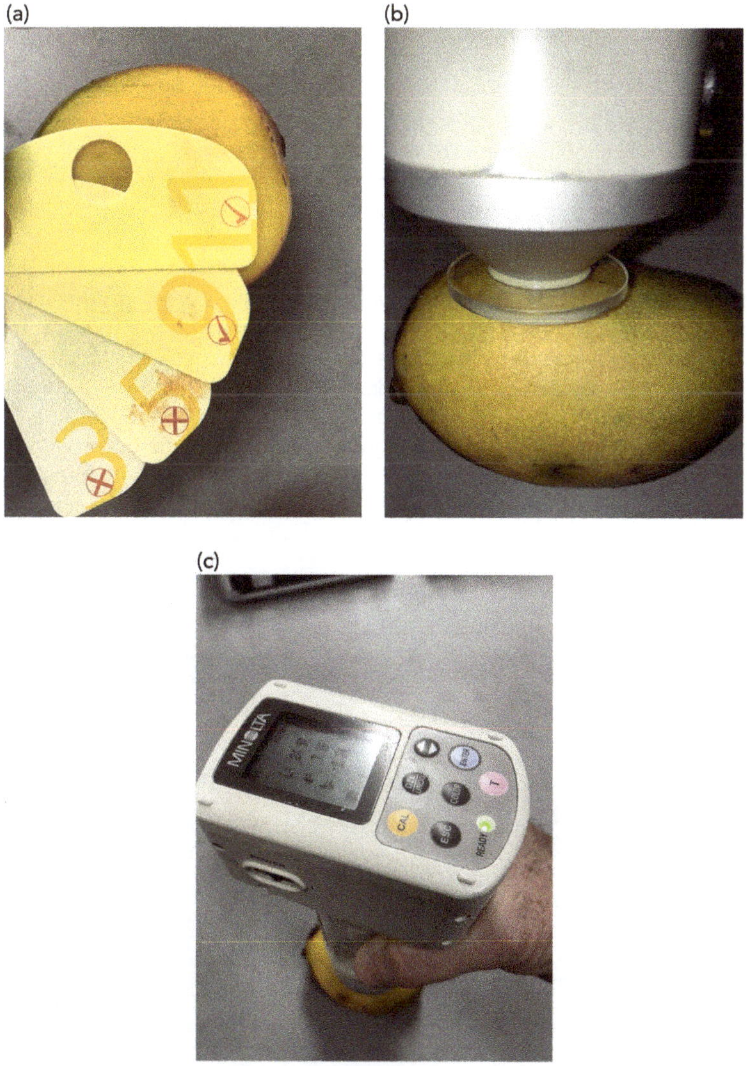

Figure 10 (a) Colour cards used in the assessment of colour of flesh of cut cheeks of mango fruit; (b) side view of colorimeter (Minolta CR400) with internal lighting for consistent estimation of colour; (c) top view of colorimeter. *Source* CQUniversity.

for a lower cost colorimeter which assesses a reasonable surface area (e.g. 20 cm^2) for the fruit industry.

Visible short-wave spectroscopy of intact fruit has also been used in the non-invasive estimation of flesh colour, using an optical geometry that ensures passage of light through flesh (Subedi et al., 2007). Hand-held spectrometers allow measurement of fruit infield. However, the calibration between spectra and flesh colour requires adjustment between growing districts, likely due to change in skin properties with growing district (Subedi et al., 2007). The technique involves passage of light through skin and flesh, and changes in skin properties can impact the assessment of flesh colour.

2.5 Monitoring DM

Fruit DM is typically assessed by weight loss of a tissue sample following 24–48 h of oven drying, typically at 65°C. Domestic dehydrator units can be used, with use of a three decimal place (i.e. 1-mg resolution) balance recommended. Microwave drying can be used, but care is required to achieve a consistent level of drying without loss of volatiles or charring.

A non-invasive technique allows repeated monitoring of individual fruit on tree, removing the sampling error associated with destructive sampling. The use of short-wave near-infrared spectroscopy for non-invasive assessment of DM in intact mango has been advocated for several decades. For example, Guthrie and Walsh (1997) reported a multiple linear regression model to predict 'Kensington Pride' mango fruit DM with $R^2 = 0.96$, and RMSEP = 0.79. Saranwong et al. (2004) reported the use of a PLS model to predict DM in a validation set of hard green 'Mahajanka' mangoes, with $R^2 = 0.92$, SEP = 0.41, bias = 0.07, while Subedi et al. (2007) reported that a multi-cultivar mango model predicted independent populations of fruit with acceptable accuracy ($R^2 = 0.79$, RMSEP = 0.97).

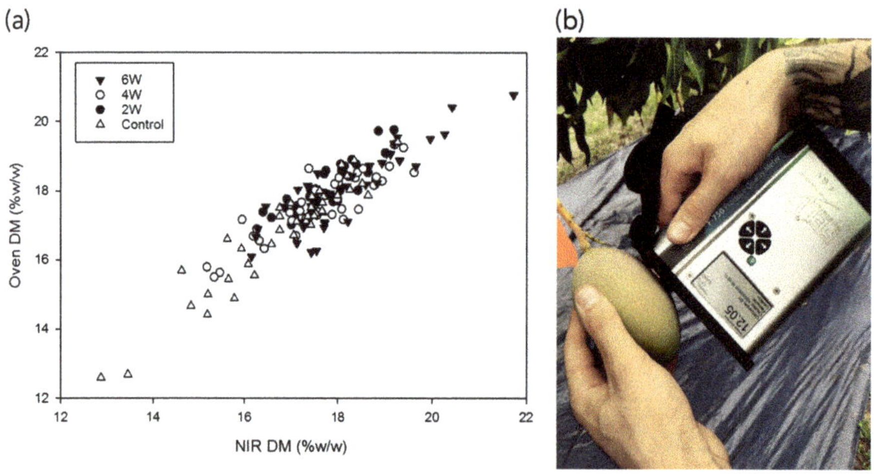

Figure 11 (a) Oven and NIR measurements of DM of fruit for trees denied water for zero, two, four and six weeks before harvest. (b) Handheld short-wave near-infrared spectrometer in use in a mango orchard. *Source* CQUniversity.

This technique relies on the absorption of light, with absorbance bands around 740, 840 and 960 nm, associated with overtones of O-H bond stretching, and at 910 nm associated with an overtone of C-H bond stretching. In essence, the technique is assessing DM in terms of water, sugar and starch content. In practice, calibrations (partial least squares regression) between spectra of intact fruit (typically using the wavelength range 720–975 nm) and oven DM values routinely achieve a $R^2 = 0.85$ and a root mean square of error of prediction (RMSEP) = 0.7%DM (Fig. 11).

2.6 Manipulating DM

Given the ability to monitor fruit DM, the grower also requires tools to manipulate fruit to meet market DM specifications. Agronomic treatments that alter carbohydrate or water allocation to the fruit are candidate treatments. For example, Yeshitela et al. (2004) reported manual and chemical fruit thinning to result in increased SSC in ripened fruit (15.1–16.3° cf. control 13.7 % SSC). Simmons et al. (1998) reported that girdling of mango branches eight weeks following flowering to achieve a single fruit per girdled branch impacted both fruit weight (441, 363, 533 and 697 g/fruit for control, 30, 60 and 120 leaves/fruit treatments) and fruit DM (14.4, 16.4, 16.3, 14.6%, respectively). Zhao et al. (2013) reported fruit bagging to result in increased fruit carbohydrate content (e.g. 114.6 and 99.5 mg total sugars/g FW, in bagged and non-bagged fruit, respectively). Fruit carbohydrate concentration can be increased using deficit irrigation (DI), partial root-zone drying (PRD) and regulated deficit irrigation (RDI), for example, as reported for kiwi fruit (Miller et al., 1998), apple (Mpelasoka et al., 2001) and musk melon (Long et al., 2006). In general, fruit size is decreased if the water deficit is applied early in fruit development results, while storage reserve levels are increased if the deficit is imposed in late development. For mango fruit, Simmons et al. (1998) reported denial of irrigation water for 56 days following panicle emergence, 56 days before harvest and 14 days before harvest to result in harvest DM of 17.4, 14.4 and 12.7%, respectively. Nagle et al. (2010) also reported fruit DM to be increased when water was denied from flowering (24.6% compared to control at 21.4%),

Table 1 DM of fruit at harvest for a number of treatments over several years and locations (Anderson et al., 2017). Water denial periods refer to the number of weeks before harvest that irrigation was discontinued

	Harvest DM%				
Treatment	Trial 1	Trial 2	Trial 3	Trial 4	Trial 5
Control	15.5	17.7	18.8	17.1	16.5
Thinning	17.0				
Trunk girdling					
Fruit bagging					
Water denial – 2 weeks					17.6
Water denial – 4 weeks		19.4	18.8		17.2
Water denial – 6 weeks			19.4		18.3
Water denial – 8 weeks		19.4	20.2		
Water denial – 10 weeks				18.5	

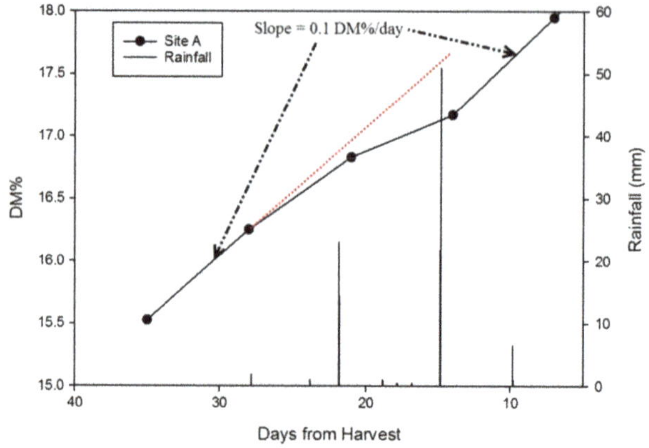

Figure 12 Time course of DM accumulation and rainfall. Modified from Anderson et al. (2017).

extending the result of Spreer et al. (2007). These treatments have also been explored by Anderson et al. (2017), who noted increased DM with thinning and water denial, but not trunk girdling (possibly too mild a girdling treatment) or fruit bagging (Table 1).

Of course, irrigation denial can only be effective in the absence of rain. In the example provided (from Anderson et al., 2017) in Fig. 12, rainfall events of approximately 20 and 50 mm at 23 and 16 days before harvest caused a temporary decrease in the rate of DM accumulation.

In summary, there is motive to increase DM to improve ripened soluble solids level and eating experience, requiring instrumentation to monitor fruit DM infield and agronomic practices to manipulate DM in the growing crop.

3 Monitoring quantity

3.1 On-tree monitoring

A rule of thumb best practice for estimation of orchard fruit load is to (manually) count all fruit on every 20th tree. Thus, in a block of 1000 trees, 50 trees would be assessed. Of course, the number of trees (n) required for a reliable estimate depends on the variation [standard deviation (SD)] on fruit load per tree – if there was no variability, a single tree could be assessed (Eqn. 1).

$$n = \left(t.\frac{SD}{e} \right)^2 \quad (1)$$

where *t* is the *t* statistic for a given probability (e.g. 1.96 for a 95% probability, n > 30) and *e* is the desired level of accuracy.

For example, the orchard described by Stein et al. (2016) had tree fruit count per tree ranging from 1 to 442, with an average of 102 and SD of 81 fruit/tree. Thus, if an error on the estimate of mean fruit count per tree of 10 fruit per tree is accepted, the number of trees to be assessed is $(1.96 \times 81/10)^2 = 252$. Note that this number is independent of population size, but is dependent on a reliable estimate of SD. Thus categorising trees in classes based on fruit load can reduce sampling effort. For example, if in the example, trees could be categorised to three types, each with SD on fruit load of 20, then n = (1.96 × 20/10)2 = 16 trees should be sampled in each category or 48 trees in total. Measures of trunk diameter, canopy volume (by LiDAR, light detection and ranging) and canopy health (by spectral indices such as NDVI, as suggested by Robson et al., 2016) are attributes of potential value for stratification of fields to regions that are more homogenous in maturity/DM, thus requiring less sampling.

Manual counting is tedious and requires continuous concentration. The counter must adopt a pattern – for example, counting from a distinguishable branch and moving in one direction around the tree. Counting fruit load on more than every 20th tree becomes an unrealistic task, especially for larger trees. Fruit load estimation using machine vision is yet to see commercial adoption, but capacity is rapidly advancing (Fig. 13). For example, Payne and Walsh (2014) reviewed the use of various technologies (colour, thermal and colour imaging, LiDAR) and Payne et al. (2014) reported on the use of night colour imagery to improve contrast between fruit and background. Quereshi et al. (2016) employed this imagery with machine vision features of K-nearest neighbour pixel classification and contour segmentation and a method based on super-pixel over-segmentation and classification using support vector machines. More recently, Stein et al. (2016) presented a method to localise every piece of fruit in a mango orchard using tracking of fruit between multiple images ('multi-view') of the canopy to locate fruit occluded by foliage or other fruit in a single frame image of the tree, coupled with use of a faster regional convolutional

Figure 13 Machine vision estimation of fruit load infield is a developing technology. (a) Farm utility mounted machine vision rig used in flowering and fruit and (b) night imaged mango canopy. *Source* CQUniversity.

neural network (R-CNN) detector for fruit detection and a LiDAR-generated mask for each canopy, with all fruit assigned to a tree. The 'multi-view' is a form of stereovision, allowing each fruit to be assigned a 3D location.

3.2 In-packhouse monitoring

In almost all supply chains, mango fruit will pass through a packhouse after harvest. The packhouse allows for functions such as cleaning and application of protective treatments, as well as for categorisation and quantification of fruit (Fig. 14). Fruit are typically singulated onto a cup conveyor and are sorted on the basis of weight using a load cell (situated under the conveyor, weighing every cup and fruit), with human sorting of visual defects (e.g. skin

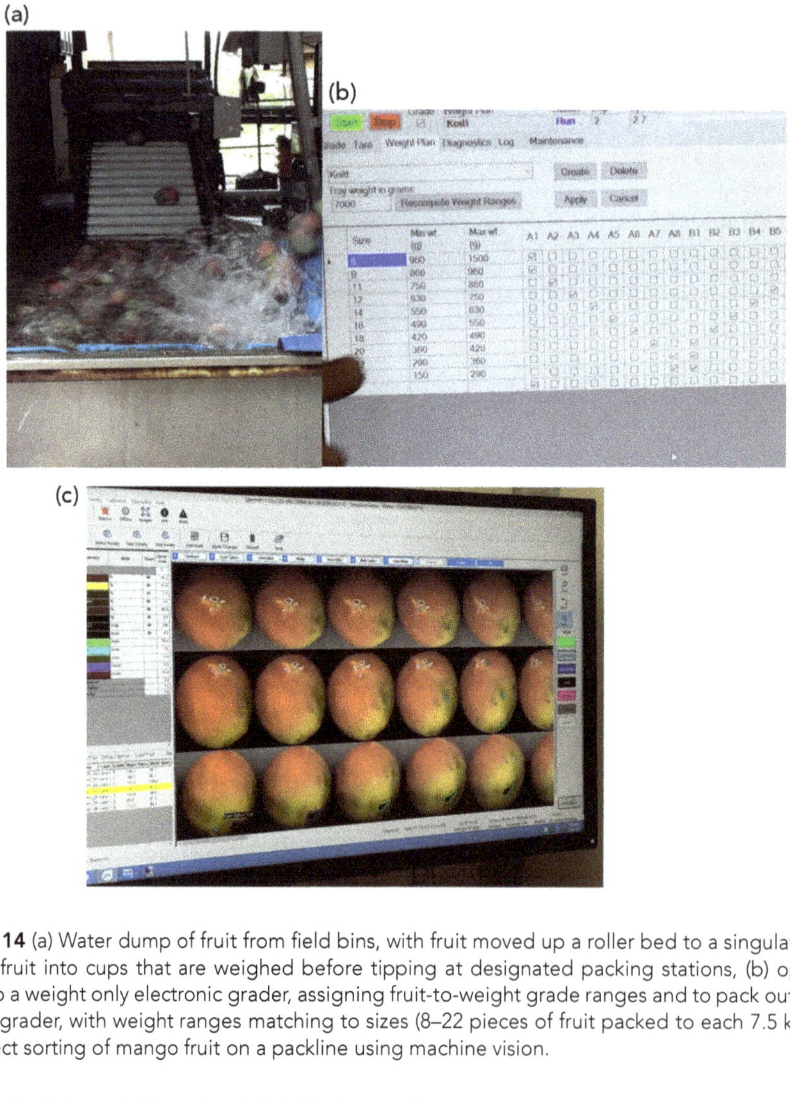

Figure 14 (a) Water dump of fruit from field bins, with fruit moved up a roller bed to a singulator that places fruit into cups that are weighed before tipping at designated packing stations, (b) operator input to a weight only electronic grader, assigning fruit-to-weight grade ranges and to pack out points on the grader, with weight ranges matching to sizes (8–22 pieces of fruit packed to each 7.5 kg tray), (c) defect sorting of mango fruit on a packline using machine vision.

rub, sun bleach, lenticel damage). Equipment is also available for machine vision-based sorting of fruit on colour and external defects.

4 Monitoring ripeness

Following grading, mango fruit consignments are typically ripened to a determined level before distribution. Measurements useful to monitoring and control of the ripening process include (i) temperature, (ii) skin colour, (iii) firmness and (iv) ethylene level. For apple in storage, commercially available systems such as Harvest Watch (www.harvestwatch.net/) allow detection of the onset of storage disorders through measurement of chlorophyll fluorescence; however, this technology has not been applied to mango given that the fruit is not stored for long periods.

4.1 Monitoring temperature

Fruit and store temperatures are typically logged using a thermistor, but infrared thermometers (which measure infrared wavelengths around 5 μm emitted from an object) have found favour, being non-contact and available in imaging as well as point modes. However, the technique is a surface measurement. If fruit core temperature is required, a thermocouple must be inserted into the fruit. An interesting development is that of logging and cloud-based cold chain decision support systems (e.g. XSense, www.bt9-tech.com).

4.2 Monitoring colour

Skin pigmentation can be visually assessed, with the use of colour charts to help maintain consistency of assessment (Fig. 15a). An instrumentation-based assessment of skin chlorophyll content is possible, for example, the handheld DA meter (Fig. 15b) uses alternating LED illumination at 680 and 720 nm to calculate an index related to fruit chlorophyll content. This index decreases during fruit ripening and could be used as a ripening stage index (Fig. 15c), although use of the instrument at a constant temperature is recommended, given shift in LED peak output wavelength with temperature (Hayes et al., 2017).

4.3 Monitoring firmness

Flesh firmness is currently routinely assessed using hand feel, with some attempt to standardise assessment by use of reference latex balls of different firmness (Fig. 16a). A penetrometer is recommended for more accurate, although destructive, measurement. However, handheld penetrometers are subject to operator variation, for example, in speed of insertion. To reduce this variation, use of a motorised stand is recommended (Fig. 16b).

In commercial practice, the destructive nature of a penetrometer limits sampling effort. Non-destructive alternatives exist, although not adopted by industry to date. The extent of deformation under a set load can be assessed, mimicking the mechanism of hand feel (e.g. the Agrosta14 (http://www.agro-technologies.com/ang/produits/crisp_fruits.htm). Fruit acoustic properties also change as flesh firmness (and Young's modulus of the material) changes. This can be assessed by either a change in the velocity of a pressure (acoustic) wave travelling through the fruit or by the resonant frequency of the fruit (Fig. 16c) (e.g. Subedi and Walsh, 2008). Both systems rely on a sharp but gentle tap of the fruit, producing a pressure wave within the fruit. As this pressure wave reaches the fruit

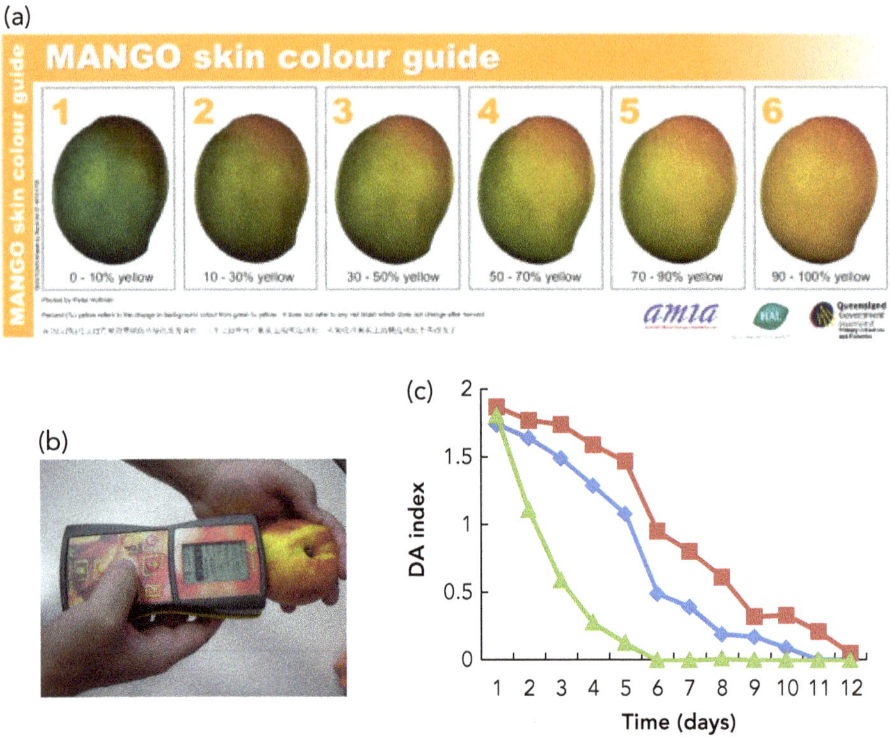

Figure 15 (a) Colour chart of fruit at different stages of ripeness (*Source* Australian Mango Industry Association), (b) DA meter and (c) change in DA values of three ripening fruit from a common lot (*Source* CQUniversity).

surface, it causes vibration in the surrounding air, detectable by an audio microphone. There will be a dominant (resonant) frequency. To estimate velocity of the pressure wave, microphones are placed at two distances from the impact point.

4.4 Monitoring ethylene and CO_2

Mango fruit benefit from a controlled ripening process, with the estimation of ethylene and CO_2 desirable. The concentration of these gases can be monitored using a colorimetric gas detection tube (Diels-Alder reaction, Dewar et al., 1986; e.g. the Kitagawa ethylene detector tube, Fig. 17a). Typically, a gas aspirating pump is used to draw air through the tube. Similar devices exist for CO_2. Alternatively, an electrochemical cell can be used in the estimation of ethylene, and infrared absorption for the estimation of CO_2, for example, the Felix Instruments F960, Fig. 17b.

5 Decision support systems

A family farm might have fewer than ten blocks, with each block of fewer than 1000 trees, involving several varieties to give a spread of harvest windows and marketing opportunities.

(a)

Stage	Description
Hard	No give under strong thumb pressure
Rubbery	Slight give under strong thumb pressure
Sprung	fruit deforms by 2-3 mm under moderate thumb pressure
Firm soft	fruit deforms by 2-3 mm under slight thumb pressure
Soft	whole fruit deforms with slight hand pressure

(b) (c)

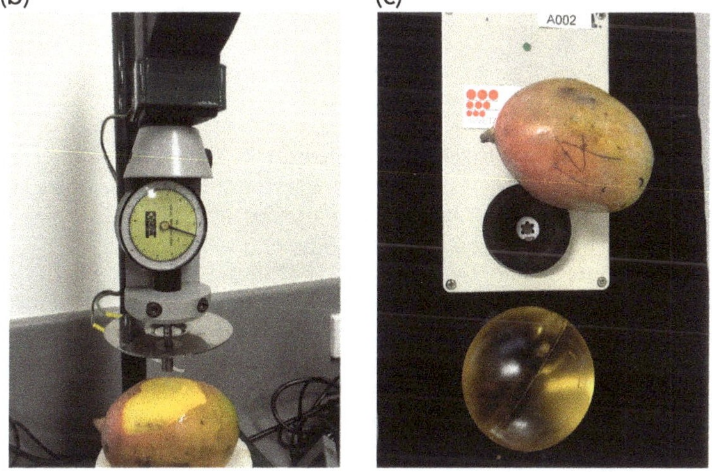

Figure 16 Measurement of fruit firmness: (a) hand feel description, (b) penetrometer on motorised stand and (c) acoustic frequency device.

This operation is small enough for the farm manager to keep a mental track of individual block maturity and other issues. In comparison, a corporate farming group may be dealing with a single cultivar under Plant Variety Rights, with a number of large operations dispersed across different geographic locations to provide an extended harvest window. Each operation can have >70 blocks of 1000 trees each, with harvest at a given location occurring over a tight window. In this circumstance, there is a clear benefit for the use of decision support aids.

As described earlier, tools to measure a range of parameters relevant to mango crop quality and quantity exist. However, for a farm manager to utilise this range of information in a timely fashion, the information must be 'digested' and presented in a visual format that enables quick comprehension. A good DSS will present information in a simplified and easily interpretable way, assisting management decisions. An example DSS for mango production is the 'Mango maturity' Web app (www.fruitmaps.info; for a demonstration version, enter 'demo' in all security checks) (Walsh, 2017), described in the following text.

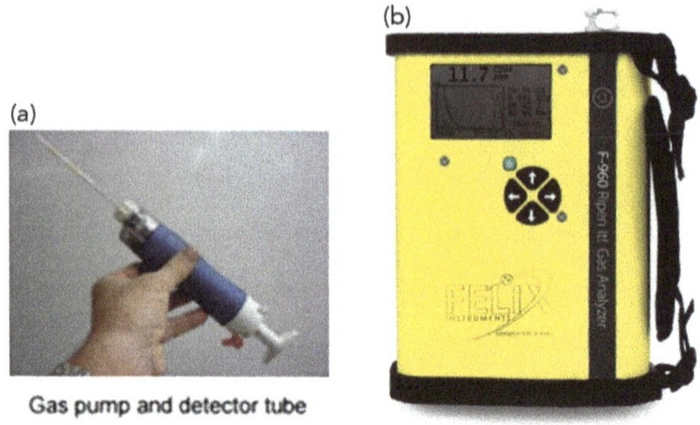

Figure 17 Ethylene and CO_2 gas concentration measurement: (a) colorimetric detection (Kitsgawa detection tube) and (b) electrochemical – IR (Felix F-960). (*Source* http://www.kitagawa-america.com/admin/efiles/PR11346.pdf; https://felixinstruments.com/).

The first requirement of a farm management app is a tool for the user to enter management boundaries, that is, tree blocks, with display to a satellite or aerial map. More sophisticated management may require several layers of boundaries, for example, blocks defined by harvest units, planting age, irrigation zones or soil type.

Ideally all data inputs should be automated. For example, to simplify the estimation of heat units, a temperature sensor installed on farm can log data to the Web app. Alternatively, temperature data could be sourced from the Web, either from a nearby publically available weather station or by a payment service offering interpolated data for the farm site. Given input of time of flowering events, a graphic that conveys maturation information easily to the grower is a graph of accumulated heat units to date for the current year together with a prediction of the date of fruit maturation based on a projection of the current year's temperature, relative to the historical temperature average and the mango cultivar (Fig. 7b).

Fruit count data (estimated after stone hardening stage, when fruit drop decreases) associated with each flowering event and geolocation records can be uploaded to the Web app for calculation of fruit load per block given information on number of trees per block. Fruit number associated with a given flowering event can be summed across orchard blocks to provide expected farm fruit numbers by harvest date. This system relies on manual estimates at present, but there is potential to have information input from machine vision estimates in the future.

The Web app utilises fruit DM readings from the F-750 Produce Quality Meter, with its geolocation records. A minimum input of a survey of fruit across a block, followed by one follow-up measurement to allow an estimate of a rate of DM increase, is suggested. The 'Farm DM' tab displays a map showing location of individual records, with colour change if the measurement exceeds the user-selected target value (Fig. 18). The block colour changes when the sampled fruit values fit the '% of fruit above target' criterion.

The 'Block DM detail' tab displays data of a single block, with current block mean DM, rate of increase of DM across time and the estimated date to achieve the user-defined criteria (for example 90% of fruit above 15% DM) given that rate of increase (Fig. 19). Another tab displays a list of all blocks of the farm, ordered by DM level and recommended date of harvest (i.e. date at which user set criterion on DM level is met), from the user set rate

Figure 18 DSS Web app for mango maturity, with display of fruit DM by location, colour coded for individual fruit above (blue) and below (red) a specification (15% DM), and block performance relative to a specification (90% of fruit above 15% DM) (fail in red, just passed, yellow).

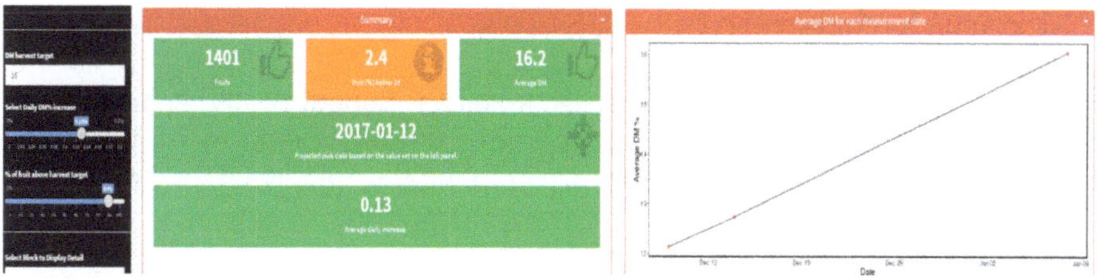

Figure 19 DSS Web app for mango maturity, with display of individual block data, including current mean DM, rate of weekly increase in DM and estimated date to achieve the criterion set in the left panel. User feedback on number of fruit sampled is provided, based on SD of data (thumbs up/down symbol).

of increase. Thus this example of a DSS offers support around the decision to harvest (i.e. estimation of harvest maturity) based on heat sums, DM and crop load information.

6 Future trends and conclusion

Bill Gates of Microsoft is attributed with saying 'We always overestimate the change that will occur in the next two years and underestimate the change that will occur in the next ten'. Over the next decade, technology and social trends from the broader society will sweep into the mango industry, in the form of satellite or drone-based inventories of the crop, autonomous vehicles and autonomous harvesting, variable rate spraying and fertiliser driven by better understanding of spatial variability or yet unforeseen applications.

Likely nascent applications include:

- Improvement of heat sum models from input of a daily average of maximum and minimum temperature to more complete temperature profile data and sunshine hours, as for apple (Li et al., 2015).
- Machine vision record of early flowering trees to enable variable rate spraying, with sprays turned off for non-flowering trees.
- Machine vision record of early flowering trees to inform selective harvest, with earlier flowering trees having fruit that will mature earlier.
- Machine vision guided selective fruit harvest.
- Matching of predicted harvest maturities to markets.

7 Where to look for further information

The International Society for Horticultural Science's 'Mango Symposium' is held every second year, with presentations published in the *Acta Horticulturae* series (see http://www.ishs.org/mango). This Symposium and the associated working group is an excellent source of expertise. Scientific publications relevant to the material of the current chapter is found in a wide range of journals, but key publications are *Computers and Electronics in Agriculture*, *Precision Agriculture*, *Postharvest Biology* and Technology and *Scienia Horticulturae*.

8 Acknowledgements

We acknowledge input of team members A. Koirala, P. Subedi and N. Anderson and of M. Matzner of Acacia Hills Farms and support of Horticulture Industry Australia project ST15005. We thank D. Swain for assistance in initiating the RStudio DSS software.

9 References

Anderson, A., Subedi, P. P. and Walsh, K. B. (2017). Manipulation of mango fruit dry matter content to improve eating quality. *Scientia Horticulturae* 226: 316–21.

Campbell, T. (2015). *Mango Quality Standards*. MG14504 final report. Horticulture Innovation, Australia.

Dewar, M. J., Olivella, S. and Stewart, J. J. (1986). Mechanism of the Diels-Alder reaction: Reactions of butadiene with ethylene and cyanoethylenes. *Journal of the American Chemical Society* 108: 5771–9.

Dhameliya, S., Kakadiya, J. and Savant R. (2016). Volume estimation of mango. *International Journal of Computer Applications* 143(12): 11–16. ISSN: 0975–8887. http://www.sciencedirect.com/science/article/pii/S0021863498902852.

Diehl, D. C., Nicole, L., Sloan, N. L., Bruhn, C. M, Simonne, A. H., Brecht, J. K. and Mitcham, E. J. (2013). Exploring produce industry attitudes: Relationships between postharvest handling, fruit flavor, and consumer purchasing. *HortTechnology* 23: 642–50.

Green, A. E., McAneney, K. J. and Astill, M. S. (1990) An instrument for measuring kiwifruit size. *New Zealand Journal of Crop and Horticultural Science* 18: 115–20, doi:10.1080/01140671.1990.10428080.

Guthrie, J. and Walsh, K. B. (1997) Non-invasive assessment of pineapple and mango fruit quality using near infra-red spectroscopy. *Australian Journal of Experimental Agriculture* 37: 253–63.

Hayes, C. J., Walsh, K. B. and Greensill, C. V. (2017) Light-emitting diodes as light sources for spectroscopy: Senistivity to temperature. *Journal of Near Infrared Spectroscopy*. http://journals.sagepub.com/doi/pdf/10.1177/0967033517736164#articleCitationDownloadContainer.

Henriod R., Sole D., Wright C. and Campbell T. (2015) Mango Quality Standards by Terry Campbell. *Horticulture Industry Australia Project MG14504 Final Report*. HIA, Sydney, Australia.

Jordan, R. B., Seelye, R. J. and McGlone V. A. (2001) A sensory based alternative to Brix/acid ratio. *Food Technology* 55(6): 36–44.

Koirala, A., Wang, Z., McCarthy, C. and Walsh, K. B. (2017). *An In-field Sizing Tool for Mango Fruit*. International mango Symposium, Nanning, China.

Ledger, S. Barker, L., Hofman, P., Campbell, J., Jones, V., Holmes, R. and Campbell, T. (2012). *Mango Ripening Manual*. Queensland Dept. of Agriculture, Fisheries and Forestry, Brisbane, Australia. https://static1.squarespace.com/static/53b0ef57e4b04ed3debabc4f/t/582a62939de4bb86345 9c7ef/1479172774823/Mango+Ripening+Manual.pdf. Accessed date: 13 January 2017.

Li, M., Chen, M., Zhang, Y., Fu, C., Xing, B., Li, W. and Yang, X. (2015) Apple fruit diameter and length estimation by using the thermal and sunshine hours approach and its application to the digital orchard management information system. *PLoS ONE* 10(4): e0120124. http://doi.org/10.1371/journal.pone.0120124

Long, R. L., Walsh, K. B., Midmore, D. J. and Rogers, G. (2006). Irrigation scheduling to increase muskmelon fruit biomass and soluble solids concentration. *HortScience* 41 (2): 367–9.

Miller, S., Smith, G., Boldingh, H. and Johansson, A. (1998) Effects of water stress on fruit quality attributes of kiwifruit, *Annals of Botany* 81(1): 73–81.

Mitcham, E. J. and McDonald, R. E. (1993) Respiration rate, internal atmosphere, and ethanol and acetaldehyde accumulation in heat-treated mango fruit. *Postharvest Biology and Technology* 3: 77–86.

Moreda, G. P., Ortiz-Cañavate, J., García-Ramos, F. J. and Ruiz-Altisent, M. (2009). Non-destructive technologies for fruit and vegetable size determination – a review. *Journal of Food Engineering* 92(2): 119–36. ISSN: 0260–8774.

Morandi, B., Manfrini, L., Zibordi, M., Massimo Noferini, M., Fiori, G. and Grappadelli, L. C. (2007). A Low-cost device for accurate and continuous measurements of fruit diameter. *HortScience* 42: 1380–2. http://hortsci.ashspublications.org/content/42/6/1380.full

Mpelasoka, B., Behboudian, M. and Green, S. (2001). Water use, yield and fruit quality of lysimeter-grown apple trees: Responses to deficit irrigation and to crop load. *Irrigation Science* 20(3): 107–13.

Nagle, M., Mahayothee, B., Rungpichayapichet, P., Janjai, S. and Müller, J. (2010) Effect of irrigation on near-infrared (NIR) based prediction of mango maturity, *Scientia Horticulturae* 125(4): 771–4.

Obenland, D., Collin, S., Mackey, B., Sievert, J., Fjeld, K. and Arpaia, M. L. (2009) Determinants of flavor acceptability during the maturation of navel oranges *Postharvest Biology and Technology* 52: 156–63 DOI:10.1016/j.postharvbio.2009.01.005

Payne, A. and Walsh, K. (2014) Chapter 16: Machine vision in estimation of crop yield. In: *Plant Image Analysis: Fundamentals and Applications* (S. Dutta-Gupta and Y. Ibaraki (Eds)). CRC Press, Taylor & Francis Group, pp. 330–72. http://www.crcpress.com/product/isbn/9781466583016

Payne, A., Walsh, K. B., Subedi, P. and Jarvis, D. (2014) Estimating mango crop yield using image analysis using fruit at 'stone hardening' stage and night time imaging. *Computers and Electronics in Agriculture* 100: 160–7.

Quereshi, W. S., Payne, A., Linker, R., Cohen, O and Dailey M. N. (2016) Machine vision for counting fruit on mango tree canopies. *Precision Agriculture* 18(2): 224–44. doi: 10.1007/s11119-016-9458-5

Robson, A. J., Petty, J., Joyce, D. C., Marques, J. R. and Hofman, P. J. (2016). High resolution remote sensing, GIS and Google Earth for avocado fruit quality mapping and tree number auditing. *Acta Horticulturae* 1130: 589–95.

Sadegaonkar, V. D. and Wagh, K. H. (2013). Quality inspection and grading of mangoes by computer vision and image analysis. *International Journal of Engineering Research and Applications* 3(5): 1208–12. ISSN: 2248–9622. www.ijera.com.

Saranwong, S, Sornsrivichai, J and Kawano, S 2004, Prediction of ripe-stage eating quality of mango fruit from its harvest quality measured nondestructively by near infrared spectroscopy. *Postharvest Biology and Technology* 31 (2): 137–45.

Simmons, S., Hofman, P., Whiley, A. and Hetherington, S. (1998). Effects of preharvest calcium sprays and fertilizers, leaf: Fruit ratios, and water stress on mango fruit quality. *ACIAR Proceedings* 81: 19–2

Spreer, W., Nagle, M., Neidhart, S., Carle, R., Ongprasert, S. and Müller, J. (2007). Effect of regulated deficit irrigation and partial rootzone drying on the quality of mango fruits (*Mangifera indica* L., cv.'Chok Anan'). *Agricultural Water Management* 88(1): 173–80.

Spreer, W. and Müller, J. (2011). Estimating the mass of mango fruit (Mangifera indica, cv. Chok Anan) from its geometric dimensions by optical measurement. *Computers and Electronics in Agriculture* 75(1): 125–31.

Stein, M., Bargoti, S. and Underwood, J. (2016). Image based mango fruit detection, localisation and yield estimation using multiple view geometry. *Sensors* 16: 1915.

Subedi, P., Walsh, K. B. and Owens, G. (2007). Prediction of mango eating quality at harvest. *Postharvest Biology and Technology* 43(3): 326–34.

Subedi, P. P and Walsh, K. B. (2008). Non – invasive measurement of fresh fruit firmness. *Journal of Postharvest Biology and Technology* 51: 297–304. doi:10.1016/j.postharvbio.2008.03.004

Walsh, K. B. (2015). Nondestructive assessment of fruit quality. In: *Advances in Postharvest Fruit and Vegetable Technology. CRC Press Series on Contemporary Food Engineering* (R. B.H. Wills and J. B. Golding (Eds)). Elsevier, CRC Press, Taylor & Francis Group, pp. 40–61.

Walsh, K. B. (2017) Fruit maturity: Developing an application to assist with the decision to pick. *Mango Matters* http://www.industry.mangoes.net.au/resource-collection/?tag=Dry+Matter Accessed date: 12 April 2017.

Wang, Z., Verma, B., Walsh, K. B. Subedi, P. and Koirala, A. 2016. Automated mango flowering assessment via refinement segmentation. In: *2016 International Conference on Image and Vision Computing New Zealand (IVCNZ)*. IEEE Palmerston North, pp. 1–6. IEEE Catalog Number: CFP1667E-ART ISBN: 978-1-5090-2748-4 Online ISSN: 2151-2205, https://opus.lib.uts.edu.au/bitstream/10453/80091/1/proceedings%20IVCNZ%202016%20-%20front%20matter.pdf, doi:10.1109/IVCNZ.2016.7804426. http://ieeexplore.ieee.org/stamp/stamp.jsp?tp=&arnumber=7804426&isnumber=7804412.

Whiley, A. W., Hofman, P. J., Christiansen, H., Marques, R., Stubbings, B. and Whiley, D. G. 2006, Development of Pre and Postharvest Protocols for Production of Calypso Mango/Anthony Whiley. Final report/Horticulture Australia; HAL FR02049. Horticultural Australia, Sydney.

Yeshitela, T., Robbertse, P. and Fivas, J. (2004). Effects of fruit thinning on 'sensation'mango (Mangifera indica) trees with respect to fruit quantity, quality and tree phenology. *Experimental Agriculture* 40(4): 433–44.

Zhao, J. J., Wang, J. B., Zhang, X. C., Li, H. L. and Gao, Z. Y. (2013). Effect of bagging on the composition of carbohydrate, organic acid and carotenoid contents in mango fruit. *Acta Horticulturae* 992: 537–54.

Chapter 12

Understanding post-harvest deterioration in mangoes

Apiradee Uthairatanakij and Pongphen Jitareerat, King Mongkut's University of Technology Thonburi, Thailand; and Robert E. Paull, University of Hawaii at Manoa, USA

1 Introduction

Mango (*Mangifera indica* L.) belongs to the family Anacardiaceae, also known as the cashew family, and has about 75 genera and 700 species. The family normally has fleshy fruit, with non-fleshy fruit being rare; if dry, the fruit is indehiscent. The fruit type is a drupe where the outer fleshy part surrounds a shell of hardened endocarp with one stone, though in the cashew the non-fleshy indehiscent fruit sits on a swollen receptacle. Other related fleshy fruits sold fresh are in the genus *Spondias*, the yellow mombin or hog plum (*S. mombin* L.); the ambarella or June plum (*S. cytherea or S. dulcis* Forst.); and the purple mombin (*S. purpurea* L.). Other species belonging to this family with edible fleshy fruit are the marula (*Sclerocarya birrea* spp. *caffra* Sand) of Africa and the Marian plum or Ma-praang (*Bouea macrophylla* Griff) of South east Asia. The genus *Mangifera* consists of about 69 species but not all bear edible fruit.

Post-harvest losses in the quality and quantity of mango from the field to the consumer is dependent upon field management, harvesting practices, fruit handling and packing procedures, post-harvest treatments, storage conditions, and wholesale and retail operations (FAO, 2016). These practices and processes vary widely from the developing to industrialized economies, making estimates of overall losses difficult (Parfitt et al., 2010; Ridoutt et al., 2010; Sivakumar et al., 2011). In India, mango losses during harvesting and packing were estimated at 52%, at the wholesale level (30%) and during retailing (18%) (Murthy et al., 2009). Retail losses for mango in the United States for 2011–12

http://dx.doi.org/10.19103/AS.2017.0026.15

range from 15% to 26% based on computerized store receivables and sales from national and regional supermarket chains (Buzby et al., 2016). The US losses at the retail level do not take into account pilferage and donations. The losses may also be associated with supermarket chain policy on the range of items all outlets must carry, claims policy for losses sustained by the retail outlet to central warehouse, socio-economic status of the stores locations and ethnicity in those areas. The causes most often cited for post-harvest losses are harvesting at the wrong stage of maturity, failure to minimize sap burn, mechanical injury from the field to the consumer, physiological disorders (spongy tissue, lenticels discolouration), fruit ripening and softening during marketing, post-harvest disease, chilling injury (CI), and pest damage (fruit flies, seed weevil). This chapter will highlight the potential causes of fruit deterioration through the whole post-harvest logistics chain. Three substantial reviews have appeared in the last seven years dealing with post-harvest physiology and changes, and the handling of mango (Brecht and Yahia, 2009; Yahia, 2011; Singh et al., 2013) that underlie mango post-harvest deterioration. In this chapter, we will summarize some of the same material and relate it directly to mango quantity and quality losses post harvest.

1.1 Mango fruit anatomy and development

The fruit of the mango is a drupe of variable size and shape, ranging in weight from a few grams to more than 1 kg and flattened, rounded or elongated in shape (Paull and Duarte, 2011). Fruit growth shows a simple sigmoidal growth curve in terms of length, thickness, mass and volume against days from anthesis with a brief period of slower growth during endocarp hardening. The skin of the young fruit is green; sometimes purplish; and when ripe, green, yellow, orange, yellowish red, purplish or purplish red. The Indian mango species tend to mature with a yellow and red colour combination, while the Indochina-Philippine mangoes have yellow or orange coloured fruit.

The peel (exocarp) is thick and the flesh (mesocarp) of ripe fruit is yellow or orangish yellow and juicy. The pericarp can be separated into exocarp, mesocarp and endocarp at about 14 days after anthesis, a period that occurs between 9 and 14 weeks after fruit set when the growth rate declines slightly associated with hardening of the endocarp and accumulation of starch and sugars. The endocarp is hard, with fibres that may extend into the flesh. The period from fruit set to maturity depends upon cultivar and climate and can range from 10 to 28 weeks. 'Saigon' grown in a hot climate is ready for harvest in 12 to 13 weeks. In cool areas, where mean temperatures fall below 20°C, maturation is delayed by up to 4 weeks (Whiley et al., 1988).

The cuticle of mango is thicker than that of other fruit at 14 to 21 μm (Martin and Rose, 2012). Cuticular compositional diversity exists between organs, species and cultivars. Based upon cuticular permeability studies, spatial differences exist on the surface of a single mango fruit, in addition to where the fruit occurred in the canopy (Léchaudel et al., 2013). Fruit that were shaded in the canopy had higher water conductance than sun-exposed fruit which had a thicker cuticle.

A network of latiferous canals and secretory ducts anastomoses in all directions in the exocarp and mesocarp (Joel, 1978). The latiferous canals in mango begin to disintegrate during ripening and the fruit becomes susceptible to fruit flies. The sap from the skin contains alkylresorcinols that are related to urushiol, a toxin also found in other members of the Anacardiaceae. The alkylresorcinols cause allergic skin rash on contact in sensitive individuals.

1.2 Mango post-harvest varietal differences

Hundreds of named mango cultivars exist throughout the tropics and subtropics with each country or region having their own selected cultivars with different names. Many excellent cultivars have been developed through introduction and selection by cooperation between researchers, growers, nurserymen and hobbyists. Preferences for cultivars vary in different regions of the world based on growth habit and fruit quality. The most important commercial cultivars are derived from selections among open-pollinated seedling populations. Seedlings can be produced by mass cross-pollination of a collection of mixed cultivars selected for known desirable characteristics among mono- and polyembryonic races (Whiley, 1993). Isozyme analysis has been used to verify or refute parentage of mango cultivars (Degani et al., 1990). This technique showed that 'Haden' appears to be a seedling of 'Mulgoba' and 'Zill'. Newer molecular biology techniques are being used to unscramble the parentage of mangoes and to determine the extent of genetic diversity relationship to other *Mangifera* spp. (Eiadthong et al., 2000; Pujarig and Patil, 2007; Hirano et al., 2010; Dillon et al., 2013). At least six genetic diversity groupings of the cultivated varieties exist: Thailand/Philippines, Vietnam, Indonesia/Malaysia/Malesia, Kenya/South Africa, and Caribbean/Sri Lanka/Australia/Hawaii and the Pacific/Florida (Dillon et al., 2013).

India, with more than 1000 named cultivars, has provided much of the germplasm for cultivar development. Florida has developed a large number of cultivars, using mainly Indian cultivars, that have shown wide geographic adaptability (Campbell and Campbell, 1992). Most mango varieties planted in the Americas, the Canary Islands and many African countries were developed in Florida. However, though cultivars were selected for fruit quality, this does not mean that they are most suited for post-harvest handling and long-distance shipping.

Fruit disorders associate with a cultivar take on greater importance depending upon the cultivars parentage (Wainwright and Burbage, 1989; Shivashankar, 2014). One fruit disorder, occurring in 'Mulgoba' from India is present in the background of many varieties, for example, 'Haden' (Olano et al., 2005; Pandit et al., 2009; Dillon et al., 2013) is referred to as 'internal flesh breakdown', 'jelly seed', 'stem end cavity' or 'soft nose'. 'Tommy Atkins', whose parent was 'Haden', has been reported to be especially susceptible to this disorder, although 'Kent', `Irwin', 'Sensation' and 'Carabao' and a few other commercial cultivars of importance are also susceptible (Knight et al., 2009). Certain cultivars from Thailand and the Philippines are particularly susceptible to 'lumpy tissue', which is not evident in green fruit but develops during ripening.

2 Ripening-related changes

2.1 Respiration and ethylene

As a climacteric fruit, mango shows typical respiratory and ethylene peaks as the fruit ripens (Fig. 1a). The climacteric respiratory rise and pattern of ripening vary among cultivars, growing conditions, tree management and orchard regional locations (Krishnamurthy and Subramanyam, 1970; Subramanyam et al., 1975; Medlicott et al., 1988). Mature green 'Alphonso' shows the respiratory peak on day 5 (Karmarkar and Joshi, 1941), 'Pairi' on day 9 (Krishnamurthy and Subramanyam, 1970), 'Kent' also on day 9 and 'Haden' on day 11 (Burg and Burg, 1962).

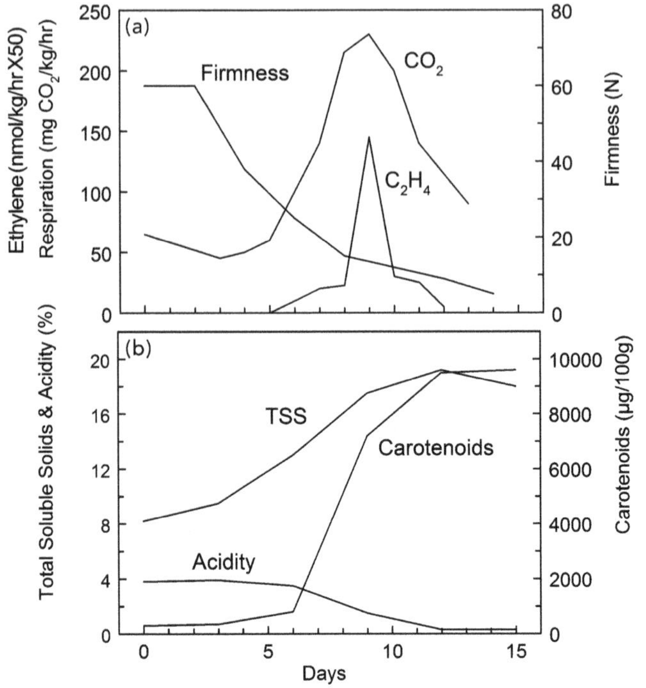

Figure 1 Patterns of changes during the ripening of mango fruit. Redrawn from various sources. (a) Respiration rate for 'Pairi' (Krishnamurthy and Subramanyam, 1970) and ethylene production for 'Kensington Pride' (Zaharah and Singh, 2011). The peak in ethylene is preceded by a rise in ACC and the enzymes ACS and ACO. Decline in fruit firmness for 'Keitt' mango (Padda et al., 2011). (b) Increase in TSS in 'Keitt' (Padda et al., 2011) and carotenoids in 'Alphonso' (Krishnamurthy and Subramanyam, 1970) and decline in titratable acidity in 'Keitt' (Padda et al., 2011).

The respiratory rise precedes the rise in ethylene production (1 to 3 µl kg^{-1} h^{-1}) (Fig. 1a) by about a few days at 20°C (Lakshminarayana, 1973; Cua and Lizada, 1990; Reddy and Srivastava, 1999; Brecht and Yahia, 2009; Zaharah and Singh, 2011). The initial slow rise in ethylene before the rapid rise would be sufficient to initiate ripening (Burg and Burg, 1962; Wills et al., 2001), although we have no details on changes in fruit sensitivity to ethylene during fruit maturation and ripening. The ethylene peak occurs just as the respiratory peak (Fig. 1a) begins to decline and both the respiratory and ethylene peaks are suppressed following nitric oxide treatment (Zaharah and Singh, 2011).

During ripening, the amount of 1-aminocyclopropane-1-carboxylic acid (ACC) increases in the peel and mesocarp while ACC oxidase (ACO) declines (Reddy and Srivastava, 1999). These changes imply new gene expression for ACC synthesis and the conversion to ethylene. The fruit peel has more ACO and ethylene and the least ACC compared to the mesocarp, indicating differential accumulation and control at the mature green stage (Lederman et al., 1997).

Exogenous ethylene treatment stimulates ripening of mature fruit (Lakshminarayana, 1973; Lakshminarayana et al., 1974; Barmore and Mitchell, 1975). The most marked changes following ethylene treatment is more uniform softening and skin degreening.

Immature fruit after ethylene treatment, however, do not develop full flavour (Chaplin, 1988).

The recommended ripening temperature is from 21°C to 24°C (Yahia, 2011; Singh et al., 2013) which presents problems for ripening fruit in the tropics where higher ambient temperatures are normally experienced. At 32 to 37°C, the fruit ripens, although it has a slightly poorer final fruit quality than fruit ripened at lower temperatures.

2.2 Fruit texture and cell wall changes

From the mature green to ripening stages, there is a dramatic decline in mesocarp starch level and its conversion to soluble sugars (Sen et al., 1985; Lima et al., 2001). Concomitant with starch breakdown is a significant change in wall structure and fruit softening (Fig. 1a). Changes in wall composition is associated with a number of carbohydrate-active enzymes, including hydrolyases, esterases and lyases (Krishnamurthy et al., 1971; Ali et al., 1990; Lazan et al., 1993; Selvaraj and Kumar, 1989; Abu-Sarra and Abu-Goukh, 1992; Yashoda et al., 2005, 2007). The activity of these enzymes increases during fruit ripening.

Initially, the focus of ripening research was on the changes in the activity of pectinases and esterases. Cell wall pectins show a reduction in molecular size and solubilization during ripening (Roe and Bruemmer, 1981; Tandon and Kalra, 1983; Lazan et al., 1986; Brinson et al., 1988; Selvaraj and Kumar, 1989; Mitchum and McDonald, 1992; Muda et al., 1995). Cell wall pectin breakdown involves exo- and endo-polygalacturonases (PG), pectin methylesterase (PME), pectin lyases, mannanases, galactosidases and arabinofuranosidases. These enzymes act synergistically with the esterases that remove methyl groups from the pectin so that the PG can breakdown the pectin chain (Ashraf et al., 1981; Yashoka et al., 2005). Paralleling these enzymatic activity is a change in the cell wall pH, enhancing the activity of these enzymes (Yashoka et al., 2006). Calcium-dependent pectic lyase has been identified to show ripening-related expression and is involved in pectin solubilization (Chourasia et al., 2006).

Hemicellulases have also been associated with cell wall breakdown and fruit textural changes in mango. The enzymes involved include β-1,4-glucanases (cellulases), galactosidase, galactanase and xylanases. The β-1,4-glucanases, referred to as cellulases because of the modified and soluble cellulose used as a substrate, are hydrolases of three types: endoglucanases, exoglucanase and glucosidase. Both endo- and exo-glucanases increase during ripening with fruit specificity and ripening-related expression and a correlation to softening (Selvaraj and Kumar, 1989; Chourasia et al., 2006). The activities of arabinanase, galactanase and mannanase peak at the climacteric stage of ripening (Bhagyalakshmi et al., 2002). At least three isoforms of α-galactosidase increase during fruit ripening (Ali et al., 1995) and parallel tissue softening (Prasanna et al., 2006). The substrates for α-galactosidase are arabinogalactans.

Expansins have also shown ripening-related expression and may play a role in wall loosening during wall softening (Sane et al., 2005; Zheng et al., 2012). Oxalic acid slows softening and expression of the mango expansin gene.

2.3 Sugar accumulation and acid metabolism

The sugar and acid contents of mango are highly dependent upon cultivar (Kapse et al., 1989; Kundu and Ghosh, 1992; Gowda et al., 1994), fruit maturity at harvest (Shashirekha and Patwardhan, 1976; Tandon and Kalra, 1983), post-harvest handling and storage conditions (Vazquez-Salinas and Lakshminarayana, 1985; Kumar et al., 1993).

The final content of sugars and acid in ripe mango fruit is a function of starch breakdown and the.flux in glycolysis and citric acid cycle (Krebs cycle). Starch is rapidly degraded and almost completely hydrolysed to simple sugars (Fig. 1b). The starch content of 'Alphonso' is ~14% at the mature green stage declining during ripening to less than 0.3% (Selvaraj et al., 1989). Total sugars in ripe mango ranges from 10% to 20% (FW basis) and this varies with cultivar and stage of ripeness (Castrillo et al., 1992; Ito et al., 1997). The majority of the simple sugars are sucrose, glucose and fructose with minor amounts of maltose and xylose (Medlicott et al., 1986; Selvaraj et al., 1989; Kumar et al., 1994; Ito et al., 1997). A minor contribution is made to the fruit sugars levels by the hydrolysis of cell wall galactans and arabinogalactans by galactosidase activity, although galactose levels decline during ripening (Seymour et al., 1990). The ratio of the non-reducing disaccharide sucrose to the reducing monosaccharides (glucose, fructose) should be interpreted with care as extraction protocols allow considerable invertase activity to occur. Sucrose is the major soluble sugars in ripe 'Keitt' fruit at 57% of the total sugars, with fructose at 28% and glucose at 15% (Medlicott and Thompson, 1985). Invertase activity could explain the decline in non-reducing sugars (sucrose) found early in ripening and the increase in the reducing sugars (glucose, fructose) in fully ripe fruit. Both neutral and acid invertase activities are present at the start of ripening; the acid invertase activity increases at the onset of ripening and declines at the later stages of ripening (Castrillo et al., 1992). Invertase activity is probably more than sufficient to invert all the sucrose, suggesting that sucrose separates from the invertase.

The activity of enzymes involved in starch breakdown and sugar interconversion such as amylase, sucrose synthase and invertase increases during fruit ripening (Fuchs et al., 1980; Tandon and Kalra, 1983; Kumar et al., 1994). The activity of glycolytic enzyme that control the flux of sugars increases during early ripening and then declines before the ripening stage. The enzymes glucose-6-phosphatase, fructose-1,6-diphosphatase (Kumar and Selvaraj, 1990) and hexokinase (6-phosphofructokinase) have maximum activity in fully ripe fruit. Pyruvate kinase activity peaks at the middle stage of ripening and then declines (Selvaraj and Kumar, 1994). A ten-fold increase occurs in sucrose phosphate synthase as sucrose accumulates (Castrillo et al., 1992).

The sugar-to-acid ratio plays an important role in flavour and this changes during fruit ripening as the acid content declines and sugar level increases (Fig. 1b). The predominant organic acid in mango is citric acid with smaller amounts of succinic, malic and tartaric acids (Shashirekha and Patwardhan, 1976; Sarker and Muhsi, 1981; Medlicott and Thompson, 1985; Kumar et al., 1993). Citric acid gradually increases during fruit development and then reaches a peak before declining during ripening (Ito et al., 1997). Succinic acid shows a similar trend to citric acid, and malic acid shows no clear trend during ripening (Lizada, 1993).

2.4 Skin degreening and carotenoid synthesis

The consumer appeal of ripe mangoes is very much dependent on the skin and flesh colour. The major change in skin colour is due to the breakdown of chlorophyll and appearance of other red and yellow anthocyanins and carotenoids (Lakshminarayana, 1980; Parikh et al., 1990; Lizada, 1993; Ketsa et al., 1999). The anthocyanins and carotenoids are synthesized and accumulate in the vacuole (Lizada, 1993). During ripening the skin colour changes from dark green or purplish to yellowish red, orangish yellow or yellow. In yellow cultivars, carotenoids and xanthophylls are the predominant pigments, whereas anthocyanins occur in the skin of some cultivars (Proctor and Creasy, 1969; Berardini et al., 2005a; Sivankalyani et al., 2016). The reddish blush on some cultivars is due to anthocyanins that decline during the ripening of 'Tommy Atkins' (Medlicott et al., 1986; Berardini et al., 2005a,b; Sivankalyani et al., 2016). The fruit mesocarp is high in carotenoids (up to 9 mg 100 g^{-1}), giving the fruit a deep yellow to orange colour.

During ripening, there is an increase in the amount (Fig. 1b) and number of carotenoids in the mesocarp (John et al., 1970; Wilberg and Rodriguez-Amaya, 1995; Chen et al., 2004). Carotenoids found include α-, β- and γ-carotenes; and the oxygenated carotenoids (xanthophylls) such as β-cryptoxanthin, lutein, zeaxanthin, violaxanthin, antheraxanthin, auroxanthin and neoxanthin (John et al., 1970; Cano and de Ancos, 1994; Mercadante et al., 1997, 1998; Setiawan et al., 2001; Ornelas-Paz et al., 2007; Vásquez-Caicedo et al., 2005). The amounts of the individual carotenoids vary with cultivar and stage of ripening (Lizada, 1993); β-carotene constituted the major carotenoid in the unripe (37.47%) and fully ripe mango (50.64%). *cis*-β-Carotene is present only in the fully ripe mango (John et al., 1970). Unripe mango contains ζ-carotene, whereas γ-carotene is present throughout ripening. The major xanthophyll present in the unripe mango is mutatoxanthin (9.44%), whereas auroxanthin constituted the major hydroxylated carotenoid of the partially ripe (5.1%) and fully ripe (10.4%) fruit. Cryptoxanthin declines to lower levels during ripening, and lutein completely disappears. Significant quantities of zeaxanthin are found in the partially ripe and fully ripe mango.

2.5 Aroma and flavour

Each cultivar has a unique taste and flavour, with aroma playing a dominant role in consumer's appreciation of a fruit. Taste is in large measure based upon the balance in sugars to acids (Medlicott and Thompson, 1985; Selvaraj et al., 1989; Lizada, 1993; Malundo et al., 2001). Some cultivars have a slight turpentine taste that is not appreciated by some consumers. The terpeny odour is most likely due to (*Z*)-β-ocimene and (*E*)-β-ocimene which are present in high amounts in 'White Alfonso' (Munafo et al., 2014). Low fibre in the mesocarp is also a major quality factor.

The number and type of aroma volatiles released by mango varies widely with cultivar, stage of ripeness and production region (Pandit et al., 2009; Kulkarni et al., 2012). Tree-ripened mangoes have greater aroma profile than fruit harvested at mature green stage (Bender et al., 2000). Post-harvest ripened fully mature fruit contain a higher amount of aroma volatiles compared to hard fruit harvested at a less mature stage (Lalel et al., 2003a). Ripening using ethylene leads to more uniform ripening with more aromatic compounds (Lakshminarayana, 1980). Controlled atmosphere (CA) storage has little impact on volatile production (Lalel et al., 2003b).

More than 550 aroma volatiles have been reported that includes different groups of terpenes and terpenoids, esters, aromatics, ketones and lactones, aldehydes, alcohols, acids, alkanes and cycloalkanes, norisoprenoids and miscellaneous compounds (Wilson et al., 1990; Lalel et al., 2003a,b; Nair et al., 2003; Pino et al., 2005; Yahia 2011; Singh et al., 2013). Lists of these volatiles are useful as some have insect and disease resistance roles in fruit though most of these volatiles do not contribute to the aroma perceived (Munafo et al., 2014). When five cultivars ('Haden', 'White Alfonso', 'Praya Sowoy', 'Royal Special' and 'Malindi') were evaluated, 4-hydroxy-2,5-dimethyl-3(2*H*)-furanone (furaneol, strawberry furanone), that has a sweet strawberry aroma, stood out as an important common aroma compound in all cultivars. A further twenty-seven aroma-active compounds are present in at least one mango cultivar, and contributed to the unique sensory profiles of that individual cultivar. The fruity note in the mango aroma profile is associated with ethyl butanoate, ethyl 3-methylbutanoate, ethyl 2-methylpropanoate and ethyl 2-methylbutanoate. The coconut-like odour is associated with the lactones such as γ-octalactone and δ-octalactone (Pino and Mesa, 2006; Pino, 2012).

Aroma-related monoterpenes and sesquiterpenes (MacLeod and De Troconis, 1982; Engel and Tressl, 1983; Pino et al., 2005; Munafo et al., 2014) and some fatty acids (MacLeod et al., 1988) are also important for mango aroma. Monoterpenes are important volatiles in New World cultivars while oxygenated volatile compounds such as esters, furanones and lactones are critical to Old World cultivars (Engel and Tressl, 1983). Terpene hydrocarbons are important contributors to Florida (New World) cultivars such as 'Keitt', 'Kent' and 'Tommy Atkins' (Malundo, 1996). The oxygenated terpenes are qualitatively and quantitatively present among Indian cultivars and non-Indian cultivars such as 'Jaffna', 'Willard', 'Parrot', 'Kensington Pride' (Lalel et al., 2003a) and several Colombian cultivars (Quijano et al., 2007). The oxygenated volatile compounds found in the Florida cultivars are ethanol, acetaldehyde and hexanal (MacLeod and Snyder, 1985). Indian Old World mangoes have a unique flavour attributed to the monoterpene (*Z*)-ocimine (Engel and Tressl, 1983; Lizada, 1993). Ripening of 'Alphonso' is characterized by the *de novo* appearance of lactones and furanones in the blend of monoterpenes (Pandit et al., 2009). Monoterpenes quantitatively are most prevalent (57 to 99%), with (*Z*)-ocimene found in the highest concentrations. The pineapple-like aroma detected in 'Haden', 'White Alfonso', 'Royal Special' and 'Malindi' is due to (*E*,*Z*)-1,3,5-undecatriene and/or (*E*,*Z*,*Z*)-1,3,5,8-undecatetraene (Munafo et al., 2014).

2.6 Phenols and Terpenes

Mature green mangoes are astringent and this astringency declines during fruit ripening. Astringency is associated with the phenolic content (Lakshminarayana et al., 1970; Selvaraj and Kumar, 1989; Kim et al., 2007). The phenolic content of the peel is higher than in the mesocarp (Lakshminarayana et al., 1970). The phenols identified in 'Ataulfo' from Mexico are closely related to Philippines cultivars 'Carabao' and 'Manila' containing chlorogenic acid (28–301 mg/100 g DW), gallic acid (94.6–98.7 mg/100 g DW), vanillic acid (16.9–24.4 mg/100 g DW) and protocatechuic acid (0.48–1.1 mg/100 g DW) (Palafox-Carlos et al., 2012). The antioxidant contribution of the four phenolic acids increases during ripening, with gallic acid making the greatest contribution.

The toxic phenols found in the primary and secondary phloem resin ducts of Anacardiaceae include alkylresorcinols, alkylcatechols and biflavonoids. These toxic compounds are thought to play a role in disease (*Alternaria*, anthracnose) and pest (fruit flies) resistance (Joel, 1978; Cojocaru et al., 1986; Konno, 2011). The alk(en)lyresorcinol (phenolic lipids) are found in mango with the moiety linked mostly with C15 to C19 alk(en)yl chain with varying degrees of unsaturation. The alkylresorcinols in mango are analogues to the saturated and unsaturated cathechol-based urushiols of other Anacardiaceae species such as poison ivy and poison sumac that induce contact dermatitis in susceptible people (Aquilar-Ortigoza and Sosa, 2004; Kozubek and Tyman, 2005; Knodler et al., 2009). The alkylresorcinols are found in the peel and mesocarp and may play a role in anthracnose susceptibility (Hassan et al., 2009). However, alkylresorcinols may not be a deciding factor or the only factor involved in disease resistance during ripening. Little changes occur in their concentration during ripening though fungal resistance decreases, with significant difference occurring between varieties (Kienzle et al., 2013).

The total polyphenol content in mango skin (4,066 mg (gallic acid equivalents)/kg DW) (Saleh and El-Ansari, 1975; Berardini et al., 2005b) increases during ripening (Ajila et al., 2007). The skin polyphenolic constituents include mangiferin, quercetin, rhamnetin, ellagic acid, kaempferol and their related conjugates (Masibo and He, 2008). The polyphenols in the mesocarp include mangiferin, gallic acids (*m*-digallic and *m*-trigallic acids), gallotannins, quercetin, isoquercetin, ellagic acid and β-glucogallin (Schieber et al., 2001). Gallic acid (6.8 mg/kg) is the major polyphenol present in mango flesh, followed by mangiferin (4.4 mg/kg) and the six hydrolysable tannins and four minor compounds, OH-benzoic acid, *m*-coumaric acid, coumaric acid and ferulic acid (Schieber et al., 2001; Berandini et al., 2004; Kim et al., 2007). Phenolic compounds are both free and also bound to the cell wall components. The chemical distribution of phenolics across varieties varies widely (Schulze-Kaysers et al., 2015).

Polyphenol oxidase (PPO) catalyses the oxidation of mono- and diphenols to *o*-quinones, which polymerize to produce brown pigments (Robinson et al., 1993). PPO activity increases slightly from the mature green stage to the half-ripe stage, then declines in 'Banganapalli', 'Dasheri', 'Fazli' and 'Langra' mangoes, and continually decreases in 'Alphonso', 'Suvarnarekha' and 'Totapuri' mangoes (Selvaraj and Kumar, 1989). PPO isolated from 'Haden' mango is active towards the *o*-diphenolic compounds, followed by chlorogenic acid, but not with monophenols (Park et al., 1980). However, Prabha and Patwardhan (1986) found that gallic acid is the substrate of PPO in the fruit mesocarp and ellagic acid in mango peel.

3 Pre-harvest handling, environment and deterioration

3.1 Internal breakdown

'Internal flesh breakdown', 'jelly seed', 'stem end cavity' and 'soft nose' are different manifestations of similar conditions and these disorders occur in a number of cultivars. The lower half of the fruit is most affected with no apparent external symptoms (Shivashankar, 2014). The disorders usually appear during the initial stages of maturity, with a loss of firmness of the mesocarp near the endocarp, which becomes jelly-like and translucent with advancing ripeness (Fig. 2a). The condition may be related to calcium deficiency with high nitrogen increasing the disorder (Malo and Campbell, 1978; Wainwright and Burbage, 1989; Burdon et al., 1991). The spatial and temporal difference in symptom development has led to classifying them as separate disorders (Raymond et al., 1998).

The stem end cavity condition involves the formation of a cavity near the peduncle due to the breakdown of the vascular tissues between the seed and the peduncle (Raymond et al., 1998). The tissues turn necrotic and brown, leading to the formation of a cavity (Fig. 2a), which enlarges in more advanced stages and the stem end appears spongy to application of even a slight pressure (Brown et al., 1981; Chaplin, 1988). Little latex is seen from the peduncle when the fruit is harvested. The disorder is first observed in 'Tommy Atkins' and 'Van Dyke' when the fruit is 10% of its final weight and when 'Irwin' is nearly 70% of its final size. The cause is unknown (Shivashankar, 2014).

Figure 2 Fruit post-harvest quality problems: (a) jelly seed condition with some stem end cavity formation; (b) 'lumpy' fruit with starchy granulation; (c) brown-black gummy exudation sometimes associated with 'lumpy' tissue affected fruit; (d) fruit splitting and sunburn; (e) pruned trees that are less than 2.5 m tall and allow all management practice to be carried out from the ground including fruit thinning, bagging and harvest; (f) severely damaged fruit with impact injury, shin scald and poor grading; and (g) individual fruit in the market showing abrasion injury from rubbing against another fruit.

3.2 'Lumpy tissue'

Some Thai and Philippine cultivars are susceptible to 'lumpy tissue', which is not evident in green fruit but develops during ripening. The mesocarp contains white starchy lumps and the fruit surface develops indentations (Fig. 2b). Internal fruit necrosis first appears as a brown area in the mesocarp and endocarp of rapidly growing fruit. This later extends to the skin and a brownish black gummy exudation occurs (Fig. 2c). These areas then collapse and are surrounded by corky tissue. The disorder has been associated with pre-harvest calcium and boron deficiency (Rossetto et al., 2000; Sharma and Singh, 2009; Shivashankar, 2014).

3.3 Fruit cracking

Internal necrosis and fruit cracking are caused by the deficiency of boron (Saran and Kumar, 2011). Mature fruit cracking can also occur due to a rapid change in fruit water status after a period of drought (Fig. 2d), most often in low-fibre varieties (Shivashankar, 2014) and is similar to cracking that occurs in mature cherry and tomato fruit (Peet and Willits, 1995).

3.4 Sunlight and sunburn

Mango fruit exposed to direct sunlight can reach higher than 40°C and be 10°C higher than shaded fruit (Léchaudel et al., 2013). High temperatures due to direct exposure to sunlight with low humidity can lead to heat injury and sunburn. Sunburn can also occur on harvested fruit left exposed to direct sunlight. Sunburn appeared as a bleaching of the skin on the exposed side of the fruit (Nguyen et al., 2004). The sunburnt area usually appear creamy white and slightly sunken, but in severe cases, the affected area can be black. The exposed heat-stressed peel has increased hydrogen peroxide, malondialdehyde (MDA) and ascorbate contents, as well as superoxide dismutase (SOD), monodehydroascorbate reductase (MDHAR) and ascorbate POX (APX) activities, regardless of the maturity stage.

Sunburn is more severe as the percentage of yellow skin increases during ripening (Bally, 2007). Nitrogen greatly influences fruit quality, having a major influence on post-harvest disease, skin colour and sunburn susceptibility (Bally, 2007). A significant exponential relationship exists between the incidence and severity of sunburn and applied-N and exocarp-N at harvest. Fruit bagging can effectively reduce sunburn injuries as well as insect pest attack, disease incidence, mechanical injuries and bird damage.

3.5 Fruit thinning, bagging and source sink relationships

Fruit bagging (Fig. 2e) has been shown when coupled to fruit selection and thinning early in fruit development to reduce culling percentage and have a higher output of blemish-free fruit (Kitagawa et al., 1992; Mathooko et al., 2011; Sharma et al., 2014). The thinned fruit may be spayed with fungicide before bagging. Bagged fruit have less disease at harvest (Hofman et al. 1997). However, fruit stored in plastic bags deteriorated faster after harvest compared to unbagged fruit. Bagging 'Harumanis' mango with different colour bags does not affect mango mesocarp, carotenoids, fruit firmness and eating quality, that is, total soluble solids (TSS), titratable acidity and vitamin C (Hofman et al., 1997; Ding et al., 2010). Black bags are used to enhance the gold colour of Jin-Hwang (Jin Huang) and other mango varieties by reducing the skin chlorophyll level during fruit development (Watanawan et al., 2008; Chonhenchob et al., 2011).

Bagging fruit influences post-harvest behaviour of mango, especially water loss during post-harvest storage that is higher in bagged fruit (Hofman et al., 1997; Joyce et al., 1997). This difference in the post-harvest response to bagging, especially fruit covered by plastic bags, might represent changes in the structure and/or composition of the cuticle and/or lenticels. Lenticel discolouration is reduced 30 to 55% by bagging and 39 to 43% by carnauba wax-based coating applied the day before harvest (Nguyen, 2015; Nguyen et al., 2016) compared to the controls. These results suggest that exposing some mango varieties to water either at the pre- or post-harvest stages leads to lenticel discoloration.

4 In-harvest handling, environment and deterioration

4.1 Mechanical injury

The high moisture content and soft texture of ripe mango fruit make them highly susceptible to mechanical injury (Fig. 2f). This injury can occur at any stage from pre-harvest to post-harvest. Mechanical injury occurs because of improper cultural practices, poor harvesting practices, unsuitable containers and careless handling (Fig. 2g). Skin scarring is common when fruit rub against other fruit (Fig. 2f) or tree stems. This pre-harvest injury is greatly reduced by cultural practices such as fruit thinning, culling of young fruit and bagging of the selected fruit (Fig. 2e).

During harvest and transport to the packing area, injury is caused by abrasion, impact and compression (Santos et al., 2002). Abrasion is caused by scratching and rubbing against rough surfaces such as in wooden field boxes. Abrasion injury is greatly reduced in packed cartons when the fruit are carefully wrapped with paper or expanded polyethylene (PE) sleeves (Fig. 3a). Impact is due to poor handling when fruit are dropped into collection bins and tipped out of bins at the packing shed. Impact injured fruit has an accentuated rate of weight loss, skin darkening, mesocarp browning, fungi infection, translucent (water-soaked) flesh adjacent to impact area and reduced post-harvest life (Santos et al., 2002). Compression injury occurs when fruit are stacked in deep containers or bins. 'Carabao' at the mature green stage should not be stacked more than five deep to avoid compression injury (Valerio and Esguerra, 2001). Compression injury becomes a major problem for fruit at advanced stages of ripeness. Dropping and throwing of fruit and over-packing often leads to bruising during culling and packing. 'Honey Gold' mangoes show under-skin browning when subject to compression injury. Using soft liners in combination with a delay in temperature reduction after harvest significantly reduces under-skin browning in the post-harvest handling of 'Honey Gold' fruit (Marques et al., 2016).

Disease and pest control is of greater concern in mechanically injured fruit. Impact and abrasion injury breaks the protective fruit cuticle and create potential sites for pathogen colonization and foci for disease development later in the handling system (Santos et al., 2002; Yahia, 2011).

4.2 Sap injury

Sap burn is one of the most important problems in mango handling that severely reduces fruit quality and its market value (Fig. 3b). The injury is caused by fruit skin contact with the

Figure 3 Post-harvest problems: (a) fruit wrapped in expanded polyethylene mesh sleeves for mechanical protection, (b) sap burn injured fruit with subsequent disease development, (c) lenticel spotting caused by a number of post-harvest treatments, (d) early stage of anthracnose development on ripening fruit, (e) stem end rot and (f) fruit with both mango seed weevil and fruit fly damage, to the right of the fruit an adult seed weevil can be seen.

sap exuded from the cut or broken peduncle and reduces consumer acceptance because of the browning and blackening of the skin after lenticel penetration. This injury can lead to secondary disease development. The Australian cultivar 'Kensington' is highly susceptible to sap burn (Lovey et al., 1992), while 'Irwin' is less susceptible. The sap component in 'Kensington', thought to cause the burn, is a major non-aqueous terpene component, terpinolene, which also causes sap burn in 'Irwin'. The predominant terpene (6.8%) in 'Irwin' is car-3-ene that does not cause sap burn and potentially imparts the characteristic flavour. Maqbool et al. (2009) similarly reported wide variations in susceptibility to sap burn, with 'Chaunsa' being more than eleven times more susceptible compared to 'Sindhri'. PPO and POX of peel are also involved in sap injury (John et al., 2002). The sap is present in the latiferous ducts of the fruit and is not interconnected with the stem ducts. When the peduncle is broken, the latex can move significantly and increase injury.

In addition to cultivar differences, the degree of injury from sap varies with time of harvest, stage of fruit maturity and growing conditions. Growing location has a significant effect on the composition of mango sap, but the effect appeared to be more related to differences in fruit maturity (Amin et al., 2008; Hassan et al., 2009; Barman et al., 2015). The amount of sap is also greater in early morning harvest than on fruit harvested later in the day (Maqbool et al., 2007, 2009). However, the severity of sap burn increases as daytime proceeded. The minimum sap burn injury occurs in fruit harvested and desapped in the morning hours (7 a.m.), whereas maximum sap injury was observed at noon.

Harvesting with the stem attached, draining with the pedicel down and washing are effective means of control. Post-harvest treatments are crucial to limit injury from the sap and to market a high-quality product. 'Langra' mango fruit de-sapped with 1% sodium hydroxide had about eleven fold less sap burn injury than the unwashed control (Barman et al., 2015). Combined anti-sap chemicals (1% calcium hydroxide) with a hot water quarantine treatment (HWQT: 48°C, 60 min) show less sap burn injury with no effect on fruit quality during storage of 'Samar Bahisht Chaunsa' (Jabbar et al., 2012). Fruit de-stemming and de-sapping in 0.5% lime solution with a 2 to 3 min dip followed with washing in tap water had less skin browning and sap burn to fruit washed with tap water without added lime (Mazhar et al., 2011). Adding a detergent (Tween 80) to 1.0% calcium hydroxide significantly reduced sap burn injury (Maqbool et al., 2009).

4.3 Lenticel spotting

Post-harvest practices to reduce sap burn are reported to increase lenticel spotting (discolouration) (Bezuidenhout et al., 2005). Washing mango fruit in a calcium hydroxide solution to prevent sap burn causes both red and black lenticel spotting. Red lenticel spotting is a physiological process that involves anthocyanin synthesis in response to low temperature and water entry into the lenticels. Black lenticel spotting is a physical process that involves the entry of water into the lenticels and the subsequent collapse and discolouration of the sub-lenticellular cells. The absence of a cork cambium and cork cells in the mango lenticel (Du Plooy, 2006) is possibly one of the most important reasons for lenticel discolouration (Fig. 3c).

Post-harvest handling and treatment such as a combination of water washes in the orchard and hot water spray applied over rollers without brushes in the packinghouse significantly reduce the severity of lenticel discolouration by 50–60% (Feygenberg et al., 2014). The USDA phytosanitary hot water treatment reduces the potential for subsequent red lenticel spotting during cold storage, but increased black lenticel spotting (de Assis et al., 2009). The induction of lenticel spotting caused by lowering storage temperature occurs in the Thai cultivars 'Nam Dok Mai See Thong' (Kaewprawet et al., 2008) and Philippines 'Carabao' mango (Rodeo and Esguerra, 2013). Moreover, irradiation induces lenticel discolouration in 'B74' mango cultivar (Hofman et al., 2009; Li et al., 2016) and 'Nam Dok Mai' mango (Jitareerat et al., 2009). Also, brushed 'Sufaid Chaunsa' mangoes irradiated at 700 Gy had significantly higher lenticels spotting (Malik et al., 2013). Since several post-harvest handling practices increase lenticel sensitivity, research is needed to develop acceptable handling protocols to facilitate mangoes distribution and marketing and minimizing lenticel spotting.

5 Post-harvest handling, environment and deterioration

5.1 Temperature

Typically, for each successive 10°C decrease in temperature, the rate of biological processes such as respiration is halved (Ryan, 1991). The respiration rate of 'Fajri' mango fruit stored at 12 and 14°C is 1 to 2 times higher than fruit kept at 10°C. In addition, fruit stored at 14°C showed significantly higher fruit skin shrivelling, fruit softening and disease incidence severity compared to those stored at 10 and 12°C (Raza et al., 2013). Baloch and Bibi (2012) indicated that the ripening rate of mango 'Langra' and 'Samar Bahisht Chaunsa' is increased and the post-harvest life decreased with an increase in storage temperature evaluated by changes in skin colour, TSS and carotenoids. Moreover, rapid removal of field heat enhances the quality and prolongs the post-harvest life of mango fruit (Baloch et al., 2011).

Mango is susceptible to CI when stored below 13°C (Singh and Zaharah, 2015). The symptoms of CI in mangoes are greyish brown or darkening of skin, slight discolouration of the mesocarp and abnormal ripening (Chidtragool and Ketsa, 2010; Miguel et al., 2011). CI severity varies with cultivars, 'Choke Anan' shows less severity than Nam Dok Mai No. 4 (Chongchatuporn et al., 2013). Siriphanich and Kunyamee (2010) also found that 'Nam Dok Mai' mango develop CI in both the peel and the mesocarp, whereas 'Chok Anan' cultivar developed CI only on the peel. No CI was found in 'Hongsawadee' mango stored at 5°C. Wongmetha et al. (2013) reported 'Irwin' mango treated with 5 μL L^{-1} 1-MCP for 12 h, coated with 0.5% chitosan or packed in PE bag and then stored at 10°C still showed CI symptoms. The peel browning of mango stored at low temperature correlates with the activity of phenylalanine ammonia lyase (PAL) and not with total free phenolics in the peel (Chidtragool et al., 2011). Storage at chilling temperatures promotes free radical accumulation that results in CI development.

The antioxidant capacity varies during storage and ripening. 'Nam Dok Mai' mango mesocarp and peel antioxidant capacity increases during 6 days of storage and thereafter declined at both 15 and 25°C (Mekwatanakarn and Chairat, 2013). The antioxidant capacity of mango mesocarp stored at 15°C is about two-fold higher than that stored at 25°C. Agillon and Lizada (2010) reported that total phenols increase with symptom expression and the susceptibility of CI is associated with a high ratio of saturated/unsaturated fatty acids.

Post-harvest treatments that strengthen the antioxidant defence system can reduce CI symptoms during cold storage. Treatment with 0.1 mM methyl jasmonate and 1 mM salicylic acid (SA) with 'Nam Dok Mai No. 4' (Junmatong et al., 2015), 5 mM oxalic acid solution for 10 min with 'Zill' (Li et al. 2014) and 2 mM SA solutions for 5 min with 'Chausa' (Barman and Asrey, 2014) reduced CI symptom development. Li et al. (2014) suggested that the application of oxalic acid improved chilling tolerance by elevating proline accumulation. Proline accumulation is associated with an increase in Δ1-pyrroline-5-carboxylate synthetase (P5CS) activity and a decrease in proline dehydrogenase (PDH) activity in both the peel and the flesh.

5.2 Relative humidity

Moisture loss due to storage of pre-climacteric mango fruit at low relative humidity hastens ripening. A linear negative relationship exists between water loss and green life of mango (MacNish et al., 1997). Mango fruit ('Manila') at mature green stage when treated with moist (95% relative humidity) hot forced air (43°C, 2.5 m s^{-1} for 220 min) delays colour development, fruit softening, weight loss and β-carotene biosynthesis (Ornelas-Paz and Yahia, 2014). Most recommendations suggest storage at 90 to 95% RH.

5.3 Storage atmosphere

CA storage in combination with optimum storage temperature prolongs storage life and maintains fruit quality including aroma volatiles, though the response depended upon the cultivar (Singh and Zaharah, 2015). 'Ewase' mango fruit stored with 7% CO_2 had lower respiration rate and better retention of fruit firmness (Dorria et al. 2015). Storage with 3% O_2 in combination with 6% CO_2 at 13°C extends the storage life of the Australian mango cultivars 'Kensington Pride' and 'R2E2' up to six weeks (Singh and Zaharah, 2013). 'Palmer' mango fruit stored in low oxygen concentrations (1%, 5% and 10% O_2) had significantly lower respiration rate, lower soluble pectin and total soluble sugars, but did not differ in colour when compared with fruit stored in 15 and 21% O_2 (Teixeira and Durigan, 2011). Mango 'Mahajanaka' stored in 5% O_2 and 10% CO_2 for 2 days at 25°C and then transferred to ambient condition had reduced respiration and ethylene production, whereas storage in 100% O_2 stimulated ethanol accumulation (Nimitkeatkai et al. 2008).

5.4 Packaging

Packaging's role is to protect products for distribution, storage and marketing (Soroka, 2014). Fresh produces including mango are perishable crops and rapidly deteriorate after harvesting. Thus, packaging can both physically protect the fruit and potentially be used to modify the storage atmosphere (MA). MA is most frequently used to prolong post-harvest life of fruits and vegetable for both local and distant markets. An MA of ~5% CO_2 and ~ 10% O_2 created in the Xtend® film is effective in reducing CI and the level of sap of 'Tommy Atkins' and 'Keitt' mangoes stored at 12°C (Pesis et al. 2000). Kumar et al. (2013) also found that 'Baneshan' mango at the M2 stage of maturity (8 to 9% TSS) packed in Xtend® bags had better retention of quality and extended post-harvest life up to 28 days at 12.5°C. Mature green mangoes (cvs. 'Alphonso' and 'Banganapalli') individually shrink wrapped with semi-permeable Cryovac films® (D-955; 15 µm thickness) can similarly be stored for up to 5 weeks at 8°C. 'Espada' mango fruit packed in PVC and stored at room temperature kept the peel and mesocarp colour for 13 days, compared with 9 days for non-packaged fruit stored at 12°C and 80–90% RH (Galli et al., 2013). Wanikanukul et al. (2007) reported that using high gas-permeable packaging (oxygen transfer rate (OTR) >15,000 cc/m^2/day) delays ripening and disease incidence of 'Nam Dok Mai See Thong' mango stored at 13°C. The post-harvest life of 'Nam Dok Mai' mangoes is extended to 40 days when packed in non-perforated ethylene-absorbing highly gas-permeable film (HNPE), 35 days with highly gas-permeable films (HNP) and 30 days with micro-perforated highly gas-permeable films (HMP) compared to 20 days with the control and 5 days with common non-perforated PE film (Boonruang et al., 2012). Packages that contain activated carbon (25% w/w) blended with old corrugated container pulp delays 'Nam Dok Mai' softening, colour change, the TSS increase and titratable acidity decrease, and extended post-harvest life up to 20 days at 20°C (Rachtanapun et al. 2010). The storage life of 'Nam Dok Mai See Thong' stored at 5°C and packed in PE terephthalate box with 1.5 cm^2 m^2 perforated holes is 28 days without CI (Kumpoun and Uthaibutra 2013). Ullah et al. (2012) reported that 'Alphonso' mango packed in PE bags ripened more slowly as indicated by skin colour, texture and TSS changes than fruit packed in 0.037 mm PE bags and stored at ambient condition. At ambient temperature, CO_2 accumulated and CO_2 injury occurred.

MA packaging not only maintains post-harvest quality but also reduces storage disorder including CI. Chidtragool and Ketsa (2013) found that MA packaging reduced CI in 'Nam Dok Mai' mango fruit stored at 4°C and proposed that it was due to increase in peel total free phenolics content while decreasing peel PAL and PPO activities. 'Alphonso' cultivar wrapped with LD-935 film have higher total antioxidant capacity and DPPH radical scavenger (Rao and Shivashankara 2014) that could contribute to the reduced CI.

A combination of MA packaging and post-harvest treatment effectively enhances the post-harvest life of mango. Active packaging film that release SO_2 suppresses the decline in fruit firmness and TSS increase, minimizing weight loss of 'Dashehari' mango fruit during 21 days storage at room temperature (Xu et al. 2012). Mango 'Baneshan' treated with 1-MCP at 1000 ppb and packed in Xtend® bags had better quality and post-harvest life than fruit stored in polypropylene during storage at 12.5°C (Kumar et al. 2015). Ramayya et al. (2012) observed that dipping fruit in hot water at 40°C for 40 min, followed by packaging under 50% CO_2 (balance nitrogen) in bags made of orientated polypropylene maintained fruit quality better than mangoes packed under 25 and 75% CO_2 that showed spoilage at 10°C. In addition, 'Dashehari' mango treated with Bavistin (500 ppm) and wax emulsion (6%) and held in perforated polythene bag showed a maximum TSS (17.5%) and minimum titratable acidity (0.26%) (Singh et al. 2015).

5.5 Diseases

Production locality, cultivar and nutrition (nitrogen content of fruit peel) have a significant effect on post-harvest disease development (Fiaz et al., 2016). Several fungal pathogens have been reported as pre- and post-harvest diseases of mango, including anthracnose (*Colletotrichum gloeosporioides*), stem end rot (*Lasiodiplodia theobromae*) and fruit rot (*Dothiorella dominicana, D. mangiferae, Phomopsis mangiferae, Aspergillus niger, A. flavus, Rhizopus stolonifer, Pestalotia* sp., *Macrophomina phaseolina, Penicllium purpurogenum, Penicillium* sp., *Cladosporium* sp., *Alternaria* sp., *Nigrospora* sp., *Pestalotia* sp., *Chaetomium* sp., *Helminthosporium* sp., *Papularia* sp., *Periconia* sp., *Phomopsis* sp. and *Phoma* sp. (Sangchote, 1987; Butharasart, 1988; Bhuvaneswari and Rao, 2001; Gutiérrez-Martínez et al., 2012, Freeman et al., 2014; Rehman et al., 2015; Fiaz et al., 2016). Significant variation exists between different growing areas or localities in disease incidence and severity (Fiaz et al., 2016). *C. gloeosporioides*, the causal agent of anthracnose, is the most serious problem for mango worldwide that leads to high fruit losses during marketing (Fig. 3d) (Akem, 2006; Diedhiou et al. 2014). Stem end and fruit rots (Fig. 3e) are caused by a number of fungi and are the second most important cause of losses. These rot organisms enter through the broken fruit peduncle or breaks in the cuticle often caused by abrasion injuries.

C. gloeosporioides normally infects mango fruit in the orchard (Fig. 3d). After spore germination and cuticle penetration (appressorium formation), fungi growth stops and the hyphae becomes quiescence (dormant) until the fruit starts to ripen. Germination of spores (conidia) and appressorium formation requires free water or relative humidity above 95% (Jeffries et al. 1990; Arauz, 2000). Conidia can survive for 1 to 2 weeks at humidity as low as 62%. In pre-harvest disease management practices that include fungicide sprays, use of biocontrol agents and fruit bagging can significantly retard pre- and post-harvest disease development.

Numerous fungicides have been evaluated over the years to control anthracnose in the field with varying levels of success as the fungi after infection and appressorium formation on the young fruit is quiescence until the onset of ripening (Prusky, 1996; Arauz, 2000). Many post-harvest fungicides have little effect on quiescent (latent) infection, such as anthracnose. Systematic fungicides (benomyl, thiabendazole, thiophanate-methyl, carbendazim, prochloraz, imazalil) have strong inhibitory effects against quiescent infections. These fungicides are able to act systematically and penetrate the wax and cuticle of plant surface to reach and inhibit the pathogen (Phillips, 1975; Sharma et al., 1994; Davidse et al. 1988; Barkai-Golan, 2001). Thiabendazole was best at inhibiting *C. gloeosporioides* growth and reducing anthracnose disease on mango. Non-systemic fungicides (captan, ziram, chlorothalonil) are less effective against anthracnose (Banik et al. 1998; Gajbhiye et al. 2000; Barkai-Golan, 2001; Sangchote 2013). The efficacy of fungicides can be improved by mixing the fungicides with wax to ensure a more uniform fruit coating (Fonseca et al., 2001; Shivarama Reddy and Thimma Raju, 1989).

For organic mango production fungicide use is limited, with biocontrol agents and bagging being alternatives (Chuang and Ann 1997). Natural compounds are known to be a source of antifungal agents, such as chitosan, essential oils, plant extracts, acetaldehyde, ethanol and ozone. Vivekananthan et al. (2004) suggested that spraying the biological control agent (*Pseudomonas fluorescens* FP7) incorporated with chitin significantly reduced anthracnose incidence both in the orchard and after harvest. Chitosan are widely used as a coating on fruit and it has protective mechanisms that are both direct (Prasothong et al., 2011; Wang et al. 2015; de Oliveira Junior et al. 2012) and indirect (Jitareerat et al., 2007; Baker and Orlandi, 1995; Zeng et al., 2010; Sánchez-Domínguez et al., 2011; El Hadrami et al., 2010). Bee propolis (Pastor et al., 2010; Ben-Hur Mattiuz et al. 2015), *Areca catechu* L. pericarp extract, fernenol, arundorin, mixtures of stigmasterol and beta-sitosterol (Yenjit et al. 2010), essential oil from *Lippia scaberrima* (Regnier et al. 2008), eucalyptus wood vinegar extract (Naychawna and Nalumpang, 2014), fumigating with 60 ppm hexanol for 24 hours (Denumbunchachai and Sangchote, 2013) and ethanol dipping (Gutiérrez-Martínez et al. 2012) also suppress anthracnose disease development.

Compounds that are generally recognized as safe (GRAS) have been used since the 1920s for post-harvest disease control. These compounds either rapidly degrade or have low or no toxic residues on the fruit (Bakai-Golan, 2001). Mango anthracnose control has been shown with bicarbonates, carbonates, calcium chloride, acetic acid, oxalic acid, benzoates and sorbates (Denumbunchachai and Sangchote 2013; Koslanund et al., 2015; Palou et al., 2016). Sodium salts of the bicarbonates and carbonates that are alkaline in solution and if applied at 40 to 50°C are more effective (Palou et al., 2016).

Physical treatments such as pre-harvest bagging can improve post-harvest fruit quality, extending post-harvest life and helping post-harvest disease control. Fruit bagging with a pre-bagging fungicide application is widely used (Fig. 2e) and significantly reduces disease incidence (Sangchote, 2013). The incidence of anthracnose declines from 80% to 50% after fruit are covered with white paper bags (Hofman et al., 1997). Chiangsin and Sangchote (2012) studied brown waterproof kraft bags with black liners (carbon bag) on disease occurrence and severity. New unused carbon bags and re-used carbon bag (one season) were compared with unbagged fruit. No significant difference in the disease incidence (65-70%) and disease severity (3.6-5.5%) was found between the new and re-used carbon bag, but they were significantly different from non-bagged fruit (92% unbagged, 17% bagged).

Post-harvest heat treatment (hot water, hot air and vapour heat) (Prusky et al. 1999; Sopee and Sangchote, 2003; Fallik, 2004; Sripong et al. 2015), gamma irradiation (Johnson et al., 1990; Jitareerat et al. 2006;2010), UV-C irradiation (Zainuri et al. 2005; Sripong et al. 2015; Suktawee et al. 2011), short-wave infrared (Lonsdale and Droomer, 1994) and radio frequency (Muenmanee, 2015) have been used to evaluate mango post-harvest disease control. Hot water treatments are widely preferred, as they are easy to apply, leave no residues and are economical at controlling anthracnose (Lonsdale, 1993; Lurie, 1998; Sunsuwan and Sardsud, 2003; Fallik, 2004; Sui et al., 2016; Sangchote 2013). The efficacy of hot water treatment on post-harvest disease control of mango can be enhanced when coupled with other treatments such as chemicals (Phongmee and Wisaratranon, 1990; Lonsdale et al. 1993), organic salts (Dessalegn et al. 2013), ethanol (Bautista-Baños et al., 2006; Gutiérrez-Martínez et al. 2012; Sangchote and Seehahai, 2008), food additives (Zhu et al., 2007; Pongpisutha et al. 2012), chitosan (Faoro et al., 2008; Djioua et al. 2010), biocontrol agents (Villiers and Korsten, 1994; Govender et al. 2005), UV-C (Sripong et al. 2015), gamma irradiation (Gagnon et al., 1993) and fungicides (Phongmee and Wisaratranon, 1990). Vapour heat used for insect control may also help to retard fungal decay from wound infection (Barkai-Golan, 2001). Hot water treatment may increase the rate of ripening (Jacobi and Giles, 1997; Sripong et al. 2015).

Biocontrol agents (antagonists) such as yeast and bacteria have potential for mango post-harvest disease control (Chuang and Ann, 1997; Barkai-Golan, 2001; Bhuvaneswari and Rao, 2001; Govender et al. 2005; Kefialew and Ayalew, 2008; Zheng et al. 2013; Bautista-Rosales et al., 2014). The mode of action of antagonists includes the following interactions: 1) secretion of antimicrobial compounds; 2) competition for nutrients at the wound site; 3) direct effect of the antagonist that may produce and secrete cell wall degradation enzymes and pathogen inhibitory volatiles; and 4) induction of plant defence mechanism.

Natural biochemical disease resistance compounds can also be induced by abiotic and biotic elicitors. The elicitors act in addition to preformed inhibitory compounds. High anthocyanin and flavonoids in red fruit have been correlated with reduced mango disease incidence (Sivankalyani et al. 2016). Antifungal compounds such as phenolic acid, coumaric acid and resorcinol (Hassan et al., 2007) and plant defence-related enzymes such as chitinase, β-1,3-glucanase, PAL and POX (Jitareerat et al. 2007; Lin et al., 2011) are involved in resistance to *C. gloeosporioides* infection. Neither of the resistance-inducing elicitors nitric oxide (NO) or SA have antifungal property against *C. gloeosporioides*. The application of NO and SA on mango fruit elicits the activity of enzymes associated with plant defence, such as chitinase (CHI), β-1,3-glucanase (GLU), PAL, cinnamate hydroxylase (C4H), 4-coumarate:CoA ligase (4CL) and POD. The elicitors also enhance generation rate of superoxide radicals and hydrogen peroxide levels (Zeng et al., 2006; Zhu et al., 2008; Hu et al., 2014; Zhang et al., 2013; Junmatong et al., 2015). Hot water treatment combined with UV-C irradiation also elicits as a stress response and increases the key plant defence-related genes (*PAL, POD, CHI, GLU*) in both peel and mesocarp of mango resulting in anthracnose disease suppression (Sripong et al., 2015).

6 Mango modification

6.1 Ethylene response (1-MCP)

The application of 1-methylcyclopropene (1-MCP) prolongs the post-harvest life of a number of mango cultivars: 'Kent' (Ngamchuachit et al., 2014; Osuna-Garcia et al., 2009, 2015), 'Baneshan' (Kumar et al., 2015), 'Keitt' (Ngamchuachit et al., 2015), 'Alphonso' (Sonune et al., 2011; Burondkar et al., 2013), 'Irwin' (Wongmetha et al., 2013) and 'Nam Dok Mai Sri Thong' (Chaiprasart and Hansawasdi, 2009). The application of 1-MCP delays mango fruit ripening and softening by suppressing autocatalytic ethylene production, the respiratory climacteric peak, and delaying the activities of fruit softening enzymes such as exo-PG, endo-PG and endoglucanases (EGase) (Razzaq et al., 2016; Chourasia et al., 2008); reducing ACC oxidase activity (Wang et al., 2009) and ascorbic acid losses (Islas-Osuna et al., 2010); and delaying the breakdown of cellulose and hemicellulose (Chourasia et al., 2008). Applying 1-MCP also impacts the free radical production with reduced levels of H_2O_2 and lipid peroxidation (Singh and Dwivedi, 2008) and the activities of antioxidant enzymes including SOX, CAT and APX (Wang et al., 2009), leading to reduction in fruit softening. The efficacy of 1-MCP in delaying fruit ripening depends on growing area, fruit maturity (Ambuko et al. 2013, 2016), storage temperature (Faasema et al., 2014), and 1-MCP concentrations and exposure time (Watkins, 2008).

6.2 Waxing and wraps

Waxing or coatings has been used to reduce weight loss, maintain quality, add gloss and extend post-harvest life of fruits and vegetables. The coating acts as a barrier to water loss and gas exchange between the fruit and the atmosphere. Waxing 'Manila' mango maintained fruit firmness for up to 18 days at 13°C and reduced weight loss and shrivel (Vasquez-Celestino et al., 2016). Gum arabic, carnauba wax, chitosan, aloe vera gel and their combinations all reduce the respiration rate and in addition reduce weight loss in 'Choke Anan' mango (Intalook et al., 2006; Khaliq et al., 2016). Chitosan combined with aloe vera gel showed the greatest reduction in weight loss and anthracnose disease incidence (Intalook et al., 2006). The post-harvest life of treated mango at 25 and 13°C were 12 and 28 days, respectively, whereas non-coated fruit was 6 and 20 days. Similar results were reported by Noiwan (2006), 1% chitosan coating delayed the ripening 'Mahajanaka' mango fruit. Carboxymethyl cellulose- and shellac-based coatings also delay weight loss, the changes of peel and mesocarp colour, and maintain fruit firmness (Rachtanapun et al., 2008).

6.3 Irradiation

Ionizing technique has been used for disinfestation and to maintain the quality of fruit during storage. Insect disinfestation is required by some importing countries to prevent the spread of fruit flies and mango seed weevil that can cause significant damage (Fig. 3f). UV-C irradiation effectively maintained fruit firmness and delays the increase in ripening index of 'Kaew Kamin' mango fruit (Promyu and Supapvanich, 2016). The decrease in polyphenolic substances in mango peel and mesocarp during storage can be controlled by UV-C treatment (Chatha et al., 2014). UV-C is also effective in minimizing the symptoms of CI in 'Tommy Atkins' (Miguel et al., 2016). Mango fruit pre-treated with 5 kJ m^{-2} UV-B for 4 h before 10 days storage at 6°C had significantly lower CI index, ion leakage, MDA content, and NO levels than that of the control during fruit ripening at ambient temperature (Ruan et al., 2015). UV-B treatment by reducing oxidative stress alleviated oxidative damage or enhanced chilling tolerance and delayed fruit ripening of mangoes by enhancing antioxidant compounds and antioxidative enzymes or triggering endogenous NO generation in the fruit (Jiang et al., 2015; Ruan et al., 2015).

The response of mango to gamma irradiation varies with cultivar and dosage (Hofman et al., 2010; San et al., 2016). Santos et al. (2015) revealed that low dose of gamma irradiation (0.25, 0.35 and 0.45 kGy) reduced the severity of the pathogen (*Lasiodiplodia theobromae*) on ripening mango with no effect on TSS, ascorbic acid and titratable acidity. 'Nam Dok Mai See Thong' treated with gamma irradiation at 1.0 and 1.5 KGy had reduced anthracnose and stem end rot, but the fruit showed peel damage (Fig. 2f) (Wisutiamonkul and Suwanagul, 2010). Yadav et al. (2013) observed higher TSS, reducing sugars and ascorbic acid content and minimum acidity in fruit treated with 0.40 kGy and stored at 12°C compared to non-irradiated fruit stored at ambient conditions. Chatha et al. (2014) found that the decrease in polyphenolic substances in mango peel and mesocarp during storage can be controlled using gamma irradiation, especially with lower doses (0.5 KGy). Mahto and Das (2013) reported that gamma irradiation injury occurred in 'Dushehri' mango treated with 6 to 10 kGy and in 'Fazli' with 1 to 10 kGy. 'Sindhri', 'Samar Bahisht Chaunsa', 'Sufaid Chaunsa' and 'Kensington Pride' subjected to 0.5 to 1 kGy had significantly less peel colour development compared to lower doses (Malik et al., 2012; San et al., 2016). The lower doses did not affect firmness, lenticels' injury, weight loss, biochemical and organoleptic properties or reduce disease development (Malik et al., 2012; San et al., 2016). 'Kensington Pride' fruit exposed to 0.5 and 1.0 kGy had a 58 and 80% reduction in emission of α-terpinolene volatiles, respectively (San et al., 2016).

Under certain conditions gamma irradiation induces fruit softening in 'Chok Anan', but not 'Nam Dok Mai' mangoes (Uthairatanakij et al., 2006). Silva et al. (2012) found that 'Tommy Atkins' mango fruit showed no alterations in the cell wall structure, middle lamella and plasma membrane when analysed immediately after irradiation at 0.5 and 1.0. kGy. Mesocarp cell wall structure, middle lamella and the plasma membrane showed changes after storage for 20 days at 12°C, followed by 5 days at 21°C, following irradiation at 0.5 and 1.0. kGy.

6.4 Chemical treatments

Sodium nitrophenolate (SNP) at 1 mM for 10 min helps to maintain quality and extend the post-harvest life of 'Nam Dok Mai Si Thong' mango fruit (Tran et al., 2015). Barman et al. (2014) also observed that dipping in 1.5 mM SNP reduced the incidence of CI by 1.5- to 1.7-fold compared to the control. NO has been shown to inhibit mango fruit softening (Zaharah and Singh, 2011). Fumigation of 'Kensington Pride' mango with 20 μL L^{-1} NO for 2 h at 21°C suppressed ethylene biosynthesis through the inhibition of the ACS and ACO activities, leading to decrease in ACC content. NO fumigated fruit also maintained higher mesocarp firmness due to decrease in the activities of exo-PG, endo-PG and EGase during cool storage and ripening. Oxalic acid has been shown to reduce fruit softening and exo-PG activity during ripening at ambient conditions and in cold storage (Razzaq et al., 2015). Oxalate treatments of fruit stored at 25°C reduces fruit decay, delays the fruit ripening (Zheng et al., 2012) by increasing the activities of PPO and POXe and also induces total phenolic content in the peel. 'Samar Bahisht Chaunsa' mango fruit treated with oxalic acid have enhanced activities of antioxidative enzymes such as SOD, CAT and POX (Razzaq et al., 2015). Ozone at 10 µl L^{-1} significantly decreases respiration rate and reduces ethylene production resulting in lower *L** and *b** values, but did not affect fruit firmness, TSS and titratable acidity in 'Nam Dok Mai No. 4' mango fruit (Tran et al., 2015).

6.5 Post-harvest light treatment

The blush associated with anthocyanin accumulation in the skin is a desirable consumer trait. Harvested mango exposed to blue light induces skin blush in mango fruit, thus enhancing their commercial value (Cao et al., 2016). To obtain skin blush, blue light-emitting diodes were applied to harvested mango fruit skin when stored at 12°C.

7 Future trends and conclusion

Mango presents some unique challenges in post-harvest handling to avoid deterioration in fruit quality. It is critical that the pre-harvest production practices focus on being able to harvest blemish-free fruit of suitable shape, size and colour to meet consumer demands with wide seasonal availability. The pre-harvest practices that show the most promise are those that include cultivar planted, and higher planting densities with shorter tree (<2.5 m) that allows all tree management operations to be done from the ground such as pruning, fruit thinning and bagging. Harvesting at the correct stage of maturity for the market being served is essential, along with the avoidance of all possible mechanical injury during handling and packing. Post-harvest disease incidence and severity is frequently a pre-harvest associated issue and spray programmes and bagging greatly assisting in reducing post-harvest disease. A number of post-harvest market requirements such as insect disinfestation place additional stress on mango fruit and need to be carefully applied to avoid quality loss. The use of 1-MCP and fruit coating help in maintaining quality and post-harvest life. To achieve high-quality fruit and avoid deterioration post-harvest requires an integrated programme from the field to the consumer, all processes must be cost-effective, environmentally sound and ensure layers of protection to allow marketing of a safe high-quality fruit.

8 References

Abu-Sarra, A. F. and Abu-Goukh, A. A. 1992, 'Changes in pectinesterase, polygalacturonase and cellulase activity during mango fruit ripening', *Journal of Horticultural Science*, vol. 67, pp. 561–8.

Agillon, A. B. and Lizada, M. C. C. 2010, 'Responses of 'Carabao' Mango (*Mangifera indica*) Fruit to Chilling Stress', *Acta Horticulturae*, vol. 877, pp. 467–74.

Aguilar-Ortigoza, C. J. and Sosa, V. 2004, 'The evolution of toxic phenolic compounds in a group of Anacardiaceae genera', *Taxon*, vol. 53, no. 2, pp. 357–64.

Ajila, C. M., Naidu, K. A., Bha, S. G. and Rao, U. P. 2007, 'Bioactive compounds and antioxidant potential of mango peel extract', *Food chemistry*, vol. 105, 982–8.

Akem, C. N. 2006, 'Mango anthracnose disease: Present status and future research priorities', *Plant Pathology Journal*, vol. 5, pp. 266–73.

Ali, Z. M., Abu, S. and Lazan, H. 1990, 'Glycosidases in ripening mango', *Transactions of the Malaysian Society of Plant Physiology*, vol. 1, pp. 11–14.

Ali, Z. M., Armuga S. and Lazan, H. 1995, 'B-Galactosidase and its significance in ripening mango fruit', *Phytochemistry*, vol. 38, pp. 1109–14.

Ambuko, J., Githiga, R. W., Hutchnison, M. J., Gemma, H. and Owino, O. 2013, 'Effect of maturity stage and cultivar on the efficacy of 1-MCP treatments in mango fruits', *Acta Horticulturae*, vol. 1007, pp. 39–48.

Ambuko, J., Onsongo, N. K., Hutchinson, M. J. and Owino, W. O. 2016, 'Response of "Apple" mango fruit to 1-methylcyclopropene as affected by agro-ecological zone and maturity stage', *Acta Horticulturae*, vol. 1111, pp. 347–54.

Amin, M., Malik, A. U., Mazhar, M. S., Din, I. U., Khalid, M. S. and Ahmad, S. 2008, 'Mango fruit desapping in relation to time of harvesting', *Pakistan Journal of Botany*, vol. 40, pp. 1587–93.

Arauz, L. F. 2000, 'Mango Anthracnose: Economic Impact and Current Options For Integrated Management', *Plant Disease*, vol. 84, pp. 600–11.

Ashraf, M., Khan, N., Ahmed, M. and Elahi, M. 1981, 'Studies on the pectinesterase activity and some chemical constituents of some Pakistani mango varieties during storage and ripening', *Journal of Agriculture and Food Chemistry*, vol. 29, pp. 526–8.

Baker, C. J. and Orlandi, E. W. 1995, 'Active oxygen in plant pathogenesis', *Annual Review of Phytopathology*, vol 33, pp. 299–321.

Bally, I. S. E. 2007, 'The effect of preharvest nutrition and crop load on fruit quality and postharvest disease in mango (*Mangifera indica* L.)', PhD Thesis University of Queensland, https://www.researchgate.net/profile/ian_bally/publication/43495891_the_effect_of_preharvest_nutrition_and_crop_load_on_fruit_quality_and_postharvest_disease_in_mango_Mangifera_indica_L/links/5408d9520cf2187a6a6ea432.pdf Accessed 2016 July 15.

Baloch, M. K. and Bibi, F. 2012, 'Effect of harvesting and storage conditions on the post harvest quality and shelf life of mango (*Mangifera indica* L.) fruit', *South African Journal of Botany*, vol. 83, pp. 109–16.

Baloch, M. K., Bibi, F. and Jilani, M. S. 2011, 'Quality and shelf life of mango (*Mangifera indica* L.) fruit: As affected by cooling at harvest time', *Scientia Horticulturae*, vol 130, no. 3, pp. 642–6.

Banik, AK, Kaiser, SAKM and Dhua, R. S. 1998, 'Evaluation of some systemic and non-systemic fungicides against Botryodiplodia theobromae, the cause of dieback disease of mango (*Mangifera indica* L.)', *Journal of Soils and Crops*, vol. 8, pp. 199–222.

Barkai-Golan, R. 2001, *Postharvest Diseases of Fruits and Vegetables: Development and Control*. Elsevier B.V. Amsterdam, The Netherland. Pp. 418

Barman, K. and Asrey, R. 2014, 'Salicylic acid pre-treatment alleviates chilling injury, Preserves bioactive compounds and enhances shelf life of mango fruit during cold storage', *Journal of Scientific and Industrial Research*, vol. 73, no. 11, pp. 713–8.

Barman, K., Asrey, R., Pal, R. K., Jha, S. K. and Bhatia, K. 2014, 'Post-harvest nitric oxide treatment reduces chilling injury and enhances the shelf-life of mango (Mangifera indica L.) fruit during low-temperature storage', *Journal of Horticultural Science and Biotechnology*, vol. 89, no. 3, pp. 253–60.

Barman, K., Asrey, R., Pal, R. K., Jha, S. K. and Sharma, S. 2015, 'Influence of different desapping agents on the incidence of sapburn, ripening behaviour and quality of mango', *Journal of Food Science and Technology*, vol. 52, no. 1, pp. 161–70.

Barman, K., Patel, V. B., Singh, R. R., Prasad, M. and Kumari, P. 2015, 'Effect of harvesting time and desapping on Sapburn and quality in mango (Mangifera indica) cv. Langra', *Indian Journal of Agricultural Sciences*, vol. 85, no. 7, pp. 945–9.

Barmore, C. R. and Mitchell, E. F. 1975, 'Ethylene pre-ripening of mangoes prior to shipment', *Proceedings of the Florida State Horticultural Society*, vol. 88, pp. 469–71.

Bautista-Baños, S., Hernández-Lauzardo, A. N., Velázquez-del Valle, M. G., Hernández-López, M., Ait Barka, E., Bosquez-Molina, E. and Wilson, C. L. 2006, 'Chitosan as a potential natural compound to control pre and postharvest diseases of horticultural commodities', *Crop Protection*, vol. 25, pp. 108–18.

Bautista-Rosales, P. U., Calderon-Santoyo, M., Servín-Villegas, R., Ochoa-Alvarez, N. A., Vazquez-Juarez, R. and Ragazzo-Sanchez, J. A. 2014, 'Biocontrol action mechanisms of *Cryptococcus laurentii* on *Colletotrichum gloeosporioides* of mango', *Crop Protection*, vol. 65, pp. 194–201.

Ben-Hur Mattiuz, B. H., Ducamp-Collin, M. N., Mattiuz, C. F. M., Vigneault, C., Marques, K. M., Sagoua, W. and Montet, D. 2015, 'Effect of propolis on postharvest control of anthracnose and quality parameters of 'Kent' mango', *Scientia Horticulturae*, vol. 184, pp. 160–8.

Bender, R. J., Brecht, J. K., Baldwin, E. A. and Malundo, T. M. M. 2000, 'Aroma volatiles of mature-green and tree-ripe Tommy Atkins mangoes after controlled atmosphere vs. air storage', *HortScience*, vol. 35, pp. 684–6.

Berardini, N., Carle, R. and Schieber, A. 2004, 'Characterization of gallotannins and benzophenone derivatives from mango (*Mangifera indica* L. 'Tommy Atkins') peels, pulp and kernels by high-performance liquid chromatography/electrospray ionization mass spectrometry', *Rapid Communications in Mass Spectrometry*, vol. 18, pp. 2208–16.

Berardini, N., Fezer, R., Conrad, J., Beifuss, U., Carle, R. and Schieber, A. 2005a, 'Screening of mango (Mangijera indica L) cultivars for their contents of flavonol O- and xanthone C-glycosides, anthocyanins and pectin', *Journal of Agriculture and Food Chemistry*, vol. 53, pp. 1563–70.

Berardini, N., Kniidler, M., Schieber, A. and Carle, R. 2005b, 'Utilization of mango peels as a source of pectin and polyphenols', *Journal of Food Science and Emerging Technologies*, vol. 6, pp. 442–52.

Bezuidenhout, J. L. J., Robbertse, P. J. and Kaiser, C. 2005, 'Anatomical investigation of lenticel development and subsequent discolouration of "Tommy Atkins" and "Keitt" mango (*Mangifera indica* L.) fruit', *Journal of Horticultural Science and Biotechnology*, vol. 80 no. 1, pp. 18–22.

Bhagyalakshmi, N., Prabha, T. N., Yashodha, H. M., Prasanna, V., Jagadeesh, B. K. and Tharanathan, R. N. 2002, 'Biochemical studies related to textural regulation during ripening of banana and mango fruit', *Acta Horticulturae*, vol. 575, pp. 717–24.

Bhuvaneswari, V. and Rao, M. S. 2001, 'Evaluation of Trichoderma viride antagonistic to post-harvest pathogens on mango', *Indian Phytopathology*, vol. 54, no. 4, pp. 493–4.

Boonruang, K., Chonhenchob, V., Singh, S. P., Chinsirikul, W. and Fuongfuchat, A. 2012, 'Comparison of various packaging films for mango export', *Packaging Technology and Science*, vol. 25, no. 2, 107–18.

Brecht, J. K. and Yahia, E. M. 2009, 'Postharvest Physiology', in Litz, R. E. (ed.), *The Mango: Botany, Production and Uses*, 2nd Edition, CAB International, Wallingford, United Kingdom, pp. 484–528.

Brinson, K., Dey, P. M., John, M. A. and Pridham, J. B. 1988, 'Post-harvest changes in *Mangifera indica* mesocarp cell walls and cytoplasmic polysaccharides', *Phytochemistry*, vol. 27, pp. 719–23.

Brown, B. I., Scott, K. J. and Chaplin, G. R. 1981, 'Spongy stem end in mango', *Queensland Fruit Veg News*, vol. 52, pp. 497.

Burdon, J. N., Moore, K. G. and Wainwright, H. 1991, 'Mineral distribution in mango fruit susceptible to the physiological disorder soft-nose', *Scientia Horticulturae*, vol. 48, pp. 329–36.

Burondkar, M. M., Mane, A. V., Mote, M. S. and Jadhav, B. B. 2013, 'Ripening behavior, storage and quality of 'Alphonso' mangoes as influenced by 1-methyl cyclopropene', *Acta Horticulturae*, vol. 992, pp. 597–606.

Burg, S. P. and Burg, E. A. 1962, 'Role of ethylene in fruit ripening', *Plant Physiology*, vol. 37, pp. 179–89.

Buzby, J. C., Bentley, J. T., Padera, B., Campuzano, J. and Ammon, C. 2016, 'Updated Supermarket Shrink Estimates for Fresh Foods and Their Implications for ERS Loss-Adjusted Food Availability Data', EIB-155, U.S. Department of Agriculture, Economic Research Service, June 2016, 32 p. http://www.ers.usda.gov/media/2105755/eib155.pdf. Accessed 08 July 2016.

Campbell, R. J. and Campbell, C. W. 1992, 'Commercial Florida mango cultivars', *Acta Horticulturae*, vol. 341, pp. 55–9.

Cano, M. P. and de Ancos, B. 1994, 'Carotenoid and carotenoid ester composition in mango fruit as influenced by processing method', *Journal of Agricultural and Food Chemistry*, vol. 42, pp. 2737–42.

Castrillo, M., Kruger, N. J. and Whatley, F. R. 1992, 'Sucrose metabolism in mango fruit during ripening', *Plant Science*, vol. 84, pp. 45–51.

Chaiprasart, P. and Hansawasdi, C. 2009, 'Effect of 1-methylcyclopropene on the shelf life of mango (*Mangifera indica* linn.) cv. nahm-dawg-mai-sri-tong', *Acta Horticulturae*, vol. 820, pp. 725–9.

Chaplin, G. R. 1988, 'Advances in postharvest physiology of mango', *Acta Horticulturae*, vol. 231, pp. 639–48.

Chatha, Z. A., Ahmad, A., Zahoor, T. and Raza, A. 2014, 'Comparative effect of gamma irradiation, UV-C and hot water on antioxidant potential of mango (*Mangifera indica* L.) fruit', *Pakistan Journal of Nutrition*, vol. 12, no. 11, pp. 1019–23.

Chen, J. P., Tai, C. Y. and Chen, H. 2004, 'Improved liquid chromatographic method for determination of carotenoids in Taiwanese mango (*Mangifera indica* L.)', *Journal of chromatography A*, vol. 1054, pp. 261–8.

Chiangsin, R. and Sangchote, S. 2012, 'Effect of Carbon Bag on Anthracnose Disease and Mango Fruit Quality After Harvest', *Agricultural Science Journal (Thailand)*, vol. 43, no. 3, (Suppl.), pp. 548–51.

Chidtragool, S. and Ketsa, S. 2013, 'Modified atmosphere reduces chilling injury of mango fruit during low temperature storage', *Acta Horticulturae*, vol. 992, pp. 513–20.

Chidtragool, S., Ketsa, S., Bowen, J., Ferguson, I. B. and van Doorn, W. G. 2011, 'Chilling injury in mango fruit peel: Cultivar differences are related to the activity of phenylalanine ammonia lyase', *Postharvest Biology and Technology*, vol 62, no. 1, pp. 59–63.

Chongchatuporn, U., Ketsa, S. and van Doorn, W. G. 2013, 'Chilling injury in mango (*Mangifera indica)* fruit peel: Relationship with ascorbic acid concentrations and antioxidant enzyme activities', *Postharvest Biology and Technology*, vol. 86, pp. 409–17.

Chonhenchob, V., Kamhangwong, D., Kruenate, J., Khongrat, K., Tangchantra, N., Wichai, U. and Singh, S. P. 2011, 'Preharvest bagging with wavelength selective materials enhances development and quality of mango (Mangifera indica L.) cv. Nam Dok Mai# 4', *Journal of the Science of Food and Agriculture*, vol. 91, no. 4, 664–71.

Chourasia, A., Sane, V. A. and Nath, P. 2006, 'Differential expression of pectate lyase during ethylene-induced postharvest softening of mango (*Mangifera indica* var. Dashehari)', *Physiologia Plantarum*, vol. 128, pp. 546–55.

Chourasia, A., Sane, V. A., Singh, R. K. and Nath, P. 2008, 'Isolation and characterization of the MiCel1 gene from mango: Ripening related expression and enhanced endoglucanase activity during softening', *Plant Growth Regulation*, vol. 56, no. 2, pp. 117–27.

Chuang, T. Y. and Ann, P. J. 1997, 'Biological control of mango anthracnose', *Plant Protection Bulletin, Taipei*, vol. 39, pp. 227–40.

Cojocaru, M., Droby, S., Glotter, E., Goldman, A., Gottlieb, H. E., Jacoby, B. and Prusky, D. 1986, '5-(12-Heptadecenyl)-resorcinol, the major component of the antifungal activity in the peel of mango fruit', *Phytochemistry*, vol. 25, no. 5, pp. 1093–5.

Cua, A. U. and Lizada, M. C. C. 1990, 'Ethylene production in the 'Carabao' mango (*Mangifera indica* L.) fruit during maturation and ripening', *Acta Horticulture*, vol. 269, pp. 169–79.

Davidse, L. C., Gerritsma, O. C., Ideler, J., Pie, K. and Velthuis, G. C. 1988, 'Antifungal modes of action of metalaxyl, cyprofuram, benalaxyl and oxadixyl in phenylamide-sensitive and phenylamide-resistant strains of Phytophthora megasperma f. sp. medicaginis and Phytophthora infestans', *Crop Protection*, vol. 7, no. 6, pp. 347–55.

De Assis, JS, Self, G and Caron, VC 2009, 'Effects of postharvest handling on lenticel spotting of 'Tommy Atkins' mangoes from northeast Brazil', *Acta Horticulturae*, vol. 820, pp. 681–8.

De Oliveira Junior, E. N., de Malo, I. S. and Franco, T. T. 2012, 'Changes in hyphal morphology due to chitosan treatment in some fungal species', *Brazilian Archives of Biology and Technology*, vol 55, no. 5, pp. 637–46.

De Villiers, E. E. and Korsten, L. 1994, 'Biological treatments for the control of mango postharvest diseases', *South African Mango Growers' Association*, vol. 14, pp. 48–51.

Degani, C., El-Batsri, R. and Gazit, S. 1990, 'Enzyme polymorphism in mango', *Journal of the American Society for Horticultural Science*, vol. 115, pp. 844–7.

Denumbunchachai, W. and Sangchote, S. 2013, 'Induced resistance of mango after harvest with generally regarded as safe (GRAS) chemicals against anthracnose disease', *Agricultural Science Journal*, vol. 44, no. 3 (Suppl.), pp. 14–17.

Diedhiou, P. M., Diallo, Y., Faye, R., Mbengue, A. A. and Sene, A. 2014, 'Efficacy of different fungicides against mango anthracnose in Senegalese Soudanian agroclimate', *American Journal of Plant Sciences*, vol. 5, no. 15, pp. 2224.

Dillon, N. L., Bally, I. S. E., Wright, C. L., Hucks, L., Innes, D. J. and Dietzgen, R. G. 2013, 'Genetic diversity of the Australian national mango genebank'. *Scientia Horticulturae*, vol. 150, pp. 213–26.

Ding, P., Syakirah, M. N., Erkan, M. and Aksoy, U. 2010, 'Influence of fruit bagging on postharvest quality of 'Harumanis' mango (Mangifera indica)', *Acta Horticulturae*, vol. 877, pp. 169–74.

Djioua, T., Charles, F., Freire Jr., M., Filgueiras, H., Ducamp-Collin, M. and Sallanon, H. 2010, 'Combined effects of postharvest heat treatment and chitosan coating on quality of fresh-cut mangoes (*Mangifera indica* L.)', *International Journal of Food Science and Technology*, vol. 45, pp. 849–55.

Dorria, M. A., Aml, R. M. Y. and Kamel, H. M. 2015, 'Postharvest high carbon dioxide and hot water treatments for maintaining quality of Ewase mango fruits', *International Journal of ChemTech Research*, vol. 8, no. 12, pp. 75–86.

Du Plooy, G. W. 2006, *Aspects of mango (Mangifera indica L.) fruit rind morphology and chemistry and their implication for postharvest quality*, (Doctoral dissertation, University of Pretoria, Pretoria, South Africa). http://www.repository.up.ac.za/handle/2263/23814 Accessed 21 July 2016.

Eiadthong, W., Yonemori, K., Kanzaki, S., Sugiura, A., Utsunomiya, N. and Subhadrabandhu, S. 2000, 'Amplified fragment length polymorphism analysis for studying genetic relationships among Mangifera species in Thailand', *Journal of the American Society for Horticultural Science*, vol. 125 no. 2, pp. 160–4.

El Hadrami, A., Adam, L. R., El Hadrami, I. and Daayf, F. 2010, 'Chitosan in plant protection', *Marine drugs*, vol. 8, no. 4, pp. 968–7.

Engel, K. H. and Tressl, R. 1983, 'Studies on the volatile components of two mango varieties', *Journal of Agriculture and Food Chemistry*, vol. 31, pp. 796–801.

Faasema, J., Alakali, J. S. and Abu, J. O. 2014, 'Effects of Storage Temperature on 1 Methylcyclopropene Treated Mango (*Mangnifera Indica*) Fruit Varieties', *Journal of Food Processing and Preservation*, vol. 38, no. 1, pp. 289–95.

Fallik, E. 2004, 'Prestorage hot water treatments (immersion, ringing and brusing)', *Postharvest Biology and Technology*, vol. 32, pp. 125–34.

FAO, 2016, 'Prevention of post-harvest food losses: Fruits, vegetables and root crops', *Agriculture and Consumer Protection*, http://www.fao.org/docrep/t0073E02.htm Accessed 13 June 2016.

Faoro, F., Maffi, D., Cantu, D. and Iriti, M. 2008, 'Chemical-induced resistance against powdery mildew in barley: the effects of chitosan and benzothiadiazole', *BioControl*, vol. 53, pp. 387–401.

Feygenberg, O., Keinan, A., Kobiler, I., Falik, E., Pesis, E., Lers, A. and Prusky, D. 2014, 'Improved management of mango fruit though orchard and packinghouse treatments to reduce lenticel discoloration and prevent decay', *Postharvest Biology and Technology*, vol. 91, pp. 128–33.

Fiaz, M., Malik, A. U., Amin, M., Khan, A. S., Rehman, A., Alam, M. W., Hofman, P. J. and Johnson, P. 2016. Production locality influences postharvest disease development and quality in mangoes. *Acta Horticulturae*, vol. 1111, pp. 369–376

Fonseca, M. J. de O., Salomao, L. C. C., Cecon, P. R. and Puschmann, R. 2001, 'Use of fungicides and wax in storage and postharvest quality of Haden mango (*Mangifera indica* L.) in cold storage', *Revista Brasileira de Armazenamento*, vol. 26, no. 2, pp. 25–35.

Freeman, S., Shtienberg, D., Maymon, M., Levin, A. G. and Ploetz, R. C. 2014, 'New insights into mango malformation disease epidemiology lead to a new integrated management strategy for subtropical environments', *Plant Disease*, vol. 98, no. 11, pp. 1456–66.

Fuchs, Y., Pesis, E. and Zauberman, C. 1980, 'Changes in amylase activity, starch and sugar contents in mango fruit pulp', *Scientia Horticulturae*, vol. 13, pp. 155–60.

Gagnon, M., Lacroix, M., Pringsulaka, V., Latreille, B., Jobin, M., Nouchpramool, K., Prachasitthisak, Y., Charoen, S., Adulyatham, P., Lettre, J. and Grad, B. 1993, 'Effect of gamma irradiation combined with hot water dip and transportation from Thailand to Canada on biochemical and physical characteristics of Thai mangoes (Nahng Glahng Wahn variety)', *Radiation Physics and Chemistry*, vol. 42, pp. 283–7.

Gajbhiye, V. T., Singh, R. H., Kumar, R. and Singh, C. P. 2000, 'Efficacy of postharvest treatments of carbendazim against stem end rot/anthracnose disease and its persistence in mango fruit', *Annals of Plant Protection Sciences*, vol. 8, pp. 230–2.

Galli, J. A., Soares, M. B. B., Martins, A. L. M. and Galli, J. C. 2013, 'Storage of 'Espada' mango fruits in modified atmosphere and cooling: effects on conservation', *Fruits*, vol. 68, no. 4, pp. 291–302.

Govender, V., Korsten, L. and Sivakumar, D. 2005, 'Semi-commercial evaluation of Bacillus licheniformis to control mango postharvest diseases in South Africa', *Postharvest Biology and Technology*, vol. 48, pp. 254–8.

Gowda, I. N. D., Ramanjaneya, K. H., Iyer, C. P. A., Subramanyam, M. D. and Dinesh, M. R. 1994, 'Physico-chemical and processing quality of four new mango hybrids in comparison to two commercial cultivars', *Journal of Food Science and Technology (Mysore)*, vol. 31, pp. 385–8.

Gutiérrez-Martínez, P., Osuna-López, S. G., Calderón-Santoyo, M., Cruz-Hernández, A. and Bautista-Baños, S. 2012, 'Influence of ethanol and heat on disease control and quality in stored mango fruits', *LWT-Food Science and Technology*, vol. 45, pp. 20–7.

Hassan, M. K., Dann, E. K., Irving, D. E. and Coates, L. M. 2007, 'Concentrations of constitutive alk(en) ylresorcinols in peel of commercial mango varieties and resistance to postharvest anthracnose', *Physiological and Molecular Plant Pathology*, vol. 71, pp. 158–65.

Hassan, M. K., Irving, D. E., Dann, E. K., Coates, L. M. and Hofman, P. J. 2009, 'Sap properties and alk(en)ylresorcinol concentrations in Australian-grown mangoes', *Annals of Applied Biology*, vol. 154, no. 3, pp. 419–27.

Hirano, R., Htun, O. T. and Watanabe, K. N. 2010, 'Myanmar mango landraces reveal genetic uniqueness over common cultivars from Florida, India, and Southeast Asia', *Genome*, vol. 53, pp. 321–30.

Hofman, P. J., Smith, L. G., Joyce, D. C., Johnson, G. I. and Meiburg, G. F. 1997, 'Bagging of mango (*Mangifera indica* cv. 'Keitt') fruit influences fruit quality and mineral composition', *Postharvest Biology and Technology*, vol. 12, pp. 83–91.

Hofman, P. J., Marques, J. R., Taylor, L. M., Stubbings, B. A., Ledger, S. N. and Jordan, R. A. 2010, 'Skin damage to two new mango cultivars during irradiation and cold storage', *Acta Horticulturae*, vol. 877, pp. 475–82.

Hu, M., Yang, D., Huber, J. D., Jiang, Y., Li, M., Gao, Z. and Zhang, Z. 2014, 'Reduction of postharvest anthracnose and enhancement of disease resistance in ripening mango fruit by nitric oxide treatment', *Postharvest Biology and Technology*, vol. 97, pp. 115–22.

Intalook, W., Huber, J. and Sargent, S. A. 2006, 'Coating material effect on postharvest quality changes of mango fruit', *Postharvest Biology and Technology*, vol. 28, pp. 247–57.

Islas-Osuna, M. A., Stephens-Camacho, N. A., Contreras-Vergara, C. A., Rivera-Dominguez, M., Sanchez-Sanchez, E., Villegas-Ochoa, M. A. and Gonzalez-Aguilar, G. A. 2010, 'Novel postharvest treatment reduces ascorbic acid losses in mango (*Mangifera indica* L.) var. Kent', *American Journal of Agricultural and Biological Science*, vol. 5, no. 3, pp. 342–9.

Ito, T., Sasaki, K. and Yoshida, Y. 1997, 'Changes in respiration rate, saccharide and organic acid content during the development and ripening of mango fruit (*Mangifera indica* L. 'Irwin') cultured in a plastic house', *Journal of the Japanese Society for Horticultural Science*, vol. 66, pp. 629–35.

Jabbar, A., Malik, A. U., Maqbool, M., Amin, M., Saeed, M. and Hameed, R. 2012, 'Anti-sap chemicals and hot water quarantine treatment effects on storage life and fruit quality of mango cv. Samar Bahisht Chaunsa', *Pakistan Journal of Botany*, vol. 44, no. 2, pp. 757–64.

Jacobi, K. K. and Giles, J. E. 1997, 'Quality of `Kensington' mango (*Mangifera indica* Linn.) fruit following combined vapour heat disinfestation and hot water disease control treatments', *Postharvest Biology and Technology*, vol. 12, pp. 285–92.

Jeffries, P., Dodd, J. C., Jeger, M. J. and Plumbley, R. A. 1990, 'The biology and control of *Colletotrichum* species on tropical fruit crops', *Plant Pathology*, vol. 39, no. 3, pp. 343–66.

Jiang, Z., Zheng, Y., Qiu, R., Yang, Y., Xu, M., Ye, Y. and Xu, M. 2015, 'Short UV-B exposure stimulated enzymatic and nonenzymatic antioxidants and reduced oxidative stress of cold-stored mangoes', *Journal of Agricultural and Food Chemistry*, vol. 63, no. 51, pp. 10965–72.

Jitareerat, P., Paumchai, S. and Kanlayanarat, S. 2007, 'Effect of chitosan on ripening, enzymatic activity, and disease development in mango (*Mangifera indica*) fruit', *New Zealand Journal of Crop and Horticultural Science*, vol. 35, pp. 211–8.

Jitareerat, P., Mohavey, C., Uthairatanakij, A. and Phochanachai, S. 2006. Effect of Irradiation on Fungal Growths and Their Pathogenesis on Mangoes cv. Nam Dok Mai. *International Symposium 'New Frontier of Irradiated food and Non-Food Products'* 22–23 September 2005, KMUTT, Bangkok, Thailand. http://filing.fda.moph.go.th/library/e-learning/fcd/ISNF/poster/16.pdf Accessed 26 July 2016.

Joel, D. M. 1978, 'The secretory ducts of mango fruits, a defense system effective against Mediterranean fruit fly', *Israel Journal of Botany*, vol. 27, pp. 44–5.

John, J., Subbarayan, C. and Cama, H. R. 1970, 'Carotenoids in 3 Stages of Ripening of Mango', *Journal of Food Science*, vol. 35, pp. 262–5.

John, K., Bhat, S. G. and Rao, U. P. 2002, 'Involvement of Peroxidase and Polyphenol Oxidase in Mango Sap Injury', *Journal of Food Biochemistry*, vol. 26, no. 5, pp. 403–14.

Johnson, G. I., Boag, T. S., Cooke, A. W., Izard, M., Panitz, M. and Sangchote, S. 1990, 'Interaction of post harvest disease control treatments and gamma irradiation on mangoes', *Annals of Applied Biology*, vol. 116, no. 2, pp. 245–57.

Joyce, D. C., Beasley, D. R. and Shorter, A. J. 1997, 'Effect of preharvest bagging on fruit calcium levels, and storage and ripening characteristics of 'Sensation' mangoes', *Australian Journal of Experimental Agriculture*, vol. 37, pp. 383–9.

Junmatong, C., Chomkitichai, W., Boonyakiat, D., Uthaibutra, J. and Saengnil, K. 2015, 'Reduction of free radical content and chilling injury in 'Nam Dok Mai No. 4' mango fruit with methyl jasmonate during low temperature storage', *Acta Horticulturae*, vol. 1088, pp. 107–12.

Junmatong, C., Faiyue, B., Rotarayanont, S., Uthaibutra, J., Boonyakiat, D. and Saengnil, K. 2015, 'Cold storage in salicylic acid increases enzymatic and non-enzymatic antioxidants of Nam Dok Mai No. 4 mango fruit', *ScienceAsia*, vol. 41, no. 1, pp. 12–21.

Kaewprawet, S., Sardsud, V. and Chantrasri, P. 2008, 'Effects of low temperatures on lenticel of mango fruit cv. Nam Dok Mai See Thong', *Agricultural Science Journal*, vol. 39, no. 3 (Suppl.), pp. 307–10.

Kapse, B. M., Rane, D. A. and Khedkar, D. M. 1989, 'Correlation between bio-chemical parameters and organoleptic evaluation in mango varieties', *Acta Horticulturae*, vol. 231, pp. 756–62.

Karmarkar, D. V. and Joshi, B. M. 1941, 'Respiration studies of the alphonse mango', *Indian Journal of Agricultural Science*, vol. 11, pp. 993.

Kefialew, Y. and Ayalew, A. 2008, 'Postharvest biological control of anthracnose (*Colletotrichum gloeosporioides*) on mango (Mangifera indica)', *Postharvest Biology and Technology*, vol. 50, pp. 8–11.

Ketsa, S., Phakawatmongkol, W. and Subhadrabhandhu, S. 1999, 'Peel enzymatic activity and colour changes in ripening mango fruit', *Journal of Plant Physiology*, vol. 154, pp. 363–6.

Khaliq, G., Mohamed, M. T. M., Ghazali, H. M., Ding, P. and Ali, A. 2016, 'Influence of gum arabic coating enriched with calcium chloride on physiological, biochemical and quality responses of mango (*Mangifera indica* L.) fruit stored under low temperature stress', *Postharvest Biology and Technology*, vol. 111, pp. 362–9.

Kienzle, S., Carle, R., Sruamsiri, P., Tosta, C. and Neidhart, S. 2013, 'Occurrence of alk (en) ylresorcinols in the fruits of two mango (*Mangifera indica* L.) cultivars during on-tree maturation and postharvest storage', *Journal of Agricultural and Food Chemistry*, vol. 62, no. 1, pp. 28–40.

Kim, Y., Brecht, J. K. and Talcott, S. T. 2007, 'Antioxidant phytochemical and fruit quality changes in mango (*Mangifera indica* L.) following hot water immersion and controlled atmosphere storage', *Food Chemistry*, vol. 105, pp. 1327–34.

Kitagawa, H., Manabe, K., Esguerra, E. B. 1992, 'Bagging of fruit on the tree to control disease', *Acta Horticulturae*, vol. 321, pp. 871–5.

Knight, Jr., R. J., Campbell, R. J. and Maguire, I. 2009, 'Important mango cultivars and their descriptors', *The Mango: Botany, Production and Uses*, 2nd edition. Wallingford, UK: CABI, pp. 42–66.

Knodler, M., Reisenhauer, K., Schieber, A. and Carle, R. 2009, 'Quantitative determination of allergenic 5-alk (en) ylresorcinols in mango (*Mangifera indica* L.) peel, pulp, and fruit products by high-performance liquid chromatography', *Journal of Agricultural and Food Chemistry*, vol. 57, no. 9, pp. 3639–44.

Konno, K. 2011, 'Plant latex and other exudates as plant defense systems: roles of various defense chemicals and proteins contained therein', *Phytochemistry*, vol. 72, no. 13, pp. 1510–30.

Koslanund, R., Niumpiboon, W. and Aousabuy, P. 2015, 'To control of anthracnose disease of mango fruit by oxalic acid', *Agricultural Science Journal*, vol. 46, no. 3/1 (Suppl.), pp. 311–4.

Kozubek, A. and Tyman, J. H. 2005, 'Bioactive phenolic lipids', *Studies in Natural Products Chemistry*, vol. 30, pp. 111–90.

Krishnamurthy, S. and Subramanyam, H. 1970, 'Respiratory climacteric and chemical changes in the mango fruit *Mangifera indica* L', *Journal of the American Society for Horticultural Science*, vol. 95, pp. 333–7.

Krishnamurthy, S., Patwardhan, M. V. and Subramanyam, H. 1971, 'Biochemical changes during ripening of the mango fruit', *Phytochemistry*, vol. 10, pp. 2577–81.

Kulkarni, R. S., Chidley, H. G., Pujari, K. H., Giri, A. P. and Gupta, V. S. 2012, 'Geographic variation in the flavour volatiles of Alphonso mango', *Food Chemistry*, vol. 130, no. 1, pp. 58–66.

Kumar, R. and Selvaraj, Y. 1990, 'Fructose 1, 6-bisphosphatase in ripening mango (*Mangifera indica* L.) fruit', *Indian Journal of Experimental Biology*, vol. 28, pp. 284–6.

Kumar, S., Das, D. K., Singh, A. K. and Prasad, U. S. 1993, 'Changes in non-volatile organic acid composition and pH during maturation and ripening of two mango cultivars', *Indian Journal of Plant Physiology*, vol. 36, pp. 107–11.

Kumar, S., Das, D. K., Singh, A. K. and Prasad, U. S. 1994, 'Sucrose metabolism during maturation and ripening of mango cultivars', *Plant Physiology and Biochemistry*, vol. 21, pp. 27–32.

Kumar, A. K., Babu, J. D., Pratap, M. 2015, 'Effect of 1-MCP and xtend bags on shelf life of mango (*Mangifera indica* L.) 'Baneshan' during extended cold storage', *Acta Horticulturae*, vol. 1088, pp. 179–84.

Kumar, A. K., Babu, J. D. and Pratap, M. 2013, 'Effect of 1-MCP and Xtend bags on shelf life of mango (*Mangifera indica* L.) 'Baneshan' during extended cold storage', *Acta Horticulturae*, vol. 1088, pp. 179–84.

Kumpoun, W. and Uthaibutra, J. 2013, 'Storage life extension at 5° C of 'Nam Dok Mai See Thong' mango by perforated package', *Acta Horticulturae*, vol. 989, pp. 143–8.

Kundu, S. and Ghosh, S. N. 1992, 'Studies on physico-chemical characteristics of mango cultivars grown in the laterite tract of West Bengal', *Haryana Journal of Horticultural Sciences*, vol. 21 no. 3/4, pp. 129–39.

Lakshminarayana, S. 1973, 'Respiration and ripening patterns in the life cycle of the mango fruit', *Journal of Horticultural Science*, vol. 48, pp. 227–33.

Lakshminarayana, S. 1975, 'Relation of time of harvest on respiration, chemical constituents and storage life of mangoes', *Proceedings of the Florida State Horticultural Society*, vol. 88, pp. 477–80.

Lakshminarayana, S. 1980, 'Mango', in Nagy, S. and Shaw, P. E. (eds), *Tropical and Subtropical Fruits*, AVI, Westport, Conn., pp. 184–257.

Lakshminarayana, S., Shety, M. S. and Krishnaprasad, C. A. 1974, 'Accelerated ripening of Alphonso mangos by the application of ethrel', *Tropical Science*, vol. l7, pp. 95–101.

Lakshminarayana, S., Subhadra, N. V. and Subramanyam, H. 1970, 'Some aspects of the developmental physiology of the mango fruit', *Journal of Horticultural Science*, vol. 45, pp. 133–42.

Lalel, H. I. D., Singh, Z. and Tan, S. C. 2003a, 'Maturity stage at harvest affects fruit ripening, quality and biosynthesis of aroma volatile compounds in 'Kensington Pride' mango', *Journal of Horticultural Science and Biotechnology*, vol. 78, pp. 225–33.

Lalel, H. J., Singh, Z. and Tan, S. C. 2003b, 'Aroma volatiles production during fruit ripening of 'Kensington Pride' mango', *Postharvest Biology and Technology*, vol. 27 no. 3, pp. 323–36.

Lazan, H., Ali, Z. M., Wah, L. K., Voon, J. and Chaplin, G. R. 1986, 'The potential role of polygalacturonase in pectin degradation and softening of mango fruit', *ASEAN Food Journal*, vol. 2, pp. 93–8.

Lazan, H., Ali, Z. M., Soh, J. and Talkan, Z. 1993, 'The biochemical basis of differential ripening in mango', *Acta Horticulturae*, vol. 341, pp. 500–9.

Léchaudel, M., Lopez-Lauri, F., Vidal, V., Sallanon, H. and Joas, J. 2013, 'Response of the physiological parameters of mango fruit (transpiration, water relations and antioxidant system) to its light and temperature environment', *Journal of Plant Physiology*, vol. 170, no. 6, pp. 567–76.

Lederman, I. E., Zauberman, G., Weksler, A., Rot, I. and Fuchs, Y. 1997, 'Ethylene-forming capacity during cold storage and chilling injury development in "Keitt" mango fruit', *Postharvest Biology and Technology*, vol. 10, pp. 107–12.

Li, P., Zheng, X., Liu, Y. and Zhu, Y. 2014, 'Pre-storage application of oxalic acid alleviates chilling injury in mango fruit by modulating proline metabolism and energy status under chilling stress', *Food Chemistry*, vol. 142, pp. 72–8.

Li, G. Q., Gupta, M., San, A. T., Joyce, D. C., Hofman, P. J., Macnish, A. J. and Marques, J. 2016. Postharvest treatment effects on 'B74' mango fruit lenticel discolouration after irradiation. *Acta Horticulturae*, vol. 1111 pp. 385–392

Lima, L. C. D. O., Chitarra, A. B. and Chitarra, M. I. F. 2001, 'Changes in amylase activity starch and sugars contents in mango fruits pulp cv. Tommy Atkins with spongy tissue', *Brazilian Archives of Biology and Technology*, vol. 44, pp. 59–62.

Lin, J., Gong, D., Zhu, S., Zhang, L. and Zhang, L. 2011, 'Expression of PPO and POD genes and contents of polyphenolic compounds in harvested mango fruits in relation to Benzothiadiazole-induced defense against anthracnose', *Scientia Horticulturae.*, vol. 130, pp. 85–9.

Lizada, C. 1993, 'Mango', in Seymour, G. B., Taylor, J. E. and Tucker, G. A. (eds), *Biochemistry of Fruit Ripening*, Chapman and Hall, London, pp. 255–71.

Lonsdale, JH 1993, 'Strategies for the control of postharvest diseases of mango', *South African Mango Growers' Association*, vol. 13, pp. 109–16.

Lonsdale, J. H. and Droomer, L. 1994, 'The application of short wave infra-red radiation to control postharvest decay of mangoes', *South African Mango Growers' Association*, vol. 14, pp. 59–61.

Lovey, B. R., Robinson, S. P., Brophy, J. J. and Chacko, E. K. 1992, 'Mango sapburn components of fruit sap and their role in causing skin damage', *Australian Journal of Plant Physiology*, vol. 19, pp. 449–57.

Lurie, S. 1998, 'Postharvest heat treatment: Review', *Postharvest Biology and Technology*, vol. 14, pp. 257–69.

MacLeod, A. J. and de Troconis, N. G. 1982, 'Volatile flavour components of mango fruit', *Phytochemistry*, vol. 21, pp. 2523–6.

MacLeod, A. J. and Snyder, C. H. 1985, 'Volatile components of two cultivars of mango from Florida', *Journal of Agricultural and Food Chemistry*, vol. 33, pp. 380–4.

MacLeod, A. J., MacLeod, G. and Snyder, C. H. 1988, 'Volatile aroma constituents of mango (cv. Kensington), *Phytochemistry*, vol. 27, pp. 2189–93.

MacNish, A. J., Joyce, D. C. and Hetherington, S. E. 1997, 'Packaging to reduce water loss can delay ripening of mango (*Mangifera indica* L. cv. Kensington Pride) fruit', *Australian Journal of Experimental Agriculture*, vol. 37, pp. 463–7.

Mahto, R. and Das, M. 2013, 'Effect of gamma irradiation on the physico-chemical and visual properties of mango (*Mangifera indica* L.), cv. "Dushehri"and "Fazli" stored at 20 C', *Postharvest Biology and Technology*, vol. 86, pp. 447–55.

Malik, A. U., Umar, M., Hameed, R., Amin, M., Asad, H. U., Hafeez, O. and Hofman, P. J. 2013, 'Phytosanitary irradiation treatments in relation to desapping and processing types affect mango fruit quality', *Acta Horticulturae*, vol. 1012, pp. 681–92.

Malo, S. E. and Campbell, C. W. 1978, 'Studies on mango fruit breakdown in Florida', *Proceedings of the Tropical Region American Society for Horticultural Science*, vol. 22, pp. 1–15.

Malundo, T. M. M. 1996, *Application of the quality enhancement (QE) approach to mango (Mangifera indica L.) flavor research*, PhD dissertation, Department of Food Science and Technology, University of Georgia, US.

Malundo, T. M. M., Shewfelt, R. L., Ware, G. O. and Baldwin, E. A. 2001, 'Sugars and acids influence flavor properties of mango (*Mangifera indica*)', *Journal of the American Society for Horticultural Science*, vol. 126, pp. 115–21.

Maqbool, M., Malik, A. U. and Jabbar, A. 2007, 'Sap dynamics and its management in commercial mango cultivars of Pakistan', *Pakistan Journal of Botany*, vol. 39, no. 5, pp. 1565–74.

Maqbool, M., Ullah Malika, A., Ahmad, I. and Pervez, M. A. 2009, 'Sapburn management for improving mango fruit quality', *Acta Horticulturae*, vol. 820, pp. 713–20.

Marques, J. R., Hofman, P. J., Macnish, A. J. and Joyce, D. C. 2016, Evaluation of temperature management and packaging options to reduce under-skin browning in 'Honey Gold' mango fruit. *Acta Horticulturae*, vol. 1111, pp. 405–8.

Martin, L. B. and Rose, J. K. 2014, There's more than one way to skin a fruit: formation and functions of fruit cuticles. *Journal of Experimental Botany*, vol. 65, pp. 4639–4651.

Masibo, M. and He, Q. 2008, 'Major mango polyphenols and their potential significance to human health', *Comprehensive Reviews in Food Science and Food Safety*, vol. 7, pp. 309–19.

Mathooko, F. M., Kahagi, E. M., Runkuab, J. M., Onyangob, C. A., Owinob, W. O., Arauz, L. F., Fonseca, J. M. and Hewett, E. W. 2011, 'Preharvest mango (*Mangifera indica* L. 'Apple') fruit bagging controls lenticel discolouration and improves postharvest quality', *Acta Horticulturae*, vol. 906, pp. 55–62.

Mazhar, M. S., Amin, M., Malik, A. U., Campbell, J. and Johnson, P. 2011, 'Improved harvest and desapping practices affect mango fruit quality along the supply chains', *International Journal of Agriculture and Biology*, vol. 13, no. 5, pp. 776–80.

Medlicott, A. P. and Thompson, A. K. 1985, 'Analysis of sugar and organic acids in ripening mango fruit (*Mangifera indica* var. Keitt) by high performance liquid chromatography', *Journal of the Science of Food and Agriculture*, vol. 36, pp. 561–6.

Medlicott, A. P., Bhogal, M. and Reynolds, S. B. 1986, 'Changes in peel pigmentation during ripening of mango fruit (*Mangifera indica* var Tommy Atkins)', *Annals of Applied Biology*, vol. 109, pp. 651–6.

Medlicott, A. P., Reynolds, S. B., New, S. W. and Thompson, A. K. 1988, 'Harvest maturity effects on mango fruit ripening', *Trop Agric*, vol. 65, pp. 153–7.

Mekwatanakarn, W. and Chairat, R. 2013, 'Antioxidant capacity of mango pulp and peel stored at two temperatures', *Acta Horticulturae*, vol. 989, pp. 139–42.

Mercadante, A. Z. and Rodriguez-Amaya, D. B. 1998, 'Effects of ripening, cultivar differences and processing on the carotenoid composition of mango', *Journal of Agriculture and Food Chemistry*, vol. 46, pp. 128–30.

Mercadante, A. Z., Rodriguez-Amaya, D. B. and Britton, G. 1997, 'HPLC and mass spectrometric analysis of carotenoids from mango', *Journal of Agriculture and Food Chemistry*, vol. 45, pp. 120–3.

Miguel, A. C. A., Durigan, J. F., Morgado, C. M. A. and Gomes, R. F. O. 2011, 'Chilling injury in mangoes cv. Palmer', *Revista Brasileira de Fruticultura*, vol. 33, pp. 255–60.

Miguel, A. C. A., Durigan, J. F., Marques, K., Morgado, C., Ascari, M. and Ferraudo, A. S. 2016, 'Prevention of Chilling Injury In 'Tommy Atkins' mangoes Previously Stored at 5°C, Using Heat Treatment and Radiation UV (UV-C)', *Revista Brasileira de Fruticultura*, vol. 38, no. 1, pp. 53–63.

Mitcham, E. J. and McDonald, R. E. 1992, 'Cell wall modification during ripening of 'Keitt' and 'Tommy Atkins' mango fruit', *Journal of the American Society for Horticultural Science*, vol. 117, pp. 919–24.

Muda, P., Seymour, G. B., Errington, N. and Tucker, G. A. 1995, 'Compositional changes in cell wall polymers during mango fruit ripening', *Carbohydrate Polymers*, vol. 16, pp. 255–60.

Muenmanee, N. 2015, 'Effect of Radio Frequency on the control of anthracnose disease of mango cv. Nam Dok Mai Si Thong', *Agricultural Science Journal*, vol. 46, no. 3/1(Suppl.), pp. 388–91.

Munafo Jr., J. P., Didzbalis, J., Schnell, R. J., Schieberle, P. and Steinhaus, M. 2014, 'Characterization of the major aroma-active compounds in mango (*Mangifera indica* L.) cultivars Haden, White Alfonso, Praya Sowoy, Royal Special, and Malindi by application of a comparative aroma extract dilution analysis', *Journal of Agricultural and Food Chemistry*, vol. 62. no. 20, pp. 4544–51.

Murthy, D. S., Gajanana, T. M., Sudha, M. and Dakshinamoorthy, V. 2009, 'Marketing and post-harvest losses in fruits: its implications on availability and economy, *Marketing*, vol 64, no. 2, pp. 259–75.

Nair, S., Singh, Z. and Tan, S. C. 2003, 'Aroma volatiles emission in relation to chilling injury in 'Kensington Pride' mango fruit', *The Journal of Horticultural Science and Biotechnology*, vol. 78, pp. 866–73.

Naychawna, W. and Nalumpang, S. 2014, 'Efficient of wood vinegar extracts from Eucalyptus and Rubber wood for controlling *Collectotrichum gloeospotiodes* causing anthracnose disease of mango cv. Nam Dok Mai', *Agricultural Science Journal*, vol. 45, no. 3/1(Suppl.), pp. 37–40.

Ngamchuachit, P., Sivertsen, H. K., Mitcham, E. J. and Barrett, D. M. 2015, 'Influence of cultivar and ripeness stage at the time of fresh-cut processing on instrumental and sensory qualities of fresh-cut mangos', *Postharvest Biology and Technology*, vol. 106, pp. 11–20.

Ngamchuachit, P., Barrett, D. and Mitcham, E. J. 2014, 'Effects of 1-methylcyclopropene and hot water quarantine treatment on quality of "Keitt" mangos', *Journal of Food Science*, vol. 79, no. 4, pp. C505–9.

Nguyen, H., Hofman, P., Holmes, R., Bally, I., Stubbings, B. and McConchie, R. 2004, 'Effect of nitrogen on the skin colour and other quality attributes of ripe "Kensington Pride" mango (*Mangifera indica* L.) fruit', *Journal of Horticultural Science and Biotechnology*, vol. 79, pp. 204–10.

Nguyen, T. M. 2015. Lenticel damage on 'B74' mango fruit. University of Queensland PhD thesis. Accessed 2017 Jan 21 https://espace.library.uq.edu.au/view/UQ:372944

Nguyen, T. M., Joyce, D. C., Gupta, M., Hofman, P. J., Macnish, A. J. and Marques, J. R., 2016. Pre-harvest bagging and coating treatment effects on lenticel discolouration in 'B74' mango fruit. *Acta Horticulturae*, vol. 1111, pp. 377–84.

Nimitkeatkai, H., Kakaew, P. and Kanlayanarat, S. 2008, 'Effect of low oxygen atmosphere on postharvest quality of mango cv. Mahajanaka', *Agricultural Science Journal*, vol. 39, no. 3 (Suppl.), pp. 291–4.

Noiwan, D. 2006, Effect of coating materials on storage life of mango fruit cv. Mahajanaka. Master of Science (Postharvest Technology), Postharvest Technology Institute, Chiang Mai University, 193 p.

Olano, C. T., Schnell, R. J., Quintanilla, W. E. and Campbell, R. J. 2005, 'Pedigree analysis of Florida mango cultivars', *Proceedings of the Florida State Horticultural Society*, vol. 118, pp. 192–7.

Ornelas-Paz, Jde, J., Yahia, E. .M. and Gardea, A. 2007, 'Identification and quantification of xanthophyll esters, carotenes and tocopherols in the fruit of seven Mexican mango cultivars by liquid chromatography-atmospheric pressure chemical ionization-time of flight mass spectrometry[LC-(APcl+)-MS]', *Journal of Agriculture and Food Chemistry*, vol. 55, pp. 6628–35.

Ornelas Paz, J. D. J. and Yahia, E. M. 2014, 'Effect of the moisture content of forced hot air on the postharvest quality and bioactive compounds of mango fruit (*Mangifera indica* L. cv. Manila)', *Journal of the Science of Food and Agriculture*, vol. 94, no. 6, pp. 1078–83.

Osuna-García, J. A., Pérez-Barraza, M. H., Vázquez-Valdivia, V., Beltran, J. A. 2009, 'Methylcyclopropene (1-MCP), a new approach for exporting 'Kent' mangos to Europe and Japan', *Acta Horticulturae*, vol. 820, pp. 721–4.

Osuna-Garcia, J. A., Brecht, J. K., Huber, D. J. and Nolasco-Gonzalez, Y. 2015, 'Aqueous 1-Methylcyclopropene to Delay Ripening of 'Kent' Mango With or Without Quarantine Hot Water Treatment', *HortTechnology*, vol. 25, no. 3, pp. 349–57.

Padda, M. S., do Amarante, C. V., Garcia, R. M., Slaughter, D. C. and Mitcham, E. J. 2011, 'Methods to analyze physico-chemical changes during mango ripening: A multivariate approach', *Postharvest Biology and Technology*, vol. 62, no. 3, pp. 267–74.

Palafox-Carlos, H., Yahia, E. M. and González-Aguilar, G. A. 2012, 'Identification and quantification of major phenolic compounds from mango (*Mangifera indica*, cv. Ataulfo) fruit by HPLC–DAD–MS/MS-ESI and their individual contribution to the antioxidant activity during ripening', *Food Chemistry*, vol. 135, pp. 105–11.

Palou, L., Ali, A., Fallik, E. and Romanazzi, G. 2016, 'GRAS, plant- and animal-derived compounds as alternatives to conventional fungicides for the control of postharvest diseases of fresh horticultural produce', *Postharvest Biology and Technology*, Available online 8 May 2016.

Pandit, S. S., Chidley, H. G., Kulkarni, R. S., Pujari, K. H., Giri, A. P. and Gupta, V. S. 2009, 'Cultivar relationships in mango based on fruit volatile profiles', *Food Chemistry*, vol. 114, pp. 363–72.

Parfitt, J., Barthel, M. and Macnaughton, S. 2010, 'Food waste within food supply chains: quantification and potential for change to 2050', *Philosophical Transactions of the Royal Society of London B: Biological Sciences*, vol 365, no. 1554, pp. 3065–81.

Parikh, H. R., Nair, G. M. and Modi, V. V. 1990, 'Some structural changes during ripening of mangoes (*Mangifera indica* var. Alphonso) by abscisic acid treatment', *Annals of Botany*, vol. 65, pp. 121–7.

Park, Y. K., Sato, H. H., Almeida, T. D. and Moretti, R. H. 1980, 'Polyphenoloxidase of mango (*Mangifera indica* var. Haden)', *Journal of Food Science*, vol. 45, pp. 1619–21.

Pastor, C., Sánchez-González, L., Cháfer, M., Chiralt, A. and González-Martínez, C. 2010, 'Physical and antifungal properties of hydroxylpropyl methylcellulose based films containing propolis as affected by moisture content', *Carbohydrate Polymers*, vol. 82, pp. 1174–83.

Paull, R. E. and Duarte, O. 2011, Tropical Fruit - Volume I. CAB International, Wallingford, England.

Peet, M. M. and Willits, D. H. 1995, 'Role of excess water in tomato fruit cracking', *HortScience*, vol. 30, pp. 65–8.

Pesis, E., Aharoni, D., Aharon, Z., Ben-Arie, R., Aharoni, N. and Fuchs, Y. 2000, 'Modified atmosphere and modified humidity packaging alleviates chilling injury symptoms in mango fruit', *Postharvest Biology and Technology*, vol. 19, no. 1, pp. 93–101.

Phillips, D. J. 1975, 'Detection and Translocation of Benomyl in Postharvest-treated Peaches and Nectarines', *Phytopathology*, vol. 65, pp. 255–8.

Phongmee, A. and Wisaratranon, N. 1990, 'Control of Nam Dok mai mango fruits rots with four fungicides at 52°C', in *Thai abstract*. Thai National AGRIS Centre, Office of The University Library Kasetsart University, Bangkok, Thailand, 496 p.

Pino, J. A. 2012, 'Odour active compounds in mango (*Mangifera indica* L. cv. Corazón)', *International Journal of Food Science and Technology*, vol. 47, no. 9, pp. 1944–50.

Pino, J. A. and Mesa, J. 2006, 'Contribution of volatile compounds to mango (Mangifera indica L.) aroma', *Flavour and Fragrance Journal*, vol. 21, no. 2, pp. 207–13.

Pino, J., Mesa, J., Mui'ioz, Y., Marti, M. P. and Marbot, R. 2005, 'Volatile components from mango (*Mangifera indica* L.) cultivars', *Journal of Agriculture and Food Chemistry*, vol. 53, pp. 2213–23.

Pongpisutta, R., Rattanakreetakul, C. and Bunjoeedchoo, R. 2012, 'Non-chemical fungicide treatments for anthracnose control in mango fruits cv. Nam Dork Mai See Tong', *Agricultural Science Journal*, vol. 43, no. 3(Suppl.), pp. 464–7.

Prabha, T. N. and Patwardhan, M. V. 1986, 'Endogenously oxidizable polyphenols of mango, sapota and banana', *Acta Alimentaria*, vol. 15, pp. 123–8.

Prabha, T. N. and Patwardhan, M. V. 1986, 'Endogenously oxidizable polyphenols of mango, sapota and banana', *Acta Alimentaria*, vol. 15, pp. 123–8.

Prasanna, V., Prabha, T. N. and Tharanathan, R. N. 2006, 'Multiple forms of polygalacturonase from mango (*Mangifera indica* L. cv Alphonso) fruit', *Food Chemistry*, vol. 95, pp. 30–6.

Prasothong, N., Plainsirichai, M., Bussaman, P., Luckanatinavong, V., Singhthongla, P. and Wongsawas, M. 2011, 'Effect of chitosan on anthracnose (*Colletotrichum gloeosporioides* Penz.) of mango fruit cv. Nam Dok Mai', *Agricultural Science Journal*, vol. 42, no. 3(Suppl.), pp. 228–31.

Proctor, J. T. A. and Creasy, L. L. 1969, 'The anthocyanins of the mango fruit', *Phytochemistry*, vol. 8, pp. 2108.

Promyou, S. and Supapvanich, S. 2016, 'Physicochemical changes in 'kaew kamin' mango fruit illuminated with ultraviolet-C (UV-C) during storage', *Journal of Agricultural Science and Technology*, vol. 18, no. 1, pp. 145–54.

Prusky, D. 1996, 'Pathogen quiescence in postharvest diseases', *Annual Review of Phytopathology*, 34, 413–34.

Prusky, D., Fuchs, Y., Kobiler, I., Roth, I., Weksler, A., Shalom, Y., Fallik, E., Zauberman, G., Pesis, E., Akerman, M., Ykutiely, O., Weisblum, A., Regev, R. and Artes, L. 1999, 'Effect of hot water brushing, prochloraz treatment and waxing on the incidence of black spot decay caused by Alternaria alternata in mango fruit', *Postharvest Biology and Technology*, vol. 15, pp. 165–74.

Pujarig, K. H. and Patil, B. P. 2007, 'Genetic diversity analysis of mango cultivars using inter simple sequence repeat markers', *Current Science*, vol. 93, pp. 1135–41.

Quijano, C. E., Salamanca, G. and Pino, J. A. 2007, 'Aroma volatile constituents of Colombian varieties of mango (Mangifera indica L.)', *Flavour and Fragrance Journal*, vol. 22, no.5, pp. 401–6.

Rachtanapun, P., Thanakkasaranee, S. and Soonthornampai, S. 2008, 'Application of carboxymethylcellulose from papaya peel for mango (*Mangifera Indica* L.) 'Namdokmai' coating', *Agricultural Science Journal*, vol. 39, no. 3 (Suppl.), pp. 74–82.

Rachtanapun, P., Khamthai, S. and Noiwan, D. 2010, 'Effect of active packaging from ethylene absorber paper on posthavest quality and storage life of banana cv. Kluai Hom Thong', *Agricultural Science Journal*, vol. 41 (Suppl.), pp. 223–6.

Ramayya, N., Niranjan, K. and Duncan, E. 2012, 'Effects of modified atmosphere packaging on quality of 'Alphonso' mangoes', *Journal of Food Science and Technology*, vol. 49, no. 6, pp. 721–8.

Rao, D. V. S. and Shivashankara, K. S. 2014, 'Individual shrink wrapping extends the storage life and maintains the antioxidants of mango (cvs. 'Alphonso' and 'Banganapalli') stored at 8°C', *Journal of Food Science and Technology*, vol. 52, no. 7, pp. 4351–9.

Raymond, L., Schaffer, B., Brecht, J. K. and Crane, J. H. 1998, 'Internal breakdown in mango fruit: symptomology and histology of jelly seed, soft nose and stem-end cavity', *Postharvest Biology and Technology*, vol. 13, pp. 59–70.

Raza, S. A., Khan, A. S., Malik, A. U., Amin, M., Asad, H. U. and Razzaq, K. 2013, 'Respiration rate, physico-chemical fruit quality and consumer acceptability for Fajri mango under different storage temperatures', *Pakistan Journal of Agricultural Sciences*, vol. 50, no. 4, pp. 585–90.

Razzaq, K., Khan, A. S., Malik, A. U., Shahid, M. and Ullah, S. 2015, 'Effect of oxalic acid application on Samar Bahisht Chaunsa mango during ripening and postharvest', *LWT-Food Science and Technology*, vol. 63, no. 1, pp. 152–60.

Razzaq, K., Singh, Z., Khan, A. S., Khan, S. A. K. U. and Ullah, S 2016, 'Role of 1-MCP in regulating 'Kensington Pride' mango fruit softening and ripening', *Plant Growth Regulation*, vol. 78 no. 3, pp. 401–11.

Reddy, Y. V. and Srivastava, G. C. 1999, 'Ethylene biosynthesis and respiration in mango fruits during ripening', *Indian Journal of Plant Physiology*, vol. 4, pp. 32–5.

Regnier, T., Plooy, du W., Combrinck, S. and Botha, B. 2008, 'Fungitoxicity of *Lippia scaberrima* essential oil and selected terpenoid components on two mango postharvest spoilage pathogens', *Postharvest Biology and Technology*, vol. 48, pp. 254–8.

Rehman, A., Malik, A. U., Ali, H., Alam M. W. and Sarfraz, B. 2015, 'Preharvest factors influencing the postharvest disease development and mango fruit quality', *Journal of Environmental and Agricultural Sciences*, vol. 3, pp. 42–7.

Ridoutt, B. G., Juliano, P., Sanguansri, P. and Sellahewa, J. 2010, 'The water footprint of food waste: case study of fresh mango in Australia', *Journal of Cleaner Production*, vol. 18, no. 16, pp. 1714–21.

Robinson, S. P., Loveys, B. R. and Chacko, E. K. 1993, 'Polyphenol oxidase enzyme in sap and skin mango fruit', *Australian Journal of Plant Physiology*, vol. 20, pp. 99–107.

Rodeo, A. J. D. and Esguerra, E. B. 2013, 'Low temperature conditioning alleviates chilling injury in mango (*Mangifera indica* L. cv. Carabao) fruit', *Philippine Journal of Crop Science*, vol. 38, no. 1, pp. 24–32.

Roe, B. and Bruemmer, J. H. 1981, 'Changes in pectic substances and enzymes during ripening and storage of 'Keitt' mangos', *Journal of Food Science*, vol. 46, pp. 186–9.

Rossetto, C. J., Furlani, P. R., Bortoletto, N., Quaggio, J. A. and Igue, T. 2000, 'Differential Response of Mango Varieties to Boron', *Acta Horticulture*, vol. 509, pp. 259–64.

Ruan, J., Li, M., Jin, H., Sun, L., Zhu, Y., Xu, M. and Dong, J. 2015, 'UV-B irradiation alleviates the deterioration of cold-stored mangoes by enhancing endogenous nitric oxide levels', *Food Chemistry*, vol. 169, pp. 417–23.
Ryan, M. G. 1991, 'Effects of climate change on plant respiration', *Ecological Applications*, vol. 1, pp. 157–67.
Saby, J. K., Bhat, S. G. and Prasada Rao, U. J. S. 2002, 'Involvement of peroxidase and polyphenol oxidase in mango sap-injury', *Journal of Food Biochemistry*, vol. 26, no. 5, pp. 403–14.
Saleh, N. A. M. and El-Ansari, M. A. I. 1975, 'Polyphenolics of twenty local varieties of Mangifera indica', *Planta Medica*, vol. 28, pp. 124–30.
San, A. T., Smyth, H. E., Joyce, D. C., Webb, R., Li, G. Q., Hofman, P. J. and Macnish, A. J., 2016. γ-Irradiation effects on appearance and aroma of 'Kensington Pride' mango fruit. *Acta Horticulturae*, vol. 1111 pp. 393–398
Sánchez-Domínguez, D., Ríos, M. Y., Castillo-Ocampo, P., Zavala-Padilla, G., Ramos-García, M. and Bautista-Baños, M. 2011, 'Cytological and biochemical changes induced by chitosan in the pathosystem *Alternaria alternata* –tomato', *Pesticide Biochemistry and Physiology*, vol. 99, pp. 250–5.
Sane, V. A., Chourasia, A. and Nath, P. 2005, 'Softening in mango (*Mangifera indica* cv. Dashehari) is correlated with the expression of an early ethylene responsive, ripening related expansin gene, MiExpAl', *Postharvest Biology and Technology*, vol. 38, pp. 223–30.
Sangchote, S. 1987, 'Postharvest diseases of mango fruits and their losses', *Agricultural Science Journal (Kasetsart University)*, vol. 21, pp. 81–5.
Sangchote, S. 2013, 'Integrated control of anthracnose (*Colletotrichum gloeosporioides*) of mango for export', *Acta Horticulturae*, vol. 973, pp. 55–8.
Sangchote, S. and Seehachai, W. 2008, 'Enhancement an efficacy of hot water treatment to control anthracnose of mango fruits', *Agricultural Science Journal*, vol. 39, no. 3 (Suppl.), pp. 115–8.
Santos, E. C., Santos, A. F., Silva, S. M., Silveira, I. R. B. S., Lira, M. L. S., Silva, L. P. and Alves, R. E. 2002, 'Effect of fruit dropping on postharvest quality of mango 'rosa' at two maturity stage',. *Acta Horticulturae*, vol. 645, pp. 543–9.
Santos, A. M. G., Lins, S. R. O., Silva, J. M. D. and Oliveira, S. M. A. D. 2015, 'Low doses of gamma radiation in the management of postharvest *Lasiodiplodia theobromae* in mangos', *Brazilian Journal of Microbiology*, vol. 46, no. 3, pp. 841–7.
Saran, P. L. and Kumar, R. 2011, 'Boron deficiency disorders in mango (*Mangifera indica*): Field screening, nutrient composition and amelioration by boron application', *Indian Journal Agricultural Science*, vol. 81, pp. 506–10.
Sarker, S. and Muhsi, A. A. 1981, 'A study on the content and interconversions of organic acids of mango (*Mangifera indica* L.) at various stages of fruit development', *Bangladesh Journal of Agricultural Science*, vol. 8, pp. 69–75.
Schieber, A., Stintzing, F. C. and Carle, R. 2001, 'By-products of plant food processing as a source of functional compounds—recent development', *Trends in Food Science and Technology*, vol. 12, pp. 401–13.
Schulze-Kaysers, N., Feuereisen, M. M. and Schieber, A. 2015, 'Phenolic compounds in edible species of the Anacardiaceae family–a review', *RSC Advances*, vol. 5, no. 89, pp. 73301–14.
Selvaraj, Y. and Kumar, R. 1989, 'Studies on fruit softening enzyme and polyphenoloxidase activity in ripening mango (*Mangifera indica* L) fruit', *Journal of Food Science and Technology*, vol. 26, pp. 218–22.
Selvaraj, Y. and Kumar, R. 1994, 'Enzymatic regulation in ripening mango fruit', *Indian Journal of Horticulture*, vol. 51, pp. 316–23.
Selvaraj, Y., Kumar, R. and Pal, D. K. 1989, 'Changes in sugars, organic acids, amino acids, lipid constituents and aroma characteristics of ripening mango (*Mangifera indica* L.) fruit', *Journal of Food Science and Technology*, vol. 26, pp. 306–11.
Sen, S., Chatterjee, B. K. and Roy, H. B. 1985, 'Amylase activity in ripening mango (*Mangifera indica* L.) fruit', *Indian Biology*, vol. 17, pp. 25–8.

Setiawan, B., Sulaeman, A., Giraud, D. W. and Driskell, J. A. 2001, 'Carotenoid content of selected Indonesian fruits', *Journal of Food Composition and Analysis*, vol. 14, pp. 166–9.
Seymour, G. B., N'Diaye, M., Wainwright, H. and Tucker, G. A. 1990, 'Effect of cultivar and harvest maturity on ripening of mangoes during storage', *Journal of Horticultural Science*, vol. 65, pp. 479–83.
Sharma, R. R. and Singh, R. 2009, 'The fruit pitting disorder—A physiological anomaly in mango (*Mangifera indica* L.) due to deficiency of calcium and boron', *Scientia Horticulturae*, vol. 119, pp. 388–91.
Sharma, I. M., Harender, R. and Kaul, J. L. 1994, 'Studies on post-harvest diseases of mango and chemical control of stem end rot and anthracnose', *Indian Phytopathology*, vol. 47, no. 2, pp. 197–200.
Sharma, R. R. S., Reddy, V. R., Reddy, S. V. R. and Jhalegark M. J. 2014, 'Pre-harvest fruit bagging: a useful approach for plant protection and improved post-harvest fruit quality–a review', *The Journal of Horticultural Science and Biotechnology*, vol. 89, pp. 101–13.
Shashirekha, M. S. and Patwardhan, M. V. 1976, 'Changes in amino acids, sugars and nonvolatile organic acids in a ripening mango fruit (*Mangifera indica* L. Badami variety)', *Lebensmittel-Wissenschaft und Technologie*, vol. 9, pp. 369–70.
Shivarama, R. L. and Thimma, K. R. 1989, 'Effect of prepackaging and postharvest treatments on the storage behavior of mango fruits cv. Alphonso', *Acta Horticulturae*, vol. 231, pp. 670–4.
Shivashankar, S. 2014, 'Physiological Disorders of Mango Fruit', *Horticultural Reviews*, vol. 42, pp. 314–47.
Silva, J. M., Villar, H. P. and Pimentel, R. M. 2012, 'Structure of the cell wall of mango after application of ionizing radiation', *Radiation Physics and Chemistry*, vol. 81, no. 11, pp. 1770–5.
Singh, R. and Dwivedi, U. N. 2008, 'Effect of Ethrel and 1-methylcyclopropene (1-MCP) on antioxidants in mango (*Mangifera indica* var. Dashehari) during fruit ripening', *Food Chemistry*, vol. 111, no. 4, pp. 951–6.
Singh, Z., Zaharah, S. S. 2013, 'Controlled atmosphere storage of mango fruit - An overview', *Acta Horticulturae*, vol. 992, pp. 481–92.
Singh, Z. and Zaharah, S. S. 2015, 'Controlled atmosphere storage of mango fruit: Challenges and thrusts and its implications in international mango trade', *Acta Horticulturae*, vol. 1066, pp. 179–91.
Singh, A. K., Singh, C. P., Kushwaha, P. S. and Chakraborty, B. 2015, 'Efficacy of postharvest treatments on fruit marketability and physico-chemical characteristics of 'Dashehari' mango', *Acta Horticulturae*, vol. 1066, pp. 201–6.
Singh, Z., Singh, R. K., Sane, V. A. and Nath, P. 2013, 'Mango - Postharvest Biology and Biotechnology', *Critical Reviews in Plant Sciences*, vol. 32, pp. 217–36.
Siriphanich, J. and Kunyamee, S. 2010, 'Chilling Injury in Fruit of Three Mango Cultivars and Their Relation to Hydrogen Peroxide Content and Total Antioxidant Capacity', *Acta Horticulturae*, vol. 877, pp. 523–9.
Sivakumar, D., Jiang, Y. and Yahia, E. M. 2011, 'Maintaining mango (*Mangifera indica* L.) fruit quality during the export chain', *Food Research International*, vol. 44, no. 5, pp. 1254–63.
Sivankalyani, V., Feygenberg, O., Diskin, S., Wright, B. and Alkanm, N. 2016, 'Increased anthocyanin and flavonoids in mango fruit peel are associated with cold and pathogen resistance', *Postharvest Biology and Technology*, vol. 111, pp. 132–9.
Sonune, G. P., Pujari, K. H., Sangale, K. K., Patil, R. S., Tirmali, A. M., Gokhale, N. B., Jadhav, S. N. and Shedge, M. S. 2011, 'Effect of 1-methylcyclopropene (1-MCP) on ripening behaviour, shelf life and quality of mango (Mangifera indica L.) cv. Alphonso under ambient temperature storage', *Ecology, Environment and Conservation*, vol. 17, no. 4, pp. 843–6.
Sopee, J. and Sangchote, S. 2005, 'Effect of heat treatment on the fungus *Colletotrichum gloeosporioides* and anthracnose of mango fruit', *Acta Horticulturae*, vol. 682, pp. 2049–56.
Soroka, W. 2014, *Fundamentals of Packaging Technology* (4th Edition). Institute Of Packaging Professional, ISBN-13: 978–0-615–70934-5.
Sripong, K., Jitareerat, P., Tsuyumu, S., Uthairatnakij, A., Srilaong, V., Wrongs-Aree, C., Ma, G., Zhang, L. and Kato, M. 2015, 'Combined treatment with hot water and UV-C elicits disease resistance

against anthracnose and improves the quality of harvested mangoes', *Crop Protection*, vol. 77, pp. 1–8.

Subramanyam, H., Krishnamurthy, S. and Parpia, H. A. B. 1975, 'Physiology and biochemistry of mango fruit, *Advances in Food Research*, vol. 21, pp. 223–305.

Sui, Y., Wisniewski, M., Droby, S., Norelli, J. and Liu, J. 2016, 'Recent advances and current status of the use of heat treatments in postharvest disease management systems: Is it time to turn up the heat?', *Trends in Food Science and Technology*, vol. 51, pp. 34–40.

Suktawee, S., Wongjunta, M., Kaewwihar, S., Boonrod, J. and Jitareerat, P. 2011, 'Effects of UV-B radiation and hot water on the control of anthracnose disease in mango fruit during storage', *Agricultural Science Journal*, vol. 42, no. 3(Suppl.), pp. 263–6.

Sunsuwan, P. and Sardsud, U. 2003, 'Control of Anthracnose disease in postharvest mango fruit by hot water treatment and food microorganism', *Agricultural Science Journal*, vol. 34, no. 4–6(Suppl.), pp. 45–8.

Tandon, D. K. and Kalra, S. K. 1983, 'Changes in sugars, starch, and amylase activity during development of mango fruit cv. Dashehari', *Journal of Horticultural Science*, vol. 58, pp. 449–53.

Teixeira, G. H. D. A. and Durigan, J. F. 2011, 'Storage of 'Palmer' mangoes in low-oxygen atmospheres', *Fruits*, vol. 66, no. 4, pp. 279–89.

Tran, T. T. L., Aiamla-Or, S., Srilaong, V., Jitareerat, P., Wongs-Aree, C. and Uthairatanakij, A. 2015, 'Application of nitric oxide to extend the shelf life of mango fruit', *Acta Horticulturae*, vol. 1088, pp. 97–102.

Ullah, H., Ahmad, S., Amjad, M. and Khan, M. A. 2012, 'Response of mango cultivars to modified atmosphere storage at an ambient temperature cv. Alphanso and Chounsa', *Pakistan Journal of Agricultural Sciences*, vol. 49, no. 3, pp. 293–9.

Uthairatanakij, A., Jitareerat, P. and Kanlayanarat, S. 2006, 'Effects of irradiation on quality attributes of two cultivars of mango', *Acta Horticulturae*, vol. 712, pp. 885–92.

Valerio, R. F. and Esguerra, E. B. 2001, 'Rheological properties of mango (*Mangifera indica* L.) fruits in relation to handling and transport', *Philippine Agricultural Scientist*, vol. 84, pp. 232–40.

Vásquez-Caicedo, A. L., Sruamsiri, P., Carle, R. and Neidhart, S. 2005, 'Accumulation of all-trans-β-carotene and its 9-cis and 13-cis stereoisomers during postharvest ripening of nine Thai mango cultivars', *Journal of Agricultural and Food Chemistry*, vol. 53, no.12, pp. 4827–35.

Vázquez-Celestino, D., Ramos-Sotelo, H., Rivera-Pastrana, D. M., Vázquez-Barrios, M. E. and Mercado-Silva, E. M. 2016, 'Effects of waxing, microperforated polyethylene bag, 1-methylcyclopropene and nitric oxide on firmness and shrivel and weight loss of 'Manila' mango fruit during ripening', *Postharvest Biology and Technology*, vol. 111, pp. 398–405.

Vazquez-Salinas, C. and Lakshminarayana, D. 1985, 'Compositional changes in mango fruit during ripening at different storage temperatures, *Journal of Food Science*, vol. 50, pp. 1646–8.

Vivekananthan, R., Ravi, M., Saravanakumar, D., Kumar, N., Prakasam, V. and Samiyappan, R. 2004, 'Microbially induced defense related proteins against postharvest anthracnose infection in mango', *Crop Protection*, vol. 23, pp. 1061–7.

Wainwright, H. and Burbage, M. B. 1989, 'Physiological disorders in mango (*Mangifera indica* L.) fruit', *Journal of Horticultural Science*, vol. 64, pp. 125–35.

Wang, B., Wang, J., Feng, X., Lin, L., Zhao, Y. and Jiang, W. 2009, 'Effects of 1-MCP and exogenous ethylene on fruit ripening and antioxidants in stored mango', *Plant Growth Regulation*, vol. 57, no. 2, pp. 185–92.

Wang, Q., Zuo, J. H., Wang, Q., Na, Y. and Gao, L. P. 2015, 'Inhibitory effect of chitosan on growth of the fungal phytopathogen, *Sclerotinia sclerotiorum*, and sclerotinia rot of carrot', *Journal of Integrative Agriculture*, vol. 14, no. 4, pp. 691–7.

Wanikanukul, C., Chonhenchob, V. and Chanthanawaranggoon, S. 2007, 'Effect of modified atmosphere packaging on quality and shelf life of mango cv. 'Nam Dok Mai See Thong' in high gas permeable package', *Agricultural Science Journal*, vol. 38, no. 5 (Suppl.), pp. 242–5.

Watanawan, A., Watanawan, C. and Jarunate, J. 2008, 'Bagging 'Nam Dok Mai #4' mango during development affects colour and fruit quality', *Acta Horticulturae*, vol. 787, pp. 325–8.

Watkins, C. B. 2008, 'Overview of 1-methylcyclopropene trials and uses for edible horticultural crops', *HortScience*, vol. 43, no.1, pp. 86–94.

Whiley, A. W. 1993, 'Environmental effects on phenology and physiology of mango - A review', *Acta Horticulture*, vol. 341, pp. 168–76.

Wilberg, V. C. and Rodriguez-Amaya, D. B. 1995, 'HPLC quantitation of major carotenoids of fresh and processed guava, mango and papaya', *Lebensmittei-Wissenschaft und Technologie*, vol. 28, pp. 474–80.

Wills, R. B. H., Warton, M. A., Mussa, D. M. D. N. and Chew, L. P. 2001, 'Ripening of climacteric fruits initiated at low ethylene levels', *Australian Journal of Experimental Agriculture*, vol. 41, pp. 89–92.

Wilson, C. W., Shaw, P. E. and Knight, R. J. 1990, 'Importance of some lactones and 2,5 dimethyl-4-hydroxy-3(2H)-furanone to mango (*Mangifera indica* L.) aroma', *Journal of Agriculture and Food Chemistry*, vol. 38, pp. 1556–9.

Wisutiamonkul, A. and Suwanagul, A. 2010, 'Effect of gamma irradiation on quality of mango fruit cv. Nam Dok Mai Si Thong', *Thai Journal of Agricultural Science Journal*, vol. 41, pp. 99–102.

Wongmetha, O., Ke, L. S. and Liang, Y. S. 2013, 'The changes in physico-chemical compositions of 'Irwin' mango fruit after postharvest treatments under low temperature storage', *Thai Journal of Agricultural Science*, vol. 46, no. 1, pp. 11–19.

Xu, W., Li, D., Fu, Y. and Wang, Y. 2012, 'The effect of active packaging film on the quality of 'Dashehari' mango fruits', *Advanced Materials Research*, vol. 399–401, pp. 1881–5.

Yadav, M. K., Patel, N. L. and Patel, S. R. 2013, 'Effect of irradiation and storage temperature on quality parameters of Kesar mango (*Mangifera indica* L.)', *Indian Journal of Plant Physiology*, vol. 18, no. 3, pp. 313–7.

Yahia, E. M. 2011, *Mango (Mangifera indica L.)*. Chapter 22 Woodhead Publishing Limited.

Yashoda, H. M., Prabha, T. N. and Tharanathan, R. N. 2005, 'Mango ripening—chemical and structural characterization of pectic and hemicellulosic polysaccharides', *Carbohydrate Research*, vol. 340, pp. 1335–42.

Yashoda, H. M., Prabha, T. N. and Tharanathan, R. N. 2006, 'Mango ripening: changes in cell wall constituents in relation to textural softening', *Journal of the Science of Food and Agriculture*, vol. 86, pp. 713–22.

Yashoda, H. M., Prabha, T. N. and Tharanathan, R. N. 2007, 'Mango ripening - Role of carbohydrases in tissue softening', *Food Chemistry*, vol. 102, pp. 691–8.

Yenjit, P., Issarakraisila, M., Intana, W. and Chantraprommа, K. 2010, 'Fungicidal activity of compounds extracted from the pericarp of *Areca catechu* against *Colletotrichum gloeosporioides in vitro* and in mango fruit', *Postharvest Biology and Technology*, vol. 55, pp. 129–32.

Zaharah, S. S. and Singh, Z. 2011, 'Mode of action of nitric oxide in inhibiting ethylene biosynthesis and fruit softening during ripening and cool storage of 'Kensington Pride' mango', *Postharvest Biology and Technology*, vol. 62, no. 3, pp. 258–66.

Zainuri, D., Irving, E., Dann, E. K., Coates, L. M. and Wearing, A. H. 2005, 'UV-C Irradiation activates resistance to Anthracnose in Mango Fruit', Program and Abstract. Australasian Postharvest Horticulture Conference. Royal Lakeside Novote, Rotorua, New Zealand. *New Zealand Journal of Crop and Horticultural Science* 35 (2) 285, http://www.tandfonline.com/doi/pdf/10.1080/01140670709510195 Accessed 24 July 2016.

Zeng, K., Deng, Y., Ming, J. and Deng, L. 2010, 'Induction of disease resistance and ROS metabolism in navel oranges by chitosan', *Scientia Horticulturae*, vol. 126, pp. 223–8.

Zeng, K. F., Cao, J. K. and Jiang, W. B. 2006, 'Enhancing disease resistance in harvested mango (*Mangifera indica* L. cv. 'Matisu') fruit by salicylic acid', *Journal of the Science of Food and Agriculture*, vol. 86, pp. 694–8.

Zhang, Z., Yang, D., Yang, B., Gao, Z., Li, M., Jiang, Y. and Hu, M. 2013, 'Aminobutyric acid induces resistance of mango fruit to postharvest anthracnose caused by *Colletotrichum gloeosporioides* and enhances activity of fruit defense mechanisms', *Scientia Horticulturae*, vol. 160, pp. 78–84.

Zheng, M., Shi, J., Wang, Q. and Li, Y. 2013, 'Antimicrobial effects of volatiles produced by two antagonistic *Bacillus* strains on the anthracnose pathogen in postharvest mangos', *Biological Control*, vol. 65, pp. 200–6.

Zheng, X. L., Jing, G. X., Liu, Y., Jiang, T. A., Jiang, Y. M. and Li, J. R. 2012, 'Expression of expansin gene, MiExpA1, and activity of galactosidase and polygalacturonase in mango fruit as affected by oxalic acid during storage at room temperature', *Food Chemistry*, vol. 132, pp. 849–54.

Zhu, X., Cao, J., Wang, Q. and Jiang, W. 2007, 'Postharvest infiltration of BTH reduces infection of mango fruits (*Mangifera indica* L. cv. Tainong) by *Colletotrichum gloeosporioides* and enhances resistance inducing compounds', *Journal of Phytopathology*, vol. 156, pp. 68–74.

Zhu, X., Cao, J., Wang, Q. and Jiang, W. 2008, 'Postharvest infiltration of BTH reduces infection of mango fruits (*Mangifera indica* L. cv. Tainong) by *Colletotrichum gloeosporioides* and enhances resistance inducing compounds', *Journal of Phytopathology*, vol. 156, pp. 688–74.

Chapter 13

Post-harvest storage management of mango fruit

Noam Alkan, Agricultural Research Organization (ARO), Volcani Center, Israel; and Anirudh Kumar, Agricultural Research Organization (ARO), Volcani Center, Israel and Indira Gandhi National Tribal University (IGNTU), India

1 Introduction: the mango fruit

1.1 Origins and global production

The mango (*Mangifera indica* L.) is a dicotyledonous plant of the family Anacardiaceae, generally cultivated in tropical and subtropical climatic zones (Bompard, 2009). The mango is referred to as the 'king of fruits' (particularly in India) due to its nutritional benefits, excellent exotic flavour and versatile uses. Mango is a pulpy, sweet, climacteric drupe fruit. It is mostly eaten when it is fresh and ripe. Alternatively, processed mangoes are also consumed in the form of juices, nectars, pickles, jams and fruit creams, among others. Consumer acceptance of mango fruit relies on both external and internal quality aspects (Kader, 2002). The quality of mango depends on various factors, including cultivar, maturity at harvest, pre- and post-harvest handling practices, mechanical damage, chilling injury (CI) and post-harvest rots (Kader 2008a,b; Brecht et al., 2010).

More than 150 varieties of mango are cultivated worldwide. Mango is considered to have originated from southern Asia, and more specifically from eastern India, Burma and the Andaman Islands (Subramanyam et al., 1975; Tjiptono et al., 1984). Today, more than

http://dx.doi.org/10.19103/AS.2017.0026.16

90 countries worldwide produce mango commercially. India is the world's largest mango producer with 21.54 million tonnes annually (Sekhar et al., 2013), followed by China. However, Mexico is the world's largest exporter of mango (41% of the world market) (Sauco, 2004), whereas the United States is the world's largest importer (about 30% of all mango imports) (Tharanathan et al., 2006). The global demand for fresh mango is on the rise (Prospectiva, 2020, http://www.prospectiva2020. com/).

A short production season, short storage period (2–3 weeks at 12°C) and development of post-harvest diseases leading to decay are the major constraints of the mango trade. Food deterioration is a vital issue as it depletes natural assets. Recent studies have estimated food losses at more than 33% (Gustavsson et al., 2011; Lipinski et al., 2013; Buzby et al., 2014; Okawa, 2015). Post-harvest losses of fruit and vegetables are even higher. In fact, an Australian case study of the mango fruit showed 40% loss from harvest to plate (Ridoutt et al., 2010). Therefore, there is much room for improvement with regard to uneven fruit maturation, ripening variability, post-harvest rots and compromised quality, all of which may be due to inappropriate management practices. In this chapter, we will attempt to summarize the available information on preserving overall fruit quality and reducing post-harvest losses.

1.2 Mango fruit composition and quality attributes

Mango is one of the most nutritious tropical fruits. Its fruit is fleshy and sweet, with cultivars varying in taste, size, colour, aroma, fragrance and composition (FAO, 2002). It is an excellent source of phytonutrient compounds such as provitamin A, carotenoids,

Table 1 Nutritive value of mango fruit per 100 g

Nutrients	Ripe mango	Unripe mango
Proteins (g)	0.6	0.7
Fats (g)	0.4	0.1
Minerals (g)	0.4	0.4
Fibre (g)	0.7	1.2
Carbohydrates (g)	16.9	10.1
Energy (kcal)	74	44
Vitamin C (mg)	16	3
Total carotenes (μg)	2,210	90
β-carotene (μg)	1,990	–
Potassium (mg)	205	83
Sodium (mg)	26	43
Calcium (mg)	14	10
Iron (mg)	1.3	0.33
Phosphorus (mg)	16	19

Source: Nigam et al. (2007).

vitamins (B6, C and A), phenolics (gallic acid, syringic acid, gentisyl-protocatechuic acid, mangiferin, ellagic acid and quercetin), antioxidants and dietary fibre, which are important for human nutrition and health (Sogi et al., 2012; Lemmens et al., 2013; Dorta et al., 2014: Jahurul et al., 2015; Gorinstein et al., 2011). Moreover, mango fruit contains other important nutritional and health components, such as carbohydrates, amino acids and the minerals, calcium, iron and potassium (Table 1). Some bioactive compounds found in mangoes have shown potential anti-carcinogenic, anti-atherosclerotic, anti-mutagenic and angiogenic activities (Cao and Cao, 1999). Its β-carotene and lycopene suppress tumour growth and protect cells from damage. The nutrient content of the mango fruit is affected by several factors, including cultivar, growing conditions, stage of maturity at harvest, storage conditions and interactions between pre- and post-harvest factors (Lee and Kader, 2000; Léchaudel and Joas, 2007). For example, cvs. Haden and Ataulfo contain higher amounts of β-carotene than cvs. Kent or Tommy Atkins (Ornelas-Paz et al., 2007). Unripe mango contains a higher amount of vitamin C and pectins, which are considered beneficial in reducing cholesterol. However, they are sour in taste due to the presence of various organic acids, including succinic, gallic, maleic and citric acids. During storage and fruit ripening, the amount of acid decreases (Lakshminarayanan et al., 1970).

2 Harvest operations

2.1 Maturity and harvest parameters

The fruit quality and post-harvest storage period of mango depend mainly on fruit maturity at harvest (Fig. 1). Ripe mango fruit are more susceptible to post-harvest disease and mechanical injury (Prusky et al., 2009), while immature fruit are prone to chilling stress during cold storage (Yahia, 1999).

Mango is a climacteric fruit and as such it can ripen after harvest, when harvested after its physiological maturity. However, fruits that ripen off the tree do not usually

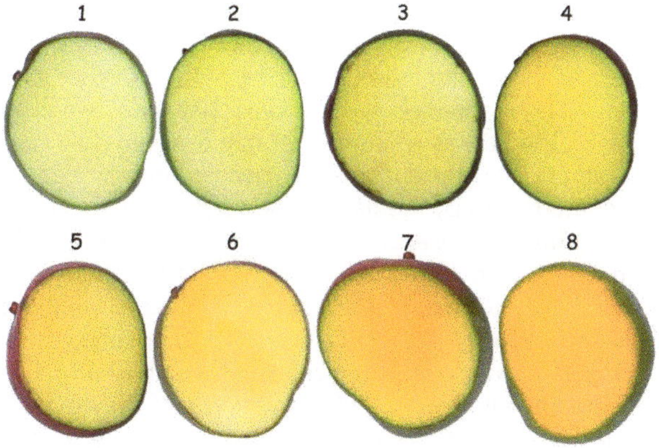

Figure 1 Inner colour of 'Shelly' mango fruit as a harvest parameter (Mango Quality Metrics, Plant Protection and Inspection Services, Ministry of Agriculture and Rural Development, Israel).

acquire their full flavour as those that ripen on the tree (Kader and Mitcham, 2008). Nevertheless, mango fruit are commonly harvested before full ripening to prolong the post-harvest storage period with minimal quality deterioration (Yahia, 1999). Application of maturity indices is extremely important for preserving fruit quality and reducing post-harvest losses. Physical, physiological and chemical parameters are generally taken into account to determine fruit maturity. Physical parameters include softening, changes in peel and pulp colour, and shoulder development (Kosiyachinda et al., 1984; Kienzle et al., 2011). Mango generally reaches maturity around 12–16 weeks after full bloom (Yahia, 1999).

Traditionally, mango harvesting time is based on appearance. Unripe mangoes are green in colour and during ripening the fruit changes from dark green to olive green and then become yellow or orange due to decreasing chlorophyll and accumulation of carotenoid pigments (Tharanathan et al., 2006). The red and purple peel colours are correlated to anthocyanin contents (Sherman et al., 2015) and can be induced by exposure to light (Sivankalyani et al., 2016a). Peel colour is not considered a good maturity index because it develops when the fruit starts to soften and, in many cases, the colour development is not uniform. On the other hand, changes in flesh colour are more uniform and advance with maturity, establishing it as an appropriate maturity index (Fig. 1, Kader, 2008a,b). The maturity of some mango cultivars, such as 'Tommy Atkins', 'Keitt', 'Kent' and 'Haden', is assessed by changes in flesh colour. For example, the inner colour of 'Shelly' mango fruit is used as a harvest parameter in Israel (Fig. 1). The recommendation for the Israeli fruit is that they be harvested with a pulp colour index of 3–7. Index values of 3–4 are considered common for harvesting for export, and values of 6–8 are considered appropriate for harvesting for ready-to-eat fruit (Fig. 1). It is impossible to obtain a uniform maturity index for all cultivars, and therefore it is essential to develop standard maturity indices for each specific cultivar.

Assessing a combination of physical indices, such as shape and size, together with flesh colour and detachment of the first fruit from the tree, is considered a better approach to judging the maturity of mango fruit (Slaughter, 2009).

During ripening, the fruit softens due to increased activity of cell wall-degrading enzymes such as pectin methylesterase and polygalacturonase (Luna-Guzman and Barrett, 2000). Simultaneously, the total soluble solids (TSS), also referred as soluble solids content (SSC), increase (Tharanathan et al., 2006). TSS, or SSC, represents mostly soluble sugars. SSC is not commonly used as a maturity index, but it can support other indices such as flesh colour. Fruit taste is mainly controlled by the ratio between sweetness and sourness. During ripening, the SSC increases while the titratable acidity (TA) decreases. The SSC TA ratio increases during ripening and indicates ripeness (Mizrach et al., 1999). Consumer acceptance is mainly determined by the flavours corresponding to sugar, acids and aroma volatile compounds (Baldwin, 2010). Aroma compounds also increase during fruit ripening. The most suitable time to harvest, from the flavour point of view, is just after the initiation of ripening (Lalel et al., 2003b). Harvested at this stage, the fruit will retain most of its flavour and aroma, while also maintaining most of its resistance to post-harvest pathogens.

Recently, new non-destructive methods have been established to determine mango quality and maturity (Ibarra-Garza et al., 2015) and some of the more important of these methods include nuclear magnetic resonance, electronic nose, hyperspectral analysis and near infrared reflection spectroscopy (Nicolai et al., 2014). Nevertheless, these technologies have not yet been integrated into commercial practice.

It is good practice to harvest mango fruit when the temperature is not high, so that the fruit will maintain reduced metabolic activity for a longer period. After harvest, the fruit should not be exposed to direct sunlight and they should be cooled as soon as possible.

2.2 Sapburn

At harvest, severing the mango fruit from the stem causes the release of resin for up to 1 h. The sap, which has a low pH, can burn the fruit surface. This burn enhances the accumulation of red and black spots and decay development on the peel, and compromises the fruit's quality (Fig. 3b). Mango resin (also termed latex) has the aroma of a ripe fruit. Terpinolene and 3-carene have been reported as sap volatiles (Robinson et al., 1993). Latex may provide protection against fruit-fly infestation (Joel, 1980) and may contribute to disease tolerance, probably because of resorcinol, which has antifungal properties (Johnson and Hofman, 2009). Thus, harvesting fruit with 1–2 cm of stem could reduce stem-end rot (SER) (Hassan et al., 2007). However, the resorcinol in mango latex can cause human skin disorders.

After harvest, the sap should be drained from the fruit or treated to minimize the incidence of sapburn. Picking fruit with long stems and de-sapping in the shade using various aids can reduce sap-burn damage. However, placing the fruit directly on the ground reduces sap-burn damage, but increases SERs (Johnson et al., 1993). Several treatments like de-sapping in the orchard by dipping in water or 0.1% NaOH were shown to reduce sapburn and are commercial practice in Israel (Holmes and Ledger, 1992). Extensive research on sapburn was conducted in Australia, which recommended de-sapping in the orchard using alkaline detergent spray or dip for at least 90 s; the alkaline detergents deactivate sap components (Bally et al., 1997). The alkaline detergent application could be applied in several commercial harvesting practices (Johnson and Hofman, 2009).

2.3 Transport to the packing house

Harvested mature mango fruit should be transported to the packing house straightaway while avoiding exposure to sunlight, preferably under cooled conditions. Rough handling during transport should be avoided.

3 Post-harvest operations: managing mango fruit diseases

Post-harvest-handling processes play a pivotal role in controlling post-harvest losses. Mango fruit has a high moisture content and soft texture, and is highly decomposable. Therefore, if not handled properly, mango fruit can deteriorate and decay rapidly. To preserve the quality and enhance fruit shelf life, post-harvest operations should include harvesting the fruit according to maturity indices, sap-burn control, washing and waxing, post-harvest disease and pest control, sorting and grading, processing if needed, packaging and cold storage, among others (Fig. 2).

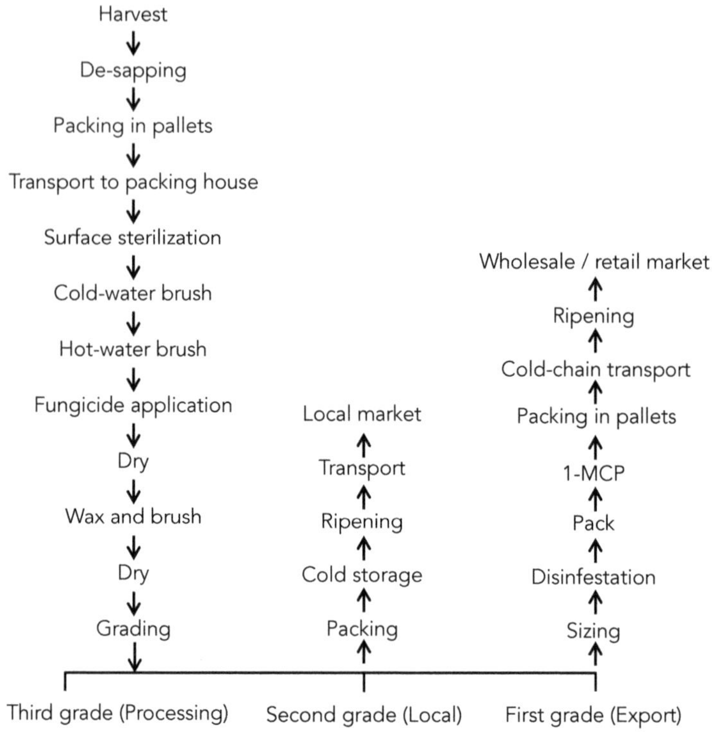

Figure 2 Post-harvest process of mango fruit handling, from orchard to market.

3.1 Overview of mango fruit diseases

Among post-harvest diseases in mango, anthracnose caused by *Colletotrichum gloeosporioides* is predominant under humid growth conditions (Fig. 3a, Alkan et al., 2015). Other common post-harvest diseases include SER, which is caused by various pathogens, mainly related to the Botryosphaeria group, such as: *Lasiodiplodia theobromae, Dothiorella dominicana, Neofusicoccum* species and *Phomopsis mangiferae* (Fig. 3a, Jonson et al., 1992). Other pathogens, such as *C. gloeosporioides* and *Alternaria alternata*, can also cause SER. Mango black spots caused by *A. alternata* and black mould rot caused by *Aspergillus niger* are more prevalent under dry environmental conditions (Fig. 3a, Prusky et al., 1993b).

Post-harvest fungal pathogens penetrate the fruit through wounds and natural openings. *Colletotrichum* can also penetrate the fruit directly via appressorium formation. These pathogens can remain quiescent for long periods, until the fruit ripens, at which time they switch to aggressive necrotrophic colonization and cause decay (Prusky et al., 2013). SER pathogens penetrate the stem through natural openings and live as endophytes in the stem until the fruit ripens. They then switch to necrotrophic colonization and cause mango SER.

Devastating post-harvest diseases, such as anthracnose, *Alternaria* black spot and SER cause huge economic losses and deterioration in fruit quality (Fig. 3a). Fruit ripening,

Figure 3 Post-harvest diseases, injuries and disorders. (a) Representative pictures of common diseases of mango fruit and their causal agents, fungal pathogen conidia and colonies: anthracnose caused by *C. gloeosporioides*, black spot caused by *A. alternata* and SER caused by *L. theobromae, Neofusicoccum mangiferae* or *P. mangiferae*. (b) Representative pictures of common injuries and disorders: sapburn, chilling injuries (red spots, black spots and pitting), jelly seed and stem-end cavity. Pictures taken by Dr. Noam Alkan, Prof. Dov Prusky and Dr. Yair Aharon.

injuries, elevated relative humidity (RH) and warm temperature favour post-harvest fungal pathogenicity and can reduce fruit resistance. Improper handling, packing and storing can injure the fruit, and the resultant openings allow the pathogens to penetrate the fruit tissue. Unripe fruit is considered to be resistant to post-harvest pathogens (Alkan and Fortes, 2015). During storage and ripening, the fruit become susceptible to various post-harvest diseases because of several physiological changes, including increased total soluble sugars, decreased acidity, increased softening, phytohormone changes, decreased phytoalexins such as resorcinol, and a general decrease in fruit resistance to pathogens (Prusky et al., 2013).

Prevention of post-harvest diseases begins with cultivar selection, cultural practices and chemical, physical and biological control. Various approaches have been tested to control the post-harvest decay and prolong the shelf life of mango without compromising fruit quality, including fruit sanitation, hot-water treatment (HWT) (Dea et al., 2010), fungicide treatments (Dang et al., 2008b; Malik and Amin, 2015), wax coating (Dang et al., 2008a), cold storage (Sivankalyani et al., 2016b), modified and controlled atmospheres (CAs), and biocontrol agents (Prusky et al., 2009; Singh and Singh, 2012b).

3.2 Pre-harvest control measures

Post-harvest diseases and pests can be reduced by various pre-harvest control measures, including the use of tolerant cultivars, orchard hygiene, manipulation of flowering and integrated management using chemical, physical and biological controls (Johnson et al., 1992; Ploetz, 2004; Akem, 2006). Chapter 5 (Osman et al., this volume) reviewed the 'Developing pest and disease-resistant mango varieties'. Chapter 19 (Ploetz et al., this volume) reviewed the 'Integrated disease management in mango cultivation'.

Several chemicals have been suggested for application in the orchard to decrease post-harvest disease. Periodic application (once a month) of copper oxychloride ($CuCl_2 \cdot 3Cu(OH)_2$), combined with mancozeb, from flowering till harvest, controls most mango post-harvest diseases (Lonsdale, 1993). Ledger (2004) recommended prochloraz application every 3–4 weeks in rotation with mancozeb and Cu during rainy periods. In Australia, azoxystrobin (Amistar®) application is recommended in the orchard to control anthracnose: one or two applications at flowering and/or early fruit set at no less than 14-day intervals, and again at 21 and 7 days before harvest (Johnson and Hofman, 2009).

3.3 Fruit sanitizers

The purpose of fruit sanitizers is to wash and kill the microorganisms on the fruit surface. Traditionally, the sanitizers consisted of water with or without chemicals (see section 'Physical control'). One of the most extensively used and studied sanitizers is chlorine (water pH 6.5–7.5; chlorine concentration 100–150 ppm). Both chlorine and sulphur dioxide have been used as fungal disinfectants (Johnson et al., 1997; Tefera et al., 2007). Different forms of chlorine, such as sodium hypochlorite, calcium hypochlorite and chlorine gas, control a wide range of post-harvest pathogens (Boyette, 1995). In the past, elevated chlorine dosages were frequently used due to the misconception that chlorine leaves no residue on the fruit.

Common alternatives to chlorine as sanitizers are ozone (O_3), oxidized water and hydrogen peroxide. Ozone and ozonated water were recognized in 1997 by the FDA as safe food disinfectants, and were proven to control post-harvest rots of various fruits, such as pineapple, banana, guava, papaya (Alothman et al., 2010; Yeoh, 2014; Romanazzi et al., 2016) and mango (Monaco et al., 2014). Recently, ozonated water has been reported as a sanitizer for mango cv. Palmer as it increases antioxidant activity (Minas et al., 2012; Lima et al., 2014, Monaco et al., 2016).

Electrolyzed water has also been suggested as a sanitizer for the industry (Colangelo et al., 2015). Electrolyzed water is produced by adding sodium chloride (as an electrolyte) to tap water and passing an electrical current through an anode or cathode to produce oxidizing (acidic) and reducing water (alkaline), respectively. The high electrolyzed water potential works against both bacteria (Pinto et al., 2015) and fungi (Guentzel et al., 2010). Hydrogen peroxide has also been recommended as an effective disinfectant against several fungi (Boyette, 1995). Fruit-sanitation techniques are very important in extending the storage life of mango.

3.4 Chemical control

Optimally, fruit sanitation eliminates all pathogens on the fruit surface. However, most of the pathogens, such as *C. gloeosporioides* and SER pathogens, may have already penetrated the fruit (Johnson et al., 1992; Alkan et al., 2015). Therefore, further treatments to control post-harvest pathogens are needed. Fungicides are the most effective treatments against fungal pathogens.

Hot benomyl and carbendazim (52°C) provide good control of SER and anthracnose (Johnson et al., 1997), but the application of benomyl after harvest has been banned. The active component of benomyl and thiabendazole in plants, carbendazim, is identical (Erwin, 1973). However, benomyl penetrates plant tissue more effectively than thiabendazole, carbendazim or thiophanate-methyl (Eckert and Joseph, 1985). Hot thiabendazole is generally effective in controlling SER, but provides poor control of anthracnose (Coates et al., 1993).

Post-harvest application of prochloraz effectively controls *C. gloeosporioides* and *A. alternata* during storage at low temperature and ripening at 20°C for cvs. Tommy Atkins, Keitt, Lilly and Haden (Prusky et al., 1999), but it does not provide good control for SER. Prochloraz is a well-recognized fungicide that is used commercially to control post-harvest diseases of mango fruit. In Australia, prochloraz at 250 ppm is applied and in Israel it is applied at 300 ppm, by overhead spray. The main disadvantage of imidazoles (i.e. prochloraz and imazalil) is that they are less effective at controlling SER pathogens than benzimidazoles (i.e. benomyl and thiabendazole) (Estrada et al., 1996).

With the appearance of various fungicide-resistant isolates, no one fungicide can provide complete protection against anthracnose, alternaria rot and SER. A combination of treatments must be applied to cope with the post-harvest pathogens. One combination offered in Australia is HWT with benomyl followed by a prochloraz spray, which provides effective control of anthracnose, SER and alternaria rot during long storage (Johnson et al., 1990b). Another combination applied in Israel includes chlorine sanitation, hot-water brushing (15–20 s) and then a spray of 50-mM hydrochloric acid (HCl), alone or in combination with prochloraz. This combination improved the control of anthracnose and alternaria rot (Prusky et al., 2006).

When fungicides are used in the packing house, fungicide containers must be cleaned periodically using approved methods.

3.5 Physical control

Growing public demand for chemical residue-free fruit has necessitated the development of alternative technologies, such as irradiation, heat treatment and cold-temperature storage. Cold storage of mango fruit (10–12°C) is one of the best ways of delaying fruit ripening and thus decreasing post-harvest decay (Sivankalyani et al., 2016b). Shortwave infrared radiation treatment reduces anthracnose in mango (Saaiman, 1996) and can also be considered for the organic market.

Heat treatment is known to reduce post-harvest diseases. Different approaches are included under this treatment option, such as hot-water dipping and rinsing, and hot water vapour and dry-air treatments (Schirra et al., 2000). There are many benefits to heat treatments, such as reduction in post-harvest decay, reduction in CI, killing of insects and pests, colour and flavour preservation and shelf-life improvement, among others (Lurie, 1998; Schirra, 2000; Fallik, 2004). Hot-water brushing at 50–60°C for 20 s after harvesting reduces decay development via both surface cleansing and induction of fruit resistance against pathogens (Prusky et al., 1996; Fallik, 2004); this method is applied in Israel. Hot-water dipping for 3–7 min has been recommended and is moderately efficient at delaying post-harvest rot (Johnson, 1994). Hot-water dips, or spray can control fungal infections such as anthracnose and alternaria rot better than SER (Johnson, 1994). Too much brushing in the hot-water brushing treatment can contribute to increased red and black spots on the peel due to lenticel damage (Luria et al., 2014). On the other hand, hot-water brushing leaves no chemical residue and could be a good alternative for organically produced mango.

3.6 Biological control

Post-harvest biological control agents might serve as an alternative to the use of synthetic chemicals and have been the focus of considerable research (Droby et al. 2016).

Numerous microbial antagonists (mostly yeast and bacteria) of post-harvest pathogens have been identified in laboratory, semi-commercial and commercial studies (Droby et al., 2009). Several of these antagonists have reached advanced levels of development and commercialization, such as *Bacillus subtilis* in South Africa (Avogreen, University of Pretoria, Pretoria, South Africa) for the control of post-harvest diseases of avocado (Demoz and Korsten, 2006), *Candida oleophila*, (Nexy, Leasafre, Lille, France) against wound pathogens on pome fruits, citrus and banana (Massart and Jijakli, 2014), *Aureobasidium pullulans* (BoniProtect, Bio-Ferm, Tulln, Austria) to control wound pathogens on pome fruit (Mounir et al., 2007), *Pantoea agglomerans* CPA-2, (Pantovital, Domca, Granada, Spain) against post-harvest pathogens of pome and citrus fruit (Torres et al., 2014), and *Metschnikowia fructicola* (Shemer, Bayer, Leverkusen, Germany) against various fruit and vegetable pathogens (Spadaro and Droby, 2016).

In South Africa, a *Bacillus licheniformis* isolate was used as a biological control agent against anthracnose on mango fruit. It was used either alone or in combination with HWT for 5 min at 45°C and with low doses of prochloraz or sodium hypochlorite (Govender et al., 2005). The yeasts *Rhodotorula minuta* (Patiño-Vera et al., 2005) and *Debaryomyces nepalensis* (Luo et al., 2015) have also been suggested as potential biocontrol agents of mango anthracnose. To date, these biocontrol products have not been widely applied commercially (Droby et al., 2016).

4 Post-harvest operations: use of ethylene, 1-MCP, modified and controlled atmospheres, waxes and edible coatings

4.1 Ethylene and 1-MCP

Mango is a climacteric fruit that ripens rapidly at ambient temperature after harvest, when it is harvested after physiological maturity. Ethylene triggers mango fruit ripening (Theologis, 1992; Singh and Singh, 2012a), but has harmful effects at high dosages, including increased senescence and increased susceptibility to post-harvest pathogens (Wills, 2005). Therefore, proper management of ethylene is important. Exposure to ethylene (100 ppm) for 12–24 h at 20–22°C results in accelerated and uniform ripening and can be applied before marketing of the ready-to-eat fruit. In contrast, another commercial approach recommends storing fresh produce at low temperature to reduce ethylene biosynthesis (Watkins, 2002).

The structure of 1-methylcyclopropene (1-MCP) is closely related to that of ethylene and therefore, it is used as a competitive inhibitor of ethylene. It binds to the ethylene receptor, thereby slowing fruit ripening (Sisler and Serek, 1997; Sisler, 2006). Application of 1-MCP has been shown to delay mango fruit-ripening parameters, including reducing respiration, delaying the formation of a climacteric peak, delaying fruit softening and delaying the decrease in ascorbic acid content during storage (Alves et al., 2004). 1-MCP is used for a wide variety of fruits and vegetables, such as apple, banana, mango, papaya, peach, persimmon, kiwifruit and tomato. Apple is an excellent example of the most successful use of 1-MCP (Watkins, 2008). Application of 1-MCP at high concentrations has a negative effect on fruit flavour and aroma (Lalel et al., 2003b; McArtney et al., 2008).

1-MCP probably has a dual effect on post-harvest fruit diseases: on the one hand, ethylene is involved in plant defence responses against pathogens, but on the other, it is a major regulator of fruit ripening, and fruit ripening is correlated with susceptibility (Alkan and Fortes, 2015). Thus, the effect of 1-MCP on fruit resistance or susceptibility is probably dependent on application timing and dose. 1-MCP application should be calibrated for each cultivar according to its ethylene production (Watkins and Miller, 2004). The impact of 1-MCP is further enhanced when applied with a CA (Watkins and Nock, 2003).

4.2 Modified and controlled atmospheres

The short storage life of mango is one of the major constraints in reaching overseas markets. Modified atmosphere (MA) and CA technology can extend storage life by delaying fruit respiration, senescence and deterioration without compromising quality, thereby supporting its long-distance export (Caleb et al., 2013). MA can be achieved by packing the fruit under semi-closed conditions such as micro-perforated polyethylene or polypropylene. Fruit respiration and gaseous exchange through the package creates a MA inside the package that can extend the fruit's shelf life. Fruit respiration leads to an increase in CO_2 concentration and decrease in O_2 concentration without water condensation. A MA can delay mango ripening by delaying ethylene production, thereby enhancing resistance to post-harvest pathogens (Prusky and Keen, 1993a; Kader, 1994; Yahia, 2009; Sivakumar et al., 2011). However, an unsuitable MA can lead to anaerobic respiration (O_2 < 2% and CO_2 > 10%), resulting in the formation of acetaldehyde and ethanol, uneven ripening and tainted flavour (Lalel et al., 2003a). Moreover, it is important to consider the humidity level inside the package to avoid condensation and pathogen development (Sousa-Gallagher et al., 2013; Malik et al., 2016).

Recently, much research has been conducted in the field of nanotechnology-based packaging. Several types of active packaging that adsorb O_2, ethylene, moisture, CO_2 and odours have been developed on the basis of nanotechnology enhancement of the mechanical properties of the packing film, in addition to the incorporation of antimicrobial agents (reviewed in Mihindukulasuriya and Lim, 2014).

Whereas in a MA, fruit respiration regulates the atmospheric composition, in a CA, the atmospheric composition, that is, O_2, CO_2 and N_2 concentrations, is externally regulated, along with temperature and RH. The general CA recommendation for mango fruit storage, 3–5% O_2 and 5–8% CO_2, can increase the cold-storage period from 2–3 weeks to 3–6 weeks. The low O_2 level affects both primary and secondary metabolism (Beaudry, 1999). In cv. Tommy Atkins, CA reduces the incidence of anthracnose without affecting total phenols and antioxidants (Kim et al., 2007). When used outside the recommended range, CA can have detrimental effects (reviewed by Kader, 2004). Other developments include dynamic CA based on the fruit's ethanol content (Veltman et al., 2003), respiration quotient (Gasser et al., 2010) and chlorophyll fluorescence (Prange et al., 2013). CA technology has been used for long-distance transport of several types of fruit, including apple, pear, avocado, strawberry, banana, mango, cranberry, plum, cherry, fig, kiwifruit and melon. However, its commercial application is limited to selected fruits, such as apples, pears and avocados. Its industrial acceptance depends on costs and benefits.

4.3 Waxing and coating

Waxes and edible coatings can be regarded as types of MA. Surface coatings are used to improve the fruit's external appearance and to alter gas permeability, reduce water loss and delay ripening (Banks et al., 1993). The surface coating is semipermeable, decreasing gas diffusion and respiration (Banks et al., 1993; Dhall, 2013). Thus, fruit coating delays ripening by lowering O_2 and increasing CO_2 concentration and hence, enhances the storage life of mango fruit. In most cases, edible coatings are environmentally friendly and are used as an alternative to film packaging. Fruit coating is also less costly than CA/MA technology (Baldwin, 2005).

Wax coatings are generally emulsions of synthetic polyethylene or natural carnauba wax, beeswax and others. Edible coatings are based on polysaccharides (cellulose, chitosan) or resins (zein, shellac). The integration of both polysaccharide- and protein-based materials enhances the functionality of the coating (Gol et al., 2013). Surface coatings containing synthetic waxes, natural waxes (carnauba and beeswax) and resins (shellac) limit water loss better than those containing polysaccharides (Amarante and Banks, 2001). Formulations based on shellac result in a shinier appearance than those based on carnauba wax or polysaccharides (Hoa and Ducamp, 2008). Coating 'Tommy Atkins' mango with carnauba wax and beeswax reduced water loss and chlorophyll loss, CI and decay after cold storage (Feygenberg et al., 2005). However, improper fruit waxing can lead to undesirable effects on fruit quality, including anaerobic respiration and development of off-flavours (Amarante and Banks, 2001). Therefore, waxes should be applied with a wax applicator such as a roller brush, with a uniform flow on dry fruit to prevent uneven wax application or water–wax emulsions. The brushes should be soft and drenched with wax before the treatment.

In recent years, significant progress has been made in the edible coating field by using chitosan, cellulose, mineral oils and protein materials (Dhall, 2013). Chitosan is superior to starch and cellulose in prolonging mango fruit shelf life and enhancing its quality (Kittur et al., 2001). Another kind of coating, *Aloe vera*, was shown to have antimicrobial properties with no effect on mango quality (Sophia et al., 2015). Even though edible coatings show potential for practical use, more research is required. In addition, coating costs and regulatory issues must be considered prior to their commercial application.

5 Post-harvest operations: quarantine treatments

The world trade in mango has expanded rapidly in the last decade. Increasing trade has increased the threat of global spread of invasive pests. Therefore, a quarantine treatment is essential for mango-exporting nations. Quarantine treatments must kill all of the undesirable pests on or in the commodities. Mango fruit pests include internal pulp feeders (fruit fly), seed and fruit pulp pests (mango weevils and fruit caterpillars) and external pests (i.e. thrips, scales, mites and mealybugs). The main quarantined pest problems are caused by fruit flies (Tephritidae). Strict quarantine measures have to be implemented to export mango to those countries that have not yet been exposed to these pests (EPPO, 2007). Approvals for disinfestation treatments and market access are obtained on a country-by-country basis. Key aspects and guidelines of quarantine regulations for crop pests are covered by the International Standards for Phytosanitary Measures (ISPM, 2007). Known quarantine treatments include chemicals, radiation and cold storage or heat quarantine. The major constraints in the development of such treatments for mango are the fruit's susceptibility to irradiation, cold, heat and O_2 depletion, all of which can impair fruit quality. Cold quarantine (Kane and Marcellin, 1978) cannot be used for mango because fruit quality is compromised at the required quarantine temperature. Fumigation with ethylene dibromide (16–35 g/m^3 ethylene dibromide for 2 h at 21–26°C) was a commonly used quarantine treatment until it was banned. Methyl bromide was phased out due to its toxicity to humans and the environment. Therefore, alternatives to chemical fumigation, such as irradiation and heat treatments (hot water, vapour heat and forced hot air), are now commonly used for fresh produce (Schneider et al., 2003; Follett, 2004; Armstrong and Mangan, 2007; Johnson and Hofman, 2009).

5.1 Irradiation

Growing public demand for chemical residue-free fruit is leading to increased development of alternative technologies, such as heat and irradiation. Gamma (γ) irradiation kills microorganisms and insects by damaging their DNA, but does not alter the nutritional value of food (Ferrier, 2010) and can even be used to extend the shelf life of fresh produce (Farkas et al., 2014). γ-Irradiation is considered safe and has been approved by the FDA (at up to 1 kGy) for fresh produce, while higher doses of irradiation are not recommended for fresh fruit because they might compromise fruit texture and flavour and increase senescence, due to severe damage to proteins and DNA. Fruit-fly disinfestation can be achieved with 70–150 Gy, while other insects, such as weevil and moth, need up to 400 Gy, depending on the species (EPA, 2002). The incidence of anthracnose was reported to be reduced by irradiation at 600 Gy (Johnson et al., 1990a). Mango treated with 0.3–0.7 kGy γ-irradiation showed delayed ripening and extended shelf life (Mahto and Das, 2013).

The importance of γ-irradiation in quarantine treatments was reviewed by Johnson and Hofman (2009). International recommendations for the effective use of irradiation as a phytosanitary measure (ISPM, 2007) and the USDA-APHIS PPQ Manual on Irradiation provide generic guidelines (APHIS, 2007). The impact of irradiation depends on many factors, such as its intensity and duration, cultivar type and fruit maturity at harvest (Mitcham and Yahia, 2009).

5.2 Heat treatments

Heat treatments include a variety of approaches, such as hot-water dipping and rinsing, and hot-vapour and dry-air treatments (Mangan and Hallman, 1998; Armstrong and Mangan, 2007). There are many benefits to heat treatments, such as reduced rotting (section 'Physical control'), a decrease in chilling injuries, pest eradication and more (Lurie, 1998; Schirra, 2000; Fallik, 2004). However, the primary purpose of heat treatments is to rid the fresh produce of pest infestation, especially fruit fly. To date, heat treatment is the most common disinfestation treatment for mangoes (Armstrong and Mangan, 2007).

Hot water

HWT via dipping is the largest-volume heat treatment for fresh produce. HWT is widely used in many countries, especially in Central and South America, as a disinfestation treatment for Tephritidae fruit flies. HWT has been used as a quarantine treatment to disinfest Mediterranean, melon and oriental fruit flies in banana, papaya and mango (Armstrong, 1982, 1994; Nascimento et al., 1992). Typical treatments include 46.1°C for 65–90 min, depending on the fruit's weight, followed by quick cooling (Armstrong and Mangan, 2007). Generic guidelines for the use of hot water are provided by the USDA-APHIS PPQ Manual for Hot Water Treatment. If HWT is not applied properly, it can also cause deterioration of the fresh fruit skin and compromise fruit quality, due to rapid heat transfer (Jacobi and Gowanlock, 1995; Jacobi et al., 2001). A detailed review of HWTs is presented in Sharp (1994).

Vapour heat and forced hot air

Vapour-heat treatment (VHT) was invented specifically for Mediterranean (*Ceratitis capitata* Wiedemann) and Mexican (*Anastrepha ludens* Loew) fruit fly disinfestation (Baker, 1952). In VHT, saturated heated air is passed across the fruit in a stream. VHT protocols include heating the fruit's internal pulp to 46–47°C for 10–30 min (Mangan and Hallman, 1998; Armstrong and Mangan, 2007). VHT guidelines for treating mango fruit for the US market are provided by the USDA-APHIS PPQ Manual on Vapour Heat Treatment.

Forced hot-air treatment (FHAT) is a modification of the VAT (Brown, 1989), which includes heating the fruit to the same temperatures as with VHT but with lower RH. FHAT provides a better fruit quality than HWT or VHT (Laidlaw et al., 1996). High temperature can cause fruit injury. Therefore, gradual heating and rapid cooling are applied to prevent damage to the fruit (Armstrong, 1994). In addition, the use of CA ($O_2 < 1\%$, $CO_2 > 15\%$) can reduce the duration of the heat treatment (Armstrong and Mangan, 2007). VHT is commonly used in Australia, Thailand, the Philippines and Taiwan, whereas FHAT is commonly used for quarantine treatment in New Zealand, Fiji and Cook Islands for the disinfestation of fruit flies in mango.

6 Preparing fruit for market

6.1 Sorting and grading

After pre- and post-harvest applications such as surface sterilization, fungicide, wax and other treatments to retain fruit quality, the fruit is sorted, graded and packed. Fruit must be sorted according to size and weight prior to disinfestation with hot water to ensure effective and consistent treatment. The sorting and grading categorizes the fruit on the basis of various disorders like shape, size, weight and colour to maintain uniformity. The fruits are divided into first-, second- and third-quality grades on the basis of consumer preference and demand and into defective unmarketable fruit, which are eventually removed.

Camera-based sorting systems can separate fruit according to colour, defects and shape. The fruits are usually separated and packed into single-layer trays according to their size and quality grade. To avoid fruit losses, the damaged fruit can be used for processing (e.g. as juice). Defects include initial decay or immature fruit, or fruit with physical injuries such as cuts, scrapes and bruises that favour the development of decay later on (Brecht et al., 2010).

Technologies need to be developed for early detection of post-harvest pathogen or chilling injuries to sort for proper storage duration and temperature. For example, the volatile 1-pentanol has been shown to be specific to *Lasiodiplodia* and *Colletotrichum* colonization (Moalemiyan et al., 2007).

6.2 Fresh-cut mango

Tremendous growth in the ready-to-eat fruit market has been largely due to increasing demand for fresh, healthy and convenient prepared salads. However, cutting or wounding mango fruit activates physiological and metabolic changes, with an upsurge in respiration rate and interference in the normal functioning of tissues. This could lead to tissue browning as a result of polyphenol oxidase and peroxidase activities (Oms-Oliu et al., 2010). In addition, wounding increases water loss and ethylene synthesis, and may enhance fruit susceptibility to microbial attack (Watada et al., 1996). The reduction in firmness, flavour and colour, and browning has negative effects on consumer acceptance of fresh-cut mango. To maintain the quality of fresh-cut mango and prolong its shelf life, several post-harvest treatments can be applied, such as anti-browning agents, organic acids, firmness stabilizers (calcium-based solutions), antioxidants, anti-ripening agents (1-MCP), ozone, electrolyzed water, edible coating, irradiation, heat treatments and MA (Beaulieu and Lea, 2003; Eduardo et al., 2007; Sothornvit and Rodsamran, 2007; Rico et al., 2007; Vilas-Boas and Kader, 2007; Robles-Sánchez et al., 2009; Djioua et al., 2010; He et al., 2016). The initial stage of mango ripening is considered the appropriate time for fresh-cut processing in order to facilitate the handling step and maintain fruit quality during storage (Ngamchuachit et al., 2015). An integrated treatment approach should be considered for treating fresh-cut fruit (Rico et al., 2007).

6.3 Packaging

Fruit-packing technology is very important for protecting fruit during distribution, storage and sale. Packaging protects fruit from microbial infection and spread, dust and dirt, as well as compression damage (Appiah and Kuma, 2009). The cardboard is designed to support individual fruit against impact damage and to allow air flow between the boxes. Packing serves a number of purposes, including protection of produce, and product description and promotion. Labelling is required by many quality-assurance systems and can serve as a marketing tool. The word 'Mango' should be visible on the label. In addition, the cultivar, fruit size and class, country of origin, grower, packing house and market agent names may be required. Designed labelling should present a marketing image that reflects customer and cultural preferences.

Mango fruit are usually packed in a single layer (Yahia, 1999), by fruit count per package, subject to fruit size. Different types of packages are used for mango around the world, but the most common are carton and plastic crates. Cartons that are used for export should be new, clean, strong and unbroken. There is increasing pressure in the EU for recyclable packing material. Cartons that are recyclable should be marked as such. Returnable plastic crates are used mostly for national trade, as the return cost would make them less profitable for international trade. Poor-quality packing can cause physical injuries and bruises. Ventilation of the produce is required to prevent condensation, while overventilation can lead to fruit shrinkage and weight loss. MA packaging is used to form an ideal balance of gases inside the package. MA packaging reduces respiration activity of the product and increases its storage life (see section 'Modified and controlled atmospheres').

6.4 Cold storage/cold-chain management

Immediate cold storage after fruit harvest slows respiration rate and metabolic processes related to ripening (McGlasson et al., 1979). Thus, cold storage is probably the most effective method of prolonging the storage period of fresh fruit. Mango fruit are stored at 10–13°C for 3–4 weeks, depending on cultivar and fruit maturation. Fruit storage at higher temperature would lead to rapid ripening and the accompanying susceptibility to fungal pathogens. Storage at lower temperature would lead to chilling injuries (Sivakumar et al., 2011). CI symptoms include red and black spots on the peel, pitting, irregular ripening, carotenoid reduction, reduced aroma and flavour, electrolyte leakage, a drop in soluble solids, short storage life and increased sensitivity to fungal pathogens (Fig. 3b, Wills et al., 1981; Lederman et al., 1997; Pesis et al., 2000; Sivakumar et al., 2011). Fruit sensitivity to cold storage depends on various factors, such as fruit maturity, storage duration and packaging type (Medlicott et al., 1990). Cold-chain management is required at several steps from harvest to storage to transportation and marketing should be uninterrupted, so that mango post-harvest life can be increased and fruit quality retained (Fig. 2; Harvey, 1978).

6.5 Transport

Mango is transported by aircraft, ship, truck and train. Transportation of tropical fruit has been reviewed by McGregor (1987) and Thompson (2002). In developing countries, non-refrigerated trucks are commonly used to transport mango. However, this can exert pressure on the fruit, which can soften the tissue and lead to microbial deterioration. In addition, lack of temperature control will reduce shelf life (Mitcham and Yahia, 2009). However, it is possible to transport the fruit at night under non-refrigerated conditions (up to 3 h) if it has been precooled.

In principle, mango should be transported in the cold in pallets of single-layer cartons (Brecht et al., 2010). The fruit must be cold before transportation because refrigerated vehicles are not designed to lower fruit temperature. Refrigerated vehicles should be equipped with temperature loggers. Additional loggers for RH, atmospheric composition or fruit temperature can be added. The refrigerated truck should be clean and in sound condition and the doors must be tightly closed. Cold-chain management is extremely important, especially when fruit are exported to a distant location. To sell ready-to-eat fruit, its artificial ripening is necessary. It is better to artificially ripen the product, using high temperature or ethylene, as close as possible to the retail end to minimize physical damage and decay during fruit storage and transportation (Fig. 2).

7 Conclusions

Short shelf life and susceptibility to chilling and post-harvest diseases are the main difficulties that need to be tackled to expand the mango market worldwide. In the post-genomic era, significant progress has been made to understand the molecular biology of fruit ripening and softening of mango (Dautt-Castro et al., 2015) and mango single nucleotide polymorphisms (SNPs) (Sherman et al., 2015). Nevertheless, there is still much to be investigated. Fruit ripening, softening and susceptibility to pathogens and chilling differ from one cultivar to another. Cultivar-specific molecular markers need to be identified and correlated with post-harvest characteristics to ensure good-quality fruit that is resistant to diseases and chilling with a longer shelf life.

In addition, post-harvest handling of mango fruit has great potential for prolonging post-harvest life. The effect of various factors, such as genotype, fruit maturity, environmental conditions at the orchard, pre-harvest and post-harvest treatments, and storage conditions, has been determined to some extent, but many aspects still need to be investigated. Rational integration of post-harvest treatments, packaging technology, biosensor development and more can contribute to improving post-harvest management of mango fruit, retaining fruit quality, reducing losses and increasing shelf life.

8 Where to look for further information

The Mango: Botany, Production and Uses, second edition, 2009, CABI. Additional important information can be found on the University of Florida website http://edis.ifas.ufl.edu/hs1185 and on the Davis University website http://postharvest.ucdavis.edu/Commodity_Resources/Fact_Sheets/Datastores/Fruit_English/?uid=37&ds=798.

9 Abbreviations

CA	Controlled atmosphere
CI	Chilling injury
FHAT	Forced hot-air treatment
HWT	Hot-water treatment
MA	Modified atmosphere
RH	Relative humidity
SSC	Soluble solids content
VHT	Vapour-heat treatment

10 References

Akem, C. N. (2006). Mango anthracnose disease: present status and future research priorities. *Plant Pathology* 5:266–73.

Alkan, N., Friedlander, G., Ment, D., Prusky, D. and Fluhr, R. (2015). Simultaneous transcriptome analysis of *Colletotrichum gloeosporioides* and tomato fruits response reveals novel fungal-fruit arm and defense strategies. *New Phytologist* 205:801–15.

Alkan, N. and Fortes, A. M. (2015). Insights into molecular and metabolic events associated with fruit response to postharvest fungal pathogens. *Frontiers in Plant Science* 6:889. doi:10.3389/fpls.2015.00889.

Alothman, M., Kaur, B., Fazilah, A., Bhat, R. and Karim, A. A. (2010). Ozone-induced changes of antioxidant capacity of fresh-cut tropical fruits. *Innovative Food Science Emerging Technology* 11:666–71.

Alves, R. E., Filgueiras, H. A. C., Almeida, A. D., Pereira, M. E. C., Cocozza, F. D. and Jorge, J. T. (2004). Postharvest ripening of 'Tommy Atkins' mangoes on two maturation stages treated with 1-MCP. *Acta Horticulturae* 645:627–32.

Amarante, C. and Banks, N. H. (2001). Postharvest physiology and quality of coated fruits and vegetables. *Horticultural Reviews* 26:161–238. https://doi.org/10.1002/9780470650806.ch4.

APHIS-Animal and Plant Health Inspection Services, USDA. (2007). http://www.aphis.usda.gov.

Appiah, F. and Kuma, P. (2009). Processing and Packaging Technology. IDL, Knust. Kumasi. Ghana. pp. 121.

Armstrong, J. W. (1982). Development of a hot-water immersion quarantine treatment for Hawaiian-grown 'Brazilian' bananas. *Journal of Economic Entomology* 75(5):787–90.

Armstrong, J. W. (1994). Heat and cold disinfestation treatments. In: Paull, R. E. and Armstrong, J. W. (eds), *Insect Pests and Fresh Horticultural Products: Treatments and Responses*. CAB International, Wallingford, UK, pp. 103–19.

Armstrong, J. and Mangan, R. (2007). Commercial quarantine heat treatments. In: Tang, J., Mitcham, E., Wang, S. and Lurie, S. (eds), *Heat Treatments for Postharvest Pest Control: Theory and Practice*. Wallingford, UK: CAB International, pp. 311–40.

Baker, A. (1952). The vapor-heat process. Insects: The Yearbook of Agriculture. US Dept. Agric. US Government Printing Office, Washington DC, pp. 401–4.

Baldwin, E. (2005). Edible Coatings In: Environmentally friendly techniques for agricultural produce quality. Ben-Yehoshua, S., Taylor and Francis Group LLC. Boca Roxton, FL. USA.

Baldwin, E. A. (2010). Fruit and vegetable flavour. http://www.ne-postharvest.com/hb66/022 flavor.pdf.

Bally, I. S. E., O'Hare, T. J. and Holmes, R. J. (1997). Detrimental effects of detergent in the development of mango skin browning. *Acta Horticulturae* 455:612–21.

Banks, N. H., Dadzie, B. K. and Cleland, D. J. (1993). Reducing gas exchange of fruits with surface coatings. *Postharvest Biology and Technology* 17:215–26.

Bompard, J. M. (2009). Taxonomy and systematics. In: LITZ, R. E. (ed.), *Mango: botany production and uses*. Wallingford: CABI, pp. 19–41.

Beaudry, R. M. (1999). Effect of O_2 and CO_2 partial pressure on selected phenomena affecting fruit and vegetable quality. *Postharvest Biology and Technology* 15:293–303.

Beaulieu, J. C. and Lea, J. M. (2003). Volatile and quality changes in fresh-cut mangos prepared from firm-ripe and soft-ripe fruit, stored in clamshell containers and passive MAP. *Postharvest Biology and Technology* 30:15–28.

Boyette, M. D. (1995). Packaging Requirements for Fruits and Vegetables. In: *Postharvest Handling and Cooling of Fresh fruits, Vegetables and Flowers for Small Farm 800–4*. North Carolina Crop Extension Services. USA, pp. 17.

Brecht, J. K., Sargent, S. A., Kader, A. A., Mitcham, E. J. and Maul, F. (2010). *Mango. Postharvest Best Management Practices Manual*. National Mango Board. Davis, UC, USA.

Brown, S. A. (1989). High-temperature forced-air quarantine treatment for papayas infested with *tephritid* fruit flies (Diptera: *Tephritidae*). *Journal of Economic Entomology* 82:1667–74.

Buzby, J. C., Wells, H. F. and Hyman, J. (2014). The estimated amount, value, and calories of postharvest food losses at the retail and consumer levels in the United States. USDA. J. EIB-121. http://www.ers.usda.gov/publications/eib-economic-information-bulletin/eib-xxx.aspx.

Caleb, O. J., Mahajan, P. V., Al-Said, F. A. and Opara, U. L. (2013). Modified atmosphere packaging technology of fresh and fresh-cut produce and the microbial consequences: a review. *Food Bioprocess Technology* 6:303–29.

Cao, Y. H. and Cao, R. H. (1999). Angiogenesis inhibited by drinking tea. *Nature* 398:381.

Coates, L., Johnson, G. and Cooke, A. (1993). Postharvest disease control in mangoes using high humidity hot air and fungicide treatments. *Annual Applied Biology* 123:441-8.

Colangelo, M. A., Caruso, M. C., Favati, F., Scarpa, T., Condelli, N. and Galgano, F. (2015). Electrolysed water in the food industry as supporting of environmental sustainability. In: Vastola, A. (ed.), *The Sustainability of Agro-Food and Natural Resource Systems in the Mediterranean Basin*. Springer International Publishing, pp. 385–97.

Dang, K. T. H., Singh, Z. and Swinny, E. E. (2008a). Edible coatings influence fruit ripening, quality and aroma biosynthesis in mango fruit. *Journal of Agriculture and Food Chemistry* 56:1361–70.

Dang, K. T. H., Singh, Z. and Swinny, E. E. (2008b). Impact of postharvest disease control methods and cold storage on volatiles, color development and fruit quality in ripe 'Kensington Pride mangoes. *Journal of Agriculture and Food Chemistry* 56:10667–74.

Dautt-Castro, M., Ochoa-Leyva, A., Contreras-Vergara, C. A., Pacheco-Sanchez, M. A., Casas-Flores, S., Sanchez-Flores, A., Kuhn, D. N. and Islas-Osuna, M. A. (2015). Mango (*Mangifera indica* L.) cv. Kent fruit mesocarp de novo transcriptome assembly identifies gene families important for ripening. *Frontiers in Plant Science* 6:62. https://doi.org/10.3389/fpls.2015.00062.

Dea, S., Brecht, J. K., Nunes, M. C. N. and Baldwin, E. A. (2010). Quality of fresh-cut 'Kent' mango slices prepared from hot water or non-hot water-treated fruit. *Postharvest Biology and Technology* 56:171–80.

Demoz, B. T. and Korsten, L. (2006). *Bacillus subtilis* attachment, colonization, and survival on avocado flowers and its mode of action on stem-end rot pathogens. *Biological Control* 37: 68–74.

Dhall, R. K. (2013). Advances in edible coatings for fresh fruits and vegetables: a review. *Critical Review in Food Science and Nutrition* 53:435–50.

Djioua, T., Charles, F., Freire, M., Filgueiras, H., Ducamp-Collin, M. N. and Sallanon, H. (2010). Combined effects of postharvest heat treatment and chitosan coating on quality of fresh-cut mangoes (*Mangifera indica* L.). *International Journal of Food Science and Technology* 45:849–55.

Dorta, E., González, M., Lobo, M. G., Sánchez-Moreno, C. and De Ancos, B. (2014). Screening of phenolic compounds in by-product extracts from mangoes (*Mangifera indica* L.) by HPLC-ESI-QTOF-MS and multivariate analysis for use as a food ingredient. *Food Research International* 57:51–60.

Droby, S., Wisniewski, M., Macarisin, D. and Wilson, C. (2009). Twenty years of postharvest biocontrol research: is it time for a new paradigm? *Postharvest Biology and Technology* 52:137–45.

Droby, S., Wisniewski, M., Teixidó, N., Spadaro, D. and Jijakli, M. H. (2016). The science, development, and commercialization of postharvest biocontrol products.In press at *Postharvest Biology and Technology*122: 22–9. https://doi.org/10.1016/j.postharvbio.2016.04.006.

Eckert, J. W. and Joseph M. O. (1985). The chemical control of postharvest diseases: subtropical and tropical fruits. *Annual review of Phytopathology* 23:421–54.

Eduardo, V., Vilas-Boas, D. B. and Kader, A. A. (2007). Effect of 1-methylcyclopropene (1-MCP) on softening of fresh-cut kiwi fruit, mango and persimmon slices. *Postharvest Biology and Technology* 43:238–44.

Environmental Protection Agency (EPA) (2002). Irradiation Phytosanitary Treatment of Imported Fruits and Vegetables Federal Register (Volume 67, Number 205). https://www.federalregister.gov/articles/2002/10/23/02–27027.

EPPO (2007). EPPO (European and Mediterranean Plant Protection Organization), 2007. Data sheets on quarantine pests. *Ceratitis capitata*. http://www.eppo.org/QUARANTINE/insects/Ceratitis.

Erwin, D. C. (1973). Systemic fungicides: disease control, translocation, and mode of action. *Annual Review Phytopathology* 11:389-422.

Estrada, A. B., Jeffries P. and Dood, J. C. (1996). Field evaluation of predictive model to control anthracnose disease of mango in the Philippines. *Plant Pathology* 45:294–301.

Fallik, E. (2004). Prestorage hot water treatments (immersion, rinsing and brushing). *Postharvest Biology and Technology* 32:125–34.

FAO. (2002). Handling and Preservation of Fruits and Vegetables by combined methods for Rural Areas. FAO Technical Manual. Rome. Italy.

Farkas, J., Ehlermann, D. A. E. and Mohácsi-Farkas, C. (2014). Food Technologies: food Irradiation. *Encyclopedia of Food Safety* 3:178–86. https://doi.org/10.1016/b978-0-12-378612-8.00259-6.

Ferrier, P. (2010). Irradiation as a quarantine treatment. *Food Policy* 35:548–55.

Feygenberg, O., Hershkovitz, V., Ben-Arie, R., Jacob, S., Pesis, E. and Nikitenko, T. (2005). Postharvest use of organic coating for maintaining bio-organic avocado and mango quality. *Acta Horticulturae* 682:1057–61.

Follett, P. A. (2004). Irradiation to control insects in fruits and vegetables for export from Hawaii. *Radiation Physics and Chemistry* 71:163–6.

Gasser, F., Eppler, T., Naunheim, W., Gabioud, S. and Nising, A. B. (2010). Dynamic CA storage of apples: monitoring of the critical oxygen concentration and adjustment of optimum conditions during oxygen reduction. *Acta Horticulturae* 876:39–46.

Gol, N. B., Patel, P. R. and Rao, T. V. R. (2013). Improvement of quality and shelf life of strawberries with edible coatings enriched with chitosan. *Postharvest Biology and Technology* 85:185–95.

Guentzel, J. L., Lam, K. L., Callan, M. A., Emmons, S. A. and Dunham, V. L. (2010). Postharvest management of gray mold and brown rot on surfaces of peaches and grapes using electrolyzed oxidizing water. *International Journal of Food Microbiology* 143:54–60.

Gustavsson, J., Cederberg, C., Sonesson, U., van Otterdijk, R. and Meybeck, A. (2011). Global food losses and food waste – extent, causes and prevention. http://www.fao.org/docrep/014/mb060e/mb060e.pdf.

Gorinstein, S., Poovarodom, S., Leontowicz, H., Leontowicz, M., Namiesnik, J., Vearasilp, S., Haruenkitf, R., Ruamsukeg, P., Katricha, E. and Tashma, Z. (2011). Antioxidant properties and bioactive constituents of some rare exotic Thai fruits and comparison with conventional fruits. *In vitro* and *in vivo* studies. *Food Research International* 44:2222–32.

Govender, V., Korsten, L. and Sivakumar, D. (2005). Semi-commercial evaluation of *Bacillus licheniformis* to control mango postharvest diseases in South Africa. *Postharvest Biology and Technology* 38:57–65.

Harvey, J. M. (1978). Reduction of losses in fresh market fruits and vegetables. *Annual Review of Phytopathology* 16:321–41.

Hassan, M. K., Dann, E. K., Irving, D. E. and Coates, L. M. (2007). Concentrations of constitutive alk(en) ylresorcinols in peel of commercial mango varieties and resistance to postharvest anthracnose. *Physiological and Molecular Plant Pathology* 17:158–65.

He, J., Ren,Y., Chen, C., Liu, J., Liu, H. and Pei, Y. (2016). Defense responses of salicylic acid in mango fruit against postharvest anthracnose, caused by *Colletotrichum gloeosporioides* and its possible mechanism. *Journal of Food Safety*. DOI:10.1111/jfs.12294.

Hoa, T. T. and Ducamp, M. N. (2008). Effects of different coatings on biochemical changes of 'Cat Hoa Loc' mangoes in storage. *Postharvest Biology and Technology* 48:150–2.

Holmes, R. and Ledger, S. (1992). Handling systems to reduce mango sapburn. International Mango symposium International Society for Horticultural Science, abstracts, pp. 98.

Ibarra-Garza, I. P., Ramos-Parra, P. A., Hernández-Brenes, C. and Jacobo-Velázquez, D. A. (2015). Effects of postharvest ripening on the nutraceutical and physicochemical properties of mango (*Mangifera indica* L. cv Keitt). *Postharvest Biology and Technology* 103:45–54.

International Standards for Phytosanitary Measures (ISPM) (2007). International Standards for Phytosanitary Measures. https://www.ippc.int.

Jacobi, K. K. and Gowanlock, D. (1995). Ultrastructural studies of 'Kensington' mango (*Mangifera indica* Linn.) heat injuries. *HortScience* 30:102–3.

Jacobi, K. K., Macare, E. A. and Hertherington, E. H. (2001). Postharvest heat disinfestations treatments of mango fruit. *Scientia Horticulturae* 89:171–93.

Jahurul, M. H., Zaidul, I. S., Ghafoor, K., Al-Juhaimi, F. Y., Nyam, K. L., Norulaini, N. A., Sahena, F. and Mohd Omar, A. K. (2015). Mango (*Mangifera indica* L.) by-products and their valuable components. *Food Chemistry* 183:173–80.

Joel, D. M. (1980). Resin ducts in the mango fruit: a defence system. *Journal of Experimental Botany* 31:1707–18.

Johnson, G. I., Boag, T. S., Cooke, A. W., Izard, M., Panitz, M. and Sangchote, S. (1990a). Interaction of postharvest diseases control treatments and gamma irradiation of mangoes. *The Annals of Applied Biology* 116:245–57.

Johnson, G. I., Sangchote, S. and Cooke, A. W. (1990b). Control of stem-end rot (*Dothiorella dominicana*) and other postharvest diseases of mangoes (cv. Kensington Pride) during short- and long-term storage. *Tropical Agriculture* 67:183–7.

Johnson, G. I., Mead, A. J., Cooke, A. W. and Dean, J. R. (1992). Mango stem end rot pathogens fruit infection by endophytic colonization of the inflorescence and pedicel. *The Annals of Applied Biology* 120:225–34.

Johnson, G. I., Cooke, A. W. and Mead, A. J. (1993). Infection and quiescence of mango stem end rot pathogens. *Acta Horticulturae* 341:329–36.

Johnson, G. I. (1994). Mango, stem end rot, Part III. In: Ploetz, R. C., Zentmyer, G. A., Nishijima, W. T., Rohrbach, K. G. and Ohr, H. D. (eds), *Compendium of Tropical Fruit Diseases*. American Phytopathological Society (APS) Press, St Paul, Minnesota, pp. 39–41.

Johnson, G. I., Sharp, J. L., Mine, D. L. and Oostluyse, S. A. (1997). Postharvest technology and quarantine treatments. In: Litz, R. E. (ed.), *The Mango: Botany, Production and Uses*, Tropical Research and Education Center, USA, pp. 44–506.

Johnson, G., Hofman, P. (2009). Postharvest technology and quarantine treatments. In: The mango, botany, production and Uses, 2nd edition, Litz, R. E. (ed.), CABI international, Cambridge, UK, pp. 529–605.

Kader, A. A. (1994). Modified and controlled atmosphere storage of tropical fruits. In: Champ, B. R. and Highley, E. (eds), *Postharvest Handling of Tropical Fruit*. ACIAR Proceedings Chang Mai, Thailand, pp. 239–49.

Kader, A. A. (2002). Quality and safety factors: definition and evaluation for fresh horticultural crops. In: *Postharvest Technology of Horticultural Crops*. 3rd ed. University of California. 22:279–85.

Kader, A. (2004). Controlled atmosphere storage. In USDA Handbook 66. See www.ba.ars.usda.gov/hb66/013ca.pdf.

Kader, A. A., Mitcham, B. (2008). Optimum procedures for ripening mangoes. In: Fruit Ripening and Ethylene Management. University of California. *Postharvest Research Centre Publication Series* 9:47–8.

Kader, A. A. (2008a). Mango quality attributes and grade standards: a review of available information and identification of future research needs (report to the national mango board). Davis, CA, USA: Kader Consulting Services.

Kader, A. A. (2008b). Fresh-cut mangoes as value added product (literature review and interviews). http://mango.org/media/31003/fresh_cut_mangos_literature_review.pdf.

Kane, O. and Marcellin, P. (1978). Incidence of ripening and chilling the mitrochondria from mango fruits. *Plant Physiology* 61:634–8.

Kienzle, S., Sruamsiri, P., Carle, R., Sirisakulwat, S., Spreer, W. and Neidhart, S. (2011). Harvest maturity specification for mango fruit (*Mangifera indica* L.'Chok Anan') in regard to long supply chains. *Postharvest Biology and Technology* 61:41–55.

Kim, Y., Brecht, J. K. and Talcott, S. T. (2007). Antioxidant phytochemical and fruit quality changes in mango (*Mangifera indica* L.) following hot water immersion and controlled atmosphere storage. *Food Chemistry* 105:1327–34.

Kittur, F. S., Saroja, N., Habibunnisa, S. and Tharanathan, R. N. (2001). Polysaccharide based composite coating formulations for shelf life extension of fresh banana and mango. *European Food Research and Technology* 213:306–11.

Kosiyachinda, S., Lee, S. K. and Poernomo, E. (1984). Maturity indices for harvesting of mango. In: Mendoza, D. B. Jr. and Wills, R. B. H. (eds), *Mango: Fruit Development and Postharvest Physiology*. Kuala Lumpur, Malaysia: ASEAN Food Handling Bureau, pp. 33–8.

Laidlaw, W. G., Armstrong, J. W., Chan, H. T. and Jr. Jang, E. B. (1996). The effect of temperature profile in heat-treatment disinfestation on mortality of pests and on fruit quality. Proceedings of the International Conference on Tropical Fruits: Global Commercialization of Tropical Fruits, pp. 343–52.

Lakshminarayanan, S., Subhadra, N. V. and Subramanyam, H. (1970). Some aspects of developmental physiology of mango fruit. *The Journal of Horticultural Science and Biotechnology* 47–53.

Lalel, H. J. D., Sing, Z. and Tan, S. C. (2003a). The role of ethylene in mango fruit aroma volatiles biosynthesis. *The Journal of Horticultural Science and Biotechnology* 78:485–95.

Lalel, H. J. D., Singh, Z., and Tan, S. C. (2003b). Elevated levels of CO_2 in controlled atmosphere storage affects shelf life, fruit quality and aroma volatiles of mango. *Acta Horticulturae* 628:407–13. https://doi.org/10.17660/actahortic.2003.628.51

Léchaudel, M. and Joas, J. (2007). An overview of preharvest factors influencing mango fruit growth, quality and postharvest behaviour. *Brazilian Journal of Plant Physiology* 19:287–98. https://doi.org/10.1590/s1677-04202007000400004

Lederman, I., Zauberman, G., Weksler, A., Rot, I. and Fuchs, Y. (1997). Ethylene-forming capacity during cold storage and chilling injury development in 'Keitt' mango fruit. *Postharvest Biology and Technology* 10:107–12.

Ledger, S. N. (2004). Control of Postharvest Diseases in Mango. Department of Primary Industries and Fisheries (DPI&F) note. Department of Primary Industries and Fisheries, Brisbane, Australia.

Lee, S. K. and Kader, A. A. (2000). Preharvest and postharvest factors influencing vitamin C content of horticultural crops. *Postharvest Biology and Technology* 20:207–20.

Lemmens, L., Tchuenche, E. S., Loey, A. M. V. and Hendrickx, M. E. (2013). Beta-carotene isomerization in mango puree as influenced by thermal processing and high-pressure homogenization. *European Food Research and Technology* 236:155–63.

Lima, G. P. P., Machado, T. M., de Oliveira, L. M., da Silva Borges, L., de Albuquerque Pedrosa, V., Vanzani, P. and Vianello, F. (2014). Ozonated water and chlorine effects on the antioxidant properties of organic and conventional broccoli during postharvest. *Scientific Agriculture* 71:151–6. https://doi.org/10.1590/s0103-90162014000200010

Lipinski, B., Hanson, C., Lomax, J., Kitinoja, L., Raite, W. and Searchinger, T. (2013). *Reducing Food Loss and Waste*. http://www.wri.org/sites/default/files/reducing_food_loss_and_waste.pdf.

Luna-Guzman, I. and Barrett, D. M. (2000). Comparison of calcium chloride and calcium lactate effectiveness in maintaining shelf stability and quality of fresh-cut cantaloupes. *Postharvest Biology and Technology* 19:61–72.

Lonsdale, J. H. (1993). Strategies for the control of postharvest diseases of mango. *South African Mango Growers' Association Yearbook* 13:109–16.

Luo, S., Wan, B., Feng, S. and Shao, Y. (2015). Biocontrol of postharvest anthracnose of mango fruit with *debaryomyces nepalensis* and effects on storage quality and postharvest physiology. *Journal of Food Science* 80:2555–63.

Luria, N., Sela, N., Yaari, M., Feygenberg, O., Kobiler, I., Lers, A. and Prusky, D. (2014). De-novo transcriptome assembly of mango fruit peel reveals mechanisms of mango response to hot water treatment. BMC Genomics. doi:10.1186/1471–2164-15–957.

Lurie, S. (1998). Postharvest heat treatments. *Postharvest Biology and Technology* 14:257–69.

Mahto, R. and Das, M. (2013). Effect of gamma irradiation on the physico-chemical and visual properties of mango (*Mangifera indica* L.), cv. Dushehri and Fazli stored at 20°C. *Postharvest Biology and Technology* 86:447–55.

Malik, A. U. and Amin, M. (2015). ASLP Mango Value Chain project R&D report. University of Agriculture Faisalabad.

Malik, A., Iram, S. and Rasool, A. (2016). Current status of mango pre and post-harvest diseases with respect to environmental factors. *International Journal of Agronomy and Agricultural Research* 8:101–10.

Mangan, R. L. and Hallman, G. H. (1998). Temperature treatments for quarantine security: new approaches for fresh commodities. In: Hallman, G. H. and Denlinger, D. L. (eds), *Temperature Sensitivity in Insects and Application in Integrated Pest Management*. Westview Press, Boulder, Colorado, pp. 201–34.

Massart, M. H. Jijakli. (2014). *Pichia anomala* and *Candida oleophila* in biocontrol of postharvest diseases of fruits: 20 years of fundamental and practical research. In: Prusky, D. and Gullino, M. L. (eds), *Plant Pathology in the 21st Century*, vol. 7, The Netherlands, Springer, pp. 111–22.

McArtney, S. J., Obermiller, J. D., Schupp, J. R., Parker, M. L. and Edgington, T. B. (2008). Preharvest 1-methylcyclopropene delays fruit maturity and reduces softening and superficial scald of apples during long-term storage. *HortScience* 43:366–71.

McGlasson, W., Scott, K. and Mendoza, D. (1979). The refrigerated storage of tropical and subtropical products. *International Journal of Refrigeration* 2:199–206.

McGregor, B. M. (1987). Tropical Products Transport Handbook. United States Department of Agriculture (USDA), Agricultural Handbook No. 668. USDA, Washington, DC.

Medlicott, A., Sigrist, J. and Sy, O. (1990). Ripening of mango following low-temperature storage. *Journal of the American Society for Horticultural Science* 115:430–4.

Mihindukulasuriya, S., Lim, L. T. (2014). Nanotechnology development in food packaging: a review. *Trends in Food Science and Technology* 40:149–67.

Minas, I. S., Tanou, G., Belghazi, M., Job, D., Manganaris, G. A., Molassiotis, A. and Vasilakakis, M., (2012). Physiological and proteomic approaches to address the active role of ozone in kiwifruit post-harvest ripening. *Journal of Experimental Botany* 63:2449–64.

Mitcham, E. and Yahia, E. (2009). *Mangoes*. National Mango Board. USDA. USA.

Mizrach, A., Flitsanov, U., Schmilovitch, Z. and Fuchs, Y. (1999). Determination of mango physiological indices by mechanical wave analysis. *Postharvest Biology and Technology* 16:179–86.

Moalemiyan, M., Vikram, A. and Kushalappa, A. C. (2007). Detection and discrimination of two fungal diseases of mango (cv. keitt) fruits based on volatile metabolite profiles using GC/MS. *Postharvest Biology and Technology* 45:117–25.

Monaco, K. D. A., Costa, S. M., Uliana, M. R. and Lima, G. P. P. (2014). Sanitizers effect in mango pulp and peel antioxidant compounds. *Food and Nutrition Sciences* 5:929–35.

Monaco, K. D. A., Costa, S. M., Minatel, L. O., Correa, C. R., Calero, F. A., Vianellod, F. and Lima, G. P. P. (2016). Influence of ozonated water sanitation on postharvest quality of conventionally and organically cultivated mangoes after postharvest storage. *Postharvest Biology and Technology* 120:69–75.

Mounir, R., Durieux, A., Bodo, E., Allard, C., Simon, J. P., Achbani, E. H., El Jaafari, S., Douira, A. and Jijakli, M. H. (2007). Production, formulation and antagonistic activity of the biocontrol like-yeast *Aureobasidium pullulans* against *Penicillium expansum. Biotechnology Letters* 29:553–9.

Nascimento, A. S., Malavasi, A., Morgante, J. S. and Duarte, A. A. (1992). Hot-water immersion treatment for mangoes infested with *Anastrepha fraterculus, A. obliqua*, and *Ceratitis capitata* (Diptera: Tephritidae) in Brazil. *Journal of Economic Entomology* 85:456–60.

Ngamchuachit, P., Sivertsen, H. K., Mitcham, E. J. and Barrett, D. M. (2015). Influence of cultivar and ripeness stage at the time of fresh-cut processing on instrumental and sensory qualities of fresh-cut mangoes. *Postharvest Biology and Technology* 106:11–20.

Nicolaï, B. M., Defraeye, T., De Ketelaere, B., Herremans, E., Hertog, M. L., Saeys, W., Torricelli, A., Vandendriessche, T., Verboven, P. (2014). Nondestructive measurement of fruit and vegetable quality. *Annual Review of Food Science and Technology* 5:285–312.

Nigam, S., Bhatt, D. and Jha, A. (2007). Different product of mango: the king of fruits. *Processed Food Industry* 10:32-40.

Okawa, K. (2015). Market and trade impacts of food loss and waste reduction. OECD Food, Agriculture and Fisheries. Papers, No. 75, OECD Publishing. http://dx.doi.org/10.1787/5js4w29h0wr2-en.

Oms-Oliu, G., Rojas-Graü, M. A., González, L. A., Varela, P., Soliva-Fortuny, R., Hernando, I. H. M., Munuera, I. P., Fiszman, S. and Martín-Belloso, O. (2010). Recent approaches using chemical treatments to preserve quality of fresh-cut fruit: a review. *Postharvest Biology and Technology* 57:139–48.

Ornelas-Paz, J. D., Yahia, E. M. and Gardea-Bejar, A. (2007). Identification and quantification of xanthophyll esters, carotenes, and tocopherols in the fruit of seven Mexican mango cultivars by liquid chromatography-atmospheric pressure chemical ionization-time-of-flight mass spectrometry. *Journal of Agriculture and Food Chemistry* 55:6628–35.

Patiño-Vera, M., Jiménez, B., Balderas, K., Ortiz, M., Allende, R., Carrillo, A. and Galindo, E. (2005). Pilot-scale production and liquid formulation of *Rhodotorula minuta*, a potential biocontrol agent of mango anthracnose. *Journal of Applied Microbiology* 99:540–50. https://doi.org/10.1111/j.1365-2672.2005.02646.x

Pesis, E., Aharoni, D., Aharon, Z., Ben-Arie, R., Aharoni, N. and Fuchs, Y. (2000). Modified atmosphere and modified humidity packaging alleviates chilling injury symptoms in mango fruit. *Postharvest Biology and Technology* 19:93–101.

Pinto, L., Ippolito, A. and Baruzzi, F. (2015). Control of spoiler Pseudomonas spp. on fresh cut vegetables by neutral electrolyzed water. *Food Microbiology* 50:102–8.

Ploetz, R. C. (2004). The major diseases of mango: strategies and potential for sustainable management. *Acta Horticulturae* 645:137–50.

Prange, R. K., Wright, A. H., DeLong, J. M. and Zanella, A. (2013). History, current situation and future prospects for dynamic controlled atmosphere (DCA) storage of fruits and vegetables, using chlorophyll fluorescence. *Acta Horticulturae* 1012:905–15.

Prusky, D., Falik, E., Kobiler, I., Fuchs. Y., Zauberman, G., Pesis, E., Roth, I., Weksler, A., Akerman, M. and Ykutiely, O. (1996). Hot water brush: a new method for the control of post-harvest disease caused by alternaria rot in mango fruits. *V International Mango Symposium* 455:780–5. https://doi.org/10.17660/actahortic.1997.455.99

Prusky, D., Fuchs, Y., Kobiler, I., Roth, I., Weksler, A., Shalom, Y., Fallik, E., Zauberman, G., Pesis, E., Akerman, M., Ykutiely, O., Weisblum, A., Regev, R. and Artes, L. (1999). Effect of hot water brushing, prochloraz treatment and waxing on the incidence of black spot decay caused by *Alternaria alternata* in mango fruits. *Postharvest Biology and Technology* 15:165–74.

Prusky, D. and Keen, N. T. (1993a). Involvement of preformed antifungal compounds in the resistance of subtropical fruits to fungal decay. *Plant Disease* 77:114–19.

Prusky, D., Kobiler, I., Zauberman, G. and Fuchs, Y. (1993b). Preharvest conditions and postharvest treatments affecting incidence of decay in mango fruits during storage. *Acta Horticulturae* 341:307–20. https://doi.org/10.17660/actahortic.1993.341.34

Prusky, D., Kobiler, I., Akerman, M. and Miyara, I. (2006). Effect of acidic solutions and acidic prochloraz on the control of postharvest decay caused by *Alternaria alternata* in mango and persimmon fruit. *Postharvest Biology and Technology* 42:134–41.

Prusky, D., Kobiler, I., Miyara, I. and Alkan, N. (2009). Fruit Diseases. In: Litz, R. E. (ed.), *The Mango, Botany, Production and Uses*, 2nd edition. CABI International, Cambridge, UK, pp. 210–31.

Prusky, D., Alkan, N., Fluhr, R. and Tesfaye, M. (2013). Quiescent and necrotrophic lifestyle choice during postharvest disease development. *Annual Review of Phytopathology* 51:155–76.

Rico, D., Martín-Diana, A. B., Barat, J. M. and Barry-Ryan, C. (2007). Extending and measuring the quality of fresh-cut fruit and vegetables: a review. *Trends in Food Science and Technology* 18:373–86.

Ridoutt, B. G., Juliano, P., Sanguansri, P. and Sellahewa, J. (2010). The water footprint of food waste: case study of fresh mango in Australia. *Journal of Cleaner Production* 18:1714–21.

Robinson, S. P., Loveys, B. R. and Chacko, E. K. (1993). Polyphenol oxidase enzymes in the sap and skin of mango fruit. *Australian Journal of Plant Physiology* 20:99–107.

Robles-Sánchez, R. M., Islas-Osuna, M. A., Astiazarán-García, H., Vázquez-Ortiz, F. A., Martín-Belloso, O., Gorinstein, S. and González-Aguilar, G. A. (2009). Quality index, consumer acceptability, bioactive compounds, and antioxidant activity of fresh-cut "Ataulfo" mangoes (*Mangifera indica* L.) as affected by low-temperature storage. *Journal of Food Science* 74:126–34. https://doi.org/10.1111/j.1750-3841.2009.01104.x

Romanazzi, G., Sanzani, S. M., Bi, Y., Tian, S. P., Martinez, P. G. and Alkan, N. (2016). Induced resistance to control postharvest decay of fruit and vegetables. *Postharvest Biology and Technology* 122:82–94.

Saaiman, W. (1995). Short wave infra-red radiation as an alternative to the hot water bath for the control postharvest disease in mango. South African Mango Growers' Association Yearbook (South Africa).

Sauco, V. G. (2004). Mango production and world market: current situation and future prospects. *Acta Horticulturae* 1:107–16.

Schirra, M., D'Hallewin, G., Ben-Yehoshua, S. and Fallik, E. (2000). Host–pathogen interactions modulated by heat treatment. *Postharvest Biology and Technology* 21:71–85.

Schneider, S. M., Rosskopf, E. N., Leesch, J. G., Chellemi, D. O., Bull, C. T. and Mazzola, M. (2003). United States department of agriculture - Agricultural research service research on alternatives to methyl bromide: pre-plant and post-harvest. *Pest Management Science* 59:814–26.

Sekhar, C., Selvarajan, M., Pounraj, A. and Praghadeeswaran, M. (2013). Production and export of mango in India: A paradigm to the developing nations. *American International Journal of Research in Humanities, Arts and Social Sciences* 13:78–84.

Sharp, J. L. (1994). Hot water immersion. In: Sharp, J. and Hallman, G. (eds), *Quarantine Treatments for Pests and Food Plants.* Westview Studies in Insect Biology, pp. 133–48.

Sherman, A., Rubinstein, M., Eshed, R., Benita, M., Ish-Shalom, M., Sharabi-Schwager, M., Rozen, A., Saada, D., Cohen Y. and Ophir, R. (2015). Mango (*Mangifera indica* L.) germplasm diversity based on single nucleotide polymorphisms derived from the transcriptome. *BMC Plant Biology* 15:277.

Singh, S.P. and Singh, Z. (2012a). Postharvest oxidative behaviour of 1-methylcyclopropene treated Japanese plums (*Prunus salicina* Lindell) during storage under controlled and modified atmospheres. *Postharvest Biology and Technology* 74:26–35.

Singh, Z. and Singh, S. P. (2012b). Mango. In: Rees, D. and Orchard, J. (eds), *Crop Post-harvest: Science and Technology.* Volume 3: Perishables, UK: Blackwell Pub., pp. 108–42.

Sisler, E. and Serek, M. (1997). Inhibitors of ethylene responses in plants at the receptor level: recent developments. *Physiologia Plantarum* 100:577–82.

Sisler, E.C. (2006).The discovery and development of compounds counteracting ethylene at the receptor level. *Biotechnology Advances* 24:357–67.

Sivakumar, D., Jiang, Y. and Yahia, E.M. (2011). Maintaining mango (*Mangifera indica* L.) fruit quality during the export chain. *Food Research International* 44:1254–63.

Sivankalyani, V., Feygenberg, O., Diskin, S. and Alkan, N. (2016a). Increased anthocyanin and flavonoids in mango fruit peel are associated with cold and pathogen resistance. *Postharvest Biology and Technology* 111:132–9.

Sivankalyani, S., Sela, N., Feygenberg, O., Zemach, H., Maorer, D. and Alkan, N. (2016b). Transcriptome dynamics in mango fruit peel reveals mechanisms of chilling stress. Submitted to Frontiers in Plant Science.

Slaughter, D. C. (2009). Nondestructive maturity assessment methods for mango: A review of literature and identification of future research needs. http://www.mango.org/media/55728/nondestructive_maturity_assesments_methods for mangoes PDF.

Sogi, D. S., Siddiq, M., Roidoung, S. and Dolan, K. D. (2012). Total phenolics, carotenoids, ascorbic acid, and antioxidant properties of fresh-cut mango (*Mangifera indica* L., cv. Tommy Atkin) as affected by infrared heat treatment. *Journal of Food Science* 77:1197–202.

Sophia, O., Robert, G. M. and Ngwela, W. J. (2015). Effect of Aloe Vera gel coating on postharvest quality and shelf life of mango (*Mangifera indica* L.) fruits Var. 'Ngowe'. *Journal of Horticulture and Forestry* 7:1–7.

Sothornvit, R. and Rodsamran, P. (2007). Effect of a mango film on quality of whole and minimally processed mangoes. *Postharvest Biology and Technology* 47:407–15.

Sousa-Gallagher, M. J. and Mahajan, P. V. (2013). Integrative mathematical modeling for MAP design of fresh-produce: theoretical analysis and experimental validation. *Food Control* 29:444–50.

Spadaro, D. and Droby, S. (2016). Development of biocontrol products for postharvest diseases of fruit: the importance of elucidating the mechanisms of action of yeast antagonists. *Trends in Food Science and Technology* 47:39–49.

Subramanyam, H., Krishnamurthy, S. and Parpia, H. A. B. (1975). Physiology and biochemistry of mango fruit. *Advances in Food Research* 21:223–305. https://doi.org/10.1016/s0065-2628(08)60092-0
Tefera, A., Seyoum, T. and Woldetsadik, K. (2007). Effect of disinfection, packaging, and storage environment on the shelf life of mango. *Biosystems Engineering* 96:201–12.
Tharanathan, R. N., Yashoda, H. M., Prabha, T. N. (2006). Mango (*Mangifera indica* L.), "The King of Fruits"- An overview. *Food Review International* 22:95–123.
Theologis, A. (1992). One rotten apple spoils the whole bushel: the role of ethylene in fruit ripening. *Cell* 70(2):181–4.
Thompson, J. F. (2002). Transportation. In: Kader, A. A. (ed.), *Postharvest Technology of Horticultural Crops*, 3rd ed. University of California, Oakland, California, pp. 259–69.
Tjiptono, P., Lam, P. E. and Mendoza, D. B. (1984). Status of the mango industry in ASEA. ASEAN Food Handling Bureau, pp. 1–11.
Torres, R., Solsona, C., Viñas, I., Usall, J., Plaza, P. and Teixidó, N. (2014). Optimization of packaging and storage conditions of a freeze-dried *Pantoea agglomerans* formulation for controlling postharvest diseases in fruit. *Journal of Applied Microbiology* 117:173–84.
Veltman, R. H., Verschoor, J. A. and van Dugteren, J. H. R. (2003). Dynamic control system (DCS) for apples (*Malus domestica* Borkh. cv 'Elstar'): optimal quality through storage based on product response. *Postharvest Biology and Technology* 27:79–86.
Vilas-Boas, E. V., de, B. and Kader, A. A. (2007). Effect of 1-methylcyclopropene (1-MCP) on softening of fresh-cut kiwifruit, mango and persimmon slices. *Postharvest Biology and Technology* 43:238–44.
Watada, A., Ko, N. and Minott, D. (1996). Factors affecting quality of fresh-cut horticultural products. *Postharvest Biology and Technology* 9:115–25.
Watkins, C. B. (2002). Ethylene synthesis, mode of action, consequences and control. In: Knee, M. (ed.), *Fruit Quality and its Biological Basis*, Sheffield, UK: Sheffield Academic Press, pp. 180–222.
Watkins, C. B. and Nock, J. F. (2003). 1-Methylcyclopropene (1-MCP)-New York update and recommendations. Apple handling and Storage. Proceeding of Storage Workshop, Dept Hort. Publ., 22. Ithaca, New York: Cornell Univ., pp. 18–55.
Watkins, C. B. and Miller, W. B. (2004). A summary of physiological processes or disorders in fruits, vegetables, and ornamental products that are delayed or decreased, increased or unaffected by application of 1-methylcyclopropene (1-MCP). *HortScience* 39:816–16.
Watkins, C. B. (2008). Overview of 1-methylcyclopropene trials and uses for edible horticultural crops. *HortScience* 43:86–94.
Wills, R., Lee, T., Graham, D., McGlosson, W. and Hall, E. (1981). Post-harvest – an introduction to the physiology and handling of fruits and vegetables. Granada, CT: AVI.
Wills, R. (2005). Minimizing the harmful effects of ethylene on the quality of fruit and vegetables. In: Ben-Yehoshua, S. (ed.), *Environmentally Friendly Technologies for Agricultural Produce Quality*. London, UK: Taylor and Francis, pp.133–48.
Yahia, E. M. (1999). Postharvest handling of mangoes. Technical Report. Agricultural Technology Utilization and Transfer Project, Giza, Egypt. pp. 57–61.
Yahia, E. M. (2009). Modified and controlled atmospheres for the storage, transportation, and packaging of horticultural commodities. Boca Raton, FL: CRC Taylor and Francis.
Yeoh, W. K., Ali, A. and Forney, C. F. (2014). Effects of ozone on major antioxidants and microbial populations of fresh-cut papaya. *Postharvest Biology and Technology* 89: 56–8.

Chapter 14

The nutritional and nutraceutical/functional properties of mangoes

Laurent Urban, University of Avignon, France; Mônica Maria de Almeida Lopes; and Maria Raquel Alcântara de Miranda, Federal University of Ceará, Brazil

1 Introduction

Mangoes are recognized as a major source of bioactive compounds with potential health-promoting activities (thereafter named phytochemicals). We review here the potential health benefits of mango fruits that can be derived from what we know about the biological effects of vitamin C, carotenoids and phenolic compounds and from what we learned from more specific studies performed on cell and animal models. An up-to-date list of the pre- and post-harvest factors influencing the concentrations in phytochemicals of the pulp of mango fruits is presented. There are not many observations about pre-harvest treatments. By contrast, the literature is especially abundant on post-harvest treatments, including, besides coating techniques, exposure to ionizing radiations; electrical fields; visible, ultraviolet (UV) or infrared (IR) radiations, either in a 'traditional' form or under the form of pulsed light (PL); heat or cold; high pressure and chemicals such as salicylic acid. It is quite obvious from this list that almost all these treatments may be considered as stressing or mimicking stress (salicylic acid is involved in biotic stress signalling). We shall therefore evoke some of the current hypotheses about the stimulating effect of stress on the secondary metabolism.

It is important to keep in mind that all parts of mango trees have been used mainly in traditional South Asian, South American and African medicine (Masibo and He, 2008). However, in this chapter, we shall rather put emphasis on the pulp of either entire fruits or fresh cuts, which represent the forms that are currently available to consumers. For reviews

http://dx.doi.org/10.19103/AS.2017.0026.17

about the potentials and uses of by-products of the food industry, see Jahurul et al. (2015), Masibo and He (2008) and Ribeiro and Schieber (2010).

2 Health benefits of mango fruits

Mango pulp and crude natural extracts from the peel or the pulp are rich in various phytochemicals, including phenolics such as gallic acid-*O*-hexoside, syringic acid hexoside, gallic acid (GA; Septembre-Malaterre et al., 2016) and glycosylated xanthine (mangiferin) (Oliveira et al., 2016); flavonols such as quercetin, naringenin, kaempferol, apigenin, luteolin and chrysin (Pierson et al., 2014); carotenoids such as β-carotene (Mercadante et al., 1997) and vitamin C (Tian et al., 2010) (Table 1). Most of these compounds have antioxidant properties and contribute to the antioxidant capacity of mango fruits, measured by oxygen radical absorbance capacity (ORAC), ferric reducing antioxidant power (FRAP), (2.2-diphenyl-1-picrylhydrazyl) (DPPH) and 2,2-azinobis (3-ethyl-benzothiazoline-6-sulphonic acid) (ABTS) tests.

Table 1 Range of contents for the major classes of phytochemicals and major compounds found in mango pulp

Phytochemicals	Range of contents	References
Vitamin C	9.8 to 186 mg 100 g^{-1} fresh weight	Nisperos-Carriedo et al. (1992) Vinci et al. (1995) Franke et al. (2004) Gil et al. (2006) Reyes and Cisneros-Zevallos (2007) Ribeiro et al. (2007) Corral-Aguayo et al. (2008) Manthey and Perkins-Veazie (2009) Tian et al. (2010)
Total phenolics	9 to 208 mg 100 g^{-1} fresh weight	Gil et al. (2006) Ribeiro et al. (2007) Manthey and Perkins-Veazie (2009)
Gallic acid	69 mg 100 g^{-1} fresh weight	Schieber et al. (2000)
Mangiferin	3.0 to 19.4 mg kg^{-1} dry basis	Berardini et al. (2005)
Total carotenoids	1159-ca. 3000 mg 100 g^{-1} fresh weight	Mercadante et al. (1997) Hulshof et al. (1997) Ben-Amotz and Fishier (1998) Setiawan et al. (2001) Pott et al. (2003) Chen et al. (2004) Ornelas-Paz et al. (2007) Ribeiro et al. (2007) Veda et al. (2007) Corral-Aguayo et al. (2008)
β-carotene	0.55–3.21 mg 100 g^{-1} fresh weight	
Antioxidant capacity	6.12 to 81.39	Shi et al. (2015)

2.1 Theoretical health benefits of mango phytochemicals

Vitamin C, also known as ascorbate, is a vital micronutrient for humans. A lack of vitamin C hampers the activity of a range of enzymes and may lead to scurvy in humans (Olmedo et al., 2006). Unlike most animals, humans are unable to synthesize their own vitamin C, and they must therefore find it in plants, in particular, fruits and vegetables (FAVs). In addition to its involvement in the production of collagen, ascorbic acid serves as a cofactor in several vital enzymatic reactions, including those involved in the synthesis of catecholamines, carnitine and cholesterol, and in the regulation of transcription factors controlling the expression of important genes of the metabolism (Arrigoni and De Tullio, 2002). Ascorbic acid is present in three forms: ascorbate, monodehydroascorbate and dehydroascorbate which corresponds to the oxidized form of ascorbate. In most cellular functions, ascorbate acts as an electron donor, but it may also act directly to scavenge reactive oxygen species (ROS) generated by cellular metabolism. Due to the role of ascorbate in protecting cells against oxidative stress and the involvement of ROS in neurodegenerative disorders (Alzheimer's and Parkinson's diseases) or inflammatory response (atherosclerosis), it is strongly suggested that vitamin C plays a positive role in the prevention of heart, chronic inflammatory and neurodegenerative diseases (Ames et al., 1993).

Gallic acid is a phenolic acid, 3, 4, 5-trihydroxybenzoic acid. Its dimeric derivative is known as ellagic acid. Both GA and ellagic acid exist either in the free or in the bound form. The bound forms of GA and of ellagic acid are called gallotannins (GTs) and ellagitannins, respectively. These two molecules can react with one another to form diagallic acid, which is an ester. GA is the major compound among the phenolic acids found in the mango pulp (Schieber et al., 2000). GA was shown to have antioxidant, anti-inflammatory, antimicrobial, anti-mutagenic, anticancer and free-radical-scavenging properties (Madsen and Bertelsen, 1995).

Mangiferin is a xanthone, C-2-β–D–glucopyranosyl-1, 3, 6, 7-tetrahydroxyxanthone. Xanthones are thought to be among the most powerful antioxidants known (Masibo and He, 2008). Mangiferin is a pharmacologically active phytochemical, with antioxidant, anti-inflammatory, antimicrobial, anti-atherosclerotic, anti-allergenic, analgesic and immunomodulary properties. See Masibo and He (2008) for a detailed review.

Quercetin is a flavonoid, often occurring in plants as glycosides such as rutin. The predominant flavonol glycoside found in mango pulp is quercetin 3-galactoside, followed by quercetin 3-arabinoside according to Schieber et al. (2000). It is, however, important to keep in mind that there are many sources of quercetin in FAVs and that quercetin supply by mango pulp exists. Other flavonol glycosides, like kaempferol, were found only in trace amounts in mango (Schieber et al., 2000). Quercetin has been found to have antihistamine, anti-inflammatory and anticancer properties, and helps prevent cardiovascular diseases. Also quercetin is thought to contribute to protect against chronic diseases such as diabetes, obesity, atherosclerosis and heart disease.

On absorption, quercetin is metabolized mainly to isorhamnetin, tamarixetin and kaempferol. Kaempferol is a strong antioxidant. It inhibits monocyte chemoattractant protein MCP-1 which plays a role in the initial steps of atherosclerotic plaque formation. See Chen and Chen (2013) for a recent review about the numerous health benefits of kaempferol.

Carotenoids endowed with provitamin A activity are vital components of the human diet. Vitamin A is involved in hormone synthesis, immune responses, and the regulation of cell growth and differentiation (Combs, 1995). It can be produced within certain tissues from carotenoids such as β-carotene which is present in the pulp of mango fruits. A carotenoid-deficient diet can lead to night blindness and premature death. Carotenoid-rich diets are correlated with a significant reduction in the risk for certain cancers, coronary heart disease and several degenerative diseases. Carotenoids have demonstrated anticancer and anti-mutagenic properties (Krinsky and Johnson, 2005). Underlying mechanisms are not well understood, but the dietary importance of carotenoids is discussed, at least in part, in terms of antioxidant properties (Beutner et al., 2001; Combs, 1995; Krinsky and Johnson, 2005). Carotenoids are known for their capacity to efficiently quench $^1O_2$ singlet oxygen by energy transfer (Baltschun et al., 1997). $^1O_2$ is a particularly active ROS, capable of damaging DNA (Sies and Menck, 1992) and provoking genetic mutations (Devasagayam et al., 1991). Eventually, $^1O_2$ can damage lipids and membranes (Kalyanaraman et al., 1987). β-carotene is an efficient antioxidant, capable of inhibiting strongly the formation of peroxide. β-carotene is prone to degradation after ingestion, but its breakdown product seems to have interesting properties that may explain the cancer preventative activity (Linnewiel et al., 2009). When lipophilic antioxidants such as lutein or lycopene are associated with hydrophilic antioxidants such as rutin, a supra-additive protection of low-density lipoprotein occurs (Mildel et al., 2007). When rutin is associated with ascorbic acid, a synergetic protection also occurs.

2.2 Mango antioxidants

As said before, many of the above-mentioned compounds are antioxidants. When the compounds endowed with antioxidant capacity are adequately released from the food matrix and absorbed in the small intestine (see below), they are believed to protect the different body tissues against oxidative stress, conferring different health benefits (Das et al., 2012; Palafox-Carlos et al., 2011). Table 2 summarizes the effects of pre- and post-harvest factors on the antioxidant properties of mango pulp.

The antioxidant capacity of phenolics is attributed mainly to the number and localization of their hydroxyl groups and their interactions with dietary fibres (DFs) which could not only limit their absorption, but also prevent these groups from stabilizing free radicals (Palafox-Carlos et al., 2011). Velderrain-Rodríguez et al. (2016) evaluated the effect of DF present in 'Ataulfo' mangoes on bioaccessibility (see below) of phenolics and antioxidant capacity in an '*in vitro*' digestion model that simulates the conditions of the human gastrointestinal tract. They concluded that DF did not represent a major limitation to the bioaccessibility of phenolics (2.48 mg GAE 100 g^{-1} FW) in mangoes.

There is, however, a debate. It would certainly be wrong to attribute the health benefits of FAVs solely to the antioxidant properties of the phytochemicals they supply, even if many of the diseases they contribute to prevent involve oxidative and inflammatory stress. The term 'antioxidant paradox' is used to refer to the observation that giving large doses of dietary antioxidant supplements to human consumers demonstrates, in most studies, little or no preventative or therapeutic effect even for diseases in which ROS are important (Halliwell, 2013). The explanation for the positive effects of phytochemicals may be that, besides their antioxidant properties, many of them may act as elicitors that alter transcription, among others activate Nrf2, a transcription factor that binds to the antioxidant response element in the promoter region of genes coding for enzymes involved in protective mechanisms (Surh and Na, 2008). In addition, there is now mounting evidence that phytochemicals, such as phenolics, may act indirectly through the mediation of the gut microbiota (Anhê et al., 2015).

Table 2 Influence of pre- and post-harvest factors on antioxidant activities in mango pulp, measured by different methods. See the text for a definition of the ripening stages (RS)

Mango (Cultivar)	Factors	Major findings	Antioxidant activity assay	References
'Keitt'	Ripening stage	High correlation between antioxidant activity and total phenolics. Ripening stages RS2 and RS6 characterized by high concentrations of phytochemicals	ORAC[1]	Ibarra-Garza et al. (2015)
'Samar Bahisht Chaunsa'	Ripening stage	Increase in antioxidant activity up to day 7	DPPH[2]	Razzaq et al. (2013)
'Ataulfo'	Ripening stage	Ripening stages RS2 and RS3 characterized by an increased antioxidant capacity	FRAP[3] and DPPH	Palafox-Carlos et al. (2012)
28 mango genotypes	Genetic factors	Total antioxidant potency composite index varied among all genotypes from 6.12 to 81.39, and was significantly correlated with total phenolics	ABTS[4], FRAP, MCC[5], ABTS, SRSA[6] and DPPH	Shi et al. (2015)
'Tommy Atkins'	Cultural practice (biodynamic culture, organic and conventional systems)	Highest antioxidant activity observed in mature green and ripe mango fruits from biodynamic culture	DPPH	Maciel et al. (2010)
'Chok Anan'	Arabic gum plus calcium chloride	High DPPH radical scavenging activity maintained during low temperature storage	DPPH	Khaliq et al. (2016)
'Tainung'	UV-B exposure	Correlation between DPPH scavenging activity and total polyphenols and vitamin C	DPPH and FRAP	Jiang et al. (2015)
'Nam Dok Mai'	Cold storage and salicylic acid	Increased levels of vitamin C acid, total phenolics and antioxidant activity	DPPH and ABTS	Junmatong et al. (2015)

(Continued)

Table 2 *(Continued)*

Mango (Cultivar)	Factors	Major findings	Antioxidant activity assay	References
'Amrapali'	High pressure	Retention of 92% of total phenolics and 90% of antioxidant activity	DPPH and ABTS	Kaushik et al. (2014)
'Kent'	Fresh cut dipped in antioxidant solutions	Higher antioxidant activity observed as a consequence of the treatment using ascorbic acid at 1%	DPPH and ABTS	Robles-Sánchez et al. (2009)

[1] ORAC: oxygen radical absorbance capacity
[2] DPPH: (2.2-diphenyl-1-picrylhydrazyl)
[3] FRAP: ferric reducing antioxidant power
[4] ABTS: 2,2-azinobis (3-ethyl-benzothiazoline-6-sulphonic acid)
[5] MCC: metal chelating capacity
[6] SRSA: superoxide radical scavenging activity

2.3 Bioaccessibility and bioavailability

The term 'bioaccessibility' refers to the proper release of nutrients or specific phytochemicals within the food matrix as influenced by the conditions of the gastrointestinal tract (Saura-Calixto et al., 2007). Bioavailability refers to the total amount that is released and absorbed, reaching the bloodstream, where bioactive compounds are delivered to the different body tissues (Manach et al., 2005).

Although there is compelling evidence that vitamins and secondary metabolites are essential for human health, because they act either as antioxidants or through other mechanisms, many questions remain unresolved. Biologically active substances found in FAVs always come as part of a mixture in the diet. In a mixture, metabolites may have potentiating, antagonizing or synergistic effects (Raskin and Ripoll, 2004). Moreover, health benefits may be influenced by other ingredients such as DF, monounsaturated fatty acids, agents stimulating the immune system, minerals and even ethanol (Halliwell, 2007). Then, there is the issue of bioavailability of biologically active substances, which is affected by several factors such as tannin and lignin concentrations that differ greatly from one species to another.

Tannins have anti-feeding effects, due to their protein-binding properties, whereas lignin decreases the digestibility of plant material. Besides, it is now established that not all individuals respond identically to bioactive food components because of the existence of genetic profiles that modulate the responses. Finally, little is known about the dynamics of food components after they are ingested and then metabolized in the body.

On the basis of the above-mentioned questions and debates, it appears unfortunately not sufficient to demonstrate that mangoes supply large amounts of compounds theoretically known for being antioxidants and for delivering health benefits from studies made on other plants, to jump to the conclusion that consuming mangoes actually supplies such benefits to real human consumers. For that, studies performed on cell models and animal models, or, even much better, clinical studies are required. Table 3 summarizes the positive health effects of mango as they may be derived from studies on cells and animals, and from clinical studies.

Table 3 Potential positive health effect of mango (*Mangifera indica* L.) evaluated in clinical studies, mammalian cells or animal models

Mangoes (pulp/juice/ whole fruit or extract)	Type of trial	Study details	Major experimental findings	References
Mango pulp ('Keitt')	Clinical study	11 healthy volunteers (21–38 years) consumed 400 g day^{-1} of pulp during 10 days. Metabolites of gallic acid (GA) in urine excreted over a 12 h period were quantified	Seven metabolites of GA were identified in the urine of healthy volunteers, and two microbial metabolites were found to be significantly more excreted	Barnes et al. (2016)
Whole and fresh cut mango ('Ataulfo')	Clinical study	During 30 consecutive days, 30 normolipidaemic volunteers (20–50 years) received daily 200 g of whole mango or fresh cut mango	Hypertriglyceridemia was prevented	Robles-Sánchez et al. (2011)
Mango (juice, pulp and dried fruit)	Clinical study	Healthy volunteers (24–25 years) were served breakfast daily, which included bread, yogurt and one of the three forms of mango fruit (juice, fresh and dry slice). Blood samples were collected three times: during fasting, 4 and 8 h after the test meals	The diet contributed efficiently to improve the vitamin A status	Gouado et al. (2007)
Mango (purified mangiferin)	Human umbilical vein endothelial cells (HUVEC)	HUVEC were treated with different concentrations of mangiferin (10 µg mL^{-1} or 20 µg mL^{-1}), and incubated at 37°C in 5% CO_2 atmosphere for 24 h	Purified mangiferin showed protective effect on H_2O_2-treated HUVEC in a dose-responsive manner	Luo et al. (2012)
Mango extracts ('Ataulfo' and 'Haden')	Human SW-480 colon cancer cells	Cells were treated with mango polyphenolics (5 and 10 mg of GAE L^{-1}) for 24 h and harvested for flow cytometer analysis	Polyphenolics exerted protection of normal colon cells (CCD-18Co) by lowering the ROS generation in a dose-dependent manner	Norato et al. (2010)

(Continued)

Table 3 *(Continued)*

Mangoes (pulp/juice/ whole fruit or extract)	Type of trial	Study details	Major experimental findings	References
Mango extracts ('Irwin')	Human hepatoma cell line (HepG2)	Cell lines were incubated with mango extracts for 24 h to allow cell attachment before exposure to varying concentrations of mango polyphenolics	Peel extract exhibited significant anti-proliferative and antioxidant effect against all tested cancer cell lines when compared to pulp in a dose-dependent manner	Kim et al. (2010)
Mango juice ('Ubá')	Male Wistar rats ($n = 32$)	The biometry and biochemical parameters were evaluated in four experimental groups. Peroxisome proliferator-activated receptor gamma (PPAR-g), lipoprotein lipase and fatty acid synthase expression, tumour necrosis factor-a and interleukin-10 (IL-10), as well as histomorphology of the epididymal adipose tissue were determined	Mango juice showed modulatory effects on both inflammation and adipogenesis, indicating potential to prevent and combat obesity	Natal et al. (2016)
Mango (peel and pulp)	Male Wistar rats ($n = 20$)	The experimental groups were fed with either mango pulp or mango peel. An hour after feeding, 150 μL of 50% ethanol was administered orally. An hour after the ethanol was given, blood was drawn from the heart	Peel and pulp decreased the mouse plasma ethanol levels and increased the activities of alcohol dehydrogenase and acetaldehyde dehydrogenase activities	Kim et al. (2011)
Mango pulp ('Ataulfo')	Female rats ($n = 116$)	Mango was administered in the drinking water (0.02–0.06 g mL^{-1}) during both short-term and long-term (LT) periods to rats treated or not with *N*-methyl-*N*-nitrosourea (MNU). The plasma antioxidant capacity was evaluated by the FRAP method	The plasma antioxidant capacity (FRAP assay) tended to increase in a dose-dependent manner in the LT rats not treated with MNU	García-Solís et al. (2008)

2.4 Cell studies

Luo et al. (2012) investigated the beneficial effects of purified mangiferin on human umbilical vein endothelial cells (HUVEC) under H_2O_2-induced stress. Luo et al. (2012) concluded that mangiferin at a concentration of either 10 or 20 μg mL^{-1} showed substantial protective effects on HUVEC with survival rates substantially improved at 0.0625 mmol H_2O_2 L^{-1}, 0.125 mmol H_2O_2 L^{-1} and 0.25 mmol H_2O_2 L^{-1}. Their observations show that purified mangiferin exerts protective effects against oxidative stress, but the underlying mechanism remains unclear requiring further investigations.

Norato et al. (2010) evaluated the anti-carcinogenic effects of polyphenols from different mango varieties in human SW-480 colon cancer cells and non-cancer CCD-18Co colon cells. Polyphenolics present in 'Ataulfo' and 'Haden' inhibited the growth of SW-480 colon cancer cells. SW-480 gene regulation included induction of apoptosis in addition to cell cycle arrest in the G2/M phase. 'Ataulfo' and 'Haden' polyphenolics exerted also protection of non-cancer CCD-18Co colon cells by lowering the ROS generation in a dose-dependent manner.

Kim et al. (2010) evaluated the antioxidant and anti-proliferative properties of pulp and peel mango in a human hepatoma cell line (HepG2). They observed that peel extract has anti-proliferative and antioxidant effects in all tested cancer cell lines in a dose-dependent manner. They moreover observed that the effects were correlated with phenolic and flavonoid contents.

2.5 Studies on animals

Recently, Natal et al. (2016) studied the effect of 'Ubá' mango juice on adiposity and inflammation in male obese Wistar rats after a high-fat diet. Their findings show that mango juice improves the gene expression related to adiposity and inflammation, and also decreased several biochemical, cytological and biometrical markers in obese rats to levels similar to the ones found in control rats. Natal et al. (2016) concluded that 'Ubá' mangoes have potential as a functional food to prevent and combat obesity.

Kim et al. (2011) investigated the ameliorating effect of mangoes on plasma ethanol levels using a mouse model. ^{1}H-NMR (spectroscopy) was employed to investigate the differences in metabolic profiles of mango fruits, and mouse plasma samples fed with mango fruit. Results confirm that mango samples (peel and pulp) remarkably decreased the mouse plasma ethanol levels and increased the activities of alcohol dehydrogenase and acetaldehyde dehydrogenase, reducing hangover symptoms in treated animals.

2.6 Clinical studies

Gouado et al. (2007) evaluated the bioavailability of carotenoids (α- and β-carotene and lycopene) present in mango and papaya consumed in three forms (juice, fresh and dried). Two groups of seven healthy volunteers, each were submitted to three types of meal treatments (juice, fresh and dried fruit). All the treatments lasted only one day during which blood samples were collected three times: during fasting (T0), 4 h and 8 h after the test meal. A comparison between the three forms revealed that papaya and mangoes consumed in the form of juice or fresh fruit lead to the highest bioavailability values (Gouado et al., 2007).

Robles-Sánchez et al. (2011) evaluated the influence of intake of 'Ataulfo' mangoes, either whole or under the form of fresh cuts, on lipids and antioxidant capacity of healthy adults' plasma. Thirty normolipidaemic volunteers were randomly divided into two groups (whole mangoes and fresh cuts). During 30 consecutive days these volunteers received daily 200 g of whole mangoes or fresh cuts. Lipid levels and antioxidant capacity in plasma were determined at the onset of the trial, as well as 15 and 30 days after. Serum triglycerides were reduced by 37% and 38%, respectively, after 30 days of supplementation with whole mangoes and fresh cuts. Very low-density lipoprotein cholesterol levels were reduced in a similar proportion. Both treatments increased plasma antioxidant capacity measured by ORAC and TEAC methods. The authors suggested that addition of mango fruit to generally accept healthy diets could have a beneficial effect preventing hypertriglyceridaemia, and that fresh cut processing does not affect the beneficial properties of mango.

Barnes et al. (2016) conducted a clinical study pilot with healthy volunteers that consumed 400 g day^{-1} of mango pulp (cv. Keitt) for 10 days. They characterized and quantified seven metabolites of GA in urine excreted over a 12 h period. A significant increase in the excretion of pyrogallol-*O*-sulphate and deoxypyrogallol-*O*-sulphate were observed between days 1 and 10, increasing from 28.5 to 55.4 mg L^{-1} and from 23.6 to 47.7 mg L^{-1}, respectively. Additionally, the *in vitro* hydrolysis of GTs was monitored at physiological pH and temperature conditions. After 4 h a shift in composition from relativity high to low molecular weight GTs was observed. Seven metabolites of GT were identified in the urine of healthy volunteers, and two microbial metabolites were found to be excreted in excess following 10 days of mango consumption. Mango GTs were also found to release free GA in conditions similar to the intestines. GTs may serve as a pool of pro-GA compounds that can be absorbed or can undergo microbial metabolism (Barnes et al., 2016).

3 Increasing phytochemical concentrations in mango fruits

Enough evidence has been accumulated through epidemiologic and clinical studies, first about the global benefits of FAVs in the human diet and second about the dietary effects of the phytochemicals they supply, especially vitamins and secondary metabolites. On the basis of such undisputed evidence, even in the absence of precise recommendations, it makes sense to encourage people to consume more FAVs. Unfortunately, the five-a-day campaigns in developed countries to persuade people to eat at least five portions of FAVs every day have proven to be a relative failure so far, and the situation is no better in developing countries. Taking these facts into account, it appears reasonable trying to improve the current situation by encouraging, besides the consumption of FAVs, the consumption of foods and food supplements with enhanced concentrations in phytochemicals. Within this view, the proposition to produce mangoes with increased concentrations in phytochemicals can be considered (Poiroux-Gonord et al., 2010).

We shall review now successively the effects of genetic factors and of the ripening stage, before considering the environmental levers that can be used before and after harvest.

3.1 Genetic factors and fruit-to-fruit variability (see Table 1)

Manthey and Perkins-Veazie (2009) compared five varieties of mango ('Ataulfo', 'Haden', Keitt', 'Kent' 'and 'Tommy Atkins') from four countries. They observed that vitamin C ranged from 11 to 134 mg 100 g^{-1} of pulp puree, and that β-carotene varied from 5 to 30 mg kg^{-1} among the five varieties. Total phenolic content ranged from 19.5 to 166.7 mg of GA equivalents (GAEs) 100 g^{-1} of puree. The varieties 'Haden', 'Keitt', 'Kent' and 'Tommy Atkins' had similar total phenolic contents with an average of 31.2 ± 7.8 mg GAE 100 g^{-1} of puree, whereas the 'Ataulfo' variety contained substantially higher amounts. In contrast, the country of origin and harvest dates had far less influence on these parameters. 'Ataulfo' mangoes contained significantly higher amounts of mangiferin and ellagic acid than the other four varieties. Large fruit-to-fruit variations in the concentrations of these compounds were observed within sets of mangoes of the same cultivar with the same harvest location and date. Since phytochemicals are often endowed with strong antioxidant properties, the large differences observed in concentrations in phytochemicals as a consequence of genetic factors should be reflected in large differences in total antioxidant capacity. Shi et al. (2015) evaluated 28 mango cultivars for phytochemical content and antioxidant capacity. Using a total antioxidant potency composite index, they found that antioxidant capacity varied indeed strongly among all the genotypes, from 6.12 to 81.39.

While the strong effects of genetic factors tend to overshadow other effects, fruit-to-fruit variability points again towards the importance of pre-harvest factors. Indeed the within-tree variability and the tree-to-tree variability arguably originate from differences in water availability among trees, differences in photosynthesis and carbon gains, as well as differences in photo-oxidative stress, as a consequence of differences in fruit position in the canopy and of drought-associated differences in stomatal conductance and sugar accumulation in leaves close to fruits. Moreover differences in carbon status and in oxidative stress necessarily translate into differences in ripening stages which again have a tremendous influence on concentrations in phytochemicals of fruits.

3.2 Effect of the harvesting stage

According to Kondo et al. (2005), contents in total phenolics and in vitamin C of the mango pulps increase and decrease, respectively, during fruit growth, from 14 to 56 days after full bloom (DAFB). Contents then appear relatively stable from 56 to 84 DAFB. But there are contradictory observations. Wenkam (1979) observed a threefold decrease in vitamin C content of 'Haden' fruits from green to ripe. Kim et al. (2007) observed similarly that phenolic compounds decrease as fruit ripens. More specifically, GA and, in general, total hydrolysable tannins were found to decrease by 22% and 57%, respectively (Kim et al., 2007).

The picture is different when considering carotenoids. Observations on 'Cogshall' fruits indicate that it is advantageous for the carotenoid content to delay the harvesting time (Joas et al., 2012). The majority of carotenoids in mango fruit are isomers of violaxanthin and β-carotene, whose contents will effectively increase in relation to maturity (Mercadante and Rodriguez-Amaya, 1998). The highest content in carotenoids is achieved for fruits well-exposed to sunlight, harvested at the latest maturity stage, and allowed to ripe at 20°C.

Since the harvesting stage has such a high impact on the phytochemical content of mango fruits, it seems desirable to acquire parameters to assess harvesting maturity, more relevant and precise than the usual criteria, such as colour changes. Chlorophyll fluorescence has been proposed as a credible tool to reduce biases associated with growth conditions (Lechaudel et al., 2010).

Maciel et al. (2010) evaluated the potential for cultural practices to influence contents in total phenolics and flavonoids, as well as antioxidant activity of mangoes by comparing at three maturation stages of 'Tommy Atkins' fruits from biodynamic culture, organic farming and conventional growing. They observed the highest antioxidant activity in mature green and ripe fruits from biodynamic culture, and the highest antioxidant activity in unripe fruits from organic farming, which they related to the flavonoid content of fruits. Interestingly, the observations of Maciel et al. (2010) are consistent with observations made by Oliveira et al. (2013) of positive effects of organic farming on contents in vitamin C and phenolics of tomato fruits.

It was observed under conventional farming that the highest carotenoid content is obtained in mangoes from girdled branches at 25 leaves per fruit (high fruit load – low carbon supply to fruits) rather than in fruits from girdled branches at 100 leaves per fruit (low fruit load – high supply of carbon to fruits) (Table 4) (Joas et al., 2012). This observation does not support the common view that carotenoid biosynthesis competes for carbohydrate supply with structural growth or sugar build-up, and thus contrasts with other reports showing that carbohydrate limitation negatively impacts carotenoids. Poiroux-Gonord et al. (2012a; 2012b) suggested that early changes in carbohydrate availability in fruiting branches of *Citrus* could provide a signal for carotenoid regulation, independent of the substrate availability *per se*. More specifically, early carbon limitation may increase the potential for fruit carotenoid synthesis and accumulation, and enhance final carotenoid content through enhancing plastid storage capacity (Poiroux-Gonord et al., 2012a; Poiroux-Gonord et al., 2012b). So far, such observations have not been made in mango even though the observations of Joas et al. (2012) suggest that such a mechanism may well be at play in mango.

Exposure to light may also influence carotenoid content of mangoes. There are substantial differences in carotenoid content between shaded and well-exposed mango fruits (Joas et al., 2012), and in vitamin C between the shaded and the well-exposed sides of fruits (Léchaudel et al., 2013). Clearly, training and pruning practices which aim at increasing light penetration to fruits could be used as a lever to increase concentrations in carotenoids and in vitamin C of mango fruits.

Table 4 Major findings about the effects of pre-harvest factors influencing the concentrations in phytochemicals of the pulp of mangoes at the time of harvest

Technique used	Cv.	Major effects observed (expressed on a fresh weight basis)	Reference
Leaf-to-fruit ratio (25 vs. 100)	'Cogshall'	+67% carotenoids when fruits are allowed to ripe at 20°C after harvest	Joas et al. (2012)
Light exposition of fruits (well exposed vs. shaded)	'Cogshall'	Global positive trend for carotenoids	Joas et al. (2012)

3.3 Stress as a lever to increase phytochemical content

Global climate change entails many threats and challenges for the majority of crops. Even though increasing temperatures stimulate mango photosynthesis up to ca. 40°C (Schaffer et al., 2009), a reduction in yield must be expected as a consequence of the stressing conditions associated with the climate change (Normand et al., 2015). Fruit crops will certainly suffer from the increased extension of drought conditions among others; however, yield is arguably not as important for fruit as for grain crops or oil crops. Yield does matter for fruit crops, but quality criteria are as important if not more important (Ripoll et al., 2014). Fruits are expected to supply health benefits and to bring hedonistic pleasures associated with specific aromatic compounds. We may thus distance ourselves from the dominant deleterious effect of stress on crop performance and consider the potential benefits (Ripoll et al., 2014). Fruits from stressed trees may in particular display a higher content in health-promoting phytochemicals.

Here is a brief summary of the physiological mechanisms behind the idea that stress may be beneficial when it comes to synthesis and accumulation of phytochemicals in fruits. For more details, see the following articles: Fanciullino et al. (2013), Poiroux-Gonord et al. (2013) and Ripoll et al. (2014).

Photosynthesis represents the major source of ROS in green plants (Asada, 1999). The mechanism is now well understood. Most stresses that result in the inhibition of the Calvin cycle, may result in an excess of energy entering the system under the form of photons when compared to the quantity of energy used by photochemistry, and a decrease in reoxidation of NADPH. This leads to the formation of singlet oxygen $^1O_2$ at the level of photosystem II, PSII (Apel and Hirt, 2004) and to the transfer of electrons to molecular oxygen, that is, the formation of superoxide O_{2-} at photosystem I (PSI). Plants are well prepared to cope with such conditions. Non-photochemical quenching allows for the dissipation of excess excitation energy in the light-collecting antennae of PSI, while the photosynthetic electron transport rate of stressed plants is reallocated from photosynthesis to photorespiration and the Mehler reaction at PSI. The glycolate oxidase and the Mehler peroxidase reactions lead to the production of substantial amounts of H_2O_2 (Noctor et al., 2002; Smirnoff, 1993) (a less reactive ROS than $^1O_2$ and O_{2-}), in peroxisomes and chloroplasts, respectively. Besides catalase (CAT) there are several enzymes and enzymatic systems to eliminate H_2O_2.

Ultraviolet light, like other stresses, may be at the origin of photo-oxidative stress, that is, the production of ROS associated with the functioning of the photosynthetic machinery (Urban et al., 2016). Similarly, drought is at the origin of photo-oxidative stress or exacerbates it by reducing stomatal conductance, the amount of CO_2 feeding the Calvin cycle, and therefore, reoxidation of NADPH (Grassmann et al., 2002). Eventually, a decrease in translocation of sugars synthetized in leaves inhibits the Calvin cycle and reduces reoxidation of NADPH.

According to Fanciullino et al. (2013) and Poiroux-Gonord et al. (2013), oxidative stress can originate both in the fruit and in the leaf, from which it propagates to nearby fruit through an unknown signal. Activation of redox-sensitive systems upregulates the transcription of genes involved in biosynthetic pathways, leading to higher levels of corresponding proteins and higher levels of secondary metabolites. Redox changes also regulate the activity of the enzymes of the biosynthetic pathways. The increase in ROS can stimulate production of secondary metabolites indirectly by promoting fruit development and, as far as carotenoids are concerned, conversion of chloroplasts into chromoplasts.

Besides regulated deficit irrigation, among stressing conditions that could be used in mango trees, girdling, a common practice among mango growers, could be a powerful technique to force accumulation of sugars in leaves by suppressing phloem connections and therefore provoke photo-oxidative stress in leaves (Urban and Léchaudel, 2005; Urban and Alphonsout, 2007) and the subsequent accumulation of phytochemicals in fruits.

4 Pre- and post-harvest factors influencing bioactive compounds of mango fruits

4.1 Ripening after harvest

Ripening climacteric fruits like mangoes, even when they are detached from the parent plant, undergo physiological, biochemical and molecular changes that directly affect their quality traits (Osorio and Fernie, 2013; Prasanna et al., 2007).

Ibarra-Garza et al. (2015) evaluated the effects of post-harvest ripening (6 ripening stages, RS) on the nutraceutical and physico-chemical properties of 'Keitt' mangoes. Based on measurements of antioxidant activity and of vitamin C, total phenolics and carotenoids they found that the optimum ripening stages were RS2 and RS6. Razzaq et al. (2013) evaluated the impact of ripening duration on the antioxidant capacity of 'Samar Bahisht Chaunsa' mangoes. They observed an increase in dismutase superoxide (SOD) activity during ripening and found that total antioxidant activity peaked at day 7. Palafox-Carlos et al. (2012) observed that antioxidant capacity of 'Ataulfo' mangoes was tightly associated with contents in phenolics and flavonoids, and increased from RS2 (20 to 30% yellow colour surface) to RS3 (70–80%).

4.2 Post-harvest handling

Post-harvest handling, processing and storage conditions have the potential to strongly influence the content in phytochemicals of mango fruits as well as their antioxidant potential. They may help to maintain positive characteristics and even boost them.

Khaliq et al. (2016) observed that gum arabic (GA) (10% *w/v*) coating enriched with calcium chloride (CA) (3% *w/v*) maintained high DPPH radical scavenging activity of mango fruits (cv Chok Anan) stored at low temperature (Table 2). They found, moreover, that GA either alone or in combination with CA effectively inhibited the loss of phenolics and ascorbic acid.

Junmatong et al. (2015) studied the effect of long-term cold storage (5°C) combined with salicylic acid (SA) at 1 mM on antioxidants of 'Nam Dok Mai' mangoes. They observed that SA-treated mango fruits exhibited significantly higher levels of vitamin C, total phenolics and antioxidant activity.

In another study, Kaushik et al. (2014) examined the effect of high-pressure processing (100 to 600 MPa for 1 s to 20 min) on colour, biochemical and microbiological characteristics of mango pulp (cv. Amrapali). They observed a retention of 92 % of total phenolics and of 90 % for antioxidant activity in treated mangoes.

The impact of processing of fresh mangoes was evaluated by (Gil et al., 2006) by comparing fresh cut and whole fruits, and found that fresh cutting resulted in less than 5% loss in vitamin C, 10 to 15% loss in carotenoids and no significant loss in total phenolics after 6 days at 5°C. In another study, Robles-Sánchez et al. (2009) demonstrated that dipping cubes of 'Kent' mangoes in ascorbic acid (1%) increased antioxidant activity.

The most spectacular effects of phytochemicals were observed with light. Gil et al. (2006) observed strong positive effects of visible light on carotenoids content of 'Ataulfo' mangoes. Jiang et al. (2015) evaluated the effect of exposure of 'Tainung' mangoes to ultraviolet B (UV-B) radiation at 5 kJ m^{-2} delivered over 4 h. They found that the UV-B associated strong increase in antioxidant compounds (vitamin C and phenolics) was highly correlated to reduced ROS level (H_2O_2 and O_{2-}) and to increased activities of SOD and CAT. While exposure to UV-C light or to infrared (IR) radiation may lead to contrasting effects according to the considered phytochemicals, PL which encompasses UV-C, UV-B, UV-A, visible and near-IR radiations was found to be at the origin of strong positive effects for vitamin C, phenolics and carotenoids as well. Apparently the boosting effect of PL on phytochemical content of mango fruits can be observed on fresh cuts (Charles et al., 2013) and on entire fruits (Lopes et al., 2015). We attached to this review a case study about the effect of PL on phytochemical content of 'Tommy Atkins' entire mangoes (see below).

5 Case study: low fluence PL to enhance mango phytochemical content

The phenomenon known as hormesis refers to physiological stimulation of beneficial responses by low levels of stressors, thus in theory, it may be used as a promising tool in the food industry leading to healthier products by enhancing phytochemical levels of either whole or fresh cut produce (González-Aguilar et al., 2010c; Bravo et al., 2012). However, little is known about the physiological basis of the accumulation of phytochemicals as a response to a post-harvest stress. In a specific case of study, Lopes et al. (2015) tested the hypothesis that a hormetic dose of PL (100–1100 nm) was capable to induce biochemical changes in the tissues (peel and pulp) of 'Tommy Atkins' mangoes.

Physiologically mature 'Tommy Atkins' mangoes were submitted to PL at a fluence of 0.6 $J.cm^{-2}$ (2 pulses of 0.3 $J.cm^{-2}$ each) and then, stored for 7 days at 20°C. Fruit pulp and peel were separated and evaluated for H_2O_2 content (Sergiev et al., 2001) as an indicator of oxidative stress, vitamin C content (Strohecker and Henning, 1967), SOD (Giannopolitis and Ries, 1977), CAT (Beers and Sizer, 1952) and ascorbate peroxidase (Nakano and Asada, 1981) activities as indicators of antioxidant metabolism. Total carotenoids were measured as described by Lichtenthaler and Buschmann (2001), and the results were expressed as mg kg^{-1}. Polyphenolic pigments as total anthocyanins and yellow flavonoids were evaluated as described by Francis (1982), with results expressed as mg kg^{-1}. Total phenol content of mangoes was measured by a colorimetric assay using Folin–Ciocalteu reagent as described by Larrauri et al. (1997) and Obanda et al. (1997) and expressed as GAE mg kg^{-1}. Phenylalanine ammonia lyase (PAL) activity was assayed as described by Mori et al. (2001) and El-Shora (2002), with slight modifications. PAL-specific activity was expressed as μmol *trans*-cinnamic acid h^{-1} mg^{-1} P.

After 7 days of storage, the PL treatment made at the onset of the storage period had resulted in a +350% increase in total carotenoid content in the pulp, when compared to the control. In the peel, the PL had similarly resulted in a +90% increase in carotenoid content when compared to the control. Storage time and PL did not influence the anthocyanin and yellow flavonoid contents in the peel whereas PL resulted in a +21% increase in anthocyanins in the pulp and, similarly, a +42% increase in yellow flavonoid after 7 days of storage. PL also strongly helped to reduce the storage-associated decrease in vitamin C content in the pulp (Table 5).

Table 5 Major findings about the effects of post-harvest techniques on the concentrations in phytochemicals of the pulp of mangoes either under the form of entire fruits or under the form of fresh cuts

Technique used	Entire fruits/ fresh cuts (cv)	Major effects observed (expressed on a fresh weight basis if not specified otherwise)	References
High pressure (100 to 600 Mpa, 1 s to 20 min)	Pulp ('Amrapali')	+85% vitamin C +92% total phenolics	Kaushik et al. (2014)
Electron beam ionizing radiation (3.1 kGy)	Entire fruits ('Tommy Atkins')	−54% vitamin C −74% carotenoids (vs. control after 18 days storage)	Reyes and Cisneros-Zevallos (2007)
High electric field (150 kV/m of electric field for 45 min)	Entire fruits ('Irwin')	Decrease in β-carotene, no clear effect on total phenolics, quercetin and vitamin C (vs. control after 20 days storage)	Shivashankara et al. (2004)
Light exposure (4 to 5 μmol photons m^{-2} s^{-1})	Fresh cuts ('Ataulfo')	+228% total carotenoids	Gil et al. (2006)
Low fluence pulsed light (2 pulses amounting to 0.6 J cm^{-2})	Entire fruits ('Tommy Atkins')	+60% vitamin C +30% total phenolics +350% total carotenoids (vs. control after 7 days storage)	Lopes et al. (2015)
Pulsed light (2 pulses amounting to 8 J cm^{-2})	Fresh cuts ('Kent')	+55% vitamin C (nmol g^{-1} DM) +350% total carotenoids (vs. control after 7 days storage)	Charles et al. (2013)
UV-B light (0.5 J cm^{-2} delivered over 4 hours)	Entire fruits ('Tainung', cold-stored)	Strong positive effect on vitamin C, quercetin, kaemferol and gallic acid	Jiang et al. (2015)
UV-C light (2.46 and 4.93 J cm^{-2})	Entire fruits ('Haden')	+14 to 25% total phenolics +20 to 80% total flavonoids (vs. control after 15 days storage)	Gonzalez-Aguilar et al. (2007a)
UV-C light (1 to 10 mn, using 15 W lamps at a 15 cm distance)	Fresh cuts ('Tommy Atkins')	Down to −71% vitamin C down to −67% β-carotene up to +28% total phenolics up to +50% total flavonoids (vs. control after 15 days storage)	Gonzalez-Aguilar et al. (2007b)

(Continued)

Table 5 *(Continued)*

Technique used	Entire fruits/ fresh cuts (cv)	Major effects observed (expressed on a fresh weight basis if not specified otherwise)	References
UV-C light ($4.93\ J\ cm^{-2}$)	Entire fruits ('Nam Dok Mai')	Negative effect on total phenolics (on a DM basis) (vs. control after 15 days storage)	Safitri et al. (2015)
IR treatment (5, 10 and 15 min)	Fresh cuts ('Tommy Atkins')	Down to −80% vitamin C contrasted effects on carotenoids up to +247% total phenolics (vs. control after 16 days storage)	Sogi et al. (2012)
Osmo-dehydrofreezing (sucrose, glucose and maltose at 45%)	Pulp ('Tainung')	Increase in phenolics like *p*–hydroxybenzoic acid, quercetin, *p*-coumaric acid and sinapic acid	Gil et al. (2006)
Hot water immersion at 50°C for 30 min followed by cooling for 15 min	Fresh cuts ('Tommy Atkins)	Slight positive but NS effect on total carotenoids	Djioua et al. (2009)
Cold shock (0°C for 4 h)	Entire fruits ('Wacheng')	Slight positive but NS effect on vitamin C +70% total phenolics (vs. control after 12 days storage)	Zhao et al. (2006)
Hot water immersion at 46.1°C for 70 to 110 min	Entire fruits ('Tommy Atkins')	Negative effect on phenolics (vs. control after 4 days storage)	Kim et al. (2009)
Hot water immersion at 50°C for 60 min	Entire fruits ('Tommy Atkins')	No significant difference in total carotenoids (vs. control after 16 days storage at 5°C)	Talcott et al. (2005)
Salicylic acid (2 mM for 5 min)	Entire fruits ('Chausa')	+27% total phenolics NS effect on carotenoids (vs. control after 30 days)	Barman and Asrey (2014)
Salicylic acid or oxalic acid (2 and 5 mM for 10 min)	Entire fruits ('Zill')	Higher % of vitamin C under the reduced form (vs. control after 30 days storage at 5°C)	Ding et al. (2007)

H_2O_2 content was evaluated as an indicator of oxidative stress, and found to increase in response to PL, but only in the pulp. SOD activity increased by 58% in both tissues of PL-treated mangoes, when compared to the control. CAT activity was also enhanced

(+104%) by PL, but only in the pulp. The concomitant increase in H_2O_2 levels and activities of antioxidant enzymes provides evidence that PL was at the origin of oxidative stress, that is, the production of ROS, and the subsequent triggering of antioxidant mechanisms (Jaleel et al., 2009).

Moreover, PAL activity was significantly enhanced by PL treatment in both mango tissues which is consistent with the higher total phenolic content found in PL-treated mango tissues. PAL has been commonly used as an indicator of stressful conditions (Sreelakshmi and Sharma, 2008) and together with higher H_2O_2 levels and activities of the antioxidant enzymes, the increase in PAL activity suggests that PL induced an oxidative imbalance in cells of mango pulp and peel.

In conclusion, it may be stated that PL positively affected the post-harvest physiology of 'Tommy Atkins' mangoes by strongly stimulating synthesis and accumulation of health-promoting phytochemicals, namely carotenoids and phenolics, while limiting the decrease in vitamin C normally observed during storage. PL emerges from our trial as an exceptionally potent technique for improving quality of mangoes that clearly deserves more attention from the scientific community in the future.

6 Future trends and conclusion

Mango fruits represent an outstanding source of phytochemicals. There is, however, a large variability in contents and therefore quality, due to the influence of numerous genetic factors and environmental factors before harvest, not to mention the roles of harvesting time, and of ripening and storage conditions and duration. Control of quality is of paramount importance to the mango industry. It is also a major health issue for consumers and stakeholders. There is therefore the need both to develop a better understanding of the way factors and their interactions influence synthesis and accumulation of phytochemicals in mangoes, and to develop innovative tools and techniques, endowed with a potent ability to drive the secondary metabolism in the fruits. We specifically advocate here for more studies to be conducted about light, especially about PL after harvest on both entire fruits and fresh cuts. In addition to more observations about doses and possibly repetitions of doses, what we need now is to go further, beyond descriptive studies, for acquiring insight in the physiological mechanisms involved in light sensing and signalling in relationship with the metabolic pathways of ascorbate, phenolic compounds and carotenoids.

7 Where to look for further information

For those who are interested in developing an integrated view of the way environmental factors and their interactions influence synthesis and accumulation of phytochemicals in fruits, we recommend to read the article of Fanciullino et al. (2013). The article is focusing on carotenoids but most of the rationales developed in this review paper also apply to phenolic compounds and, to a lesser extent, to vitamin C. We also strongly advise readers to increase their awareness of the existing debates about the concept of antioxidant and about the mechanisms of action of phytochemicals after ingestion by human consumers. The articles of Halliwell (2013) and of Anhê et al. (2015) are must-reads in our view.

8 References

Ames, B. N., Shigenaga, M. K. and Hagen, T. M. (1993). Oxidants, antioxidants, and the degenerative diseases of aging. *Proceedings of the National Academy of Sciences of the United States of America* 90:7915–22.

Anhê, F. F., Roy, D., Pilon, G., Dudonné, S., Matamoros, S., Varin, T. V., Garofalo, C., Moine, Q., Desjardins, Y. and Levy, E. (2015). A polyphenol-rich cranberry extract protects from diet-induced obesity, insulin resistance and intestinal inflammation in association with increased Akkermansia spp. population in the gut microbiota of mice. *Gut* 64:872–83.

Apel, K. and Hirt, H. (2004). Reactive oxygen species: Metabolism, oxidative stress, and signal transduction. *Annual Review of Plant Biology* 55:373–99.

Arrigoni, O. and De Tullio, M. C. (2002). Ascorbic acid: much more than just an antioxidant. *Biochimica et Biophysica Acta (BBA) – General Subjects* 1569:1–9.

Asada, K. (1999). The water-water cycle in chloroplasts: Scavenging of active oxygens and dissipation of excess photons. *Annual Review of Plant Physiology and Plant Molecular Biology* 50:601–39.

Baltschun, D., Beutner, S., Briviba, K., Martin, H. D., Paust, J., Peters, M., Rover, S., Sies, H., Stahl, W., Steigel, A. and Stenhorst, F. (1997). Singlet oxygen quenching abilities of carotenoids. *Liebigs Annalen-Recueil* 1887–93.

Barman, K. and Asrey, R. (2014). Salicylic acid pre-treatment alleviates chilling injury, preserves bioactive compounds and enhances shelf life of mango fruit during cold storage. *Journal of Scientific & Industrial Research* 73:713–18.

Barnes, R. C., Krenek, K. A., Meibohm, B., Mertens-Talcott, S. U. and Talcott, S. T. (2016). Urinary metabolites from mango (*Mangifera indica* L. cv. Keitt) galloyl derivatives and in vitro hydrolysis of gallotannins in physiological conditions. *Molecular Nutrition & Food Research* 60(3): 542–50. DOI:10.1002/mnfr.201500706.

Ben-Amotz, A. and Fishier, R. (1998). Analysis of carotenoids with emphasis on 9-cis β-carotene in vegetables and fruits commonly consumed in Israel. *Food Chemistry* 62:515–20.

Berardini, N., Fezer, R., Conrad, J., Beifuss, U., Carle, R. and Scieber, A. (2005). Screening of mango cultivars for their contents of flavanol *O* and xanthone *C*-glycosides, anthocyanins and pectins. *J Agric. Food Chem* 52:1563–70.

Beutner, S., Bloedorn, B., Frixel, S., Blanco, I. H., Hoffmann, T., Martin, H. D., Mayer, B., Noack, P., Ruck, C., Schmidt, M., Schulke, I., Sell, S., Ernst, H., Haremza, S., Seybold, G., Sies, H., Stahl, W. and Walsh, R. (2001). Quantitative assessment of antioxidant properties of natural colorants and phytochemicals: carotenoids, flavonoids, phenols and indigoids. The role of β-carotene in antioxidant functions. *Journal of the Science of Food and Agriculture* 81:559–68.

Charles, F., Vidal, V., Olive, F., Filgueiras, H. and Sallanon, H. (2013). Pulsed light treatment as new method to maintain physical and nutritional quality of fresh-cut mangoes. *Innovative Food Science & Emerging Technologies* 18:190–5.

Chen, A. Y. and Chen, Y. C. (2013). A review of the dietary flavonoid, kaempferol on human health and cancer chemoprevention. *Food Chemistry* 138:2099–107.

Chen, J., Tai, C. and Chen, B. (2004). Improved liquid chromatographic method for determination of carotenoids in Taiwanese mango (*Mangifera indica* L.). *Journal of Chromatography A* 1054:261–8.

Combs, G. F. (1995). *The Vitamins: Fundamental Aspects in Nutrition and Health*. Academic Press, San Diego.

Corral-Aguayo, R. D., Yahia, E. M., Carrillo-Lopez, A. and Gonzalez-Aguilar, G. (2008). Correlation between some nutritional components and the total antioxidant capacity measured with six different assays in eight horticultural crops. *Journal of Agricultural and Food Chemistry* 56:10498–504.

Das, L., Bhaumik, E., Raychaudhuri, U. and Chakraborty, R. (2012). Role of nutraceuticals in human health. *Journal of Food Science and Technology* 49(2):173–83. DOI:http://dx.doi.org/10.1007/s13197-011-0269-4. PMid:23572839.

Devasagayam, T. P. A., Steenken, S., Obendorf, M. S. W., Schulz, W. A. and Sies, H. (1991). Formation of 8-hydroxy(deoxy)guanosine and generation of a strand breaks at guanine residues in DNA by singlet oxygen. *Biochemistry* 30:6283–9.

Ding, Z.-S., Tian, S.-P., Zheng, X.-L., Zhou, Z.-W. and Xu, Y. (2007). Responses of reactive oxygen metabolism and quality in mango fruit to exogenous oxalic acid or salicylic acid under chilling temperature stress. *Physiologia Plantarum* 130:112–21.

Djioua, T., Charles, F., Lopez-Lauri, F., Filgueiras, H., Coudret, A., Freire, Jr. M., Ducamp-Collin, M.-N. and Sallanon, H. (2009). Improving the storage of minimally processed mangoes (*Mangifera indica* L.) by hot water treatments. *Postharvest Biology and Technology* 52:221–6.

Fanciullino, A. L., Bidel, L. P. R. and Urban, L. (2013). Carotenoid responses to environmental stimuli: integrating redox and carbon controls into a fruit model. *Plant, Cell & Environment*:n/a-n/a. DOI: 10.1111/pce.12153.

Francis, F. (1982). Analysis of anthocyanins. In P. Markakis (ed.), *Anthocyanins as food colors*, pp. 181–207.

Franke, A. A., Custer, L. J., Arakaki, C. and Murphy, S. P. (2004). Vitamin C and flavonoid levels of fruits and vegetables consumed in Hawaii. *Journal of Food Composition and Analysis* 17:1–35.

García-Solis, P., Yahia, E. M. and Aceves, C. (2008). Study of the effect of 'Ataulfo' mango (*Mangifera indica* L.) intake on mammary carcinogenesis and antioxidant capacity in plasma of N-methyl-N-nitrosourea (MNU)-treated rats. *Food Chemistry* 111: 309–15.

Gil, M. I., Aguayo, E. and Kader, A. A. (2006). Quality changes and nutrient retention in fresh-cut versus whole fruits during storage. *Journal of Agricultural and Food Chemistry* 54:4284–96.

Gonzalez-Aguilar, G., Zavaleta-Gatica, R. and Tiznado-Hernandez, M. (2007a). Improving postharvest quality of mango 'Haden' by UV-C treatment. *Postharvest Biology and Technology* 45:108–16.

Gonzalez-Aguilar, G. A., Villegas-Ochoa, M. A., Martinez-Téllez, M., Gardea, A. and Ayala-Zavala, J. F. (2007b). Improving antioxidant capacity of fresh-cut mangoes treated with UV-C. *Journal of Food Science* 72:S197–202.

Gouado, I., Schweigert, F. J., Ejoh, R. A., Tchouanguep, M. F. and Camp, J. V. (2007). Systemic levels of carotenoids from mangoes and papaya consumed in three forms (juice, fresh and dry slice). *European Journal of Clinical Nutrition* 61, 1180–8.

Grassmann, J., Hippeli, S. and Elstner, E. F. (2002). Plant's defence and its benefits for animals and medicine: role of phenolics and terpenoids in avoiding oxygen stress. *Plant Physiology and Biochemistry* 40:471–8.

Halliwell, B. (2007). Dietary polyphenols: Good, bad, or indifferent for your health? *Cardiovascular Research* 73:341–7. DOI:10.1016/j.cardiores.2006.10.004.

Halliwell, B. (2013). The antioxidant paradox: less paradoxical now? *British Journal of Clinical Pharmacology* 75:637–44.

Hulshof, P. J., Xu, C., van de Bovenkamp, P., Muhilal, A. and West, C. E. (1997). Application of a validated method for the determination of provitamin A carotenoids in Indonesian foods of different maturity and origin. *Journal of Agricultural and Food Chemistry* 45:1174–9.

Ibarra-Garza, I., Ramos-Parra, P. A., Hernandéz-Brenes, C. and Jacobo-Velázquez, D. A. (2015). Effects of postharvest ripening on the nutraceutical and physicochemical properties of mango (*Mangifera indica* L. cv Keitt). *Postharvest Biology and Technology* 103:45–55.

Jahurul, M., Zaidul, I., Ghafoor, K., Al-Juhaimi, F. Y., Nyam, K.-L., Norulaini, N., Sahena, F. and Omar, A. M. (2015). Mango (*Mangifera indica* L.) by-products and their valuable components: A review. *Food Chemistry* 183:173–80.

Jaleel, C. A., Riadh, K., Gopi, R., Manivannan, P., Ines, J., Al-Juburi, H. J., Chang-Xing, Z., Hong-Bo, S. and Panneerselvam, R. (2009). Antioxidant defense responses: physiological plasticity in higher plants under abiotic constraints. *Acta Physiologiae Plantarum* 31:427–36.

Jiang, Z., Zheng, Y., Qiu, R., Yang, Y., Xu, M., Ye, Y. and Xu, M. (2015). Short UV-B exposure stimulated enzymatic and nonenzymatic antioxidants and reduced oxidative stress of cold-stored mangoes. *Journal of Agricultural and Food Chemistry* 63:10965–72. DOI:10.1021/acs.jafc.5b04460.

Joas, J., Vulcain, E., Desvignes, C., Morales, E. and Léchaudel, M. (2012). Physiological age at harvest regulates the variability in postharvest ripening, sensory and nutritional characteristics of mango (*Mangifera indica* L.) cv. Coghshall due to growing conditions. *Journal of the Science of Food and Agriculture* 92:1282–90.

Junmatong, C., Faiye, B., Rotarayanont, S., Uthaibutra, J., Boonyakiat, D. and Saengnil, K. (2015). Cold storage in salicylic acid increases enzymatic and non-enzymatic antioxidants of Nam Dok Mai No. 4 mango fruit. *Scientia Asia* 41:12–21.

Kalyanaraman, B., Feix, J. B., Sieber, F., Thomas, J. P. and Girotti, A. W. (1987). Photodynamic action of merocyanine-450 on artificial and natural cell-membranes – involvment of singlet molecular oxygen. Proceedings of the National Academy of Sciences of the United States of America 84:2999–3003.

Kaushik, N., Kaur, B. P., Rao, S. and Mishra, H. N. (2014). Effect of high pressure processing on color, biochemical and microbiological characteristics of mango pulp (*Mangifera indica* cv. Amrapali). *Innovative Food Science and Emerging Technologies* 22:40–50.

Khaliq, G., Mohamed, M. T. M., Ghazali, H. M. Ding, P. and Ali, A. (2016). Influence of gum arabic coating enriched with calcium chloride on physiological, biochemical and quality responses of mango (*Mangifera indica* L.) fruit stored under low temperature stress. *Postharvest Biology and Technology* 111:362–9.

Kim, S. H., Moon, J. Y., Kim, H., Lee, D. S., Cho, M. and Choi, H. K. (2010). Antioxidant and antiproliferative activities of mango (*Mangifera indica* L.) flesh and peel. *Food Chemistry* 121:429–35.

Kim, S. H., Cho, S. K., Min, T. S., Kim, Y., Yang, S. O., Kim, H. S., Hyun, S. H., Kim, H. A., Kim, Y. S. and Choi, H. K. (2011). Ameliorating effects of mango (*Mangifera indica* L.) fruit on plasma ethanol level in a mouse model assessed with ^{1}H-NMR based metabolic profiling. *Journal of Clinical Biochemistry and Nutrition* 48:214–21.

Kim Y., Brecht J. K., Talcott S. T. (2007) Antioxidant phytochemical and fruit quality changes in mango (*Mangifera indica* L.) following hot water immersion and controlled atmosphere storage. *Food Chemistry* 105:1327–34.

Kim, Y., Lounds-Singleton, A. J. and Talcott, S. T. (2009). Antioxidant phytochemical and quality changes associated with hot water immersion treatment of mangoes (*Mangifera indica* L.). *Food Chemistry* 115:989–93.

Kondo, S., Kittikorn, M. and Kanlayanarat, S. (2005). Preharvest antioxidant activities of tropical fruit and the effect of low temperature storage on antioxidants and jasmonates. *Postharvest Biology and Technology* 36:309–18.

Krinsky, N. I. and Johnson, E. J. (2005). Carotenoid actions and their relation to health and disease. *Molecular Aspects of Medecine* 26:459–516.

Larrauri, J. A., Rupérez, P. and Saura-Calixto, F. (1997). Effect of drying temperature on the stability of polyphenols and antioxidant activity of red grape pomace peels. *Journal of Agricultural and Food Chemistry* 45:1390–3.

Léchaudel, M., Urban, L. and Joas, J. (2010). Chlorophyll fluorescence, a nondestructive method to assess maturity of mango fruits (Cv. 'Cogshall') without growth conditions bias. *Journal of Agricultural and Food Chemistry* 58:7532–8. DOI:10.1021/jf101216t.

Léchaudel, M., Lopez-Lauri, F., Vidal, V., Sallanon, H. and Joas, J. (2013). Response of the physiological parameters of mango fruit (transpiration, water relations and antioxidant system) to its light and temperature environment. *Journal of Plant Physiology* 170:567–76.

Lichtenthaler, H. K. and Buschmann, C. (2001). Chlorophylls and carotenoids: Measurement and characterization by UV-VIS spectroscopy. *Current Protocols in Food Analytical Chemistry.*

Linnewiel, K., Ernst, H., Caris-Veyrat, C., Ben-Dor, A., Kampf, A., Salman, H., Danilenko, M., Levy, J. and Sharoni, Y. (2009). Structure activity relationship of carotenoid derivatives in activation of the electrophile/antioxidant response element transcription system. *Free Radical Biology and Medicine* 47:659–67.

Lopes, M. M. A., Miranda, M. R. A., Moura, C. F. H. and Filho, J. E. (2012). Bioactive compounds and total antioxidant capacity of cashew apples (*Anacardium occidentale* L.) during the ripening of early dwarf cashew. *Ciência e Agrotecnologia* 36:325–32.

Lopes, M. M., Silva, E. O., Canuto, K. M., Silva, L. M., Gallao, M. I., Urban, L., Ayala-Zavala, J. F. and Miranda, M. R. A. (2015). Low fluence pulsed light enhanced phytochemical content and antioxidant potential of 'Tommy Atkins' mango peel and pulp. *Innovative Food Science & Emerging Technologies* 33, 216–24.

Luo, F., Ly, Q., Zhao, Y., Hu, G., Huang, G., Zhang, J., Sun, C., Li, X. and Chen, K. (2012). Quantification and purification of mangiferin from Chinese mango (*Mangifera indica* L.) cultivars and its protective effect on human umbilical vein endothelial cells under H_2O_2-induced Stress. *International Journal of Molecular Sciences* (13):11260–74. DOI:10.3390/ijms130911260.

Maciel, L. F., Oliveira, L. S., Bispo, E. S. and Miranda, M. P. S. (2011). Antioxidant activity, total phenolic compounds and flavonoids of mangoes coming from biodynamic, organic and conventional cultivations in three maturation stages. *British Food Journal* 13(9):1103–13.

Madsen, H. L. and Bertelsen, G. (1995). Spices as antioxidants. *Trends in Food Science & Technology* 6:271–7.

Manach, C., Williamson, G., Morand, C., Scalbert, A. and Rémésy, C. (2005). Bioavailability and bioefficacy of polyphenols in humans. I. Review of 97 bioavailability studies. *The American Journal of Clinical Nutrition* 81(1 Suppl):230S–42S. PMid:15640486.

Manthey, J. A. and Perkins-Veazie, P. (2009). Influences of harvest date and location on the levels of β-carotene, ascorbic acid, total phenols, the in vitro antioxidant capacity, and phenolic profiles of five commercial varieties of mango (*Mangifera indica* L.). *Journal of Agricultural and Food Chemistry* 57:10825–30.

Masibo, M. and He, Q. (2008). Major mango polyphenols and their potential significance to human health. *Comprehensive Reviews in Food Science and Food Safety* 7:309–19.

Mercadante, A. Z. and Rodriguez-Amaya, D. B. (1998). Effects of ripening, cultivar differences, and processing on the carotenoid composition of mango. *Journal of Agricultural and Food Chemistry* 46:128–30.

Mercadante, A. Z., Rodriguez-Amaya, D. B. and Britton, G. (1997). HPLC and mass spectrometric analysis of carotenoids from mango. *Journal of Agricultural and Food Chemistry* 45:120–3.

Mildel, J., Elstner, E. F. and Grabmann, J. (2007). Synergetic effects of phenolics and cpartenoids on human low-density lipoprotein oxidation. *Molecular Nutrition & Food Research* 51:956.

Molan, A. L., Liu, Z. and Plimmer, G. (2014). Evaluation of the effect of blackcurrant products on gut microbiota and on markers of risk for colon cancer in humans. *Phytotherapy Research* 28: 416–22.

Natal, D. I. G., Moreira, M. E. C., Militão, M. S., Benjamin, L. A., Dantas, M. I. S., Ribeiro, S. M. R. and Martino, H. S. D. (2016). Ubá mango juices intake decreases adiposity and inflammation in high-fat diet-induced obese Wistar rats. *Nutrition* (in press) 1–8.

Nisperos-Carriedo, M. O., Buslig, B. S. and Shaw, P. E. (1992). Simultaneous detection of dehydroascorbic, ascorbic, and some organic acids in fruits and vegetables by HPLC. *Journal of Agricultural and Food Chemistry* 40:1127–30.

Noctor, G., Veljovic-Jovanovic, S., Driscoll, S., Novitskaya, L. and Foyer, C. H. (2002). Drought and oxidative load in the leaves of C-3 plants: a predominant role for photorespiration? *Annals of Botany* 89:841–50.

Noratto, G. D., Bertoldi, M. C., Krenek, K., Talcott, A. T., Stringheta, P. C. and Mertens-Talcott, S. (2010). Anticarcinogenic effects of polyphenolics from mango (*Mangifera indica*) varieties. *Journal of agricultural and Food Chemistry* 58:4104–12. DOI:10.1021/jf903161g.

Normand, F., Lauri, P.-E. and Legave, J.-M. (2015). Climate change and its probable effects on mango production and cultivation, [ISHS Mango symposium. *Acta Horticulturae* 1075:21–31.

Obanda, M., Owuor, P. O. and Taylor, S. J. (1997). Flavanol composition and caffeine content of green leaf as quality potential indicators of Kenyan black teas. *Journal of the Science of Food and Agriculture* 74:209–15.

Oliveira, B. G., Costa, H. B., Ventura, J. A., Kondratyuk, T. P., Barroso, M. E. S., Correia, R. M., Pimentel, E. F., Pinto, F. E., Endringer, D. C. and Romão, W. Chemical profile of mango (*Mangifera indica* L.) using electrospray ionisation mass spectrometry (ESI-MS). *Food Chemistry* 149:253–63.

Oliveira, A. B. Moura, C. F. H., Filho, E. G., Marco, C. A., Urban, L. and Miranda, M. R. A. (2013). The impact of organic farming on quality of tomatoes is associated to increased oxidative stress during fruit development. *PLoS ONE* 8(2):e56354 DOI:doi:10.1371/journal.pone.0056354.

Olmedo, J. M., Yiannias, J. A., Windgassen, E. B. and Gornet, N. K. (2006). Scurvy: a disease almost forgotten. *International Journal of Dermatology* 45:909–13.

Ornelas-Paz, J. D. J., Yahia, E. M. and Gardea-Bejar, A. (2007). Identification and quantification of xanthophyll esters, carotenes, and tocopherols in the fruit of seven Mexican mango cultivars by liquid chromatography-atmospheric pressure chemical ionization-time-of-flight mass spectrometry [LC-(APcl+)-MS]. *Journal of Agricultural and Food Chemistry* 55:6628–35.

Osorio, S. and Fernie, A. R. (2013). Biochemistry of fruit ripening. In: Seymour, G. B., Poole, M., Giovannoni, J. J. and Tucker, G. A. (eds), *The Molecular Biology and Biochemistry of Fruit Ripening*. Blackwell Publishing Ltd., Iowa, pp. 1–19.

Palafox-Carlos, H., Ayala-Zavala, J. F. and González-Aguilar, G. A. (2011). The role of dietary fiber in the bioaccessibility and bioavailability of fruit and vegetable antioxidants. *Journal of Food Science* 76(1):R6–15. DOI:http://dx.doi.org/10.1111/j.1750-3841.2010.01957.x. PMid:21535705.

Palafox-Carlos, H., Yahia, E., Islas-Osuna, M. A., Gutierrez-Martinez, P., Robles-Sánchez, M. and González-Aguilar, G. A. (2012). Effect of ripeness stage of mango fruit (*Mangifera indica* L., cv. Ataulfo) on physiological parameters and antioxidant activity. *Scientia Horticulturae* 135:7–13.

Poiroux-Gonord, F., Fanciullino, A.-L., Poggi, I. and Urban, L. (2012a). Carbohydrate control over carotenoid build-up is conditional on fruit ontogeny in clementine fruits. *Physiologia Plantarum*:n/a-n/a. DOI:10.1111/j.1399–3054.2012.01672.x.

Poiroux-Gonord, F., Fanciullino, A.-L., Bert, L. and Urban, L. (2012b). Effect of fruit load on maturity and carotenoid content of clementine (*Citrus clementina* Hort. ex Tan.) fruits. *Journal of the Science of Food and Agriculture* 92:2076–83. DOI:10.1002/jsfa.5584.

Poiroux-Gonord, F., Bidel, L. P. R., Fanciullino, A.-L., Gautier, H., Lauri-Lopez, F. and Urban, L. (2010). Health benefits of vitamins and secondary metabolites of fruits and vegetables and prospects to increase their concentrations by agronomic approaches. *Journal of Agricultural and Food Chemistry* 58:12065–82. DOI:10.1021/jf1037745.

Poiroux-Gonord, F., Santini, J., Fanciullino, A.-L., Lopez-Lauri, F., Giannettini, J., Sallanon, H., Berti, L. and Urban, L. (2013). Metabolism in orange fruits is driven by photooxidative stress in the leaves. *Physiologia Plantarum* 149, 175–87. DOI:10.1111/ppl.12023.

Pott, I., Breithaupt, D. E. andCarle, R. (2003). Detection of unusual carotenoid esters in fresh mango (*Mangifera indica* L. cv. 'Kent'). *Phytochemistry* 64:825–9.

Prasanna, V., Prabha, T. N. and Tharanathan, R. N. (2007). Fruit ripening phenomena: an overview. *CRC Critical Reviews in Food Science* 47:1–19.

Raskin, I. and Ripoll, C. (2004). Can an apple a day keep the doctor away? *Current Pharmaceutical Design* 10:3419–29.

Razzaq, K., Khan, A. S., Malik, A. U. and Shahid, M. (2013). Ripening period influences fruit softening and antioxidative system of 'Samar Bahisht Chaunsa' mango. *Scientia Horticulturae* 160:108–14.

Reyes, L. F. and Cisneros-Zevallos, L. (2007). Electron-beam ionizing radiation stress effects on mango fruit (*Mangifera indica* L.) antioxidant constituents before and during postharvest storage. *Journal of Agricultural and Food Chemistry* 55:6132–9.

Ribeiro, S. M. R. and Schieber, A. (2010). *Bioactive foods in promoting health: fruit s and vegetables.* Bioactive Compounds in Mango (*Mangifera indica* L.), Academic Press, London.

Ribeiro, S. M. R., Queiroz, J. H., de Queiroz, M. E. L. R., Campos, F. M. and Sant'Ana, H. M. P. (2007). Antioxidant in mango (*Mangifera indica* L.) pulp. *Plant Foods for Human Nutrition* 62:13–17.

Ripoll, J., Urban, L., Staudt, M., Lopez-Lauri, F., Bidel, L. P. and Bertin, N. (2014). Water shortage and quality of fleshy fruits, making the most of the unavoidable. *Journal of Experimental Botany*:eru197.

Robles-Sánches, M., Astiazarán-García, H., Belloso, O. M., Gorinstein, S., Parrilla, E. A., Rosa, A. L. de., Yepiz-Plascencia, G. and González-Aguillar, G. A. (2011). Influence of whole and fresh-cut mango intake on plasma lipids and antioxidant. *Food Research International* 44:1386–91.

Robles-Sánchez, R. M. Islas, O., Astiazáran-Garcia, H., Vásquez-Ortiz, O., Martín-Belloso, O., Gorinstein, S. and González-Aguillar, G. A. (2009). Quality index, consumer acceptability, bioactive compounds, and antioxidant activity of fresh-cut 'Ataulfo' mangoes (*Mangifera Indica* L.) as affected by low-temperature storage. *Journal of Food Science* 74(3): S126–S134. DOI: doi:10.1111/j.1750-3841.2009.01104.x.

Safitri, A., Theppakorn, T., Naradisorn, M. and Setha, S. (2015). Effects of UV-C irradiation on ripening quality and antioxidant capacity of mango fruit cv. Nam Dok Mai Si Thong. *Journal of Food Science and Agricultural Technology* (JFAT) 1:164–70.

Saura-Calixto, F., Serrano, J. and Goñi, I. (2007). Intake and bioaccessibility of total polyphenols in a whole diet. *Food Chemistry* 101(2):492–501. DOI:http://dx.doi.org/10.1016/j.foodchem.2006.02.006.

Schaffer, B., Urban, L., Lu, P. and Whiley, A. (2009). 6 Ecophysiology. *The Mango: Botany, Production and Uses*, p. 170.

Schieber, A., Ullrich, W. and Carle, R. (2000). Characterization of polyphenols in mango puree concentrate by HPLC with diode array and mass spectrometric detection. *Innovative Food Science & Emerging Technologies* 1:161–6.

Septembre-Malaterre, A., Stanilas, G., Douraguia, E. and Marie-Paule, G. (2016). Evaluation of nutritional and antioxidant properties of the tropical fruits banana, litchi, mango, papaya, passion fruit and pineapple cultivated in Réunion French Island. *Food Chemistry* 212:225–33.

Setiawan, B., Sulaeman, A., Giraud, D. W. and Driskell, J. A. (2001). Carotenoid content of selected Indonesian fruits. *Journal of Food Composition and Analysis* 14:169–76.

Shi, S., Ma, X., Xu, Y., Wu, H. and Wang, S. (2015). Evaluation of 28 mango genotypes for physicochemical characters, antioxidant capacity, and mineral content. *Journal of Applied Botany and Food Quality* 88:264–73. DOI:10.5073/JABFQ.2015.088.039.

Shivashankara, K., Isobe, S., Al-Haq, M. I., Takenaka, M. and Shiina, T. (2004). Fruit antioxidant activity, ascorbic acid, total phenol, quercetin, and carotene of Irwin mango fruits stored at low temperature after high electric field pretreatment. *Journal of Agricultural and Food Chemistry* 52:1281–6.

Sies, H. and Menck, C. F. M. (1992). Singlet oxygen induced DNA damage. *Mutation Research* 275:367–75.

Smirnoff, N. (1993). The role of active oxygen in the response of plants to water-deficit and desiccation. *New Phytologist* 125:27–58.

Sogi, D., Siddiq, M., Roidoung, S. and Dolan, K. (2012). Total phenolics, carotenoids, ascorbic ccid, and antioxidant properties of fresh-cut mango (*Mangifera indica* L., cv. Tommy Atkin) as affected by infrared heat treatment. *Journal of Food Science* 77:C1197–202.

Sreelakshmi, Y. and Sharma, R. (2008). Differential regulation of phenylalanine ammonia lyase activity and protein level by light in tomato seedlings. *Plant Physiology and Biochemistry* 46:444–51.

Surh, Y.-J. and Na, H.-K. (2008). NF-kB and Nrf2 as prime molecular targets for chemoprevention and cytoprotection with anti-inflammatory and antioxidant phytochemicals. *Genes & Nutrition* 2:313–17.

Talcott, S. T., Moore, J. P., Lounds-Singleton, A. J. and Percival, S. S. (2005). Ripening associated phytochemical changes in mangos (*mangifera indica*) following thermal quarantine and low-temperature storage. *Journal of Food Science* 70:C337–41.

Tian, S., Liu, J., Zhang, C. and Meng, X. (2010). Quality properties of harvested mango fruits and regulating technologies. New trends in postharvest management of fresh produce II. *Fresh Produce* 4:49–54.

Urban, L. and Léchaudel, M. (2005). Effect of leaf-to-fruit ratio on leaf nitrogen content and net photosynthesis in girdled branches of *Mangifera indica* L. *Trees-Structure and Function* 19:564–71. DOI:10.1007/s00468-005-0415-6.

Urban, L. and Alphonsout, L. (2007). Girdling decreases photosynthetic electron fluxes and induces sustained photoprotection in mango leaves. *Tree Physiology* 27:345–52.

Urban, L., Charles, F., de Miranda, M. R. A. and Aarrouf, J. (2016). Understanding the physiological effects of UV-C light and exploiting its agronomic potential before and after harvest. *Plant Physiology and Biochemistry*.

Valderrain-Rodríguez, G., Quirós-Sauceda, A. Mercado-Mercado, G., Ayala-Zavala, J., Astiazarán-García, H., Robles-Sánchez, R. M., Wall-Medrano, A., Sayago-Ayerdi S., and González-Aguillar, G. A. (2016). Effect of dietary fiber on the bioaccessibility of phenolic compounds of mango, papaya and pineapple fruits by an *in vitro* digestion model. *Food Science and Technology*. DOI:http://dx.doi.org/10.1590/1678-457X.6729.

Veda, S., Platel, K. and Srinivasan, K. (2007). Varietal differences in the bioaccessibility of β-carotene from mango (*Mangifera indica*) and papaya (*Carica papaya*) fruits. *Journal of Agricultural and Food Chemistry* 55:7931–5.

Vinci, G., Botrè, F., Mele, G. and Ruggieri, G. (1995). Ascorbic acid in exotic fruits: a liquid chromatographic investigation. *Food Chemistry* 53:211–14.

Wang, L. S. Burke, C. A., Hasson, H., Kuo, C. T., Molmenti, C. L. S., Seguin, C., Liu, P., Huang, T. H. M., Frankel, W. L. and Stoner, G. D. (2014). A phase Ib study of the effects of black raspberries on rectal polyps in patients with familial adenomatous polyposis. *Cancer Prevention Research* 7:666674.

Wenkam, N. (1979). Nutritional aspects of some tropical plant foods. *Tropical Foods: Chemistry and Nutrition* 2:341–50.

Zhao, Z., Jiang, W., Cao, J., Zhao, Y. and Gu, Y. (2006). Effect of cold-shock treatment on chilling injury in mango (*Mangifera indica* L. cv. Wacheng') fruit. *Journal of the Science of Food and Agriculture* 86:2458–62.

Chapter 15

Life cycle assessment of mango systems

Claudine Basset-Mens, Sandra Payen, Henri Vannière, Angela Braun and Yannick Biard, CIRAD, France

1 Introduction

Like all food products, fruits are under a growing scrutiny for their environmental impacts. Although often considered to have a lower environmental impact potential than most foods in western diets (Cerutti et al., 2014), fruit supply chains include a wide diversity of agricultural situations and functions which have not been covered by available studies. Moreover, export fruits are associated with a complex logistics chain including sorting, conditioning, transporting, refrigerating and waste management that also generate impacts. Both private and public institutions call today for the production of reliable reference studies on the environmental impacts of fruits at national and international levels (e.g. AGRIBALYSE, WFLDB, etc.). However, this interest concerns mostly western countries and far less developing countries in spite of important environmental challenges.

Like other fruits, mango is produced and consumed locally according to a diversity of cropping systems and practices, but is also traded globally. According to FAOSTAT, less than 5% of world's mango production is actually exported. The two main exporting regions are Asia (e.g. India, Thailand and Pakistan) for Asian and Middle East markets and America (e.g. Mexico, Brazil, Peru and Ecuador) targeting North American and European markets. Western markets impose strict trading standards and regulations that can only be achieved by well-structured and well-managed production systems (OECD, 2011; Galán Saúco, 2011). Although covering a range of situations and showing a diversity of practices, these production systems all seek for an intensive use of inputs and resources and high yields. Besides these mango exportations, about 90% of the world mango production is consumed locally, mainly in Asia which represents 75% of the

http://dx.doi.org/10.19103/AS.2017.0026.18

world production. Often present in small, traditional and family-owned farming systems, these mango cropping systems show a diversity of functions. Beyond the production of fruit, they can help farmers secure their land, earn cash or be part of a savings system and these diverse functions will affect the farmers' choices in terms of variety and practices. These cropping systems are often localized to benefit from local natural resources such as water access. However, because farmers receive less technical support and have far less cash to invest in inputs, these cropping systems are seldom optimal in terms of agronomic performance and might also be affected by resistance to pests due to long-term cropping. As a result of this great diversity and complexity of situations, mango cropping systems face contrasted challenges in terms of environmental and human health impacts.

A need for robust and global assessment tools adapted to fruit production systems in general and mango in particular is highlighted. Life cycle assessment (LCA) is the most consensual methodology available worldwide to evaluate the environmental impacts associated with a food product and is described in two ISO norms. However, it has been seldom applied to tropical fruits. Apart from the LCA for mango presented in this chapter no full LCA of mango was available in the literature. We therefore present in this chapter the state of the art for the LCA of fruits for which issues and challenges are globally the same as for mango systems and we propose a more detailed analysis of the environmental issues associated with mango systems at the end of the chapter.

The objectives of this chapter are:

- to present the LCA methodology to the mango experts as a tool to evaluate and support the eco-design of mango cropping systems and supply chains
- to present the state of the art of existing literature on LCA of fruits and associated scientific challenges
- to illustrate the use of LCA for a typical export mango case study
- to analyse the range of environmental challenges faced by mango systems over the world and the feasibility of using LCA for their environmental assessment

2 Life cycle assessment

2.1 General LCA methodology

The LCA methodology has been developed over the last decades and is the most widely recognized methodology for estimating the environmental impact of anthropogenic activities along a supply chain. After the LCA methodology was first developed for the manufacturing–packaging sector in Europe and North America in the early 1980s, it became more popular in the 1990s with methodological improvements and international harmonization regarding environmental impact assessment (Guinée et al., 2011) and first applications to the agricultural sector. This methodological strengthening were mainly supported by the Society of Environmental Toxicology and Chemistry which started playing a coordinating role in the LCA scientific community, and contributed to the harmonization of the LCA framework, terminology and methodology. The International Organization for Standardization (ISO) was involved in the LCA standardization effort in the mid-1990s and produced two international standards in 2006 which are the main references for the LCA community: ISO 14040 and ISO 14044. LCA is widely promoted and used by diverse public and private entities. For instance, it has been selected as the reference methodology to support the eco-labelling programme of the French government as part of the Grenelle II law. It is highly supported by the European Commission, Parliament and Council with Directives for the eco-design (2009/125/CE) and eco-labelling requirements for energy-related products (2010/30/UE). LCA has also become the worldwide standard for implementing environmental product declarations (ISO 14025 Type III Environmental Declarations).

2.2 LCA conceptual framework

LCA is structured in four stages (ISO series 14040:2000-6) (Fig. 1): i) definition of the study goals and scope (including definition of the boundaries of the system studied); ii) inventory of all resource flows used and substances released to the environment; iii) characterization of environmental impacts based on the inventory; and iv) interpretation of the results.

The LCA methodology is based on the following core principles:

- The life cycle perspective: the environmental impacts are quantified throughout the entire supply chain or entire 'life cycle', from raw material extraction ('cradle') to end-of-life of the product or service ('grave') – 'from-cradle-to-grave'. This global approach allows avoiding that local environmental improvement at one production stage, leads to a shifting of problem to another stage (Jolliet, 2010).
- The environmental impacts are quantified in relation to a functional unit (FU), either a product quantity (one kilogram, one piece, etc.) or a usage or service (hours, kilometres,

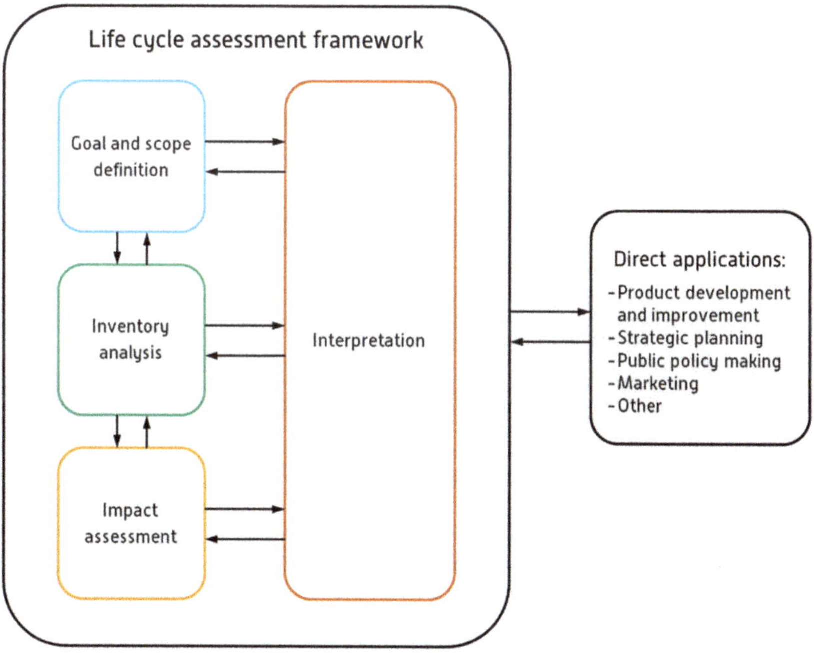

Figure 1 The general methodological framework for LCA defined by ISO 14040.

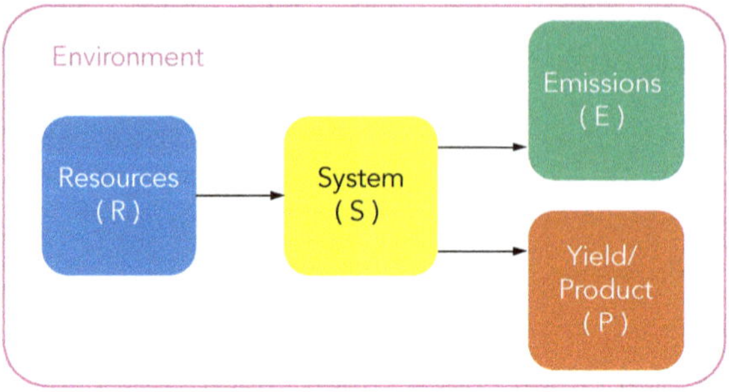

$$\text{Eco} - \text{efficiency} = \frac{\text{Impact (R; E)}}{\text{P}}$$

Figure 2 Simplified representation of the eco-efficiency concept as used in LCA. LCA results will depend on both the impacts associated with the resources used and the emissions (due to fertilizers and pesticides, for instance) and the yield of the system.

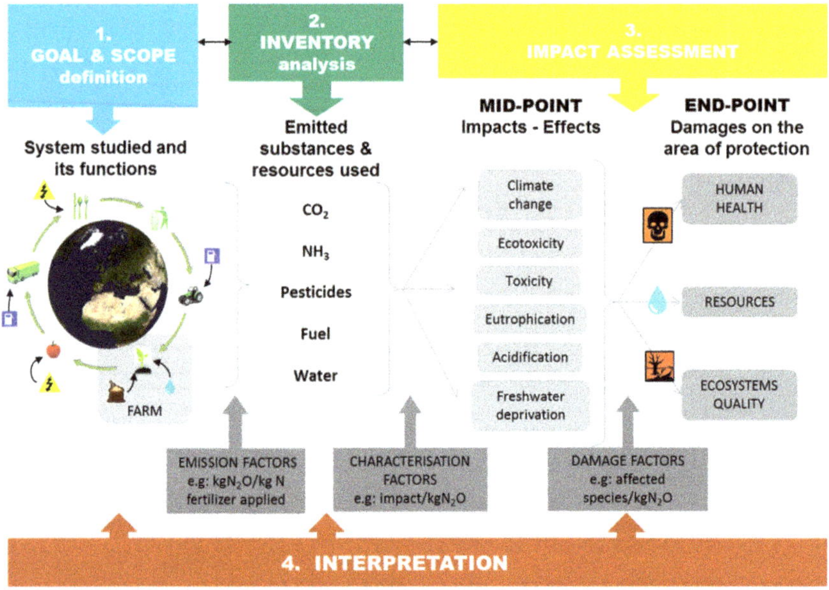

Figure 3 Description of the main principles of the four phases of LCA.

etc.). These concepts of function and associated FU enable the comparison between the environmental impacts of systems fulfilling the same function. It is often said that LCA allows evaluating the eco-efficiency of a system, that is, the impact per FU (e.g. impacts per kg of serving of fruit) (Fig. 2).

- LCA is a multi-criterion approach, considering all important environmental impacts for the studied system, such as climate change, eutrophication, toxicity and ecotoxicity, acidification, ozone layer destruction and so on. Based on an exhaustive inventory of all input and output flows among all processes of the studied system, different environmental impact indicators are calculated thanks to characterization (or equivalency) factors that convert the impacts of all flows (resource use or substance emitted) into a common impact unit. For each impact category this allows converting and aggregating the impact of all inventory flows into one single value. The most well-known equivalency unit is the kg CO_2-eq for climate change. When the IPCC model is integrated over a hundred year time horizon, 1 kg CH_4 is worse 21 kg CO_2-eq in terms of climate change potential. For other impact categories such as toxicity impacts, it will depend on the method selected. Environmental indicators can be defined either at the mid-point level or at the end-point level close to the final environmental targets, also called areas of protection (Fig. 3). The modelling of potential impacts through characterization models is an important research activity for the LCA community, and LCA practitioners benefit from the most up-to-date characterization models (e.g. ReCiPe) in the LCA software available.

The objectives of the assessment with LCA are as follows:

- to identify the most eco-efficient manner to fulfil a function (e.g. conventional vs. organic mango production)
- to highlight the greatest contributors to the environmental impacts for a given system and to identify margins of improvement.

2.3 LCA in the agri-food sector

The agricultural stage is generally a hot spot of the agri-food chains in terms of environmental impacts; for instance, the majority of food greenhouse gas emissions (40–70%) is generated during the agricultural phase (Baroni et al., 2007; Labouze et al., 2012; Garnett, 2011; Audsley et al., 2010; Smith and Gregory, 2013). Moreover, applying LCA to agri-food products is challenging, and requires specific methodological adjustments. Several general guidelines have been formulated to support the implementation of the LCA framework in general according to best available practices and methods (ILCD Handbook, 2011; PEF: EC, 2014). Sector-specific guidelines have also been developed and are already available for the agri-food sector, for instance, the methodological report from the AGRIBALYSE® project (Koch and Salou, 2015), the ENVIFOOD Protocol (Food SCP RT, 2013) and the methodological report of the World Food LCA Database (Nemecek et al., 2014). The AGRIBALYSE® programme launched by ADEME in France produced the LCA study on more than a hundred typical French agricultural products and six imported products at the farm-gate, applying a consensual methodology defined among all partners. LCA implementation in the case of tropical cropping systems raises methodological questions that will be further presented in the next section.

3 LCA of fruits

3.1 A brief history of LCA of fruits

As already mentioned, the LCA methodology was initially developed for industrial production systems, and its adaptation to agricultural production systems started in the early 1990s, with the assessment of cereals in temperate regions of Europe. This adaptation came with important challenges due to:

- the multifunctionality and diversity of farms,
- the interaction of these land-based activities with their surrounding environment making the estimation of field and livestock emissions highly complex and uncertain,
- the predominance of regional and local impacts such as eutrophication and biodiversity loss due to land use.

The application of LCA to the horticultural crops which generally use spatially structured field in rows and feature from very short (vegetables) to very long cycle (perennials) productions started only ten years ago.

Most full LCA studies on fruits including apple (Milà i Canals et al., 2006; Mouron et al., 2006a ,b; Sim et al., 2007; Alaphilippe et al., 2013; Basset-Mens et al., 2016), citrus (Sanjuan et al., 2005; Beccali et al., 2010; Pergola et al., 2013; Lo Giudice et al., 2013; Basset-Mens et al., 2016), grape (Gazulla et al., 2010), olive (Avraamides and Fatta, 2008; Cappelletti et al., 2010) and peach (Basset-Mens et al., 2016) concerned temperate or Mediterranean regions. Numerous partial LCA studies presenting only energy use balances and/or carbon footprint assessments have also been published in the last ten years (Blanke and Burdik, 2005; Ardente et al., 2006; Strapatsa et al., 2006; Mithraratne et al., 2008; Pizzigallo et al., 2008; Beccali et al., 2009; Coltro et al., 2009; Liu et al., 2010; Page et al., 2011). Only a few years ago, the life cycle approach started to be applied to fruits in tropical, subtropical and arid regions of the world. They mostly consisted of partial LCA studies on various fruits (Dwivedi et al., 2012; Graefe et al., 2013) and export banana from Ecuador (Iriarte et al., 2014; Roibas et al., 2016) and Costa Rica (Svanes and Aronsson, 2013). In 2012, Ingwersen published the first full LCA study on pineapple from Costa Rica (Ingwersen, 2012) and Trydeman Knudsen et al. (2011) published an LCA study on organic orange juice from Brazil. This year, the authors of this chapter published the first full LCA study for export mango from the Rio São Francisco Valley in Brazil in a comparative cradle-to-farm-gate LCA study for four fruits consumed in France, including apple and peach from France, clementine from Morocco and mango from Brazil (see next section) (Basset-Mens et al., 2016).

3.2 Lessons learned from LCA studies on fruits

In spite of the numerous challenges attached to the application of LCA to fruit production systems (see next section), and the diversity of goal and scope across studies, consistent trends and insights emerge from the existing literature in terms of environmental impacts of fruits.

As explained by Cerutti et al. (2011) and Bessou et al. (2013), the objectives of the available LCA studies for fruits were either to assess the environmental profile of a given crop in a specific region, or to compare different agricultural practices, organic/conventional for instance, or to compare domestic versus imported fruits. The system boundaries depended on the objectives of each study with cradle-to-farm-gate, cradle-to-processing-gate, cradle-to-market and a few cradle-to-grave LCA studies. Consistent with Cerutti et al. (2011), who identified six major groups of field operations contributing to impacts, for Basset-Mens et al. (2016), the key drivers of the eco-efficiency (environmental impacts per FU) of fruit cropping systems at the farm-gate were the yield, the fertilizer rates, the use of toxic substances and the mechanized operations for pest management, the water requirement associated with its local availability and the source of energy for irrigation. As a consequence of these drivers, among the fruits studied and for most impact categories, apple showed the least impacts compared to other fruits while citrus showed the worst impacts. In terms of hot spots from studied fruits, the farm stage appeared as a main contributor of most impacts, especially global warming, eutrophication, toxicity impacts, fossil energy/abiotic depletion and water use, bearing in mind that most LCA studies did not include all relevant impacts for fruit production systems such as water deprivation and toxicity impacts (Bessou et al., 2013). More specifically, the hot spots contributing to these impacts were the production and use of fertilizers especially for global warming and eutrophication, and the pesticide and fertilizer use for toxicity impacts. Fossil energy use and water use were predominantly related to the irrigation of fruit orchards (Bessou et al., 2013). The considerable influence of the farmers' management and the local context on LCA results is consistently acknowledged in the literature (Mouron et al., 2012; Bessou et al., 2013; Basset-Mens et al., 2016). The eco-efficiency of fruits is in the end enormously dependent on their level of intensification, agronomic performance and farming practices. Results from existing LCA studies cannot be considered representative of the extreme diversity of fruit production systems over the world and need to be complemented with more case studies.

3.3 Scientific challenges and perspectives

All four phases of the LCA methodology cause difficult challenges for an application to fruit production systems.

3.3.1 Choice of FU and system boundaries definition

The choice and definition of relevant FUs and system boundaries to evaluate fruit systems will depend on the goal and scope of the study but are not straightforward. In existing studies, a diversity of choices is observed depending on their goal and scope but also depending on a certain degree of value judgement. In comparisons across contrasted rules of production (e.g. organic/conventional), it is recommended to use several FUs: mass, land and economic value-based FUs (Cerutti et al., 2011). Regarding a mass-based FU, the number of servings per fruit and the quality should be accounted for. In the case of export fruits, the quality of the fruit was taken into account in reviewed studies through commercial grades and corresponded to different co-products of the system (Bessou et al., 2013) but the actual nutritional value of fruit is generally not accounted for in the FUs defined. Using the edible part of the fruit (or serving part) is acknowledged as the most relevant (and accessible) FU in comparative LCA studies for fruits. In terms of system boundaries, Basset-Mens et al. (2016) recommended including the post-farm-gate stages in comparative LCA for local versus imported fruits on a given market such as conditioning, sorting, transportation and also the export quality grades.

3.3.2 Diversity and complexity of agricultural practices

Many fruits, and mango in particular, are produced on perennial fruit trees which have long-term complex and evolving interactions with the ecosystem. However, this well-known complexity of perennial cropping systems in the agronomy and plant physiology disciplines is not accounted for in most LCA studies. Designing a representative average scenario constitutes one of the most difficult challenges for applying LCA to fruits (Basset-Mens et al., 2016). In the reviewed literature on LCA for fruits, the data used for agricultural practices generally lacked representativeness, consistency and transparency (Cerutti et al., 2012; Bessou et al., 2013). Many studies were based on secondary data or expertise while the rest of the studies were most often based on small samples of farms. Few studies used large samples of farms and confirmed the wide range of agricultural practices for a given product in a given region (Bessou et al., 2013). Until recently, in the vast majority of reviewed LCA studies on perennial products, very little attention was paid on integrating the perennial cropping cycle, be it through a conceptual modelling of the perennial cycle or at least through the collection of data on practices over several years. Overall, aspects related to the length of the crop cycle, the succession of immature and mature phases, the temporal variability or the biannual alternating yield, as in citrus orchards, were mostly not accounted for. Bessou et al. (2016) demonstrated that a partial modelling of the perennial crop cycle misled LCA results in two contrasted case studies of palm oil in Indonesia and small citrus in Morocco. The proposition to model orchards through a succession of stages was first proposed by Mila i Canals et al. (2006), and then validated by Cerutti et al. (2010, 2011, 2014). Bessou et al. (2013) supporting this modelling approach proposed different conceptual models to account for the orchard cycle depending on the objective of the study and the data availability. Moreover, as mentioned by Bessou et al. (2013), perennial crops in the tropics can be associated with other crops either over time when trees are still young or in agroforestry systems. The application of LCA to these complex systems induces specific methodological challenges related to their multifunctionality and the allocation of inputs, emissions and impacts. In such analysis, as explained by Bessou et al. (2013), 'a comprehensive description and quantification of costs and services of all crops within the system' should be searched for. Moreover, these authors proposed two different approaches to model agroforestry systems depending on their complexity and data availability.

3.3.3 Appropriate modelling of all field emissions

Regarding field emissions, the key substances of concern are nitrogen substances (NH_3, NO_3^-, N_2O), phosphorous substances (phosphate and particulate P), CO_2 and CH_4, pesticides and heavy metals. Their inclusion was uneven across available studies with few studies evaluating all key substances (Bessou et al., 2013). In addition, the description of the models used for their estimation most often lacked precision and transparency. When described, one could note that these models were not adapted to tropical perennials as they corresponded to measurements and parametrized equations obtained for annual crops in temperate regions. 'Research is therefore needed to produce data sets on perennial cropping systems and their field emissions in order to account better for perennial crops in existing operational models' (Bessou et al., 2013). Cerutti et al. (2011) and Bessou et al. (2013) consistently recommended using models or measurements for field emissions of nutrients and pesticides. Cerutti et al. (2011) also acknowledged the importance of calculating a nutrient balance to estimate possible nutrient surplus. Bessou et al. (2013)

emphasized the need for field experiments and modelling works on the characterization of the biomass stands (above and below ground) and the dynamics of soil cover and organic matter pools in relation with agricultural management. This would support the development of a robust framework for carbon cycle modelling in LCAs of perennial crops which is missing nowadays.

3.3.4 Modelling all important impacts

In reviewed LCA studies for fruits, important impact categories such as water deprivation and toxicity were generally not included (Bessou et al., 2013; Basset-Mens et al., 2016). The uncertainty and complexity attached to toxicity models is a key problem for the environmental assessment of fruit production systems with LCA that warrants important research effort. Moreover, although emphasized by several authors as a key environmental issue for fruit production, water consumption and water deprivation (impact on local water resource depending on local water stress) were generally not considered in LCA studies for fruits. The need for irrigation in perennial cropping systems, especially in semi-arid countries (Mediterranean and tropical), represents a specific and new issue for LCA compared to LCA dedicated until recently to annual crops in temperate regions (Bessou et al., 2013). Moreover, as for all perennial crops, the calculation of reliable and complete carbon budgets and associated climate change impacts, and the assessment of soil quality and biodiversity loss due to land use constitute key challenges for LCA studies applied to fruits. Finally, the exclusion of other dimensions of the sustainability framework such as socio-economic aspects is also an important gap highlighted by Bessou et al. (2013).

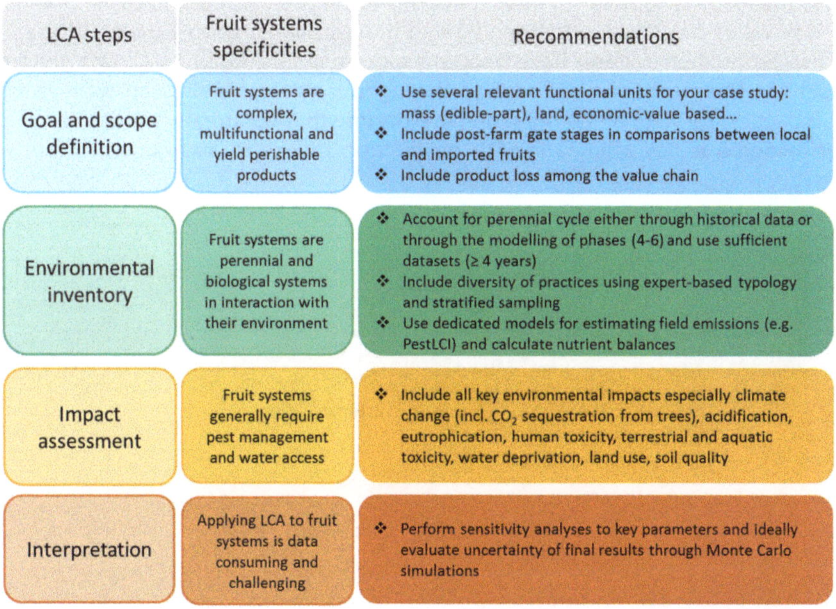

Figure 4 Up-to-date recommendations for an application of LCA to fruit life cycle systems.

3.3.5 Guidelines for applying LCA to fruits

Basset-Mens et al. (2016) called for the formalization of consensual guidelines for the comparison of fruits' environmental impacts with LCA. Up-to-date guidelines summarizing the most recent research on this topic are presented in Fig. 4.

The next section illustrates the above-mentioned challenges for LCA applied to fruits in the first LCA of mango production and exportation.

4 LCA case study: exports of mango from the Rio São Francisco Valley in Brazil

4.1 Goal and scope definition

The objective of this study was to assess the potential environmental impacts of mango produced in Brazil for export to the European market, in compliance with the AGRIBALYSE® methodology, based on the LCA framework (Fig. 5).

The FU was 1 kg of mango on the French market. Results for raw mango at the farm-gate in Brazil have been reported in Basset-Mens et al. (2016). In this chapter we also present results including the post-farm-gate stages (conditioning and transport to France). The potential impacts were assessed based on a representative mango orchard modelled over its entire life.

4.1.1 Mango from the Rio São Francisco Valley in Brazil

Brazil is the leading supplier of fresh mangoes to the EU (Gerbaud, 2016). The Rio São Francisco Valley produces more than 90% of the Brazilian mango exports. Modern and intensive production systems have been developed in this region to allow a year-round production. This production of mango throughout the year relies on well-controlled floral induction through a fine management of growth regulators and irrigation (Fig. 6). Water resource is not a limitation for production thanks to an abundant dam water access.

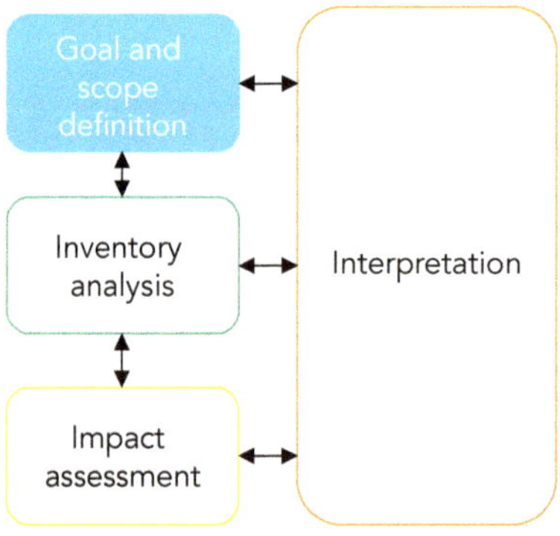

Figure 5 Goal and scope definition for Brazilian mango case study.

4.1.2 Systems studied

The system studied includes the full orchard life (from planting to uprooting), the fruit conditioning and the transportation from Brazil to France. 'From-cradle-to-market-gate', all inputs and outputs were inventoried (Fig. 7).

4.2 Inventory analysis

4.2.1 Modelling of perennial cropping systems

A representative system was modelled based on a detailed survey of eight contrasted 'Kent' and 'Tommy Atkins' mango orchards surveyed in 2012 in the Rio São Francisco Valley in Brazil (Fig. 8). Modelling perennial cropping systems is a data-demanding and time-consuming task as highlighted in the state of the art. It requires collecting or estimating agronomic management data for the whole orchard's life, modelled in three phases:

Figure 6 A well-controlled floral induction on a mango orchard, Rio São Francisco Valley, Brazil (S. Payen).

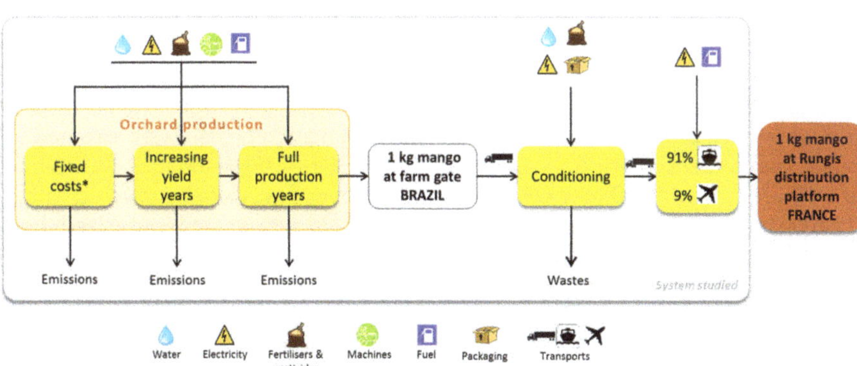

Figure 7 Mango production system from cradle-to-French market-gate. *Fixed costs include plantation, non-productive years and uprooting of trees at the end of the orchard's life.

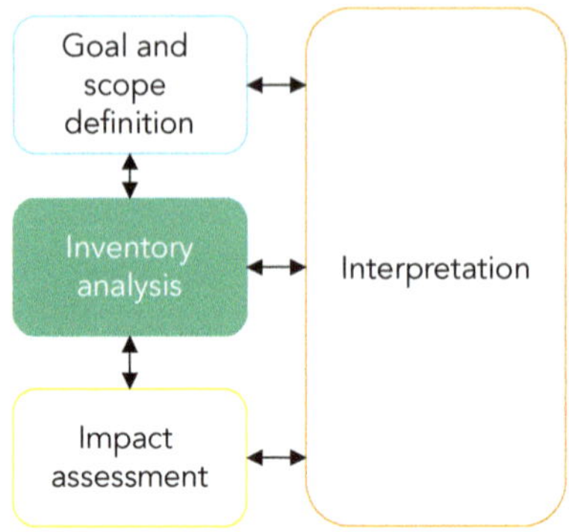

Figure 8 Inventory analysis for Brazilian mango case study.

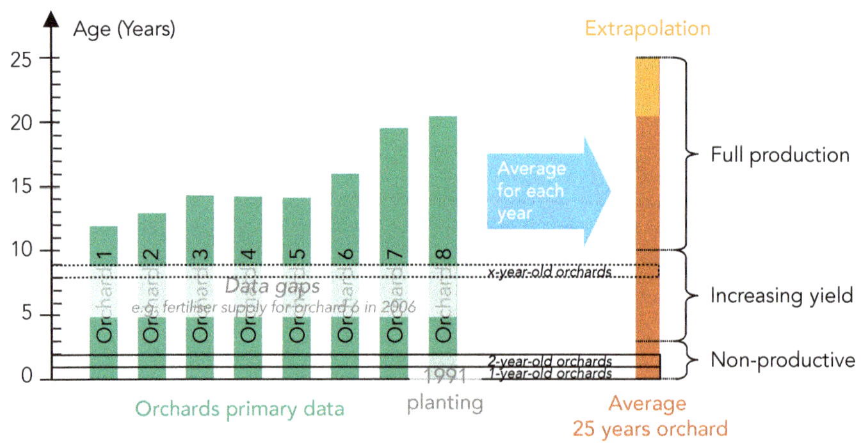

Figure 9 Modelling approach to design a representative 25-year-old mango orchard based on eight surveyed orchards.

non-productive phase (plantation, non-productive years, uprooting of trees), increasing yield years and full production years. The lifespan of a mango orchard was assumed to be 25 years. The reference period recommended by AGRIBALYSE® for strongly alternating fruits is 10 years, from 2000 to 2009. This reference period was largely covered with data collected over more than 20 years on real orchards. Field surveys were organized through a partnership with the most important exporter association of mango from the region. In spite of the tremendous amount of data collected, mostly in paper format, historic data gaps remained since most ancient orchards were planted in 1991. Only the yield was consistently available across the whole of the orchard cycle and the eight orchards (Fig. 14). Annual averages for all input and yield data available across the eight orchards were first calculated and then aggregated into average data for each phase (Fig. 9). Table 1 shows the main agronomic data for the three production phases. In addition, since the oldest orchard was 21 years old and the assumed lifetime of the mango orchard was 25,

Table 1 Main agronomic data for the three production phases of a representative mango orchard

Intervention	Unit	Non-productive years	Increasing yield phase	Full production phase
Orchard age	Years	25	–	–
Density	Trees/ha	280	–	–
Yield	t/ha	33	–	–
Fertilization				
N	kg/ha	61	190	165
P_2O_5	kg/ha	59	83	100
K_2O	kg/ha	112	297	273
Irrigation				
Water	m^3/ha	1560	6683	7999
Energy	MJ/ha	566	2387	2,946
Plant protection				
Herbicides	kg/ha	0	0	0
Insecticides	kg/ha	0.800	1.134	0.301
Fungicides	kg/ha	3.573	3.884	5.66
Total pesticides	kg/ha	4.373	5.018	5.961
Growth regulators	kg/ha	0.92	2.39	4.03

all input and yield data for the last four years were extrapolated from the average of each input and yield data for the full production phase.

4.2.2 Field emissions

Regarding the field emissions from orchards, best available emission factors as recommended by the AGRIBALYSE® method were used (Fig. 10) (Koch and Salou, 2015). Phosphate and pesticide emissions were calculated according to Nemecek and Kägi (2007). On this basis, 100% of the pesticides applied were assumed to be emitted to the soil. Nitrous oxide, carbon dioxide from urea and lime, and nitrate leaching were estimated according to IPCC (2006). Ammonia emissions were based on emission factors from EMEP/CORINAIR (2006) and nitrogen oxides according to EMEP/EEA (2009). According to IPCC (2006), nitrate leaching was considered nil because localized irrigation was used and rainfall was low (the daily irrigation (or rainfall) volume was constantly below the soil field capacity).

4.2.3 Post-farm-gate stages

Harvested mangoes are sent to packaging stations for treatment and conditioning for export. Water, electricity, chemicals (fungicide, detergent) and packaging use data were collected from two packaging stations. Packed mangoes are then exported to Europe by

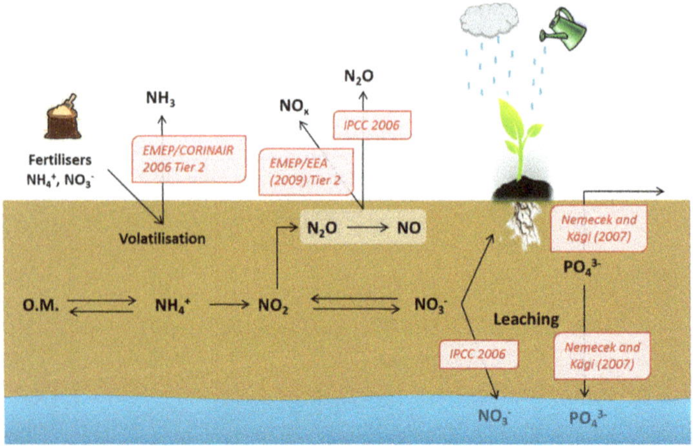

Figure 10 Field emissions and methods used for their assessment.

boat (91%) or by plane (9%) (average percentage based on the packaging stations surveyed in 2012). The energy consumption for transport and refrigeration was accounted for.

For indirect inventory data such as fertilizer manufacture or electricity generation, the Ecoinvent Life Cycle Inventory database (version 2.2) was used.

4.3 Impact assessment

4.3.1 Method

The impact assessment was performed using the ReCiPe life cycle impact assessment method (Goedkoop et al., 2008), at the mid-point level and adopting the hierarchist perspective (Fig. 11; see also Fig. 3). The impact category indicators considered were climate change (100 years; in kg $CO_{2\text{-eq}}$), terrestrial acidification (in g $SO_{2\text{-eq}}$), freshwater and marine eutrophication (in g $P_{\text{-eq}}$ and g $N_{\text{-eq}}$, respectively, based on the nutrient-limiting

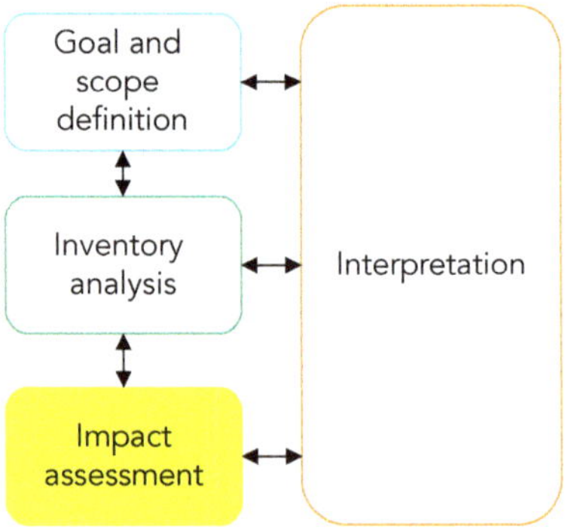

Figure 11 Impact assessment of Brazilian mango case study.

factor of the aquatic environment), human toxicity (in g 1,4-DB$_{-eq}$: 1,4-dichlorobenzene), and terrestrial and freshwater ecotoxicity (in g 1,4-DB$_{-eq}$) (see Section 2.2). The non-renewable energy consumption (fossil and nuclear; in MJ) was assessed using the cumulative energy demand method (Hischier et al., 2009). To allow comparison with published LCA studies, results were also calculated using the CML 2001 methodology (Guinée et al., 2002).

4.3.2 Comparison with other LCA studies for fruits

Since no complete LCA studies on mango could be found, we compared cradle-to-farm-gate LCA results for mango with other fruits available on the French market, assessed with the same AGRIBALYSE® methodology. We thus compared mango with apple and peach produced in France, and small citrus (clementine) imported from Morocco. Table 2 shows the results for the four fruits at the farm-gate.

Apple reveals the least impacts for most impact categories (climate change, human toxicity, terrestrial acidification, freshwater eutrophication and non-renewable energy use), followed by mango, and then peach and clementine. This ranking was explained by several factors:

- yield: higher for apple (54 t.ha^{-1}) and lower for peach and clementine (28 t.ha^{-1});
- fertilizer input: lower for apple and higher for the imported mango and clementine;
- access to water: requiring a lot of energy in water-scarce Morocco in contrast to the abundant dam water flowing into the mango orchards;
- share of fossil energy in the electricity mix of each country: lower in France (10%) and higher in Morocco (50%).

Mango showed the least impacts for terrestrial and freshwater ecotoxicity, followed by apple and peach, and then by clementine with greatest impacts (Fig. 12). This low

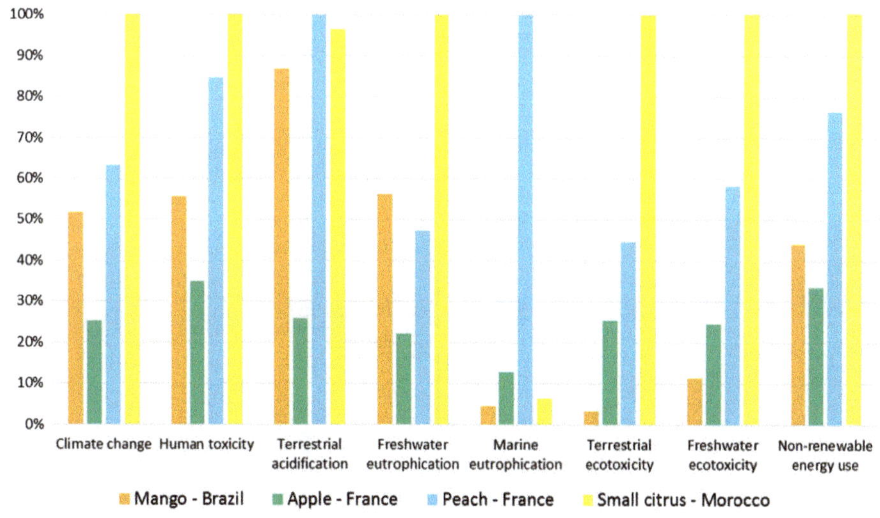

Figure 12 Cradle-to-farm-gate life cycle assessment (LCA) results per kg of raw fruit for a selection of environmental indicators (CML for apple, mango, peach and clementine). Results are expressed as a percentage of the greatest result for each impact category.

Table 2 Cradle-to-farm-gate life cycle assessment (LCA) results for a selection of environmental indicators (ReCiPe Midpoint (H) V1.06) for apple, mango, peach and clementine. Results are expressed per kilogram of raw fruit

Impact category	Climate change	Human toxicity	Terrestrial acidification	Freshwater eutrophication	Marine eutrophication	Terrestrial ecotoxicity	Freshwater ecotoxicity	Non-renewable energy use
Unit	kg CO_{2eq}	kg 1,4-DB_{eq}	kg SO_{2eq}	kg P_{eq}	kg N_{eq}	kg 1,4-DB_{eq}	kg 1,4-DB_{eq}	MJ
Mango – Brazil	1.39E-01	4.36E-02	2.05E-03	7.15E-05	8.42E-05	2.30E-04	7.06E-04	1.46E+00
Apple – France	6.78E-02	2.73E-02	6.10E-04	2.83E-05	2.33E-04	1.77E-03	1.51E-03	1.12E+00
Peach – France	1.70E-01	6.64E-02	2.36E-03	6.02E-05	1.83E-03	3.12E-03	3.59E-03	2.54E+00
Small citrus – Morocco	2.69E-01	7.83E-02	2.27E-03	1.27E-04	1.16E-04	6.99E-03	6.16E-03	3.32E+00

ecotoxicity impact for mango can be explained by the highly favourable conditions of production for mango trees in the Rio São Francisco Valley. Mango also showed the least impacts for marine eutrophication, closely followed by clementine, then by apple and finally by peach. This was explained by the use of IPCC nitrate emission factors, assuming no leaching for drip-irrigated crop under semi-arid climate (i.e. mango from Brazil and clementine from Morocco), which is in contrast with the 30% recommended for crops under a temperate climate (i.e. apple and peach from France).

4.3.3 Contribution analysis

For each impact category, the contribution of the different life cycle stages is shown in Fig. 13.

For climate change, human toxicity, freshwater eutrophication, freshwater ecotoxicity and energy consumption, the main contributor was the manufacture and transport of fertilizers (ranging from 52% contribution to climate change up to 83% contribution to freshwater eutrophication). For climate change, the second most important contributor was field emissions after application of fertilizers with 23% contribution to the total impact, due to nitrous oxide emissions. For freshwater ecotoxicity, human toxicity and freshwater eutrophication, the second most important contributor was the plant protection, with contributions of 28%, 22% and 9%, respectively. For terrestrial acidification and marine eutrophication, field emissions after application of fertilizers constituted the main contributor with 65% and 62%, respectively, mainly due to ammonia volatilization. The second most important contributor was the manufacturing and transportation of fertilizers with 25% and 20%, respectively.

For terrestrial ecotoxicity, plant protection was the main contributor, representing 80% of the total impact. The impact was mostly due to field emissions of pesticides with contributions from a wide range of substances, the most significant being ethephon, chlorothalonil, methomyl and dimethoate. No obvious hot spots could be diagnosed for mango in terms of toxic pesticide use. Finally, electricity consumption for fertigation had small contributions to all impacts: from 0% to less than 6%.

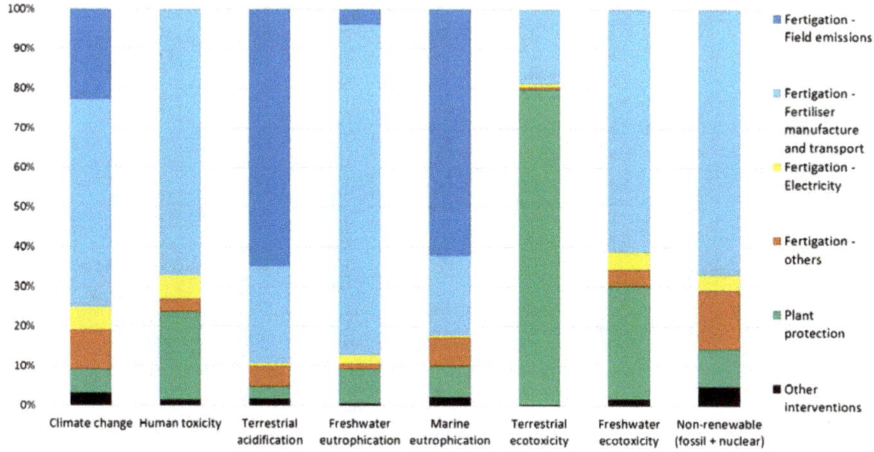

Figure 13 Contribution analysis of 1 kg of mango at the farm-gate in Brazil, ReCiPe Midpoint (H).

This contribution analysis reveals the assets of mango cultivation in the Rio São Francisco Valley in Brazil that benefits from favourable conditions of production in terms of environment and agronomic practices. First, orchards have unlimited access to water at low energy cost while benefiting from dry and warm weather; second, the most common pests for mango such as fruit flies are less problematic in Brazil than in Africa (Duyck et al., 2004). This can be partly explained by the releases of infertile insect males by planes. Furthermore, the arid climate is not favourable to diseases such as powdery mildew and anthracnose, leading to reduced anti-fungal treatments. However, the ratio kg of N fertilizer to kg of mango produced remains high at five despite the expertise of farmers. Reducing the use of fertilizers per kg fruit represents the best perspective of impact reduction for the mango production.

4.3.4 Post-farm stages

Figure 14 shows the great contribution of the transportation phase to the cradle-to-market-gate results of mangoes exported to the French market especially for climate change, terrestrial acidification, marine eutrophication and freshwater ecotoxicity. This large contribution was mostly due to 9% of mango being transported by plane.

4.4 Interpretation

In the interpretation phase, the robustness of the results is evaluated by analysing the relevance of the methods and data used compared to the goal and scope of the study (Fig. 15). For the cradle-to-farm-gate stages, the consensual method developed as part of the AGRIBALYSE programme was applied. It represents an important step forward, especially with regard to the harmonization of methodological choices, the selection of most up-to-date field emission methods and the modelling of the entire perennial cropping system through key orchard's phases. However, we completed the cradle-to-farm-gate

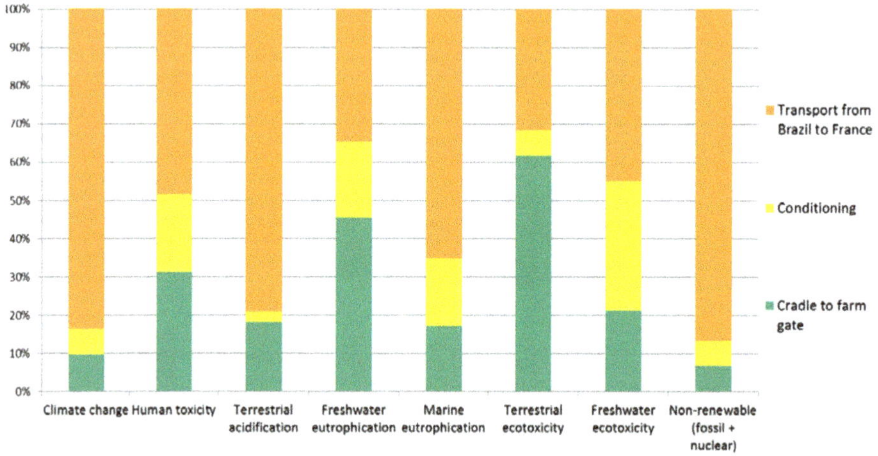

Figure 14 Contribution analysis of 1 kg of mango available on the French market, ReCiPe Midpoint (H).

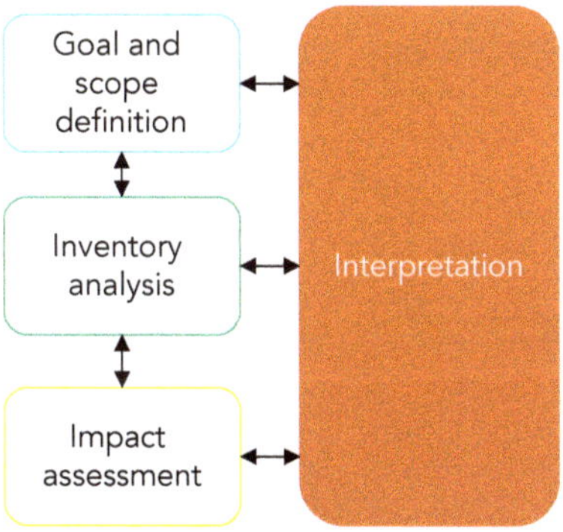

Figure 15 Interpretation for Brazilian mango case study.

LCA study by an assessment of post-farm-gate stages which revealed the important impacts of the transportation by plane. This is important for a proper comparison of local versus imported fruits. However, the quality of fruits was not accounted for in the FU in this study and would warrant further research. Regarding field emission methods, although the most up-to-date methods were used, they are not particularly valid for perennial crops under a semi-arid climate. The use of agro-hydrological models to better estimate field emissions and water flows constitutes a relevant perspective for the modelling of cropping systems in LCA. Another important gap in our study was the non-inclusion of water deprivation which potentially represents a key environmental concern for fruit and horticultural products (Payen et al., 2015).

One key difficulty of the AGRIBALYSE® programme was the design of representative systems in terms of technology, time and space. The average mango production system was based on a sample of eight orchards, with unknown statistical representation. This represents a significant bias in the assessment of a representative system, but this study constitutes a first and important step forward for the comprehensive impact assessment of mango production. As already explained, the lack of statistical representation of the average orchard is particularly questionable regarding the substances used for plant protection: since the toxicity of active substance is highly variable, not reporting the use of a highly toxic substance can drastically change the ecotoxicity results. It is important to keep in mind that designing representative practices over a period of 25 years represents a contradiction in itself and a difficult challenge, especially for pesticide treatments which follow constantly changing rules. In addition, the recent understanding that the whole orchard life cycle should be included in LCA studies of perennials leads to a very data-intensive protocol that can be at the expense of the orchard variability exploration such as in our mango case study. In our case study the important variability of agronomic data can be illustrated through the yield curves over time for the eight orchards (coloured dotted lines in Fig. 16) compared to the yield curve over time for our average scenario (black line in Fig. 16). Novel and less data-intensive strategies must be developed to allow the exploration of the variability of orchards within LCA studies for perennials, as discussed in Bessou et al. (2016).

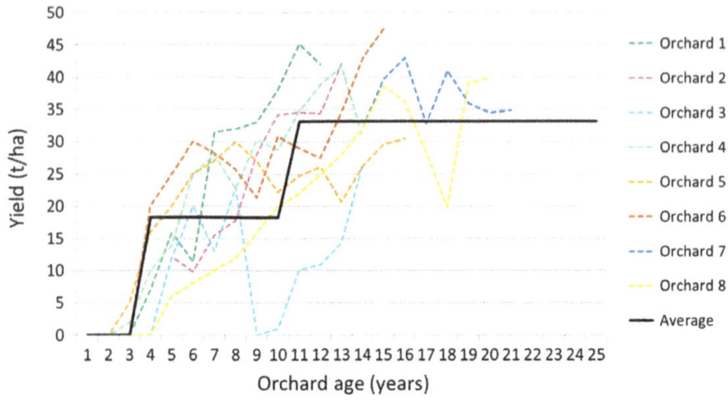

Figure 16 Yield of studied mango orchards (coloured dotted lines) and average scenario (black line).

5 Environmental challenges

As introduced in this chapter, mango cropping systems cover a wide diversity of situations, functions, practices and consequently of environmental issues (Volavi et al., 2012). Our case study presented in the previous section is a typical example of LCA study for export mango; however, its conclusions cannot be extrapolated to other situations. Although it is impossible to describe in so little words the whole diversity of situations for mango cropping systems, in this section we propose an analysis of the range of challenges associated with them from an agronomic and environmental point of view. This analysis covers both the intensive production systems dedicated to export (the minority) and the more traditional ones (the majority). Based on the eco-efficiency definition, we analyse the drivers of the yield and losses on one hand and the main operations responsible for environmental impacts on the other hand: pest management, water access and fertilization.

5.1 Yield and losses

National average yields reported by FAOSTAT reflect the share of intensive cropping systems for export and small traditional ones. In case of a majority of intensive cropping systems for export (Brazil, Peru, etc.), national average yields are 2 to 2.5 times greater than in cases where small traditional cropping systems are dominant. For more reduced cropping system samples, the yield variations can be even greater (5 to 7 tonnes vs. 25 to 40 tonnes). These yield data include the losses during the production phase due to pest attacks, climate risks and diseases. Post-harvest yield data are generally less available. Value chain analyses estimated in different countries the losses over the whole mango supply chain. For instance in different Western African countries, these post-harvest losses are very high for small producers, often ranging from 25–30% up to 60% (Van Melle, 2013; CEDEAO, 2011). Similar losses are reported by experts for small mango producers from Kenya, Nigeria or India (USAID-KAVES, 2015; Akinyemi, 2012; Prasad, 2006). Conversely, in the same region, the losses for small mango producers that are members of fruit marketing associations can be as low as 1%, being comparable to that for intensive cropping systems. In conclusion, for small mango producers, low yields at farm-gate will be associated with important post-harvest losses in most cases while intensive cropping systems will show high yield and low post-harvest losses due to proper marketing logistics. After the farm-gate, it will be necessary to estimate the fate of raw fruits including losses, fruits for local market and fruits for export. This will be critical to allocate the impacts to the actual FU studied, for example, kg export mango or kg mango for local market.

In brief, yield and losses as a critical data for calculating the eco-efficiency of systems will need to be estimated in the most reliable way as possible with sufficient sampling of orchards and post-harvest stages.

5.2 Pest management contexts and practices

In 2013, Bhushan et al. identified strong inconsistencies between, on one hand, official Indian regulations for pesticides as defined by the Central Insecticides Board and Registration Committee (CIBRC) in charge of pesticide homologation and the Food Safety and Standard Authority of India (FSSAi), in charge of defining MLRs, and, on the other hand, the technical recommendations made by extension services to farmers. In the case of mango, nine pesticides recommended by the National Horticulture Board (NHB) have never been homologated by the CIBRC. Conversely, 11 pesticides homologated for mango orchards are never recommended by the NHB in practice. Similar situations are observed in the majority of countries producing mango. In Western Africa, as part of the Permanent Interstates Committee for Drought Control in Sahel (CILSS), the Sahelian committee for pesticides (CSP) is in charge of the homologation of pesticides for the nine countries of the CILSS. As in most countries, phytosanitary companies will initiate the homologation of a pesticide in a given context, while its evaluation and validation will be done by the official competent authorities. As a consequence, the number of homologated pesticides is high for the economically important crops (e.g. 85 pesticides available for cotton) but reduced for less important ones such as mango with only five active ingredients among eight commercial pesticides and no fungicide (CSP, 2015). The risks associated with these situations are diverse: from a technical impasse that will affect the productivity of the crops when no pesticide is available for controlling a pest, to the development of resistance in case of a repeated use of a unique active ingredient and finally the misuse of pesticides (e.g. use on mango orchards of a pesticide homologated for cotton). These statements illustrate the great complexity and diversity of situations, especially for small producers and the diverse risks associated with pest management in mango producing countries.

In brief, an inadequate management of pests and diseases will lead to a yield drop, while the use of pesticides, although supposed to protect fruits, trees and associated yield, will have human health and ecotoxicity impacts, especially when excessive rates of pesticides are applied or highly toxic molecules are used. In addition, in certain regions of long-term production, resistance to pesticides may develop and lead to a technical impasse and a general degradation of the orchards' productivity.

5.3 Water access

In historical regions of mango production, the water requirements of this crop are properly covered. Rainfall of at least 750 mm, evenly spread over a period of six to seven months, allows a reasonably good mango production with no irrigation (Singh, 1990). Better than other fruit perennial crops, adult mango trees can use the underground water resources during the dry season thanks to their strong rooting system. This asset added to other adaptative features to drought allows the plantation of mango orchards in areas where the rainfalls should be limiting either in terms of quantity or of distribution. This situation is very frequent in small traditional mango plantations in Asia (46% of mango farms in Karnataka state, 4th Indian state for mango production (Bung, 2013; Thomas, 2013)) and in all Western Sudano-Sahelian Africa except in Senegal (Vannière, 2007). In such situations, mango trees will experience a water stress that will contribute to the low yield observed. If all farmers use water irrigation during the first years of the orchard, then irrigation practices will be highly contrasted from no irrigation at all to high-tech localized fertigation, or intermediate practices with more intuitive and punctual irrigation. Over the last 25–30 years, new irrigation perimeters have been developed in warm and arid regions where regional facilities supply abundant water resources to farmers. The best examples for mango production are the Rio São Francisco Valley in Brazil, the Piura region in the North of Peru and Israel. In these contexts, as demonstrated in our case study, the control of irrigation as well as all of the production factors lead to very high yields compared to that for traditional systems.

In brief, mango orchards located in the historical regions of production will not be limited by water, but in warm and arid regions, insufficient water supply will lead to a yield reduction while a supply in irrigation water will lead potentially to a water deprivation impact and to energy use for water withdrawal. This will potentially impact climate change (non-renewable energy), acidification, human toxicity and water deprivation category indicators.

5.4 Fertilization

Fertilization practices are highly diverse among mango cropping systems, as all other practices will necessitate dedicated and heavy sampling of orchards and survey for a proper characterization. In low-input systems, fertilizers will only be applied in the plantation hole of the tree before the plantation. Then, mango trees may benefit from the fertilization of associated crops planted in the inter-rows and then no further fertilization will be supplied. Conversely, in more intensive systems, fertilization is recognized as an important driver of productivity. Each region has its own fertilization references based on soil and leaf analyses, exportations by fruits harvested and sequestration in the tree trunk. Recommendations are made depending on the age of the trees, their expected yield and their nutritional status. In practice, important gaps can be observed between these recommendations and the farmers' practice. The fertilizer rates might be less than the recommendations due to cash issues or might be greater due to the behaviour of farmers, who perceive an excess of fertilization as positive for securing the yield.

In brief, fertilization as a key production factor will reduce the yield when insufficiently applied, while it will be associated with environmental impacts such as eutrophication, climate change and acidification when applied in excess or not applied.

6 Future trends and conclusion

As demonstrated above, mango production systems show highly contrasted situations of production that will diversely affect their eco-efficiency, that is, their environmental impacts per FU studied. Be it for suboptimal production reasons leading to low yield and high losses or for excessive and inadequately managed inputs, mango systems will have environmental impacts per FU and they could all greatly benefit from a proper LCA study. The key challenges that an LCA application to mango systems faces are similar to that for all tropical fruit systems and key recommendations summarized in Fig. 4 can be followed to this end. However, it is important to notice that applying LCA for small family-owned mango farms will be even more challenging than for intensive mango systems for export due to extreme data scarcity. Overall, applying LCA to mango value chains will necessitate important research effort and dedicated data collection protocols to capture the actual diversity of these complex systems.

In conclusion, although LCA application to tropical fruits is still scarce and recent, research on this topic is active especially in CIRAD which develops research projects and an LCA platform dedicated to the LCA of tropical agri-food products. LCA can support the decision of all stakeholders across mango value chains regarding the eco-design of farms, the choice of logistics systems and the sourcing of products, and contribute globally to improving their eco-efficiency. Besides environmental LCA, social LCA is emerging with the aim to evaluate the social impacts associated with a function (Feschet, 2014). When used jointly, environmental and social LCA will offer to decision-makers a consistent decision tool for designing more sustainable mango value chains.

7 Where to look for further information

LCA food conference every two years: http://lcafood2016.org/

On the AGRIBALYSE® programme: http://www.ademe.fr/en/expertise/alternative-approaches-to-production/agribalyse-program

Top five research centres on LCA for agri-food products, in general, and fruits and horticultural products, in particular:

- Agroscope: https://www.agroscope.admin.ch/agroscope/en/home/topics/environment-resources/life-cycle-assessment.html
- Pôle ELSA: http://www1.montpellier.inra.fr/elsa/?lang=en, including CIRAD, www.cirad.fr/
- SIK, Sweden: http://www.sp.se/en/units/fb/sustainable/Sidor/default.aspx
- IRTA, Spain: http://www.irta.cat/en-US/RIT/Pages/default.aspx
- NARO, Japan: http://www.naro.affrc.go.jp/english/

On the World Food LCA Database: http://www.quantis-intl.com/microsites/wfldb/

Basset-Mens, C., Vannière, H., Grasselly, D., Heitz, H., Braun, A., Payen, S., Koch, P., Biard, Y., 2016. Environmental impacts of imported and locally-grown fruits for the French market: a cradle-to-farm-gate LCA study. *Fruits* 71(2): 93–104. http://dx.doi.org/10.1051/fruits/2015050.

Bessou, C., Basset-Mens, C., Tran, T. and Benoist A., 2013. LCA applied to perennial cropping systems: a review focused on the farm stage. *Int. J. LCA*, 18:340–61.

Bessou, C., Basset-Mens, C., Latunussa, C., Vélu, A., Heitz, H., Vannière, H. and Caliman, J. P. (2016), Partial modelling of the perennial crop cycle misleads LCA results in two contrasted case studies, *Int. J. LCA*, 21, 297–310. doi: 10.1007/s11367-016-1030-z.

Hellweg, S. and Mila i Canals, L. (2014), Emerging approaches, challenges and opportunities in life cycle assessment. *Science* 344(6188), 1109-13. doi: 10.1126/science.1248361.
Jolliet, O., Saadé, M., Crettaz, P. and Shaked, S. (2010), Analyse du cycle de vie: Comprendre et réaliser un écobilan. Retrieved from http://books.google.com/books?hl=fr&lr=&id=g9S55CklsOoC&pgis=1
LCA compendium - The Complete World of Life Cycle Assessment. *Book series.* Klöpffer, W. and Curran, M. A. (eds), Springer book, ISSN 2214-3505.

8 References

Affognon, H., Mutungi, C., Sanginga, P. and Borgemeister, C. (2015), Unpacking Postharvest Losses in Sub-Saharan Africa: A Meta-Analysis, 2015, World Development, 66, 49–68. Available at: http://dx.doi.org/10.1016/j.worlddev.2014.08.002
Akinyemi, S. O. (2012), Mango Cultivation in Nigeria, in Mango, 2, Cultivation in Different Countries. STUDIUM PRESS LLC. ISBN: 1-933699-94-9, pp. 247–59.
Alaphilippe, A., Simon, S., Brun, L., Hayer, F. and Gaillard, G. (2013), Life cycle analysis reveals higher agroecological benefits of organic and low-input apple production. *Agronomy for Sustainable Development* 33, 581–92. doi: 10.1007/s13593-012-0124-7.
Alila, P. (2012), Mango in the Eastern and North Eastern States of India , in Mango, 2, Cultivation in Different Countries. STUDIUM PRESS LLC. ISBN: 1-933699-94-9, pp. 28–41.
Amin, M. N. and Siddiqui, M. (2012), Mango Cultivation in Bangladesh, in Mango, 2, Cultivation in Different Countries. STUDIUM PRESS LLC. ISBN: 1-933699-94-9, pp. 145–62.
Ardente, F., Beccali, G., Cellura, M. and Marvuglia, A. (2006), POEMS: A Case Study of an Italian Wine-Producing Firm. *Environ. Manag.*, 38(3), 350–64. doi: 10.1007/s00267-005-0103-8.
Avraamides, M. and Fatta, D. (2008), Resource consumption and emissions from olive oil production: a life cycle inventory case study in Cyprus. J. Clean. Prod., 16, 809–21.
Baroni, L., Cenci, L., Tettamanti, M. and Berati, M. (2007), Evaluating the environmental impact of various dietary patterns combined with different food production systems. *Eur. J. Clin. Nut.*, 61(2), 279–86.
Basset-Mens, C., Vannière, H., Grasselly, D., Heitz, H., Braun, A., Payen, S., Koch, P. and Biard, Y. (2016), Environmental impacts of imported and locally-grown fruits for the French market. *Fruits*, 71(2), 93–104. doi: 10.1051/fruits/2015050.
Beccali, M., Cellura, M., Iudicello, M. and Mistretta, M. (2009), Resource Consumption and Environmental Impacts of the Agrofood Sector: Life Cycle Assessment of Italian Citrus-Based Products. *Environ. Manage.*, 43, 707–24.
Beccali, M., Cellura, M., Iudicello, M. and Mistretta, M. (2010), Life cycle assessment of Italian citrus-based products. Sensitivity analysis and improvement scenarios. *J. Environ. Manag.*, 91, 1415–28. doi: 10.1016/j.jenvman.2010.02.028.
Bessou, C., Basset-Mens, C., Tran, T. and Benoist, A. (2013), LCA applied to perennial cropping systems: a review focused on the farm stage. *Int. J. LCA*, 18, 340–61.
Bessou, C., Basset-Mens, C., Latunussa, C., Vélu, A., Heitz, H., Vannière, H. and Caliman, J. P. (2016), Partial modelling of the perennial crop cycle misleads LCA results in two contrasted case studies, *Int. J. LCA*, 21, 297–310. doi: 10.1007/s11367-016-1030-z.
Bhushan, C., Bhardwaj, A. and Misra, S. S. (2013), State of Pesticide Regulations in India, Centre for Science and Environment, New Delhi. Available at: http://www.cseindia.org/userfiles/paper_pesticide.pdf
Blanke, M. and Burdick, B. (2005), Food (miles) for Thought - Energy Balance for Locally-grown versus Imported Apple Fruit. *Environ. Sci. Pollut. R. – Int.*, 12, 125–7, 3 p. doi: 10.1065/espr2005.05.252.

Bung, P. (2013), Mango cultivation industry of India: problems and prospects. In International Conference on Management and Information Systems September 22, 422–34.

Cappelletti, G. M., Nicoletti, G. M. and Russo, C. (2010), Life Cycle Assessment (LCA) of Spanish-Style Green Table Olives. *Ital. J. Food Sci.*, 22, 3–14.

CEDEAO, (2011), Document d'Orientation Stratégique pour la chaine de valeur mangue dans la Communauté Économique Des États de l'Afrique de l'Ouest. Available at: http://www.intracen.org/uploadedFiles/intracenorg/Content/About_ITC/Where_are_we_working/Multi-country_programmes/Pact_II/Strategie_CEDEAO_mangue%20Final%20draft.pdf

Cerutti, A. K., Bagliani, M., Beccaro, G. L. and Bounous, G. (2010), Application of Ecological Footprint Analysis on nectarine production: methodological issues and results from a case study in Italy. *J. Clean. Prod.*, 18, 771–6. doi: 10.1016/j.jclepro.2010.01.009.

Cerutti, A. K., Bruun, S., Beccaro, G. L. and Bounous, G. (2011), A review of studies applying environmental impact assessment methods on fruit production systems. *J. Environ. Manag.*, 92(10), 2277–86.

Cerutti, A. K., Beccaro, G. L., Bruun, S., Bosco, S., Donno, D., Notarnicola, B. and Bounous, G. (2014), Life cycle assessment application in the fruit sector: state of the art and recommendations for environmental declarations of fruit products. *J. Clean. Prod.*, 73, 125–35.

Coltro, L., Mourad, A. L., Kletecke, R. M., Mendonça, T. A. and Germer, S. P. M. (2009), Assessing the environmental profile of orange production in Brazil. *Int. J. LCA*, 14, 656–64.

Comité Sahélien des Pesticides (CSP), (2015), Liste globale des produits pesticides autorisés dans les pays membres du Comité permanent Inter-Etats de Lutte contre la Sécheresse dans le Sahel (CILSS). Available at: http://www.reca-niger.org/IMG/pdf/liste_globale_pesticides_autorises_CSP_mai_2015.pdf

Do Carmo Mouco, M. A., Silva, D. J. and De Mello Prado, R. (2012), Mango Cultivation in Brazil, , in Mango, 2, Cultivation in Different Countries. STUDIUM PRESS LLC. ISBN: 1-933699-94-9, pp. 331–45.

Duyck, P. F., David, P. and Quilici, S. (2004), A review of relationships between interspecific competition and invasions in fruit flies (Diptera: Tephritidae). *Ecol. Entomol.*, 29(5), 511–20.

Dwivedi, P., Spreen, T. and Goodrich-Schneider, R. (2012), Global warming impact of Florida's Not-From-Concentrate (NFC) orange juice. *Agr. Syst.*, 108, 104–11. doi: 10.1016/j.agsy.2012.01.006.

EMEP/CORINAIR, (2006), Air pollutant emission inventory guidebook, Technical report No 11. Ed European Environment Agency (EEA), Copenhagen, Denmark.

EMEP/EEA, (2009). Air pollutant emission inventory guidebook, Technical report No 9, European Environment Agency (EEA), Copenhagen, Denmark.

European Commission, (2009), Directive 2009/125/EC of the European Parliament and of the Council of 21 October 2009 establishing a framework for the setting of ecodesign requirements for energy-related products (Text with EEA relevance).

European Commission, (2010), Directive 2010/30/EU of the European Parliament and of the Council of 19 May 2010 on the indication by labelling and standard product information of the consumption of energy and other resources by energy-related products (Text with EEA relevance).

European Commission - Joint Research Centre - Institute for Environment and Sustainability, (2011), International Reference Life Cycle Data System (ILCD) Handbook: Recommendations for Life Cycle Impact Assessment in the European context. EUR 24571 EN. European Commission, 159. http://doi.org/10.278/33030

Feschet, P. (2014), Analyse du Cycle de Vie Sociale. Pour un nouveau cadre conceptuel et théorique. Thèse de doctorat de l'Université Montpellier 1 – Sciences économiques. 298p.

Food SCP RT. (2013), ENVIFOOD Protocol, Environmental Assessment of Food and Drink Protocol. European Food Sustainable Consumption and Production Round Table (SCP RT) - Working Group 1.

Galán Saúco, V. (2011), Worldwide mango production and market: current situation and future prospects. *Acta Horticulturae*. 992, 37–48.

Galatola, M. and Pant, R. (2014), Reply to the editorial 'Product environmental footprint – breakthrough or breakdown for policy implementation of life cycle assessment?' written by Prof. Finkbeiner (Int. J. LCA, 19(2), 266–71). *Int. J. LCA*, 19, 1356–60. doi:10.1007/s11367-013-0678-x.

Garnett, T. (2011), Where are the best opportunities for reducing greenhouse gas emissions in the food system (including the food chain)? *Food Policy*, 36, 23–32.

Gazulla, C., Raugei, M. and Fullana-i-Palmer, P. (2010), Taking a life cycle look at Crianza wine production in Spain: where are the bottlenecks? *Int. J. LCA*, 15, 330–7.

Gerbaud, P. (2016), Mango (close-up). *Fruitrop* (English ed.), 239, 38–86.

Goedkoop, M., Heijungs, R., Huijbregts, M., De Schryver, A., Struijs, J. and Van Zelm, R. (2013), ReCiPe 2008, A life cycle impact assessment method which comprises harmonised category indicators at the midpoint and the endpoint level (version 1.08), Report I: Characterisation, 126 p.

Graefe, S., Tapasco, J. and Gonzalez, A. (2013), Resource use and GHG emissions of eight tropical fruit species cultivated in Colombia. *Fruits*, 68, 303–14. doi: 10.1051/fruits/2013075.

Guinée, J. B., Gorrée, M., Heijungs, R., Huppes, G., Kleijn, R., de Koning, A., van Oers, L., Wegener Sleeswijk, A., Suh, S., Udo de Haes, H. A., de Bruijn, H., van Duin, R. and Huijbregts, M. A. J. (2002), Life cycle assessment, An operational guide to the ISO standards, Centre of Environmental Science, Leiden University, Leiden, The Netherlands.

Guinée, J., Heijungs, R., and Huppes, G. (2011), Life Cycle Assessment: Past, Present, and Future†. *Environ. Sci. Tech.*, 45(1), 90–6. Retrieved from http://pubs.acs.org/articlesonrequest/AOR-UmwbwhSaJygz8KG4gc87

Haidara, Moussa, (2012). Mali - Etude Nationale Mangue, Centre international du Commerce. http://www.intracen.org/uploadedFiles/intracenorg/Content/About_ITC/Where_are_we_working/Multi-country_programmes/Pact_II/Etude%20national%20mangue%20-%20Mali.pdf

Hischier, R., Weidema, B., Althaus, H. J., Bauer, C., Doka, G., Dones, R., Frischknecht, R., Hellweg, S., Humbert, S., Jungbluth, N., Köllner, T., Loerincik, Y., Margni, M. and Nemecek, T. (2009), Implementation of life cycle impact assessment methods, Ecoinvent report No. 3, v2.1, Swiss Centre for Life Cycle Inventories, Dübendorf, Switzerland.

Ingwersen, W. W. (2012), Life cycle assessment of fresh pineapple from Costa Rica. *J. Clean. Prod.*, 35, 152–63. doi: 10.1016/j.jclepro.2012.05.035.

IPCC, (2006), IPCC Guidelines for National Greenhouse Gas Inventories, Prepared by the National Greenhouse Gas Inventories Programme, Eggleston, H. S., Buendia, L., Miwa, K., Ngara, T. and Tanabe, K. (eds), IGES, Japan.

Iriarte, A., Almeida, M. G. and Villalobos, P. (2014), Carbon footprint of premium quality export bananas: Case study in Ecuador, the world's largest exporter. *Sci. Total Environ.*, 472, 1082–8. doi: 10.1016/j.scitotenv.2013.11.072.

ISO, (2006a), ISO - 14040 : 2006 - Environmental management - Life cycle assessment - Principles and framework, Intec. Available at: http://www.mendeley.com/research/environmental-management-life-cycle-assessment-principles-and-framework/[Accessed 11 March 2012].

ISO, (2006b), ISO - 14044: 2006 - Environmental management - Life cycle assessment - Requirements and guidelines, Geneva: Intec. Available at: http://www.iso.org/iso/iso_catalogue/catalogue_tc/catalogue_detail.htm?csnumber=38498 [Accessed 11 March 2012].

ISO, (2005), ISO/DIS 14025 - Environmental labels and declarations - Type III environmental declarations - Principles and procedures.

Javaid, A., Shoaib, A. and Khan, S. N. (2012), Mango Cultivation in Pakistan, in Mango, 2, Cultivation in Different Countries. STUDIUM PRESS LLC. ISBN: 1-933699-94-9, pp. 385–94.

Jolliet, O., Saadé, M., Crettaz, P., and Shaked, S. (2010), Analyse du cycle de vie: Comprendre et réaliser un écobilan. Retrieved from http://books.google.com/books?hl=fr&lr=&id=g9S55CklsOoC&pgis=1

Koch, P. and Salou, T. (2015), Agribalyse®: Rapport méthodologique (v. 1.2). ADEME, 116p+annexes. www.ademe.fr/agribalyse-r-rapport-methodologique

Labouze, E., Mugdal, S., Cachia, F., Petiot, C., Schloesing, E., Faninger, T. and Gremont, M. (2012), Analyse des impacts environnementaux de la consommation des ménages et des marges de manœuvre pour réduire ces impacts. Rapport de l'étude 'Outils économiques d'incitation à la consommation durable: étude de positionnement (environnementale, économique, sociologique) et élaboration de propositions', préparé pour l'ADEME, Service Économie et Prospective. ADEME, 289p.

Liu, Y., Langer, V., Høgh-Jensen, H. and Egelyng, H. (2010), Life cycle assessment of fossil energy use and greenhouse gas emissions in Chinese pear production. *J. Clean. Prod.*, 18, 1423–30.

Lo Giudice, A., Mbohwa, C., Clasadonte, M. T. and Ingrao, C. (2013), Environmental assessment of the citrus fruit production in Sicily using LCA. *Ital. J. Food Sci.*, 25, 202–12.

Mehdi, M., Ahmad, B., Yaseen, A., Adeel, A. and Sayyed, N. (2016), A comparative study of traditional versus best practices mango value chain. *Pak. J. Agri. Sci.*, 53(3), 733–42. Available at: http://www.pakjas.com.pk/papers%5C2626.pdf

Milà i Canals, L., Burnip, G. M. and Cowell, S. J. (2006), Evaluation of the environmental impacts of apple production using Life Cycle Assessment (LCA): Case study in New Zealand. *Agr. Ecosyst. Environ.*, 114, 226–38.

Mithraratne, N., McLaren, S. and Barber, A. (2008), Carbon Footprinting for the Kiwifruit Supply Chain – Report on Methodology and Scoping Study, 63p.

Mouron, P. J. P., Nemecek, T., Scholz, R. W. and Weber, O. (2006a), Management influence on environmental impacts in an apple production system on Swiss fruit farms: Combining life cycle assessment with statistical risk assessment. *Agr. Ecosyst. Environ.*, 114, 311–22.

Mouron, P. J. P., Nemecek, T., Scholz, R. W. and Weber, O. (2006b), Life cycle management on Swiss fruit farms: Relating environmental and income indicators for apple-growing. *Ecol. Econ.* 58, 561–78.

Mouron, P., Heijne, B., Naef, A., Strassemeyer, J., Hayer, F., Avilla, J., Alaphilippe, A., Hohn, H., Hernandez, J., Mack, G., Gaillard, G., Sole, J., Sauphanor, B., Patocchi, A., Samietz, J., Bravin, E., Lavigne, C., Bohanec, M., Golla, B., Scheer, C., Aubert, U. and Bigler, F. (2012), Sustainability assessment of crop protection systems: SustainOS methodology and its application for apple orchards. *Agr. Syst.*, 113, 1–15.

Namuco, L. O. (2012), Mango in the Philippines, in Mango, 2, Cultivation in Different Countries. STUDIUM PRESS LLC. ISBN: 1-933699-94-9, pp. 125–44.

Nemecek, T. and Kägi, T. (2007), Life cycle inventories of Swiss and European agricultural production systems. Final report ecoinvent V2.0 No. 15a, Agroscope Reckenholz-Taenikon Research Station ART, Swiss Centre for Life Cycle Inventories, Zürich and Dübendorf, Switzerland, Retrieved from: www. ecoinvent.ch.

Nemecek, T., Bengoa, X., Lansche, J., Mouron, P., Riedener, E., Rossi, V., and Humbert, S. (2015), World Food LCA Database - Methodological Guidelines for the Life Cycle Inventory of Agricultural Products, Version 3.0. July 2015.World Food LCA Database (WFLDB). Quantis and Agroscope, Lausanne and Zurich, Switzerland.

OECD, (2011), International Standards for Fruit and Vegetables. Mangoes E-Brochure, 70p. Available at: http://www.oecd.org/tad/code/49200564.pdf

Page, G., Kelly, T., Minor, M. and Cameron, E. (2011), Modeling Carbon Footprints of Organic Orchard Production Systems to Address Carbon Trading: An Approach Based on Life Cycle Assessment. *Hort. Sci.*, 46, 324–7.

Payen, S., Basset-Mens, C. and Perret, S. (2015), LCA of local and imported tomato: An energy and water trade-off. *Journal of Cleaner Production* 87, 139–48. doi: 10.1016/j.jclepro.2014.10.007.

Pergola, M., D'Amico, M., Celano, G., Palese, A. M., Scuderi, A., Di Vita, G., Pappalardo, G. and Inglese, P. (2013), Sustainability evaluation of Sicily's lemon and orange production: An energy economic and environmental analysis. *J. Environ. Manag.*, 128, 674–82.

Pizzigallo, A. C. I., Granai, C. and Borsa, S. (2008), The joint use of LCA and emergy evaluation for the analysis of two Italian wine farms. *J. Environ. Manage.*, 86, 396–406.

Prasad, R., Ranjan, K. R. and Sinha, A. K. (2006). AMRAPALIKA: An expert system for the diagnosis of pests, diseases, and disorders in Indian mango. *Knowledge-based Systems*, 19(1), 9–21. Available at: http://www.sciencedirect.com/science/article/pii/S0950705105000845

Roibás, L., Elbehri, A. and Hospido, A. (2016), Carbon footprint along the Ecuadorian banana supply chain: methodological improvements and calculation tool. *J. Clean. Prod.*, 112, 2441–51. doi: 10.1016/j.jclepro.2015.09.074.

Sanjuán, N., Úbeda, L., Clemente, G., Mulet, A. and Girona, F. (2005), LCA of integrated orange production in the Comunidad Valenciana (Spain). *Int. J. Agr. Res. Gov. Ecol.*, 4(2), 163–77.

Smith, P. and Gregory, P. J. (2013), Climate Change and Sustainable Food Production. *Proceedings of the Nutrition Society*, 72 (01), 21–8. doi:10.1017/S0029665112002832.

Strapatsa, A. V., Nanos, G. D. and Tsatsarelis, C. A. (2006), Energy flow for integrated apple production in Greece. *Agr. Ecosyst. Environ.*, 116, 176–80.

Svanes, E. and Aronsson, A. K. S. (2013), Carbon footprint of a Cavendish banana supply chain. *Int. J. LCA*, 18, 1450–64. doi: 10.1007/s11367-013-0602-4.

Sim, S., Barry, M., Clift, R. and Cowell, S. J. (2007), The relative importance of transport in determining an appropriate sustainability strategy for food sourcing. *Int. J. LCA*, 12, 422–31.

Singh, R. N. (1990), Mango, Indian Council of Agricultural Research, New Delhi (India), p. 21, 23.

Strebelle, J. (2013), Analyse et propositions sur la construction des marchés locaux nationaux régionaux en Afrique – Analyse complémentaire. Position des OP dans la filière mangue en Afrique de l'Ouest et au Sénégal. Coopération de l'Union européenne EuropAid - Coopération belge au Développement.http://www.csa-be.org/IMG/pdf_Ann_2_Rapport_filiere_mangue_Afrique_Occ_Senegal_24oct12_ORIGINAL.pdf

Thomas, C. G. (2012), Climat, Soil & Agronomy, in Mango, 1, Production and Processing Technology tries. STUDIUM PRESS LLC. ISBN: 1-933699-93-0, pp. 309–19.

Trydeman Knudsen, M., Fonseca de Almeida, G., Langer, V., Santiago de Abreu, L. and Halberg, N. (2011), Environmental assessment of organic juice imported to Denmark: a case study on oranges (Citrus sinensis) from Brazil. *Org. Agr.*, 1, 167–85. doi: 10.1007/s13165-011-0014-3.

USAID-KAVES, (2015), Kenyan Mango Value Chain Analysis. 53p. Available at: http://pdf.usaid.gov/pdf_docs/PA00M2SZ.pdf

Van Melle, C. and Buschmann, S. (2013), Analyse comparative des chaînes de valeur de la mangue au Bénin, au Ghana et au Burkina Faso. In: Reconstruire le potentiel alimentaire de l'Afrique de l'Ouest, Elbehri, A. (ed.), FAO/FIDA. Available at: http://www.fao.org/docrep/018/i3222e/i3222e10.pdf

Vannière, H., Didier, C., Rey, J. Y., Diallo, T. M., Kéita, S. and Sangaré, M. (2007), The mango in French-speaking West Africa: cropping systems and agronomical practices. *Fruits*, 62(3), 187–201. doi.org/10.1051/fruits:2006051

Volavi, S. G., Rajmohan, K., Govil, J. N., Peter, K. V. and Thottappily, G. (2012), Mango, 2, Cultivation in Different Countries. STUDIUM PRESS LLC. ISBN: 1-933699-94-9.

Part 4

Diseases and pests

Chapter 16

Integrated disease management in mango cultivation

Randy C. Ploetz, University of Florida, USA

1 Introduction

Mango, *Mangifera indica* L., is affected by a great number of fruit, foliar, stem and root diseases (Lim and Khoo, 1985; Singh, 1968; Cook, 1975; Snowdon, 1990; Ploetz et al., 1994; Ploetz, 2003; Ploetz and Freeman, 2009; Prakash and Srivastava, 1987; Ridgeway, 1989). In this chapter, diseases that seriously impact the crop are covered. Their significance, geographical distribution and history are outlined, and the symptoms, causal agent(s) and epidemiology of each are detailed with emphases on their management. Insect pests, nutritional disorders and other non-disease items are not considered.

Diseases of mango are caused mainly by eukaryotes (Domain Eukaryota), among which the true fungi, Eumycota (Ascomycota and Basidiomycota), are most important. Other eukaryotic pathogens of this crop include the fungus-like oomycetes (Oomycota), nematodes (Metazoa), parasitic plants and green algae (Plantae). Only two mango pathogens are prokaryotes in the Domain Eubacteria, both of which are Gram-negative γ-proteobacteria. No mango diseases are known to be caused by protozoa (Eukaryota),

http://dx.doi.org/10.19103/AS.2017.0026.19

α- and β-proteobacteria, Mollicutes and Firmicutes (Eubacteria), or nucleic acid-based plant pathogens, that is, viruses and viroids.

2 Fruit diseases: anthracnose

Diseases affect mango fruit at all stages of development. However, due to the monetary investments made prior to harvest, post-harvest diseases are of the greatest concern. This section focuses on anthracnose. Other fruit diseases are then covered in Sections 3 and 4.

Anthracnose is the most important disease of mango in humid production areas (Lim and Khoo, 1985; Dodd et al., 1997; Arauz, 2000; Ploetz, 2003, 2017; Ploetz and Freeman, 2009). Losses of fruit occur in the field, but the post-harvest losses caused by this disease are most significant. Anthracnose can also be a serious problem that affects the foliage and flowers.

2.1 Symptoms

Small fruit can develop minute brown spots and abort. The development of infections usually stops when the pathogen forms an appressorium (an infection structure). These latent infections resume development when the concentrations of fungal inhibitors known as resorcinols decline in the fruit during the ripening process. Lesions can form anywhere on larger (especially ripening) fruit, but linear smears that radiate from the stem end to the apex are common (Fig. 1). The lesions on fruit are superficial and extend into the flesh only after large portions of the fruit surface are affected. However, even superficial disease development results in aesthetic damage and rejection of fruit along the marketing chain.

Figure 1 Anthracnose on 'Edward'. Note the development of a linear smear of lesions that emanate from the pedicle and stem end (photo: R. C. Ploetz).

Necrotic flowers abscise on the panicles, leaving persistent peduncles (Fig. 2). Small, circular dark spots also develop on the pedicles and peduncles. Lesions may enlarge and coalesce to form large patches of necrotic, dark brown tissue, where the terminal portions of the peduncle form a 'shepherd's crook' (Fig. 2). The salmon- to orange-coloured fructifications of the pathogen develop on affected tissues after sufficient rainfall (Fig. 3). The lesions are dark brown on the leaves and surrounded by chlorotic haloes with irregular, rounded margins that are not delimited by veins. The lesions reach 0.5–1.0 cm in diameter on the mature leaves but can expand further on young leaves.

2.2 Aetiology

Anthracnose of mango is caused by species of *Colletotrichum*. The identification of *Colletotrichum* spp. has typically relied on morphology and host range (Freeman et al., 1998; Du et al., 2005). Currently, DNA sequences are used to identify cryptic species within the *C. acutatum*, *C. boninense*, *C. gloeosporioides* and *C. siamense* species complexes (de Souza et al., 2016; Du et al., 2005; Fitzell, 1979; Pardo-De la Hoz et al., 2016; Ploetz,

Figure 2 Blossom blight caused by *Colletotrichum gloeosporioides* s.l. (photo: R.C. Ploetz), and (inset) 'Shepherd's crook' symptom on a panicle after artificial inoculation with *Colletotrichum asianum* (photo: T. Tarnowski).

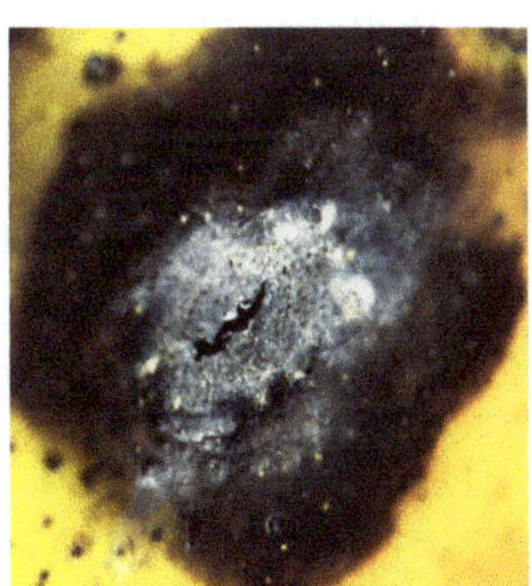

Figure 3 Sporulation of *Colletotrichum gloeosporioides* s.l. on the surface of a mango fruit (photo: R.C. Ploetz).

2003; Prakash, 1990; Shivas et al. 2016; Snowdon, 1990; Taba et al., 2004; Tarnowski, 2009; Udayanga et al., 2013; Viera et al., 2014; 2015; Afanador-Kafuri et al., 2003; Du et al., 2005; Damm et al., 2012; Weir et al., 2012).

As the ability to discern phylogenetic (cryptic) species has improved, so has our understanding of this disease on mango. *Colletotrichum gloeosporioides* s.l. is most important (e.g. *C. asianum*, *C. fructicola*, *C. gloeosporioides* (*sensu stricto*), *C. queenslandicum*, *C. theobromicola* and *C. tropicale*). Other species complexes play secondary roles; they include *C. acutatum* (*C. fioriniae* and *C. simmondsii*), *C. boninense* (*C. cliviae* and *C. karstii*), *C. siamense* (*C. dianesei* (syn. *C. melanocaulon*) and *C. endomangiferae*).

A mango biotype in the *Colletotrichum gloeosporioides* complex, *C. asianum*, may be most important. To date, *C. asianum* has been reported in Australia, Brazil, Florida (USA), Ghana, Mexico, Panama, Philippines, South Africa and Thailand (Honger et al., 2014; Lima et al., 2013; Sharma et al., 2013; Tarnowski, 2009; Udayanga et al., 2013; Weir et al., 2012). In field experiments in Florida, it was the only species in the complex that was recovered from latently infected (attached preclimacteric and asymptomatic) fruit (Tarnowski, 2009; S. Rehner, USDA-ARS, Beltsville, MD USA, 2008, pers. comm.). And in controlled experiments in which attached developing organs were protected from outside inoculum after inoculation, *C. asianum* was the only species in the complex that caused blossom blight (Fig. 2) and lesions on mango leaves. Generalist species in the complex caused disease on ripe, climacteric fruit, but not on leaves or blossoms of mango.

Colletotrichum asianum may be the only species that fits the 'typical' latent infection model for this disease (Arauz, 2000). It should be noted that previous reports, which indicated a broad host range for species in the *C. gloeosporioides* species complex, were conducted with detached leaves or fruit (e.g. de Souza et al., 2013). Studies with detached organs are poor surrogates for what would be expected in the field. For example, isolates of *Colletotrichum* that were capable of forming lesions on detached leaves of *Arabidopsis* were incapable of doing so on attached leaves of that host plant (Liu et al., 2007). Clearly, more work is needed on the host and organ specificity of different species in the *C. gloeosporioides* species complex on mango.

Colletotrichum gloeosporioides s.l. causes diseases on a large number of subtropical and tropical hosts (Jeffries et al., 1990; Freeman et al., 1998). Cultures of the fungus on potato dextrose agar (PDA) are greyish white to dark grey and usually produce an aerial mycelium that varies from a thick mat to sparse tufts (Holliday, 1980). In general, *C. gloeosporioides* s.l. produces longer and narrower conidia than *C. acutatum* s.l., as well as circular rather than lobed hyphopodia. Although hyphopodia have been used to distinguish isolates of *C. gloeosporioides* s.l. and *C. acutatum* s.l. (Du et al., 2005), they provided ambiguous results in Florida (Palmateer et al., 2006; Tarnowski, 2009).

2.3 Epidemiology

Anthracnose affects the leaves, flowers and fruit, and inocula are present all year round in the canopy (Fig. 4). Management requires an awareness of the sources of inoculum and the weather conditions that promote infection and disease development. Moist conditions and high humidity are the primary factors that affect the development of anthracnose and spread of the casual agents (Fitzell and Peak, 1984; Dodd et al., 1991). Conidia produced on branch terminals, mummified inflorescences, flower bracts and leaves are significant

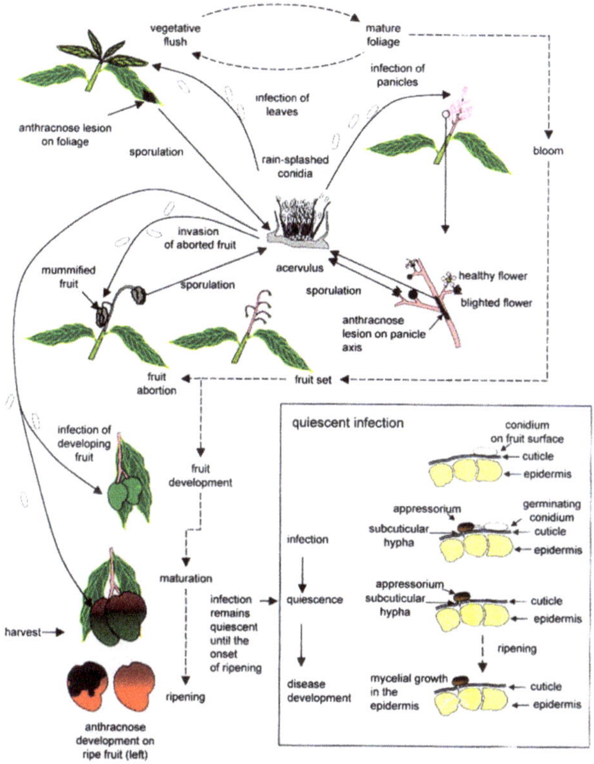

Figure 4 Anthracnose disease cycle (Arauz, 2000).

sources of inocula (Dodd et al., 1991; Fitzell and Peak, 1984). They are produced most abundantly when free moisture is available, but also at relative humidities as low as 95%. Conidia are dispersed by rain splash and infection requires free moisture (Jeffries et al., 1990).

2.4 Management

The fruit of different mango cultivars vary considerably in their resistance to anthracnose (Table 1), but even the most tolerant must be protected by fungicides in humid environments (Lim and Khoo, 1985; Jefferies et al., 1990). During commercial production, these diseases are managed using diverse chemicals, the registration of which varies in different production areas. Export quality fruit can often be produced only in arid production areas in which disease pressure is low (Arauz, 2000).

Optimum disease control relies on an integrated approach that holistically combines the best measures, which depend on the cultivar, production location and market destination. Pre-harvest management of anthracnose relies on 1) orchard sanitation (removing sources of inocula), 2) altering the time of flowering to ensure that fruit set and development occur during dry conditions and 3) an integration of these with chemical and biological measures (Johnson and Hofman, 2009).

Table 1 Susceptibility of the fruit of different mango cultivars to anthracnose

Highly susceptible	Susceptible	Moderately resistant
'Irwin'	'Brooks'	'Carrie'
'Kent'	'Bullocks Heart'	'Earlygold'
	'Fascell'	'Edward'
	'Haden'	'Florigon'
	'Lippens'	'Glenn'
	'Palmer'	'Julie'
	'Sensation'	'Keitt'
	'Zill'	'Tommy Atkins'
		'Van Dyke'

Despite its potential benefits, sanitation is often not practised due to its difficulty and expense (Akem, 2006; Prusky et al., 2009). Furthermore, flowering cannot be manipulated in all situations (Johnson and Hofman, 2009). Flowering is promoted with applications of KNO_3, although this is not effective in the subtropics or on all cultivars (e.g. Kensington Pride) and does not alter its timing on other cultivars (e.g. Kent). The growth retardant paclobutrazol is also used to alter flowering but is not registered in the United States and other areas.

Pre-harvest management of anthracnose usually relies on fungicides. Pre-harvest anthracnose control focuses solely on protecting the flowers and fruit in all but the most disease-conducive environments and on the most susceptible cultivars. In moist environments, this entails one or two fungicide applications during flowering and early fruit set, with subsequent fungicide applications before harvest; thus, applications are needed throughout the season. Experience with a given cultivar in a given environment will indicate the extent to which fungicides are needed. In general, pre-harvest fungicides are not necessary where dry conditions prevail (Arauz, 2000).

Disease forecasting

Two anthracnose forecasting models have been developed to schedule, and reduce, fungicide applications (Fitzell et al., 1984; Dodd et al., 1991). Akem (2006) noted differences between the time each model predicted was needed for infection; he suggested to use caution when a model was used in an area other than where it was developed. Forecasting would be most useful in seasonally dry situations (where infection occurs only after significant rainfall) (Arauz, 2000). Calendar-based application schedules are needed wherever regular rainfall occurs.

Fungicides

Fungicide use is constrained by the limited number of products that are available and efficacious, and the pesticide regulations that exist in the producing and destination countries. In general, copper fungicides have the widest acceptance. However, they are usually not very effective unless they are applied with other fungicides. For example, monthly applications of copper oxychloride combined with mancozeb are effective for most post-harvest diseases in South Africa (Lonsdale and Kotze, 1993). However, the registration of dithiocarbamate fungicides, such as mancozeb, varies among production areas. Another contact fungicide, chlorothalonil, is effective but phytotoxic to fruit larger than a golf ball (Fig. 5); thus, it should not be used after early fruit set.

Few systemic fungicides are available. The benzimidazoles, primarily benomyl and carbendazim, provided excellent anthracnose control before resistance to them developed (Akem, 2006). Two imidazoles, prochloraz and imazalil, are used in some countries for pre- and post-harvest anthracnose control, respectively. The imidazoles are moderately effective against anthracnose, but they are ineffective against stem-end rot, which is managed by thiabendazole. The stobilurins are effective against anthracnose and several other post-harvest diseases, but must be used sparingly to avoid the development of fungicide resistance. No more than three stobilurin applications should be made per season, preferably alternating or combining with fungicides that have a different mode of action (Brent and Hollomon, 2007). Johnson and Hofman (2009) suggested that one or two fungicide applications should be made during flowering and early fruit set, with two additional applications at 21 and 7 days prior to harvest.

Fungicide applications usually focus on reducing damage to fruit, but foliar disease control is indicated in some situations and on inflorescences in most situations. In nurseries, trees usually require protection if they are crowded or receive overhead irrigation. Since infected foliage and branch terminals are important reservoirs of inoculum, fruit set and anthracnose control on fruit are enhanced if applications are made prior to flowering (Jefferies et al., 1990). Off-season control measures are especially beneficial in production environments that receive significant rainfall.

Figure 5 Phytotoxicity caused by application of chlorothalinil on small mango fruit. Although this fungicide is effective against anthracnose on smaller fruit, it should not be used once fruit reach this size (photo: R.C. Ploetz).

Induced resistance

Increasing the natural defence responses to this disease has been investigated. Fruit anthracnose has been reduced by salicylic acid, an analogue of benzothiadiazole (BTH) (= acibenzolar-*S*-methyl = Bion), and ultraviolet (UV-C) irradiation, albeit inconsistently (Zainuri et al., 2001; Zeng and Waibo, 2005; Zainuri, 2006; Zeng et al., 2006; Karunanayake, 2007).

Resistance

The use of tolerant cultivars can reduce anthracnose severity. Even highly susceptible cultivars (e.g. 'Irwin' and 'Kent') can be produced in arid locations, but no commercial cultivar is sufficiently resistant to be produced for export in humid environments (Dodd et al., 1997). Where susceptible cultivars are produced in arid locations, unexpected rainfall (e.g. periodic El Niño events in north-western Peru) can cause serious outbreaks of anthracnose.

Previously, resorcinols (preformed antifungal inhibitors) were associated with resistance of unripe fruit to *Alternaria* black spot and anthracnose (Droby et al., 1986; Hassan et al., 2007). As fruit ripen resorcinol concentrations decline, symptom development starts. Recently, additional factors have been investigated as contributors to anthracnose resistance in fruit of different mango cultivars, including latex retention in the fruit pedicel, and the concentrations of chitinase and gallotannins in the fruit peel (Karunanayake et al., 2014, 2015).

Fruit with elevated resorcinol levels may be more resistant to anthracnose (latent infection). However, resistant lines with increased levels of resorcinols may pose risks, as some consumers develop mango dermatitis when exposed to these compounds (Knodler et al., 2009). Better understandings are needed for the influence of different fruit constituents in anthracnose susceptibility and whether anthracnose resistance is a reasonable short- or long-term breeding objective for this crop. Likewise, additional work is warranted to determine the impact of different species in the *C. gloeosporioides* species complex on different mango organs and on pre- and post-harvest disease development on fruit.

Non-fungicidal measures

Since there is a close relationship between ripening and post-harvest disease, the latter can be managed indirectly by delayed onset and reduced rates of ripening (Prusky and Keen, 1993). Mango is a climacteric fruit that undergoes profound biochemical changes as it ripens. Ripening is a fruit senescence process that is associated with, and enhanced by, increased ethylene production (Snowdon, 1990; Brecht and Yahia, 2009).

During the ripening process, ethylene levels in fruit increase naturally from <0.1 $\mu L\ kg^{-1}\ h^{-1}$ to 1–3 $\mu L\ kg^{-1}\ h^{-1}$. Mature fruit can be stored in the unripe state if the climacteric initiation of ethylene production is prevented. External sources of ethylene, such as ripening fruit, smoke and engine exhaust fumes, should be removed from storage environments. The climacteric rise in ethylene production can be delayed with refrigeration, but mango is sensitive to chilling injury and most cultivars must be stored at >10–13°C. Conversely, ripening can be initiated in unripe fruit with concentrations of exogenous ethylene as low as 0.005 $\mu L\ L^{-1}$.

Ripening and post-harvest disease can be inhibited by modified atmosphere (MA) storage (usually reduced O_2 and increased CO_2) (Brecht and Yahia, 2009; Prusky et al.,

2009; Kobiler et al., 1998). For example, when exposed to 30% CO_2 for 24 h, mangos developed less disease upon ripening. However, because fruit flavour is affected in atmospheres with <1% O_2 or >15% CO_2 and more extreme concentrations of O_2 and CO_2 are needed to affect disease development, MA is generally not used to manage anthracnose (Burg, 2004).

Hypobaric storage is superior to MA for extending the unripe, post-harvest life of mango fruit. It has been used to suppress the post-harvest development of anthracnose in papaya, although its use for the long-distance shipment of mango is hindered by technical constraints (Burg, 2004).

Heat

Hot water (most common), vapour heat and forced hot air are used to control fruit flies, which are often quarantine pests in mango fruit (Jacobi et al., 2001). For example, the Mediterranean, Mexican and related species of fruit flies must be controlled in fruit that will be sold in the United States (McGuire, 1991). An added benefit of heat treatment for fruit fly control is the reduction of anthracnose and other post-harvest decays (McGuire, 1991).

Biological control

Relatively little research has been conducted on the biological control of anthracnose. *Bacillus licheniformis* resisted desiccation and is food-safe, but caused minor reductions in disease at 10°C and 25°C, either alone or in combination with fungicides (Govender and Korsten, 2006). Reductions have also been achieved with Gram-negative bacteria and other amendments (Vivekananthana et al., 2004). To date, no biocontrol measure has been as effective as the most effective fungicides.

3 Fruit diseases: bacterial black spot (black canker)

Bacterial black spot (BBS) is a destructive leaf, stem and fruit disease in many mango-production areas (Gagnevin and Pruvost, 2001). In India, the disease is known as bacterial canker because of the cankers it causes on the stems of some cultivars (Prakash et al., 1994). This can be the most important mango disease in areas where those caused by fungi are well managed (Gagnevin and Pruvost, 2001).

Until 2013 (Yasuhara-Bell et al., 2013), BBS had been reported only in the following locations in the Eastern Hemisphere: Australia (New South Wales, Northern Territory and Queensland), Burma (Myanmar), China, the Comoros, Egypt, India (Andhra Pradesh, Bihar, Delhi, Haryana, Karnataka, Kerala, Madhya Pradesh, Maharashtra, Tamil Nadu and Uttar Pradesh), Japan (Bonin Islands and Ryukyu Archipelago), Kenya, Malaysia, Mauritius, New Caledonia, Pakistan, the Philippines, Réunion, Rodrigues, South Africa, Sudan, Taiwan, Thailand and the United Arab Emirates (Fukuda et al., 1990; Pruvost et al., 1992; Prakash et al., 1994; Gagnevin and Pruvost, 1995; Kishun, 1995; Gagnevin and Pruvost, 2001; Ah-You et al., 2007b; CABI/EPPO, 2012). Other locations in the Western Hemisphere in which the disease had been reported (Central America, Caribbean and South America) (CABI/EPPO, 2012) require confirmation (Ah-You et al., 2007a).

Ploetz and Freeman (2009) suggested that BBS should have an even wider distribution given the ease with which the BBS pathogen is disseminated in propagation materials. Beginning in 2011, BBS was reported in Burkina Faso and Ghana with reports soon after in Benin, Ivory Coast and Mali (Pruvost et al., 2011a, b, 2012, 2014; Zombre et al., 2015, 2016). Clearly, neighbouring areas in West Africa are at risk. Recently, BBS was confirmed for the first time in the Western Hemisphere in Hawaii (Yasuhara-Bell et al., 2013) and later in Florida (Sanahuja et al., 2016). The appearance of BBS in Florida poses a threat to mango production in tropical America.

3.1 Symptoms

Leaves, stems and fruit of mango are affected (Gagnevin and Pruvost, 2001; Manicom and Pruvost, 1994). Water-soaked spots on leaves are initially 1–3 mm in diameter, become raised, black and angular as they enlarge, and are delimited by veins and surrounded by chlorotic haloes. These lesions are larger and raised more than those caused by other xanthomonads that are isolated from related species in the Anacardiaceae (Fig. 6) (Ah-You et al., 2007a). Under conducive, wet conditions, BBS lesions can merge to form large necrotic patches from which bacteria ooze. Old lesions become dry, white or grey, and crack; in severe cases trees defoliate. Anthracnose lesions are not raised or as black and angular as those caused by BBS. BBS lesions on branches are dark and cracked. They develop only on highly susceptible cultivars and are often associated with wounds. The most conspicuous symptoms of BBS are bleeding, star-shaped lesions that develop on fruit (Fig. 7).

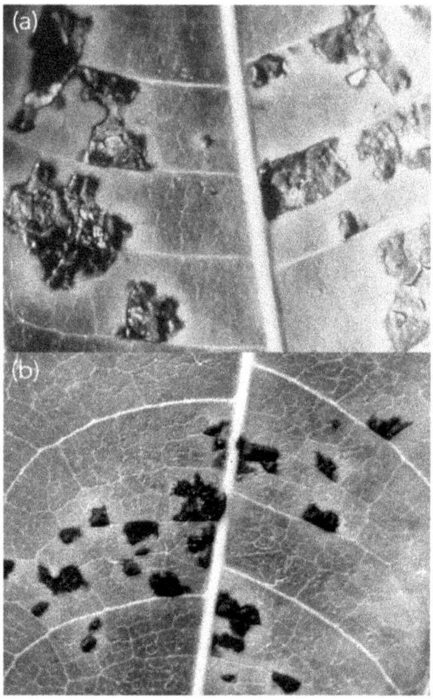

Figure 6 (a) Symptoms of bacterial black spot on the underside of a mango leaf caused by *Xanthomonas citri* pv. *mangiferaeindicae* (photo: O. Pruvost), and (b) symptoms of bacterial spot on the underside of a mango leaf caused by a yellow-pigmented xanthomonad (photo: R.C. Ploetz). The bacterial black spot lesions are larger and more raised than those caused by bacterial spot.

Figure 7 Bleeding, star-shaped bacterial black spot lesions, caused by *Xanthomonas citri* pv. *mangiferaeindicae*, on the surface of a 'Springfels' mango fruit in Florida (photo: R.C. Ploetz).

3.2 Aetiology

Diverse xanthomonads have been recovered from mango and other plants in the Anacardiaceae, although only some of these cause typical symptoms of BBS (Gagnevin and Pruvost, 2001; Ah-You et al., 2007a). Early reports that this disease was caused by *Pseudomonas mangiferaeindicae* (Patel et al., 1948) and *Erwinia mangiferae* (Steyn et al., 1974) were erroneous. The pathogen's placement in *Pseudomonas* may have been due to its production of non-pigmented colonies in culture [*Pseudomonas syringae* pv. *syringae* causes a different disease in the mango, that is, apical necrosis (see below)]. Atypically, the BBS pathogen does not produce the yellow pigment, xanthomonadin, that is characteristic for *Xanthomonas* (Midha and Patil, 2014). Cook (1975) indicated that *E. mangiferae* was a saprophyte that reached high populations in old lesions.

Pathological, cultural, biochemical, physiological, serological and genetic data indicate that strains of the pathogen from different production areas around the world are diverse (Sanders et al., 1994; Gagnevin and Pruvost, 1995; Kishun, 1995; Gagnevin and Pruvost, 2001; Pruvost et al., 2005). Genetic diversity is highest among strains from Southeast Asia (Gagnevin and Pruvost, 1995).

The pathogen has a single flagellum, is Gram-negative and rod-shaped and measures 0.4–0.5 × 1.0–1.5°µm (Manicom and Wallis, 1984). Its colonies are cream-coloured on artificial media. Yellow-pigmented xanthomonads have been recovered from mango in Brazil, Réunion, South Africa and the United States (Florida). These strains cause flat leaf lesions, do not cause fruit or stem lesions, and are not classified as pv. *mangiferaeindicae* (Ah-You et al., 2007a).

Three genetically and pathologically distinct groups of bacteria were identified from different geographic regions and hosts in the Anacardiaceae by Ah-You et al. (2007a). Group I strains from the Old World multiplied in mango and cashew (*Anacardium occidentale*), and were in AFLP group 9.5 of *X. axonopodis*, which contained strains that produce typical BBS symptoms on mango. These strains of *X. campestris* pv. *mangiferaeindicae* s.l. were re-described as *X. axonopodis* pv. *mangiferaeindicae* sensu novo (s.n.) (Ah-You et al., 2007a). Group II strains from Brazil multiplied in cashew but not in mango, and were in AFLP group 9.6 of *X. axonopodis*. They were associated with symptoms on mango that differed from those of BBS, including brown, flat lesions on leaves, and black, depressed lesions on the fruit of a few cultivars that was periodically associated with pulp rot. These strains were responsible for previous erroneous reports of BBS in Brazil (Gagnevin and Pruvost, 2001; Ah-You et al., 2007a). Group III strains were responsible for a unique syndrome on ambarella, *Spondias dulcis*, and mombin, *S. mombin*, in the French West Indies, and they were in AFLP group 9.4. Groups II and III were described as *X. axonopodis* pv. *anacardii* and *X. axonopodis* pv. *spondiae*, respectively (Ah-You et al., 2007a). With additional genetic data, Ah-You et al. (2009) subsequently re-described *X. axonopodis* pv. *mangiferaeindicae* as *X. citri* pv. *mangiferaeindicae* (Xcm). Diagnostic PCR amplicons of two housekeeping genes, *efp* and *dnaK*, can be used to identify Xcm (Bui Thi Ngoc et al., 2010).

3.3 Epidemiology and management

Xcm is an epiphytic colonist of leaves (Manicom, 1986; Pruvost et al., 1990, 2009), buds (Pruvost et al., 1993) and fruit (Pruvost and Luisetti, 1991b). Infection requires free moisture and occurs via wounds and, less often, stomata on old leaves (Gagnevin and Pruvost, 2001). High humidity (>90% RH) and moderate temperatures (25–30°C) favour disease development (Kishun and Sohi, 1983; Pruvost and Luisetti, 1991b). There is a direct relationship between the level of disease that develops on the leaves and fruit (Manicom, 1986; Pruvost et al., 1990). Thus, leaf susceptibility can be an important criterion when selecting cultivars for lower fruit susceptibility (Pruvost and Luisetti, 1991b).

Resistance to BBS varies greatly among mango cultivars (Manicom and Pruvost, 1994). Pathogen-free planting material should be utilized when new orchards are established. The pathogen moves only short distances in wind-blown aerosols (usually within orchards) (Gagnevin and Pruvost, 2001), and long-distance dissemination occurs almost entirely via infected propagation material and less frequently in surface-contaminated seeds (Manicom and Pruvost, 1994). Windbreaks should be used to reduce wounding and infected twigs should be removed from the canopy.

BBS can be difficult to control on susceptible cultivars, as the available chemicals are marginally effective (Pruvost et al., 1989). During rainy weather, applications of copper-based bactericides are recommended. Their application should focus on protecting fruit and account for the length of time fruit are exposed to wet (infective) conditions (Manicom and Pruvost, 1994). Agricultural antibiotics, such as streptomycin sulphate or nitrate, have been effective (Misra and Prakash, 1992; Viljoen and Kotze, 1972), but their long-term effectiveness is reduced by resistance that develops after continued use.

Biological control of BBS has been researched infrequently. Pruvost and Luisetti (1991a) reported little success in preliminary work. Kishun (1994) indicated that a strain of *Bacillus coagulans* from the phylloplane of the mango affected strains of the pathogen, but did not report BBS control in the field.

4 Other fruit diseases

4.1 *Alternaria* diseases

Alternaria alternata causes black spot on fruit, *Alternaria* leaf spot and lesions on inflorescences (Prusky et al., 1983; Cronje et al., 1990). These diseases are most prevalent

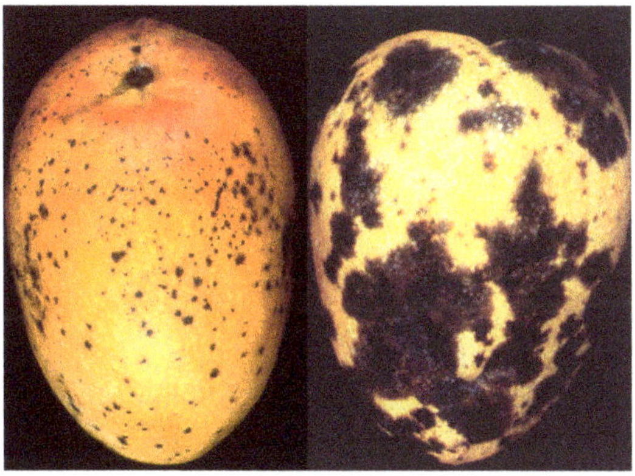

Figure 8 Initial (left) and advanced (right) symptoms of *Alternaria* rot in mango. This damage is darker and harder than that caused by anthracnose (photos: D. Prusky).

in arid environments, which apparently favour *A. alternata* over other pathogens that require higher levels of moisture (Prusky et al., 1983).

Round, dark spots, 1–3 mm in diameter, usually occur on the undersides of leaves (Prusky, 1994). Inflorescence disease can reduce fruit set when severe (Cronje et al., 1990). Lesions also develop around the lenticels on fruit (Fig. 8); they usually start near the stem end and can expand and merge to cover much of the fruit surface. Diseased areas of fruit do not soften or extend more than 1–2 mm into the flesh until late in their development. The lesion centres are depressed and are covered with olive brown spores of the pathogen under moist conditions. These lesions are more restricted, darker and harder than those caused by anthracnose.

Woudenberg et al. (2015) examined members of this large genus with genomic and transcriptomic data. They reported that *Alternaria* sect. *Alternaria* consists of only 11 phylogenetic species and one species complex; 35 morphospecies were synonymized under *A. alternata*.

Conidia of *A. alternata* are produced on single or small groups of conidiophores, which are smooth and pale to mid-olivaceous or golden brown in colour. They are 20–36 × 9–9.5 μm, simple or branched, straight to flexuous, have four to six cells and are produced in long chains. Infected leaves, twigs and inflorescences are significant sources of conidia that infect fruit, as are leaves found in the litter beneath trees. The conidia of *A. alternata* are dispersed by air currents (Prusky, 1994).

Several approaches can be used to combat these diseases. Post-harvest development of fruit disease during storage can be prevented using a combination of hot water brushing with prochloraz at 225 μg mL^{-1} (Prusky et al., 2002).

4.2 Black mildew, sooty blotch and sooty mould

Several ascomycetes produce dark, usually superficial growths on stems, leaves and fruit of mango (Lim and Khoo, 1985). These range from thin, diffuse webs of dark hyphae to opaque, felty layers, to thick crusts (Fig. 9). Presumably, these variable signs are due to the different fungi that are involved. Although these fungi are usually not important in

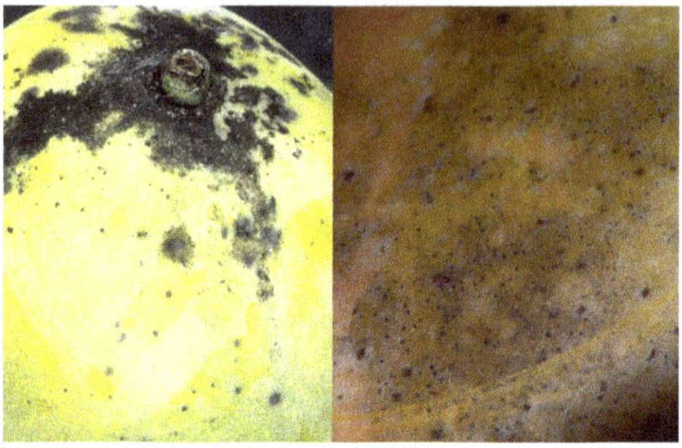

Figure 9 Symptoms of sooty mould (left) (Fig. 12.38 in Cooke et al. (2009)), and sooty blotch (photo: R.C. Ploetz) on surfaces of mango fruit.

well-maintained orchards, they can block sunlight and inhibit photosynthesis, as well as degrade the appearance and marketability of fruit.

Sooty moulds develop in the presence of aphids, mealybugs, scales and other sap-feeding insects that produce honeydew (excreta), an essential food source for these fungi. These problems dissipate after the associated insects are controlled. Black mildews and sooty blotchs grow directly on host surfaces and do not require honeydew.

Aetiology

The black or dark mildews are a group of mostly tropical obligate plant pathogens that produce two types of hyphopodia (Alexopoulos et al., 1996). Capitate hyphopodia are lobed appressoria, from which infectious haustoria are formed, whereas mucronate hyphopodia function as conidiogenous cells. *Meliola mangiferae* causes black mildew on mango (Lim and Khoo, 1985).

In contrast, the fungi that cause sooty moulds are diverse saprophytes. In Malaysia, Lim and Khoo (1985) listed coelomycetes (*Polychaeton*), hyphomycetes (*Tripospermum*) and loculoascomycetes (*Antennulariella*, *Chaetothyrium*, *Limacinula* and *Scorias*), whereas the reported agents in India were hyphomycetes (*Leptoxyphium*, *Microxyphium* and *Tripospermum*) and loculoascomycetes (*Capnodium*) (Butler and Brisby, 1931; Prakash, 1988, 18). In Pakistan, species from eight genera were associated with sooty mould, including the foliar pathogens *Aspergillus, Alternaria, Botryodiplodia, Capnodium, Cladosporium, Curvularia, Fusarium* and *Helminthosporium* spp. (Hamid and Jalaluddin, 2006). Some of the sooty mould fungi are often found in combination with each other; since they may not sporulate on plants it can be difficult to identify the species involved. In most cases, their growth is quite dense and is responsible for opaque blackish patches (Fig. 9).

Although 'sooty blotch' has been used as a synonym for 'sooty mould' (e.g. Singh, 1968), it is a distinct problem. Sooty blotch refers specifically to disease complexes that are not associated with honeydew and are caused by a diverse group of dothidiomycetes (Ascomycota) (Batzer et al., 2005; Johnson et al., 1997; Ploetz et al., 2000). The signs of sooty blotch are usually more diffuse than sooty mould, and often appear as a smoky blemish (Fig. 9). The sooty blotch agents on mango resemble those on apple, carambola and pear (Martinez et al., 2009).

Epidemiology and management

Sooty moulds are managed by controlling the associated insects with oils and/or insecticides (Lim and Khoo, 1985). In Pakistan, separate sprayings with fungicides (sulphur and mancozeb) and insecticides (malathion, diazinon and coal tar at 1 kg $tree^{-1}$) reduced the incidence of sooty mould on the foliage, whereas a mixture of fungicides and insecticides further decreased the incidence of sooty mould (Hamid and Jalaluddin, 2006). In India, sooty mould has been controlled using sulphur and parathion methyl (Prakash, 1991).

Although sooty blotch management has not been investigated on mango, these fungi have been controlled with fungicides in orchards (Williamson and Sutton, 2000) and removed from apple fruit with post-harvest washes (Batzer et al., 2002).

4.3 Stem-end rot

Stem-end rot is a post-harvest problem of mango fruit that is exceeded in importance only by anthracnose. In arid areas where anthracnose is generally not a problem, stem-end rot is the most important cause of post-harvest fruit loss (Ploetz and Freeman, 2009).

Symptoms

Different symptoms are caused by various pathogens as fruit ripen (Ploetz and Freeman, 2009). *Lasiodiplodia theobromae* s.l. (Fig. 10), *Botryosphaeria dothidea* (*Fusicoccum aesculi*) (Fig. 11) and *Neofusicoccum mangiferae* (*Fusicoccum mangiferae*) (Fig. 12) cause diffuse, water-soaked symptoms that develop under the skin and spread from the stem end with darkening. *Phomopsis mangiferae* produces dark lesions at the stem end that penetrate the flesh, although they do not spread as quickly as those caused by the former fungi. These lesions may resemble the stem-end anthracnose caused by *Colletotrichum gloeosporioides* s.l., although the latter lesions usually penetrate the flesh no deeper than 10–20 mm and are covered with the salmon pink spores of the pathogen. *Cytosphaera mangiferae* causes slow-spreading, tan lesions at the stem end, which are similar to the symptoms caused by *Aspergillus niger*; however, *C. mangiferae* produces conidiomata around the stem end. *Pestalotiopsis mangiferae* causes tan lesions that develop slowly and are eventually covered with acervuli.

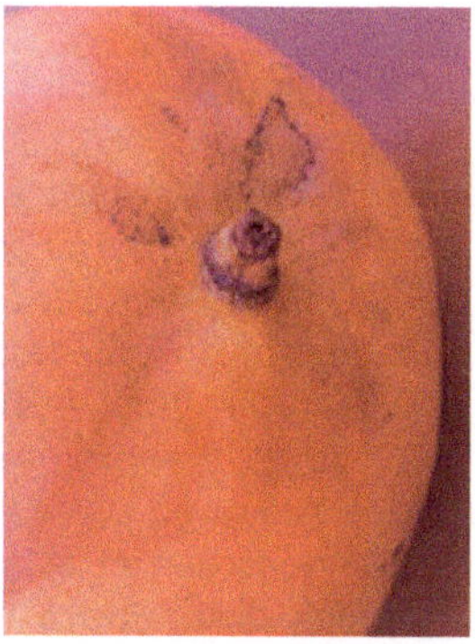

Figure 10 Symptoms of stem-end rot caused by *Lasiodiplodia theobromae* s.l. (photo: R.C. Ploetz).

Figure 11 Symptoms of stem-end rot caused by *Botryosphaeria dothidea* (*Fusicoccum aesculi*) (photo: A.W. Cooke).

Figure 12 Symptoms of stem-end rot caused by *Neofusicoccum mangiferae* (*Fusicoccum mangiferae*) (photo: A.W. Cooke).

Aetiology

The causal fungi are endophytes, many of which are in the Botryosphaeriaceae (dothidiomycetes, Ascomycota) (Johnson et al., 1992; Slippers et al., 2005; Slippers and Wingfield, 2007; Ploetz and Freeman, 2009; Marques et al., 2013; Phillips et al., 2013; Trakunyingcharoen et al., 2014, 2015). The taxonomy and nomenclature of these fungi were clarified by Phillips et al. (2013). Based on a phylogenetic analysis (ITS and EF1-α sequence data) of isolates from mango, Trakunyingcharoen et al. (2014) resolved seven clades with distinct morphological features from Australia, Brazil, Egypt, Iran, Mali, Peru, South Africa, Taiwan and Thailand. *L. theobromae* s.s. was the dominant species (Trakunyingcharoen et al., 2014, 2015). Although some of the large list of species in the family that were reported by Trakunyingcharoen et al. (2014) have not been tested for pathogenicity, six have wide geographical distributions and significantly impact this host: *B. dothidea*, *L. pseudotheobromae*, *L. theobromae* s.s., *Neofusicoccum mangiferae*, *N. parvum* and *Neoscytalidium* (*Hendersonula*) *hyalinum*.

L. theobromae is the most common and widespread cause of stem-end rot (Ploetz, 2003). The teleomorph of *L. theobromae*, formerly known as *B. rhodina*, is usually not found in nature. Crous et al. (2006) reserved the genus *Botryosphaeria* for the type species for the family, *B. dothidea* (*Fusicoccum aesculi*). *L. theobromae* produces a fluffy, greyish black mycelium on oatmeal agar (OA) and PDA (Johnson, 1994b). Conidiomata may be simple or develop into aggregated stromatic bodies (Burgess et al., 2006), and cirri of conidia may ooze from ostioles. Conidia are initially hyaline, aseptate, granular, ovoid to ellipsoid and thick-walled. The mature conidia are two-celled, measuring 26.2–27 × 14–14.4 μm, and brown-walled with numerous longitudinal striations (Phillips et al., 2013). *L. pseudotheobromae* is distinguished from *L. theobromae* based on their conidia; *L. pseudotheobromae* has larger and more ellipsoid conidia, 27.5–28.5 × 15.5–16.5 μm, than those of *L. theobromae*, which are ovoid and have a tapered base (Phillips et al., 2013; Trakunyingcharoen et al., 2014). *L. pseudotheobromae* also produces a dark pink pigment on PDA at 35°C and grows at 10°C; *L. theobromae* does neither. Phillips et al. (2013) distinguished the species with ITS and EF-1α sequence data.

Neofusicoccum parvum (*Fusicoccum parvum*), one of the most important mango pathogens, causes stem-end rot of fruit, dieback and blossom blight. It produces cottony grey mycelium and discrete pycnidia or stromatic multilocular fruiting bodies on PDA and OA, respectively (Johnson et al., 1991). Discrete, immersed pycnidia in a subcuticular pseudostroma are produced on mango. Conidia are fusiform to navicular, hyaline and unicellular, measuring 16.9–17.3 × 5.4–5.6 μm (Phillips et al., 2013). Occasionally, brown, three-celled conidia are observed. The teleomorph develops infrequently on OA and has been found on mango twigs in the tree litter in Australia (Johnson et al., 1991). Pseudothecia measuring 210 × 120 μm form beneath the epidermis on twigs. The ascostromata are hemilenticular and up to 10 mm wide on OA. The asci are eight-spored and bitunicate, while the ascospores are hyaline, single-celled, fusiform and measure 16–25 × 4.5–9.5 μm.

Neofusicoccum mangiferae causes post-harvest fruit rot and blossom blight (Lonsdale and Kotze, 1993; Saaiman, 1996). On PDA, *N mangiferae* produces a grey, felty mycelium with partly immersed conidiomata and 'pepper-spot' patterns in the pycnidial initials, and the dark grey mycelium lacks the white tufts found in similar species, such as *N. parvum* (Johnson, 1994b; Slippers et al., 2005). On mango, the fungus produces unilocular conidiomata in a subcuticular pseudostroma. Conidia differ from those produced by other *Neofusicoccum* spp. because of their shorter length and smaller length/width ratio (2–2.5) (Slippers et al., 2005). They are usually unicellular, ellipsoid to ovoid and hyaline, measuring 13.6 × 5.4 μm, although the conidia often become two- or three-celled, and are coloured light brown with darker middle cells (Phillips et al., 2013).

Management

Stem-end rot is difficult to control. Techniques that could be used to detect these pathogens in plants would be useful because they would enable the identification of pathogen-free propagation materials. The internal location and the diversity of fungi involved decrease the opportunities for controlling these diseases with fungicides (see Peterson et al., 1990). Broad-spectrum, systemic fungicides might be beneficial (see Lonsdale and Kotze, 1993) but have not been tested.

For high-value fruit, especially those destined for export, various post-harvest treatments have been beneficial. For example, Alvindia and Acda (2015) reported a 48–61% reduction in stem-end rot of 'Carabao' fruit after 20 min in 53°C water. As indicated above, hot water as practised for fruit fly eradication also reduces anthracnose. Terao et al. (2015) indicated that a low dose (< 3 kJ m^{-2}) of UV-C irradiation helped manage post-harvest diseases of mango caused by *B. dothidea*, *L. theobromae*, *A. alternata* and *C. gloeosporioides*, even though a direct impact on the pathogens was not evident. Santos et al. (2015) suggested that a dose of 0.45 kGy of gamma irradiation reduced disease caused by *L. theobromae*.

5 Foliar and floral diseases: algal leaf spot, apical necrosis and decline disorders

Diseases that affect the above-ground portions of mango are among the most important and conspicuous problems that affect this crop. Since many of the pathogens that cause foliar diseases also affect panicles, diseases of each organ are included in this section, as are the few diseases that affect the branches and trunks of mature trees.

5.1 Algal leaf spot

Algal leaf spot of mango, which is also known as red rust, is caused by *Cephaleuros virescens* and, less frequently, *C. parasiticus* (Lim and Khoo, 1985; Ponmurugan et al., 2009; Sanahuja et al., 2017). This common problem affects mango and many other plants in the tropics and subtropics (Joubert and Rijkenberg, 1971).

Cephaleuros spp. are parasitic green algae. The diseases they cause are frequently misidentified, due to their resemblance to other diseases (Brooks et al., 2015). For example, a disease of guava fruit that is caused by *C. virescens* has been erroneously reported to be caused by *Pestalotiopsis* spp. (compare Fig. 2F in Brooks et al., 2015, and Fig. 1B in Keith et al., 2006). Other diseases caused by *Cephaleuros* spp. have been mistaken for coffee rust, powdery mildew, various leaf spots and bacterial canker (Thompson and Wujek 1997; Wellman 1965).

Cephaleuros spp. produce conspicuous, orange- to rust-coloured velutinous patches on both leaf surfaces (Fig. 13) (Lim and Khoo, 1985). They are initially 5–8 mm in diameter, but

Figure 13 Symptoms of algal leaf spot on the upperside of a mango leaf (photo: R.C. Ploetz). Note that some of the algal thalli have become lichenized.

can merge to include large, irregular sections of the leaf, which later become dull greyish green and, eventually, turn to bleached patches. The twigs and branches are also affected, which causes the bark to crack.

Orange patches, that are the algal thallus, are located beneath the host cuticle. The thallus produces erect cells, some of which enlarge to produce stalked, terminal or ovoid sporangia, measuring 30 × 24 μm. The sporangia produce biflagellate zoospores, which are dispersed by rain splash and wind as the primary infective propagules.

Cephaleuros spp. require a humid environment to establish and spread (Lim and Khoo, 1985). In general, algal leaf spot is a minor problem in well-maintained orchards (Brooks et al., 2015). Pruning the canopy, mowing beneath trees and using a wider row spacing to increase the air circulation and sunlight penetration help combat the disease. In addition, appropriate fertilization and irrigation, and the control of insect pests, mites and other foliar diseases increase a tree's ability to cope with this disease. Algaecides and fungicides such as fentin acetate or those containing copper are effective.

5.2 Apical necrosis

Apical necrosis was first reported in southern Spain in 1991 and now occurs in Israel, Italy, Portugal and possibly Egypt (Cazorla et al., 1998, 2006). The disease can be quite damaging and limits production when panicles are affected.

Apical buds, leaves and panicles are susceptible (Fig. 14), but not the fruit (Cazorla et al., 1998). A dark brown to blackish necrosis develops on vegetative and floral apices (Fig. 14a and 14c), and leaf necrosis begins with blackened, water-soaked lesions, 1–3 mm in diameter, that can coalesce and expand to cover large areas (Fig. 14c). Necrosis can extend from affected buds to the petioles (Fig. 14a). A milky bacterial exudate develops often on affected apical buds but infrequently on petioles (Fig. 14b). Large portions of the canopy and much of a tree's bloom can be killed.

The causal bacterium, *Pseudomonas syringae* pv. *syringae*, affects many perennial fruit crops (Hirano and Upper, 1983; Kennelly et al., 2007). It is an epiphyte and generally not an aggressive pathogen. Apical necrosis usually develops after high populations of the

Figure 14 Symptoms caused by the apical necrosis pathogen, *Pseudomonas syringae* pv. *syringae*. (a) Extensive necrosis on young stem, apical bud, petioles and leaves; (b) bacterial exudate on necrotic stem; and (c) death of a developing floral panicle and associated leaf necrosis (photos: F.M. Cazorla, Universidade de Malaga).

bacterium are produced in the host tissues. Strains from mango produce an antimetabolite toxin, mangotoxin, which plays a role in pathogen virulence and symptom development (Arrebola et al., 2007).

Cold, wet weather and host genotype are primary factors that affect the development of apical necrosis (Cazorla et al., 1998; 2006). 'Tommy Atkins', 'Lippens' and 'Manzanillo' are highly susceptible, whereas 'Keitt' and 'Sensation' are less so.

Apical necrosis is managed in commercial orchards with copper-based products, although control failures and copper resistance have occurred in Spain and Portugal (Cazorla et al., 2002, 2006). These outbreaks have been associated with different copper resistance plasmids in the causal bacterium. Cazorla et al. (2006) determined that the plant resistance activator acibenzolar-*S*-methyl and the phosphonate derivative fosetyl-Al provided comparable control to Bordeaux mixture; they suggested that fosetyl-Al might provide a protective film against wound entry by the pathogen.

5.3 Decline disorders

Several diseases of mango have been variously termed blight, canker, decline, gummosis, twig blight, tip die back and stem bleeding (Ploetz and Freeman, 2009). They have similar symptoms and aetiologies.

Symptoms

All or some of the following symptoms have been reported: (i) marginal scorching of leaf lamina; (ii) foliar symptoms of nutritional deficiencies, particularly of iron and manganese; (iii) vascular discolouration (Fig. 15); (iv) dieback of small branches from the terminal (Fig. 15b and 15c); (v) gummosis, which is an oozing of a clear or cloudy exudate from either the terminal buds, branches, scaffold limbs or trunks; and (vi) root degeneration (Lim and Khoo, 1985; Ploetz et al., 1996a; Ploetz and Freeman, 2009).

Aetiology

Diverse biotic and abiotic factors have been suggested as causes or predisposing agents (McSorley et al., 1980; Kadman and Gazit, 1984; Schaffer et al., 1988; Ploetz and

Figure 15 Symptoms associated with mango decline: (a) internal/vascular discolouration and branch terminal death (tip dieback) may not (b) or may be associated with defoliation (c) (photos: D. Benscher).

Freeman, 2009); however, fungi have been shown to be the most common agents. They are endophytes that also cause stem-end rots on mango fruit and are generally secondary pathogens that cause disease in weakened hosts (Johnson et al., 1992; Slippers et al., 2005; Slippers and Wingfield, 2007; Ploetz and Freeman, 2009). Several species have been shown to cause all or some of the above symptoms when used to inoculate plants (Ploetz et al., 1996a). Their frequent association in affected tissues may indicate that these symptoms usually develop, or develop more severely, after multiple infections. The most important fungi are covered in Section 4.3.

L. theobromae s.l. attacks weakened trees that are predisposed by high temperatures, frost, drought, high humidity, hardpan soils, sun scorch, tar and tanglefoot (Muller, 1940; Das Gupta and Zacchariah, 1945; Alvarez-García and López-García, 1971; Acuna and Waite, 1977; Ploetz et al., 1996a). It infects wounded plants and is found in soil, on dead twigs, mummified fruit and on organic debris beneath trees (Johnson et al., 1992).

Dieback caused by *L. theobromae* s.l. has been recognized as a significant disease in India since at least the 1940s. It was the most serious disease affecting mango in the Jaipur district (Verma and Singh, 1970), whereas it affected 30–40% of plantations in the Moradabad region of Uttar Pradesh (Prakash and Srivastava, 1987). Das Gupta and Zacchariah (1945) indicated that only *L. theobromae* s.l. caused dieback. It caused a canker on mango in Indonesia (Muller, 1940) and Malaysia (Lim and Khoo, 1985), dieback in Egypt and the Sonsonate area of El Salvador (Acuna and Waite, 1977), and gummosis and dieback in Puerto Rico (Alvarez-García and López-García, 1971).

A dieback disease of mango was recognized in Niger in the early 1980s (Reckhaus and Adamou, 1987). *Neoscytalidium dimidiatum* (*Hendersonula toruloidea*) causes sudden wilting from the shoots to large branches, firing of the leaves and trunk cankers from which a clear exudate originates. Reckhaus and Adamou (1987) considered that water stress was a primary, predisposing factor in the development of this disease.

Mango decline is important in Florida, Israel and other mango-producing areas with calcareous soils (Schaffer, 1994; Ploetz et al., 1996a). The symptoms include interveinal chlorosis and marginal necrosis of the leaves, dieback of young twigs that progresses to larger branches, reduced growth of secondary roots, gummosis and vascular discolouration. Several different factors have been associated with mango decline in Florida. Schaffer et al. (1988) used the Diagnostic Recommendation and Integrated System (DRIS) to assess the nutritional status of declining and healthy 'Tommy Atkins' trees. The nutrient imbalance index was greatest for declining trees. DRIS identified Mn, Fe or both elements as the most deficient in declining trees, while the concentrations of these elements were below the critical range in two of three declining orchards. Mineral deficiencies may be predisposing factors in the development of mango decline because pathogenic fungi are recovered from symptomatic trees.

McSorley et al. (1980) detected the nematode *Hemicriconemoides mangiferae* at low, but consistent, levels in declining mango trees. It was suggested that it may have been responsible for the reduced root growth noted in affected trees and that it could play a role in the development of the disorder.

In Florida, Smith and Scudder (1951) reported that *Diplodia* sp. caused a dieback of the mango. Additional species of fungi were examined by Ploetz et al. (1996a). *A. alternata*, *Colletotrichum gloeosporioides* s.l., *Neofusicoccum parvum*, *L. theobromae* s.l. and two *Phomopsis* spp. were recovered from trees with diverse decline symptoms, which caused one or more of these symptoms in 'Keitt' and 'Tommy Atkins'. *Colletotrichum gloeosporioides* s.l., *Neofusicoccum parvum* and *L. theobromae* s.l. were most damaging and caused significant bud necrosis, tip dieback, gummosis and vascular discolouration. Symptoms caused by these fungi were distinguishable only when they sporulated on the inoculated branches.

In summary, several different fungi cause, or are associated with, decline symptoms worldwide. Most are endophytes with *Botryosphaeria* teleomorphs (Botryosphaeriaceae). Stress and wounds are usually associated with a predisposition to symptom development.

Management

The protected, interior location of the endophytic agents makes managing the decline disorders difficult. Management is usually restricted to controlling the manageable predisposing factors, such as drought stress and host nutrition. Pruning to force synchronous flushes of foliar growth and strategic applications of broad-spectrum fungicides to protect new growth might be beneficial (Johnson, 1994b). More work is needed on these topics.

6 Foliar and floral diseases: galls, scaly bark and powdery mildew

6.1 Galls and scaly bark

Gall and scaly bark disorders of the mango are usually minor problems but can cause a general loss of tree vigour.

Symptoms

In India, bark scaling develops as deep cracks along the entire rootstock portion of the plant, and these cracks may penetrate the phloem and become necrotic (Prakash and Srivastava, 1987). These symptoms resemble those of a scaly bark disorder known as 'cuarteado' in Colombia (Cook, 1975). Similar symptoms have been reported on mango seedlings in Hawaii and Oahu (Cook et al., 1971). The bark from the soil line to the first branches was rough and scaly, and 5-mm-long xylem pegs were evident when the bark was removed around the leaf scars and secondary branches.

In Mexico, a disorder known as 'nanahuate,' 'bolas' or 'buba of mango' causes galls, measuring 5–10 cm in diameter, which resemble a cauliflower and that are initially light green but become dark brown as they die (Angulo and Villapudua, 1982). The galls remain attached to trees for many years and severely affect branches until they die. Similar symptoms are found in Florida and are associated with pruning injuries. Larger galls have also been noted in Puerto Rico and mango germplasm collections in Florida, at the USDA-ARS in Miami and University of Florida in Homestead (Fig. 16) (Ploetz et al., 1996b; unpublished). The latter galls have rough, scaly exteriors, and are found on the trunks and scaffold limbs.

Aetiology

Fusarium decemcellulare causes these diseases in Florida (USA), Mexico and Venezuela (Malaguti and Reyes, 1964; Angulo and Villapudua, 1982; Ploetz et al., 1996b). The colonies produced on PDA have dark carmine red undersides. The fungus produces microconidia in

Figure 16 One-metre-wide gall on mango tree in mango germplasm collection of the University of Florida in Homestead (photo: R.C. Ploetz).

false heads or chains of branched and non-branched monophialides. Large macroconidia, measuring 92–55 × 7–5.5 μm, are produced in slimy yellow sporodochia that measure ca. 1 mm in diameter. The teleomorph of the fungus, *Albonectria rigidiuscula* (Rossman et al., 1998), has not been observed on mango.

Epidemiology

Isolates of *F. decemcellulare* from the mango are only mildly aggressive, and required wounding to infect mango. In Florida (USA), scaly bark is often associated with pruning wounds (Ploetz et al., 1996b).

Management

No pesticides have been identified that control this problem. Measures that may be helpful include the removal and destruction of the affected branches and trees in the orchard, disinfestation of pruning equipment to ensure that the pathogen is not spread during pruning operations and the use of healthy planting material in new orchards.

6.2 Powdery mildew

Powdery mildew is a widespread and important disease of leaves, panicles and fruit. The disease can reduce yields by as much as 90%, due mainly to its effect on fruit set and development (Schoeman et al., 1995).

Symptoms

Mango cultivars vary in their response to powdery mildew (Palti et al., 1974). Virtually all of the foliar, floral and fruit parts of the plant are affected in susceptible cultivars (Fig. 17). Powdery growth can cover all of the panicle tissues, resulting in a brown, shrivelled

Figure 17 Signs/symptoms of powdery mildew on a panicle of 'Glenn' (photo: R.C. Ploetz).

necrosis. Since fruit set and retention can be affected, the disease can have a profound impact on yield. Foliage can also be damaged significantly and young leaves are most susceptible. White, powdery coatings of conidia develop on either side of leaves, but on the undersides is often restricted to the midrib. Leaves become distorted and the affected areas turn purple and ultimately necrotic.

Aetiology

Powdery mildew is caused by the host-specific fungus, *Oidium mangiferae* (Prakash and Srivistava, 1987; Ploetz and Freeman, 2009). It was first described in Brazil (Berthet, 1914) and is now recognized in most mango-producing regions (Palti et al., 1974). The conidium and haustorium traits indicate that *O. mangiferae* belongs to the *Erysiphe polygoni* group (Johnson, 1994a). The pathogen was originally classified as *Erysiphe cichoracearum* by Wagle (1928), but Uppal et al. (1941) noted that it produced saccate and lobed appressoria, which are not characteristics of *Erysiphe cichoracearum*. The conidia of *Oidium mangiferae* are single-celled, hyaline and elliptical- to barrel-shaped, and measure 33–43 × 18–28 μm (Uppal et al., 1941; Palti et al., 1974). They are produced in large numbers on the host surfaces and give the affected tissues a powdery appearance (Fig. 17). The lengths of germ tubes vary depending on RH and terminate in appressoria. Globular haustoria form in host epidermal cells. The conidiophores are of the pseudoidium type (Boesewinkel, 1980).

Epidemiology

Powdery mildew is most severe during cool, dry weather. Conidia are disseminated by wind and are released on a diurnal basis (Schoeman et al., 1995). Peak spore release occurs between 1100 and 1600 h and is positively correlated with temperature and negatively correlated with RH, vapour pressure deficit and leaf wetness. Conidia germinate between 9 and 32°C (23°C is optimal) and at RHs as low as 20% (Palti et al., 1974). Thus, disease development is usually independent of RH. Infection can occur within 5–7 h and conidia are produced within five days of infection. Disease development occurs between 10 and 31°C.

Management

'Zill', 'Kent', 'Alphonso', 'Seddek' and 'Nam Doc Mai' are highly susceptible to powdery mildew; 'Haden', 'Glenn', 'Carrie', 'Zebda', 'Hendi Besenara', 'Ewaise' and 'Keitt' are moderately susceptible; and 'Sensation', 'Tommy Atkins' and 'Kensington' are slightly susceptible (Ploetz et al., 1994; Nofal and Haggag, 2006). In India, Tiwari et al. (2006) reported that 'Baigan Phalli', 'Barbalia', 'Dabari', 'Dilpasand', 'Khirama', 'Agarideeh', 'Oloor' and 'Totapari' were highly resistant and that 'Amrpali' was most susceptible.

Schoeman et al. (1995) recommended that fungicide applications for controlling this disease should start when panicles begin to change colour. Assuming an effective period of three weeks for a given application, they concluded that applications should continue every three weeks until panicle susceptibility decreases at the end of fruit set. Powdery mildew is easily controlled with sulphur, although it can burn flowers and young fruit during warm, sunny conditions (Johnson, 1994a). Foliar sprays of K_2HPO_4 and KH_2PO_4, systemic fungicides, and alternate treatments with fertilizer and systemic fungicides are also useful (Nofal and Haggag, 2006; Reuveni et al., 1998). Treatments with the fertilizers and one-half or one-quarter of the recommended rate of sterol-inhibitor fungicides, as well as kresoxim-methyl, provided protection that was comparable or superior to that with standard fungicides alone (Oosthuyse, 1998; Reuveni et al., 1998). However, some fungicides, such as dinocap, fenbuconazole and hexaconazole, can reduce pollen germination (Dag et al., 2001).

7 Foliar and floral diseases: malformation

Malformation is one of the most yield-limiting diseases on this crop (Ploetz, 2001, 2017; Freeman et al., 2014). In Egypt, which is a moderately important mango producer, 215 657 t of fruit were produced in 1998 with a value of ca. US$150 million. Based on the incidence and severity of malformation in that country, an estimated US$15 million of fruit were lost that year (Ploetz et al., 2002). Losses in more important producing countries, such as India, are undoubtedly greater.

Malformation was first described in India in 1891 (Kumar and Beniwal, 1991). It is now widely distributed and continues to spread to the remaining disease-free production areas (e.g. Crespo et al., 2012). To date, the disease has been reported in Australia, Brazil, Burma (Myanmar), China, Egypt, El Salvador, India, Israel, Malaysia, Mexico, Nicaragua, Oman, Pakistan, Senegal, South Africa, Sri Lanka, Sudan, Spain, Swaziland, Uganda and the United States (Flechtmann et al., 1973; Crookes and Rijkenberg, 1985; Liew et al., 2016; Lim and Khoo, 1985; Kumar and Beniwal, 1991; Ploetz, 2001; Ploetz and Freeman, 2009; Kvas et al., 2008; Crespo et al., 2012; Senghor et al., 2012; Sinniah et al., 2012; Zhan et al., 2012).

In general, malformation is most severe in areas with pronounced dry seasons. The mango bud mite, *Aceria mangiferae*, probably plays a significant synergistic role in malformation development, presumably as a wounding agent and disseminator of the pathogen (Gamliel-Atinsky et al., 2010). Its abundance in dry climates may be a contributing factor in the severity of malformation in these areas.

Although the aetiology of malformation was initially confused and continues to be debated by some (e.g. Ansari et al., 2013), its fungal aetiology has been understood for several decades. In 1966, *F. moniliforme* was shown to cause malformation in India (Summanwar et al., 1966). Subsequently, that pathogen, renamed *F. mangiferae* (Britz et al., 2002), has been reported from Australia, China, Egypt, Florida (USA), Israel, Malaysia, Oman, South Africa, Spain and Sri Lanka. Since the description of *F. mangiferae*, a growing list of additional species have also been described as causal agents, including *F. mexicanum* (Otero-Colina et al., 2010) and *F. pseudocircinatum* in Mexico (Freeman et al.,

2014a); *F. sterilihyphosum* in Brazil and South Africa (Britz et al., 2002); and *F. tupiense* in Brazil, Senegal and Spain (Freeman et al., 2014b; Lima et al., 2012). In addition, other described [e.g. *F. proliferatum* (Australia, China, Malaysia)] and undescribed species in the genus (Australia, Mexico, Spain) have been associated with the disease (Marasas et al., 2006; Nor et al., 2013; Zhan et al., 2010).

Identification of the different malformation pathogens relies on multigene genealogies, since no microscopic feature consistently distinguishes them from other, morphologically similar fungi in the Liseola section of *Fusarium* (Britz et al., 2002). Furthermore, only one of these pathogens has a sexual stage, *F. tupiense*, and it cannot be distinguished from other members of the *Gibberella fujikuroi* species complex (GFSC) (Lima et al., 2012). Given the time-consuming nature of identifying phylogenetic species, alternative species-specific diagnostics for these pathogens would be useful. For example, detection of new malformation outbreaks, confirmation of the pathogen-free nature of budwood and epidemiological studies on the movement of and infection by different pathogens might be made possible with rapid and accurate measures for identifying these fungi. Unfortunately, a PCR-based diagnostic for *F. mangiferae* (Zheng and Ploetz, 2002) is the only species-specific diagnostic that is available for any of these species. Additional detection measures for these species would be useful.

7.1 Symptoms

Malformation affects the vegetative and floral meristematic tissues (Fig. 18) (Ploetz, 2001). Vegetative malformation is most serious in seedlings and small plants in nurseries, especially where the seedlings are grown beneath affected trees, which is a common practice in the Middle East (Ploetz et al., 2002; Youssef et al., 2007). Vegetative malformation also occurs on mature trees. Apical and axillary buds produce misshapen shoots with shortened internodes and dwarfed leaves that are brittle and recurve towards the supporting stem (Fig. 18). The shoots may fail to expand fully, resulting in a bunched appearance, that is, the 'bunchy-top' symptom of the disease. Young nursery plants can be severely stunted.

Floral malformation is most important as affected inflorescences usually fail to set, or abort, fruit. The primary or secondary axes are shortened, which are thickened and highly branched (Fig. 18). Malformed panicles produce up to three times the normal number of flowers, which range from half to twice the normal size and have an increased proportion of male to perfect flowers (Kumar and Beniwal, 1992; Prakash and Srivastava, 1987). The malformed panicles may also produce dwarfed and distorted leaves (exhibit phyllody).

7.2 Aetiology

Historically, the aetiology of this disease has been confused (Ploetz, 2001). Despite the clear role that fungi play, mites (Narasimhan, 1954), nutritional problems (Prasad et al., 1965), physiological or hormonal imbalances (Dang and Daulta, 1982; Singh and Dhillon, 1989), viruses (Kauser, 1959) and unknown causes (Kumar and Beniwal, 1991) have all been indicted as causes of malformation.

Summanwar et al. (1966) demonstrated that a fungus, then named *F. moniliforme*, was responsible for the floral phase of this disease. Varma et al. (1974) later showed that *F. moniliforme* also caused vegetative malformation. This pathogen has had several synonyms in the literature, including *F. subglutinans*, *F. moniliforme* var. *subglutinans* and *F. moniliforme* 'Subglutinans'.

Figure 18 Symptoms caused by malformation: (a) in the panicles, there is an increase in the size and number of flowers and interspersed floral and vegetative organs (phyllody); and (b) compact or retarded growth of buds and brittle, dwarfed and recurved leaves in the vegetative shoots. The symptoms shown in (a) are on 'Haden' in Michoacan, Mexico, and are associated with *Fusarium mexicanum*, whereas (b) is on a 'Van Dyke' plant that was artificially inoculated with an isolate of *Fusarium mangiferae* (photos: R.C. Ploetz).

Since the original work in India, the above pathogen, in addition to several new species, has been described as the cause of this disease. In 2002, 29 strains of the pathogen from Egypt, Florida (USA), Israel, Malaysia and South Africa were described as a new species in the GFSC, *F. mangiferae* (Britz et al., 2002; Steenkamp et al., 2000). It was established based on β-tubulin and histone H3 DNA sequences, subtle morphological traits and malformation symptoms in mango after artificial inoculation with the species. Subsequently, *F. mangiferae* was reported in Australia (Liew et al., 2016), China (Zhan et al., 2012), India (O'Donnell et al., 1998; Zheng and Ploetz, 2002), Oman (Kvas et al., 2008) and Spain (Crespo et al., 2012). It has also been reported in Pakistan, although the identity of the pathogen is not clear since the authors used only morphological characteristics for its identification (Iqbal et al., 2006), which are insufficient for identifying this species. In summary, *F. mangiferae* is the most common and widely disseminated malformation pathogen.

Based on DNA sequence data (O'Donnell et al., 1998, 2000; Steenkamp et al., 1999, 2000), *F. mangiferae* is related to a lineage that includes *F. fujikuroi*, *F. proliferatum* and *F. sacchari* (Marasas et al., 2006); it corresponds to the 'Asian Clade' described by O'Donnell et al. (1998). *F. mangiferae* produces white, floccose mycelium on PDA with light to dark purple pigments in the agar (Leslie and Summerall, 2006). Cream-coloured sporodochia on carnation leaf agar (CLA) produce abundant thin-walled, long, slender and straight to slightly curved, three- to five-septate macroconidia, with curved apical cells and foot-shaped basal cells. Single-celled or, rarely, two-celled microconidia are produced in false heads on polyphialides with two to five conidiogenous openings on monophialides. Microconidial chains and chlamydospores are absent.

A second species, *F. sterilihyphosum*, was described based on isolates from a small area in South Africa (Britz et al., 2002). In a subsequent study, it was detected in Brazil (Ploetz, 2003) and was recently shown to cause malformation after artificial inoculation (Lima et al., 2006b). On PDA, colonies of *F. sterilihyphosum* produce white, floccose mycelium with a rose to light purple pigmentation in the agar (Leslie and Summerall, 2006). Uncommon, cream- to orange-coloured sporodochia are produced on CLA with rare, long, slender, four- to six-celled macroconidia. Single-celled microconidia are produced on the false heads of mono- and polyphialides. Distinctive sterile coiled hyphae are produced by some isolates of this species. *F. sterilihyphosum* is relatively uncommon in South Africa and Brazil where *F. mangiferae* and *F. tupiense* predominate, respectively. The latter taxon is phylogenetically distinct from *F. mangiferae* and *F. sterilihyphosum*. *F. tupiense*, which produces a unique teleomorph in the GFSC, caused malformation in pathogenicity tests in Brazil (Lima et al., 2006a,b, 2009, 2012), and was recently reported in Senegal and Spain (Crespo et al., 2016; Senghor et al., 2012). Other malformation agents include a new species in Mexico, *F. mexicanum* (Otero-Colina et al., 2010), and two previously reported species, *F. pseudocircinatum* in Australia and Mexico (Freeman et al., 2016; Liew et al., 2016), and *F. proliferatum* in Australia, China and Malaysia (Liew et al., 2016; Zhan et al., 2010; Zheng and Ploetz, 2002).

Several other taxa have been associated with mango malformation (Britz et al., 2002; Bhatnagar and Beniwal, 1977; Liew et al., 2016) but have not been tested for pathogenicity. Given the growing number of species that have been shown to cause this disease, it seems probable that additional malformation agents will be described in the future. Obviously, this complicates diagnosis and quarantine efforts (Liew et al., 2016).

PCR primer pairs have been used to identify some of the above taxa. Zheng and Ploetz (2002) developed a pair, 1-3 F/R, which amplified a 608 bp fragment from *F. mangiferae*. It has been used extensively in Egypt, Israel and Spain for diagnostic purposes (Crespo et al., 2012; Youssef et al., 2007). Another pair, 61-2F/R, was developed to diagnose *F. verticillioides* ('*Fusarium moniliforme*' in Muller et al., 1999), but it did not amplify *F. mangiferae* DNA. However, when the protocols were modified, 61-2F/R amplified a 445 bp fragment from strains of *F. sterilihyphosum* and *F. mexicanum* (Zheng and Ploetz, 2002; Rodríguez et al., 2008). It has not been tested with other malformation agents.

In summary, several different species in the GFSC have been shown to cause malformation. It is probable that other as yet unnamed species also cause this disease in different production areas.

7.3 Epidemiology

Although several species of *Fusarium* cause this disease, most research has been conducted with *F. mangiferae*. Additional work is needed to determine the extent to which the following results for *F. mangiferae* are relevant to other malformation agents.

F. mangiferae is spread by infected scions that are used for grafting and infected nursery stock (Prakash and Srivastava, 1987). Since seeds do not appear to harbour the fungus (Saeed and Schlosser, 1972; Youssef et al., 2007), seedlings should be disease-free. However, malformation can be severe when seedlings are raised in nurseries beneath affected trees (Ploetz, 2001).

The disease spreads slowly in orchards, perhaps because conidia of the pathogen die quickly when exposed to sunlight (Manicom, 1989). In Egypt and Israel, populations of *F. mangiferae* in infected panicles declined rapidly during the summer (Youssef et al., 2007). Wounding enhances infection and subsequent disease development (Ploetz, 2001).

The mango bud mite, *Aceria* (*Eriophyes*) *mangiferae*, is often observed in high numbers on malformed trees and has been proposed as a cause of this disease (Narasimhan, 1954, 1959; Nariana and Seth, 1962). Although it does not cause malformation (Ploetz, 2001), *Aceria mangiferae* is probably a vector of the pathogen. *F. mangiferae* was recovered from the body of the mite (Crookes and Rijkenberg, 1985), and was shown to adhere to its body thereby facilitating movement to infection courts in mango buds (Gamliel-Atinsky et al., 2009a, 2010). Infection of buds by the pathogen was increased in the mite's presence, presumably due to wounds it caused during feeding.

The distribution of *F. mangiferae* in affected trees suggests that vegetative and floral buds are the primary sites of infection and that systemic colonization of older, subtending tissues does not occur. Freeman et al. (1999) confirmed that buds and flower tissues of the host were primary infection sites, and that wounds provided points of entry for the pathogen. In Florida, *F. mangiferae* was restricted almost entirely to the malformed floral and vegetative tissues (Ploetz, 1994). The levels of infection were highest in malformed flowers and vegetative shoots (ca 65–85%), much lower or non-existent in asymptomatic tissues (0–11%), and rare in branches (0–4%) even when they supported malformed flowers or shoots. Remnant infections of *F. mangiferae* in scaffold branches and trunks were restricted almost exclusively to branch scars or dormant apices (Gamliel-Atinsky et al., 2009b). Reports that root infection by *F. oxysporum* (Kumar and Beniwal, 1991) or *F. mangiferae* (Abdel-Sattar, 1973; Kumar and Beniwal, 1991) cause malformation in seedling plants have not been corroborated. Roots can be infected by *F. mangiferae*, but these infections are not systemic and do not appear to result in symptom development (Youssef et al., 2007).

The localized and variable levels of infection by *F. mangiferae* in diseased and non-symptomatic tissue (Ploetz, 1994; Youssef et al., 2007) suggest that there are thresholds of infection and that malformation develops only after a sufficient proportion of the host meristem is colonized by the pathogen. This hypothesis is supported by the long latent period that exists before symptoms develop in artificially inoculated plants (Ploetz, 2001; Ploetz, unpublished data) and the hormonal perturbations that probably occur when meristematic tissues are infected by this pathogen (van Staden et al., 1989; van Staden and Nicholson, 1989). The ability of several phylogenetically distinct species to cause this disease suggests that a common, horizontally tranferred, genetic element may be associated with malformation induction.

Hormonal imbalances occur in affected tissues. Singh and Dhillon (1989) assayed the levels of indoleacetic acid (IAA), gibberellic acid (GA_3) and zeatin in malformed and healthy mango seedlings. The IAA and GA_3 levels were about ten and five times lower in malformed plants, respectively, whereas the levels of zeatin were about five times higher. Van Staden et al. (Nicholson and van Staden, 1988; van Staden et al., 1989; van Staden and Nicholson, 1989) examined the specific cytokinins produced by the mango and '*Fusarium moniliforme*' (presumably *F. mangiferae*). They determined that the cytokinin complements in healthy and malformed panicles differed qualitatively and quantitatively, whereas the pathogen was capable of producing some of the hormones and metabolites that were implicated in disease development. Recently, a transcription profile analysis of mango x *F. mangiferae* interaction indicated significant upregulation of zeatin genes (Liu et al., 2016); this and another publication help discern genetic attributes that distinguish this pathogen from other members of the GFSC (Niehaus et al., 2016).

7.4 Management

Ideally, the malformation pathogens should be kept out of uninfested areas. Awareness is key to effective exclusion efforts. New plantings should be established with pathogen-free nursery stock. Scion material should not be taken from affected orchards, and symptomatic plants in the nursery should be removed and burned. Nurseries should not be established in orchards that are affected by malformation. Control is possible after the disease is found in an orchard but it is time consuming. Cultural management has been most effective in these cases (Narisimhan, 1959; Singh et al., 1974; Manicom, 1989). Affected terminals and the subtending three nodes should be cut from the trees, removed from the field and burned.

The protected, internal location of the pathogen in affected trees makes it difficult to control this disease with chemicals. Although a diverse array of pesticides, hormones and growth regulators has been tested against malformation, few have shown much potential. Singh et al. (1994) obtained moderate control using sulphates of cobalt, cadmium and nickel in India, but it is unlikely that these toxic compounds could be used safely. Most other compounds have been less effective (Diekman et al., 1982; Chakrabarti and Ghosal, 1989). Freeman et al. (2014) described the use of prochloraz fungicide applications in conjunction with sanitation measures (as described above). Although significant reductions in malformation were reported, prochloraz is not labelled for use in some producing countries (i.e. the United States) and in some importing regions. Darvas (1987) reduced the percentage of malformed inflorescences from 96% to 48% by injecting 'Keitt' trees with the fungicide fosetyl-Al. This reduction was significant ($P < 0.05$), but the increase in fruit yield was not. Recently, more extensive applications of phosphonates (the active ingredient of fosetyl-Al) were shown to be more effective against malformation in South Africa. Nonetheless, the efficacy of intensive phosphonate applications needs to be demonstrated in other areas, and its adoption would need to address phosphonate residue tolerances that are imposed in some importing countries.

There is considerable anecdotal evidence for resistance to this disease (Prakash and Srivistava, 1987). Thus, there should be great potential for the use of resistant cultivars where grafted trees or nucellar seedlings from polyembryonic cultivars are used. Local cultivars that have consistently been deemed resistant, such as 'Bhadauran' in Brazil and India, 'Primor' in Brazil and 'Zebda' in Egypt, should receive wider attention. In other cases, note should be made of inconsistent reports (Ploetz, 2001). For example, Bastawros (1996) reported that two newly introduced cultivars in Egypt, 'Kent' and 'Keitt', were immune (0% disease), although these cultivars are susceptible in Florida (Ploetz, unpublished data). Better information is needed on resistance to this disease and its potential use in areas that are significantly impacted.

8 Foliar and floral diseases: seca and sudden decline

A disease that is known by several different names in Brazil and the Middle East routinely kills mango trees. This disease is known as seca (drying), murcha (withering), branch blight and Recife sickness in Brazil; it was first recognized in the State of Pernambuco in 1938, and is now also found in the states of Bahia, Goias, the Federal District, Rio de Janeiro and São Paulo (Ribeiro, 1997; Colosimo et al., 2000; Silveira et al., 2006). Neighbouring states in Brazil are threatened, due to the efficient movement of the pathogen(s) via infected propagation materials and pruning equipment, and a mobile beetle vector.

In 1998, a disease termed sudden decline began to kill trees in Oman (Al Adawi et al., 2003, 2006) at about the same time a similar problem known as quick decline or sudden death was observed in Pakistan (Malik et al., 2005). These diseases resembled seca in many ways. By 2007, many mango-producing areas in Oman and Pakistan were affected and uncontrolled dissemination of infected germplasm was believed to have spread the disease throughout the region.

8.1 Symptoms

Symptoms include discolouration of the vascular cambium, and exudation of an amber-coloured gum from the trunk and branches, particularly from galleries made by the beetle vector of the pathogen. Wilting, rapid death of branches and entire trees without defoliation, and a scorched appearance develops on affected trees (Fig. 19) (Al Adawi et al., 2006; Junqueira et al., 2002). The scions, rootstocks or both on grafted trees can exhibit vascular symptoms. In Oman, for example, where susceptible Omani seedlings are used as rootstocks, the disease frequently affects rootstocks (Al Adawi et al., 2006). In contrast, the disease was associated with the scion in Brazil (Colosimo et al., 2000). When symptoms begin in the canopy, they may be initiated in a branch or portion of a tree but death of the entire plant usually follows. Sudden death of the entire tree usually occurs when a rootstock infection is involved.

8.2 Aetiology

The genus *Ceratocystis* contains many tree pathogens (Kile, 1993). The wide host range of *Ceratocystis fimbriata* s.l. led Webster and Butler (1967) to propose that it was a species complex, and DNA sequences have begun to delineate some of the host-specific, often morphologically indistinct, species in the complex (van Wyk et al., 2007). A contemporary view is that *Ceratocystis fimbriata* s.s. specifically refers to the cause of black rot of the sweet potato, *Ipomoea batatas*, on which it was first described (Halsted and Fairchild, 1891). Other cryptic, monophyletic lineages of *Ceratocystis fimbriata* s.l. have been described as distinct species (Engelbrecht and Harrington, 2005; Johnson et al., 2005; van Wyk et al., 2005; van Wyk et al., 2007).

Ceratocystis fimbriata s.l. was reported to cause seca in Brazil during the 1930s (Viegas, 1960; Ribiero, 1980; Silveira et al., 2006). *Diplodia recifiensis* (= *L. theobromae*?) was indicted as the cause of Recife sickness in Brazil (Batista, 1947), but it probably plays a secondary role in the development of this disease (see below). Recently, van Wyk et al. (2011) used multigene genealogies to delineate two new seca-associated species in Brazil, *Ceratocystis mangicola* and *Ceratocystis mangivora*.

Figure 19 Symptoms of sudden wilt in Oman caused by *Ceratocystis manginecans*: (a) death of all or portions of mango trees grown on susceptible Omani rootstocks; (b) external evidence of the boring activity of the bark beetle vector of *Ceratocystis manginecans*, *Hypocryphalus mangiferae*, on the lower trunk of an affected tree; and (c) internal necrosis of the cambium (left) caused by the disease where galleries of *Hypocryphalus mangiferae* are evident (centre) (photos: R.C. Ploetz). A similar disease, seca, has been recognized in Brazil since the 1930s.

Two other species have been described on mango in the Oman Gulf region, *Ceratocystis omanensis*, which is a minor pathogen (Al Subhi et al., 2006), and the primary sudden decline agent in Oman and Pakistan, *Ceratocystis manginecans*. The latter species represents a monophyletic lineage of *Ceratocystis fimbriata* s.l., based on ITS, β-tubulin and TEF 1-α DNA sequences (van Wyk et al., 2007).

Colonies of *Ceratocystis manginecans* are greyish olive and have a banana-like odour on 2% MEA (van Wyk et al., 2007). The hyphae are smooth and segmented. The ascomatal bases are globose, black and (153–) 192–254 (–281) μm in diameter; the ascomatal necks are dark brown, lighter towards the apices, (514–) 557–635 (–673) μm long, (25–) 32–42 (–48) μm wide at the base and (14–) 16–22 (–26) μm wide at the tip; and ostiolar hyphae are hyaline, divergent and (42–) 45–59 (–69) μm long. The asci are evanescent and the ascospores are hyaline, hat-shaped, 3–4 μm long, 4–5 μm wide externally and 7–8 μm wide internally within the sheath. The primary conidiophores are phialidic, lageniform, hyaline, (72–) 81–109 (–144) μm long, 5–7 (–9) μm wide at the base, 6–8 (–9) μm wide at the broadest point and 3–6 μm wide at the tip. The secondary conidiophores are tube-like, flared at the mouth, short, hyaline, (59–) 65–77 (–84) μm long, 5–8 μm wide at the base and (5–) 6–8 μm wide at the tip. The primary conidia are hyaline, cylindrical, (15–) 23–29 (–33) μm in length and 3–6 μm wide. The secondary conidia are hyaline, barrel-shaped, (8–) 9–11 (–12) μm in length and 5–7 (–8) μm wide. The chlamydospores are brown, thick-walled, globose to sub-globose, (11–) 12–14 μm long and 9–11 (–12) μm wide.

At least two pathotypes of *Ceratocystis fimbriata* s.l. are present in Brazil (Rossetto et al., 1996; Junqueira et al., 2002; Silveira et al., 2006). Rossetto et al. (1996) evaluated 8-year-old trees of 15 cultivars with two isolates of the pathogen, IAC FITO 4905, which is pathogenic to 'Jasmim', and IAC FITO 334-1, which is not. 'São Quirino', 'Irwin', 'Edwards' and 'Van Dyke' were resistant; 'IAC 100 Bourbon' was moderately resistant; and 'Glenn', 'Joe Welch', 'Zill' and 'Haden' were susceptible to both isolates. However, 'Kent' responded in a similar manner to 'Jasmim' by resisting IAC FITO 334-1 and succumbing to IAC FITO 4905. Whether the two pathotypes represented by IAC FITO 334-1 and IAC FITO 4905 are the species reported by van Wyk et al. (2011), *Ceratocystis mangicola* and *Ceratocystis mangivora*, should be determined.

8.3 Epidemiology

Mango genotype has a profound impact on disease development, and severe epidemics occur wherever susceptible rootstocks and/or scions are used. A higher level of disease develops when trees are stressed, although it is not clear whether this is due to an increased attraction of the vector to stressed trees or reduced resistance to the disease. The associated pathogens are moved easily in infected germplasm. Pruning implements also move the pathogen, and the soil can be a long-term reservoir of inocula after it has been infested with chlamydospores of the pathogen. Insect dissemination has a particularly insidious role.

Beetles (Coleoptera: Scolytidae) are closely associated with seca in Brazil (Batista, 1947; Viegas, 1960; Piza, 1966; Ribiero, 1980). Batista (1947) indicated that the ambrosia beetle *Xyleborus affinis* was the sole vector of *Diplodia recifensis*. In contrast, Ribiero (1980) reported that a bark beetle, *Hypocryphalus mangiferae*, was the primary vector of *Ceratocystis fimbriata* s.l. This beetle produces galleries in the cambium of affected trees (Fig. 19) and was the only scolytid found on healthy and diseased trees. *Hypocryphalus mangiferae* is also the vector of *Ceratocystis manginecans* in Oman (Al Adawi et al., 2013a).

The interactions between *Hypocryphalus mangiferae* and the *Ceratocystis* pathogens of mango are incompletely understood. In olfactometer tests in Brazil, Ribiero (1980) showed that *Hypocryphalus mangiferae* was attracted to cultures of *Ceratocystis fimbriata* s.l., and larvae of the insect were raised to adulthood on the fungus. Several other species, many of which belong to the genus *Xyleborus*, were also associated with seca, although they were found only in diseased trees. The sequence of events has not been studied in Brazil and the Oman Gulf. However, *Hypocryphalus mangiferae* may contaminate its body with these pathogens while feeding in diseased trees and subsequently disseminate the pathogen to healthy trees (Ploetz et al., 2013).

Hypocryphalus mangiferae is native to some of the same areas in southern Asia where mango evolved (Atkinson and Peck, 1994; Butani, 1993; Mukherjee, 1997; Wood, 1982). Thus, the insect would have been introduced into Brazil where it would have a new encounter, rather than co-evolved relationship, with *Ceratocystis fimbriata* s.l. (van Wyk et al., 2007). Additional work is needed to understand the interactions of the vectors and pathogens in these pathosystems.

8.4 Management

Given their destructive impact, preventing the dissemination of these pathogens to new areas should be a high priority. Pathogen-free propagation material is needed whenever new plantings are established and germplasm is moved. Clean pruning implements should be used in affected areas and should be frequently disinfested with bleach, formalin or other disinfectants (Junqueira et al., 2002). Trees that have been killed by the disease should be removed and destroyed because they are significant reservoirs of the vector and pathogens.

Managing these diseases with fungicides would be a challenge, especially on susceptible cultivars. External applications of protectant or systemic fungicides would probably be ineffective given the internal location of the pathogens. In areas where partially resistant cultivars are grown, the removal and burning of affected branches and treatment of the exposed branch stubs with copper fungicides are recommended (Ribeiro et al., 1995; Ribeiro, 1997). Alternatively, injecting fungicides might be effective, as is done to control Dutch elm disease.

Genetic resistance may offer the best hope for managing these diseases. Various levels of tolerance have been observed in Brazil and resistant clones have been developed. However, pathogenic variation in the causal fungus in Brazil has hindered progress (Rossetto et al., 1996; Junqueira et al., 2002; Silveira et al., 2006). The disease responses of some genotypes vary in different production areas in the country, but 'Manga Dagua', 'Pico', 'IAC 101', 'IAC 102', 'Edwards', 'Van Dyke' and 'Carabao' resist two pathotypes that are recognized, while 'Rosa', 'Sabina', 'Sao Quirino', 'Oliveira Neto', 'Jasmim', 'Sensation',' Irwin' and 'Tommy Atkins' are generally tolerant (Ribiero, 1997; Junqueira et al., 2002). 'Espada' is also reported to be tolerant, but old trees are attacked. Seca is managed on 'Espada' by grafting onto resistant rootstocks and pruning diseased branches. Colosimo et al. (2000) reported that 'Oliveira' was the most resistant, while 'Carlota', 'Imperial', 'Extrema' and 'Pahiri' had intermediate resistance. Carvalho et al. (2004) described two new cultivars, 'IAC 103 Espada Vermelha' and 'IAC 109 Votupa', with moderate resistance to seca. 'IAC 103 Espada Vermelha' also had moderate resistance to powdery mildew but was susceptible to anthracnose, and both cultivars were susceptible to malformation. In Oman, 'Hindi Besennara', 'Sherokerzam', 'Mulgoa', 'Baneshan', 'Rose' and 'Alumpur Baneshan' developed significantly less disease when challenged with *C. maginecans* (Al Adawi, et al., 2013b).

9 Soil-borne diseases

Soil-borne diseases of mango are relatively less important than foliar and floral diseases, but they can cause significant damage to seedlings, nursery stock and mature trees. In general, the pathogens involved are different from those that cause problems above-ground.

9.1 *Phytophthora* diseases

Phytophthora palmivora (Oomycota) has caused wilt, crown rot, root rot and the death of nursery trees in Arizona, the Philippines and Thailand (Kueprakone et al., 1986; Matheron and Matejka, 1988; Tsao et al., 1994). Gumming and conspicuous bark lesions developed above-ground on these plants, whereas root and crown rots were evident at or below the ground level. Crowded conditions and excessive irrigation and rainfall exacerbate these diseases. Sanitation, the use of less-crowded conditions and reduced irrigation would be beneficial.

Damage has also been recorded on the trunks of field-grown, mature trees in the Ivory Coast (Lourd and Keuli, 1975), and on fruit in Australia, Malaysia and West Africa (Turner, 1960; Chee, 1969; Cooke, 2007). Mortality of trees has not been observed, but substantial stem cracking and bleeding does occur. In Australia, a firm chocolate-brown decay with a sweet odour was reported on 'Calypso' fruit. Fruit isolates in Australia caused leaf blight and crown canker on mango seedlings (Fig. 20).

P. palmivora has coenocytic hyphae, up to 7 µm in diameter, and papillate sporangia that measure 31–56.4 × 20.7–36.7 µm, which germinate directly with germ tubes or indirectly via motile zoospores (Erwin and Ribiero, 1996; Waterhouse, 1970). Zoospores, the primary infective propagules, require free water for movement. *P. palmivora* is heterothallic. The antheridia are amphigynous and the oogonia are spherical. Chlamydospores are also formed, which measure ca. 35 µm in diameter.

Recently, a *Phytophthora* sp. was isolated from mango trees in Andalucia in Spain, which were wilted and chlorotic, and had sparse canopies and cracked bark (Zea-Bonilla et al., 2007). The sporangia were semi-papillate and obovoid on V8 agar, and measured 51 (28–52) × 36 (22–37) µm. An isolate deposited in the Spanish Type Culture Collection, CECT 20567, caused root rot on 'Florida' and lesions on the leaves and stems of seedlings of 'Gomera 3'.

9.2 Root rot and damping off

Another oomycete, *Pythium vexans*, has been reported to cause root rot and wilt of seedlings (Lim and Khoo, 1985). In Malaysia, it caused losses of up to 30% in nurseries. Foliage wilted and was initially pale green, but later developed necrotic patches. Roots developed a wet, blackened necrosis that began in the fine roots before progressing to the larger roots and the root collar. Seedlings were often killed. Lim and Khoo (1985)

Figure 20 Symptoms induced by *Phytophthora palmivora* after artificial inoculation of (a) the stems and (b) the foliage of mango seedlings (photos: A.W. Cooke).

indicated that overcrowding, excessive moisture and the use of polybags promoted this disease.

Prakash and Singh (1980) reported that the basidiomycete *Rhizoctonia solani* caused root rot and damping off of seedlings in India. Affected tissues became soft, dark brown or black, and seedlings were ultimately girdled and collapsed. Mycelia and sclerotia of the pathogen were conspicuous on affected tissues.

9.3 *Sclerotium* rot

This disease was reported in Brazil (Almeida et al., 1979), India (Prakash and Singh, 1976) and the Philippines (Palo, 1933). The causal fungus, *Sclerotium rolfsii*, produces globular, brown sclerotia, 1.0–2.6 mm in diameter.

Symptom development begins with the formation of felty white tufts of the mycelium of the pathogen around the base of seedlings. The fungus can girdle the entire stem to a height of 5 cm or more above the soil line. Eventually, it forms high numbers of conspicuous sclerotia. Ultimately, the seedlings wilt and die. Seeds may also rot prior to germination. The disease is controlled via sanitation and disinfestation of seedbeds.

9.4 *Verticillium* wilt

Verticillium wilt of mango was first reported in Florida (USA) (Marlatt et al., 1970). It was originally attributed to *Verticillium albo-atrum* but the causal fungus formed microsclerotia, indicating that *V. dahliae* was involved.

Symptoms of the disease include firing and necrosis of leaves, usually in a portion of the canopy. Sectoral development of the disease often does not progress to other portions of the tree, which may recover. Dead leaves usually remain attached to the tree, and the xylem of affected branches is discoloured brown (Fig. 21). *Verticillium* wilt is a relatively uncommon disease that is found on land where susceptible vegetable crops, such as

Figure 21 Vascular discolouration caused by *Verticillium dahliae* (photo: R.C. Ploetz).

potato, tomato and eggplant, have recently been grown (Pohronezny and Marlatt, 1982). New mango orchards should not be planted on such sites.

10 Summary

Mango is plagued by damaging diseases that can have enormous impacts on tree health and fruit yield. Although each of the above diseases can cause significant problems, anthracnose, stem-end rot/mango decline, malformation and the *Ceratocystis*-incited diseases are most serious. For each of the latter diseases, different phylogenetic species of the pathogens have been identified within morphologically defined species complexes.

The available evidence suggests that one of the phylogenetic species that is associated with anthracnose, *C. asianum*, is most important, as it appears to be the only species that fits the latent infection model for this disease (Arauz, 2000). It also appears to be the only or primary taxon that causes leaf and blossom disease.

Anthracnose is difficult to manage in areas with significant rainfall, due mainly to the poor resistance of most cultivars to this disease and the requirement for blemish-free fruit, especially when they are destined for foreign markets. Orchard sanitation, fungicides, post-harvest treatments and the use of tolerant cultivars can reduce anthracnose severity, but no commercial cultivar is sufficiently resistant to be produced for export in humid environments.

Although new, anthracnose-resistant cultivars would be useful, little progress has been made towards their development. More research is needed on the impact of resorcinols, latex retention in the fruit pedicel and other fruit attributes on the development of anthracnose, as well as stem-end rot (Droby et al., 1986; Hassan et al., 2007; Karunanayake et al., 2014, 2015).

Several causal agents have been described for malformation. More information is needed on the response of different mango cultivars to the various taxa, as virtually all of the results to date have been for *F. mangiferae*. In general, better information is needed on which cultivars possess useful resistance and the traits that are associated with resistance. Furthermore, more work is needed to confirm the efficacy of intensive phosphonate use, as it may enable production of even the most susceptible cultivars.

The seca/sudden decline pathogens, *Ceratocystis manginecans*, *C. mangicola* and *C. mangivora* are phylogenetically, anatomically and pathologically very similar (Al Adawi et al., 2006; Van Wyk et al., 2007, 2011). Unlike the pathogens that are responsible for anthracnose and malformation, the *Ceratocystis* pathogens have narrow geographic distributions. As for malformation, awareness is key for effective exclusion of these agents. Recognizing the symptoms that they cause and how they are moved are essential. In this regard, importing scion material from infested areas is dangerous and should be avoided whenever possible.

The *Ceratocystis*-induced diseases are difficult to manage. Pathogen-free germplasm should be utilized to establish new plantings, and machetes and other tools should be routinely disinfested in the affected areas. When they are available, resistant cultivars should be considered.

There are several key needs for the improved management of these diseases. They include better fungicides and resistance for all of the important problems. Heightened awareness of the threats that are posed by malformation and the Ceratocystis wilts is necessary if they are to be excluded from (preferably), or interdicted in, new areas. To that end, diagnostic procedures that would enable the rapid and accurate detection of the different phylogenetic pathogens would be very useful. Finally, regulations that would monitor or restrict the international movement of mango germplasm should be considered in mango-production areas that remain free of the major problems.

11 Where to look for further information

There have been relatively few comprehensive publications on mango, let alone on mango diseases. Books which deal generally with the crop usually give cursory coverage of diseases (e.g. Purseglove, 1968). Reviews on diseases are included in books by Cook (1975), Cooke et al. (2009), Litz (1997, 2009), Singh (1968), Ploetz et al. (1994), Ploetz (2003) and Snowdon (1990), and major portions of books by Lim and Khoo (1985), Prakash and Srivastava (1987) and Ridgeway (1989) are devoted to these subjects.

12 Acknowledgements

The following are thanked for the figures indicated: the American Phytopathological Society (Fig. 5), David Benscher (Fig. 15), Francisco Cazorla (Fig. 14), Tony Cooke (Figs. 11–13 and 20), Dov Prusky (Fig. 9) and Tara Tarnowski (Fig. 2).

13 References

Abdel-Sattar, M. M. A. 1973. Histopathology of mango malformation. M.Sc. thesis. Al-Azhar University: Cairo.

Acuna, H. E. and Waite, B. H. 1977. La muerte regresiva del mango (*Mangifera indica* L.) en El Salvador. *Proc. Amer. Soc. Hort. Sci., Trop. Reg.* 21:15–16.

Adaskaveg, J. E. and Hartin, R. J. 1997. Characterization of *Colletotrichum acutatum* isolates causing anthracnose of almond and peach in California. *Phytopathology* 87:979–87.

Afanador-Kafuri, L., Minz, D., Maymon, M. and Freeman, S. 2003. Characterization of *Colletotrichum* isolates from tamarillo, passiflora and mango in Colombia and identification of a unique species from the genus. *Phytopathology* 93:579–87.

Ah-You, N., Gagnevin, L., Chiroleu, F., Jouen, E., Rodrigues Neto, J. and Pruvost, O. 2007a. Pathological variations within *Xanthomonas campestris* pv. *mangiferaeindicae* support its separation into three distinct pathovars that can be distinguished by amplified fragment length polymorphism. *Phytopathology* 97:1568–77.

Ah-You, N., Gagnevin, L., Pruvost, O., Myint, N. T. and Johnson, G. I. 2007b. First report in Myanmar of *Xanthomonas axonopodis* pv. *mangiferaeindicae* causing mango bacterial canker on *Mangifera indica*. *Plant Dis.* 91:1686.

Ah-You, N., Gagnevin, L., Grimont, P. A. D., Brisse, S., Nesme, X., Chiroleu, F., Bui Thi Ngoc, L., Jouen, E., Lefeuvre, P., Verniere, C. and Pruvost, O. 2009. Polyphasic characterization of xanthomonads pathogenic to Anacardiaceae and their relatedness to different *Xanthomonas* species. *Int. J. Syst. Evol. Microbiol.* 59:306–18.

Akem, C. N. 2006. Mango anthracnose disease: Present status and future research priorities. *Plant Pathol. J.* 5:266–73.

Al Adawi, A. O., Al Jabri, R. M., Deadman, M. L., Barnes, I., Wingfield, B. and Wingfield, M. J. 2013a. The mango sudden decline pathogen, *Ceratocystis manginecans*, is vectored by *Hypocryphalus mangiferae* (Coleoptera: Scolytinae) in Oman. *Eur. J. Plant Pathol.* 135:243–51.

Al Adawi, A. O., Al Sadi, B. A., Deadman, M. L., Barnes, I., Al Jabri, M. H., Wingfield, B. D. and Wingfield, M. J. 2013b. Evaluation of mango cultivars for resistance to infection by *Ceratocystis manginecans*. *Acta Hort.* 992:393–406.

Al Adawi, A. O., Deadman, M. L., Al Rawahi, A. K., Khan, A. J. and Al Maqbali, Y. M. 2003. *Diplodia theobromae* associated with sudden decline of mango in the Sultanate of Oman. *Plant Pathol.* 52:419.

Al Adawi, A. O., Deadman, M. L., Al Rawahi, A. K., Al Maqbali, Y. M., Al Jahwari, A. A., Al Saadi, B. A., Al Amri, I. S. and M. J. Wingfield. 2006. Aetiology and causal agents of mango sudden decline disease in the Sultanate of Oman. *Eur. J. Plant Pathol.* 116:247–54.

Al Subhi, A. M., Al Adawi, A. O., Van Wyk, M., Deadman, M. L. and Wingfield, M. J. 2006. *Ceratocystis omanensis*, a new species from diseased mango trees in Oman. *Mycol. Res.* 110:237–45.

Alahakoon, P. W., Brown, A. E. and Sreenivasaprasad, S. 1994. Genetic characterization of *Colletotrichum gloeosporioides* isolates obtained from mango. *Int. J. Pest Manage.* 40:225–9.

Alexopoulos, C. J., Mims, C. W. and Blackwell, M. 1996. *Introductory Mycology*. 4th edition. John Wiley & Sons: New York.

Almeida, R. T. de., Landim, C. M. U. and Teixeira, L. M. S. 1979. Occurrence of *Sclerotium rolfsii* Sacc. on cashew and mango seedlings in Ceara State. *Fitossanidade* 3:40–1.

Alvarez-García, L. A. and López-Gracía, J. 1971. Gummosis, dieback, and fruit rot disease of mango (*Mangifera indica* L.) caused by *Physalospora rhodina* (B. and C.) Cke. in Puerto Rico. *J. Agric. Univ. P. Rico* 55:435–50.

Alvindia, D. G., and Acda, M. A. 2015. Revisiting the efficacy of hot water treatment in managing anthracnose and stem-end rot diseases of mango cv. 'Carabao'. *Crop Prot.* 67:96–101.

Angulo, S. M. and Villapudua, J. R. 1982. Buba of mango (*Mangifera indica* L.) in the state of Sinaloa, Mexico. *Phytopathology* 72:171.

Arauz, L. F. 2000. Mango anthracnose: Economic impact and current options for integrated management. *Plant Dis.* 84:600–11.

Arrebola, E. Cazorla, F. M., Romero, D., Pérez-García, A. and de Vicente, A. 2007. A nonribosomal peptide synthetase gene (mgoA) of *Pseudomonas syringae* pv. *syringae* is involved in mangotoxin biosynthesis and is required for full virulence. *Mol. Plant-Microbe Interact.* 20:500–9.

Atkinson, T. H. and Peck, S. B. 1994. Annotated checklist of the bark and ambrosia beetles (Coleoptera: Platypodidae and Scolytidae) of tropical Southern Florida. *Fla. Entomol.* 77:313–29.

Bastawros, M. K. 1996. Mango malformation in Egypt. *Acta Hort.* 455:566–74.

Batista, A. C. 1947. Mal do Recife (grave doenca da Manggueira). Unpublished thesis, Pernambuco College of Agriculture. (abstracted in Review of Appl. Mycol (1948) 27:77–8).

Batzer, J. C., Gleason, M. L., Harrington, T. C. and Tiffany, L. H. 2005. Expansion of the sooty blotch and flyspeck complex on apples based on analysis of ribosomal DNA gene sequences and morphology. *Mycologia* 97:1268–86.

Batzer, J. C., Gleason, M. L., Weldon, B., Dixon, P. M. and Nutter Jr, F. W. 2002. Evaluation of postharvest removal of sooty blotch and flyspeck on apples using sodium hypochlorite, hydrogen peroxide with peroxyacetic acid, and soap. *Plant Dis.* 86:1325–32.

Berthet, J. A. 1914. Molestia cia mangueira. *Bolm Agric. (Sao Paolo)* 15:818–19.

Bhatnagar, S. S. and Beniwal, S. P. S. 1977. Involvement of *Fusarium oxysporum* in causation of mango malformation. *Plant Dis. Reptr.* 61:894–8.

Boesewinkel, H. J. 1980. The identity of mango mildew, *Oidium mangiferae*. *Phytopathol. Z.* 99:126–30.

Brecht, J. K. and Yahia, E. M. 2009. Postharvest physiology, pp. 484–529. In Litz, R. E. (Ed.) *The Mango: Botany, Production and Uses*. 2nd edition. CAB International, Wallingford Oxon, UK.

Brent, K. J. and Hollomon, D. W. 2007. *Fungicide Resistance: The Assessment of Risk*. Fungicide Resistance Action Committee. Croplife International: Brussels, Belgium.

Brooks, F., Rindi, F., Suto, Y., Shimane, M., Ohtani, S. and Green, M. 2015. The Trentepohliales (Ulvophyceae, Chlorophyta): An unusual algal order and its novel plant pathogen – *Cephaleuros*. *Plant Dis.* 99:740–53.
Britz, H., Steenkamp, E. T., Coutinho, T. A., Wingfield, B. D., Marasas, W. F. O. and Wingfield, M. J. 2002. Two new species of *Fusarium* section *Liseola* associated with mango malformation. *Mycologia* 94:722–30.
Bui Thi Ngoc, L., Verniere, C., Jouen, E., Ah-You, N., Lefeuvre, P., Chiroleu, F., Gagnevin, L. and Pruvost, O. 2010. Amplified fragment length polymorphism and multilocus sequence analysis-based genotypic relatedness among pathogenic variants of *Xanthomonas citri* pv. *citri* and *Xanthomonas campestris* pv. *bilvae*. *Int. J. Syst. Evol. Microbiol.* 60:515–25.
Burg, S. P. 2004. *Postharvest Physiology and Hypobaric Storage of Fresh Produce*. CAB International: Wallingford, Oxon, UK.
Burgess, T. I., Barber, P. A., Mohali, S., Pegg, G., de Beer, W. and Wingfield, M. J. 2006. Three new *Lasiodiplodia* spp. from the tropics, recognized based on DNA sequence comparisons and morphology. *Mycologia* 98:423–35.
Butani, D. K. 1993. *Mango: Pest Problems*. Periodical Expert Book: Agency D-42, Vivek Vihar, Delhi, India.
Butler, E. J. and Bisby, G. R. 1931. *The Fungi of India*. Scientific Monograph No.1. The Imperial Council of Agricultural Research in India: Calcutta.
CABI/EPPO. 2012. *Xanthomonas campestris* pv. *mangiferaeindicae*. Distribution Maps of Plant Diseases. Map 1135 (Edition 1).
Carvalho, C. R. L., Rossetto, C. J., Mantovani, D. M. B., Morgano, M. A., de Castro, J. V. and Bortoletto, N. 2004. Avaliaca de cultivares de mangueira selectionadas pelo Insitituto Agronimico de Campinas comparadas a outras de importancia comercial. *Rev. Bras. Fruti.* 26:264–71.
Cazorla, F. M., Arrebola, E., Olea, F., Velasco, L., Hermoso, J. M., Pérez-García, A., Torés, J. A., Farré, J. M. and de Vicente, A. 2006. Field evaluation of treatments for the control of the bacterial apical necrosis of mango (*Mangifera indica*) caused by *Pseudomonas syringae* pv. *syringae*. *Eur. J. Plant Pathol.* 116:279–88.
Cazorla, F. M., Arrebola, E., Sesma, A., Pérez-García, A., Codina, J. C., Murillo, J. and de Vicente, A. 2002. Copper resistance in *Pseudomonas syringae* strains isolated from mango is encoded mainly by plasmids. *Phytopathology* 92:909–16.
Cazorla, F. M., Torés, J. A., Olalla, L., Pérez-García, A., Farré, J. M. and de Vicente, A. 1998. Bacterial apical necrosis of mango in southern Spain: A disease caused by *Pseudomonas syringae* pv. *syringae*. *Phytopathology* 88:614–20.
Chakrabarti, D. K. and Ghosal, S. 1989. The disease cycle of mango malformation induced by *Fusarium moniliforme* var. *subglutinans* and the curative effects of mangiferin metal chelates. *J. Phytopathol.* 125:238–46.
Chee, K. H. 1969. Hosts of *Phytophthora palmivora*. *Rev. Appl. Mycol.* 48:337–44.
Colosimo, E. A., Chalita, L. V. A. S. and Demetrio, C. G. B. 2000. Tests of proportional hazards and proportional odds models for grouped survival data. *Biometrics* 56:1233–40.
Cook, A. A. 1975. *Diseases of Tropical and Subtropical Fruits and Nuts*. Hafner Press: New York.
Cook, A. A., Milbrath, G. M. and Hamilton, R. A. 1971. Woody gall and scaly bark of *Mangifera indica* in Hawaii. *Phytopathology* 61:1320.
Cooke, A. W. 2007. Phytophthora postharvest fruit rot in mango. Advisory poster produced by the Queensland Government Department of Primary Industries and Fisheries. Indooropilly, Australia.
Cooke, T., Persley, D. and Hodge, S. (Eds). 2009. *Diseases of Fruit Crops in Australai*. CSIRO Publishing: Collingwood, Victoria, Australia.
Crespo, M., Cazorla, F. M., de Vicente, A., Arrebola, E., Torés, J. A., Maymon, M., Freeman, S., Aoki, T. and O'Donnell, K. 2016. Analysis of genetic diversity of *Fusarium tupiense*, the main causal agent of mango malformation disease in southern Spain. *Plant Dis.* 100:276–86.
Crespo, M., Cazorla, F. M., Hermoso, J. M., Guirado, E., Maymon, M., Tores, J. A., Freeman, S. and de Vicente, A. 2012. First report of mango malformation disease caused by *Fusarium mangiferae* in Spain. *Plant Dis.* 96:286–7.

Cronje, C., Wehnwe, F. C. and Kotze, J. M. 1990. *Alternaria alternata* as a lesion pathogen of mango inflorescences in South Africa. *Phytophylactica* 22:117–18.

Crookes, C. A. and Rijkenberg, F. H. J. 1985. A literature review of the distribution, symptomatology, cause and control of mango blossom malformation. *S. Afr. Mango Grow. Assoc. Res. Rept.* 5:15–24.

Crous, P. W., Slippers, B., Wingfield, M. J., Rheeder, J., Marasas, W. F. O., Philips, A. J. L., Alves, A., Burgess, T., Barber, P. and Groenewald, J. Z. 2006. Phylogenetic lineages in the *Botryosphaeriaceae. Stud. Mycol.* 55:235–53.

Dag, A., Eisenstein, D. and Gazit, S. 2001. Effects of three fungicides used to control powdery mildew in mango on pollen germination and pollen-tube growth. *Proc. Interamer. Soc. Trop. Hort.* 43:123–6.

Damm, U., Cannon, P. F., Woudenberg, J. H. C. and Crous, P. W. 2012. The *Colletotrichum acutatum* species complex. *Stud. Mycol.* 73:37–113.

Dang, J. K. and Daulta, B. S. 1982. Mango malformation – A review. *Pesticides* 16:5–11.

Dann, E., Coates, L., Cooke, A. W., Smith, L. and Dean, J. 2007. Novel methods for integrated control of mango postharvest diseases, pp. 5–7. In *Delivering Mango Research*. Proceedings of the Sixth Australian Mango Conference. Gold Coast, Queensland, Australia.

Darvas, J. M. 1987. Control of mango blossom malformation with trunk injection. *S. Afr. Mango Grow. Assoc. Yearb.* 7:21–4.

Das Gupta, S. N. and Zacchariah, H. T. 1945. Studies on the diseases of *Mangifera indica*. Part V. On the die-back of mango trees. *J. Ind. Bot. Soc.* 24:101–8 (abstracted in Rev. Appl. Mycol. (1946) 25:267–8).

de Souza, A., Delphino Carbonia, L. C., Wicker, E., de Macedo Lemos, E. G. and de Goes, A. 2013. Lack of host specificity of *Colletotrichum* spp. isolates associated with anthracnose symptoms on mango in Brazil. *Plant Pathol.* 62:1038–47.

Diekman, F., Manicom, B. Q. and Coetzee, K. 1982. An attempt to control blossom malformation of mangoes with chemical sprays. *Subtropica* 3:15–16.

Dodd, J. C., Estrada, A. B., Matcham, J., Jeffries, P. and Jeger, M. J. 1991. The effect of climatic factors on *Colletotrichum gloeosporioides*, causal agent of mango anthracnose, in the Philippines. *Plant Pathol.* 40:568–75.

Dodd, J. C., Prusky, D. and Jeffries, P. 1997. Fruit diseases, pp. 257–80. In Litz, R. E. (Ed.) *The Mango: Botany, Production and Uses.* CABI: Wallingford, UK.

Du, M., Schardl, C. L., Nuckels, E. M. and Vaillancourt, L. J. 2005. Using mating-type gene sequences for improved phylogenetic resolution of *Collectotrichum* species complexes. *Mycologia* 97:641–58.

Engelbrecht, C. J. and Harrington, T. C. 2005. Intersterility, morphology and taxonomy of *Ceratocystis fimbriata* on sweet potato, cacao and sycamore. *Mycologia* 97:57–69.

Erwin, D. C. and Ribiero, O. K. 1996. *Phytophthora Diseases Worldwide*. APS Press: St. Paul.

Fitzell, R. D. 1979. *Colletotrichum acutatum* as a cause of anthracnose of mango in New South Wales. *Plant Dis. Reptr.* 63:1067–70.

Fitzell, R. D. and Peak, C. M. 1984. The epidemiology of anthracnose disease of mango: Inoculum sources, spore production and dispersal. *Ann. Appl. Biol.* 104:53–9.

Fitzell, R. D., Peak, C. M. and Darnell, R. E. 1984. A model for estimating infection levels of anthracnose disease of mango. *Ann. Appl. Biol.* 104:451–8.

Flechtmann, C. H. W., Kimati, H., Medcalf, J. C. and Ferré, J. 1973. Preliminary observations on malformation in mango (*Mangifera indica* L.) inflorescences and the fungus, insects and mite found on them. *Rev. Plant Pathol.* 52:162–3.

Freeman, S., Katan, T. and Shabi, E. 1998. Characterization of *Colletotrichum* species responsible for anthracnose diseases of various fruits. *Plant Dis.* 82:596–605.

Freeman, S., Maimon, M. and Pinkas, Y. 1999. Use of GUS transformants of *Fusarium subglutinans* for determining etiology of mango malformation disease. *Phytopathology* 89:456–61.

Freeman, S., Minz, D., Maymon, M. and Zveibil, A. 2001. Genetic diversity within *Colletotrichum acutatum sensu* Simmonds. *Phytopathology* 91:586–92.

Freeman, S., Otero-Colina, G., Rodríguez-Alvarado, G., Fernández-Pavía, S., Maymon, M., Ploetz, R. C., Aoki, T., and O'Donnell, K. 2014. First report of mango malformation disease caused by *Fusarium pseudocircinatum* in Mexico. *Plant Dis.* 8:1583.

Freeman, S., Pham, M., and Rodriguez, R. J. 1993. Molecular genotyping of *Colletotrichum* species based on arbitrarily primed PCR, A+T-rich DNA, and nuclear DNA analyses. *Exp. Mycol.* 17:309–22.

Freeman, S. and Rodriguez, R. J. 1995. Differentiation of *Colletotrichum* species responsible for anthracnose of strawberry by arbitrarily primed PCR. *Mycol. Res.* 99:501–4.

Freeman, S., Shtienberg, D., Maymon, M., Levin, A. G. and Ploetz, R. C. 2014. New insights into mango malformation disease epidemiology lead to a new integrated management strategy for subtropical environments. *Plant Dis.* 98:1456–66.

Fukuda, T., Uehara, K., Azegami, K., Tabei, H. and Nishiyama, K. 1990. Bacterial canker of mango in Japan caused by *Xanthomonas campestris* pv. *mangiferaeindicae*. *Ann. Phytopathol. Soc. Japan* 56:474–80.

Gagnevin, L. and Pruvost, O. 1995. Assessment of genomic variability in *Xanthomonas campestris* pv. *mangiferaeindicae*. Phytopathology 85:1163.

Gagnevin, L. and Pruvost, O. 2001. Epidemiology and control of mango bacterial black spot. *Plant Dis.* 85:928–35.

Gamliel-Atinsk, E., Freeman, S., Maymon, M., Belausov, E., Ochoa, R., Bauchan, G., Skoracka, A., Pena, J. and Palevsky, E. 2010. The role of eriophyoids in fungal pathogen epidemiology, mere association or true interaction? *Exp. Appl. Acarol.* 51:191–204.

Gamliel-Atinsky, E., Freeman, S., Sztejnberg, A., Maymon, M., Ochoa, R., Belausov, E. and Palevsky, E. 2009a. Interaction of the mite *Aceria mangiferae* with *Fusarium mangiferae*, the causal agent of mango malformation disease. *Phytopathology* 99:152–9.

Gamliel-Atinsky, E., Sztejnberg, A., Maymon, M., Vintal, H., Shtienberg, D. and Freeman, S. 2009b. Infection dynamics of *Fusarium mangiferae*, causal agent of mango malformation disease. *Phytopathology* 99:775–81.

Gantotti, B. V. and Davis, M. J. 1993. Pectic zymogram analysis for characterizing genetic diversity of the mango anthracnose pathogen. *Acta Hort.* 341:353–9.

Govender, V. and Korsten, L. 2006. Evaluation of different formulations of *Bacillus licheniformis* in mango pack house trials. *Biol. Cont.* 37:237–42.

Gupta, J. H. 1989. Perpetuation and epidemiology of powdery mildew of mango. *Acta Hort.* 231:528–33.

Halsted, B. D. and Fairchild, D. G. 1891. Sweet-potato black rot. *J. Mycol.* 7:1–11.

Hamid, M. and Jalaluddin, M. 2006. Recurring incidence of sooty mould of mango in Karachi and its control. *Intl. J. Biol. Biotech.* 3:561–5.

Hirano, S. S. and Upper, C. D. 1983. Ecology and epidemiology of foliar bacterial plant pathogens. *Annu. Rev. Phytopathol.* 21:243–69.

Holliday, P. 1980. *Fungus Diseases of Tropical Crops*. Cambridge University Press: Cambridge.

Iqbal, Z., Mehboob-ur-Rahman, Dasti, A. A., Saleem, A. and Zafar, Y. 2006. RAPD analysis of Fusarium isolates causing 'Mango Malformation' disease in Pakistan. *World J. Microbiol. Biotech.* 22:1161–7.

Jacobi, K. K., MacRae, E. A. and Hetherington, S. E. 2001. Postharvest heat disinfestations treatments of mango fruit. *Sci. Hort.* 89:171–93.

Jeffries, P., Dodd, J. C., Jeger, M. J. and Plumbley, R. A. 1990. The biology and control of *Colletotrichum* species on tropical fruit crops. *Plant Pathol.* 39:343–66.

Johnson, E. M., Sutton, T. B. and Hodges C. S. 1997. Etiology of apple sooty blotch disease in North Carolina. *Phytopathology* 87:88–95.

Johnson, G. I. 1992. Biology and control of stem-end rot pathogens of mango. Unpublished PhD dissertation, University of Queensland, Brisbane, Australia.

Johnson, G. I. 1994a. Powdery mildew, pp. 38–9. In Ploetz, R. C., Zentmyer, G. A., Nishijima, W., Rohrbach, K. and Ohr, H. D. (Eds), *Compendium of Tropical Fruit Diseases*. APS Press: St Paul, USA.

Johnson, G. I. 1994b. Stem-end rot, pp. 39–41. In Ploetz, R. C., Zentmyer, G. A., Nishijima, W., Rohrbach, K. and Ohr, R. D. (Eds), *Compendium of Tropical Fruit Diseases*. APS Press: St Paul, USA.

Johnson, G. I., Cooke, A. W., Mead, A. J. and Wells, LA. 1991. Stem end rot of mango in Australia: Causes and control. *Acta Hort.* 219:288–95.

Johnson, G. I., Mead, A. J., Cooke, A. W. and Dean, J. R. 1992. Mango stem end rot pathogens – Fruit infection by endophytic colonisation of the inflorescence and pedicel. *Ann. Appl. Biol.* 120:225–34.

Johnson, G. I. and Hofman, P. J. 2009. Postharvest technology and quarantine treatments, pp. 529–605. In Litz, R. E. (Ed.) *The Mango: Botany, Production and Uses*. 2nd edition. CABI: Wallingford, UK.

Johnson, J. A., Harrington T. C. and Engelbrecht C. J. B. 2005. Phylogeny and taxonomy of the North American clade of the *Ceratocystis fimbriata* complex. *Mycologia* 97:1067–92.

Joubert, J. J. and Rijkenberg, F. H. J. 1971. Parasitic green algae. *Annu. Rev. Phytopathol.* 9:45–64.

Junqueira, N. T. V., Pinto, A. C.de Q., da Cunha, M. M. and Ramos, V. H. V. 2002. Controle das doencas da mangueira, pp. 323–403. In Zambolin, L., do Vale, F. X. R., Monteiro, A. J. A. and Costa, H. (Eds) *Controle de Doencas de Plantas Fruteiras vol. 1*. Universidade Federal de Vicosa: Vicosa.

Kadman, A. and Gazit, S. 1984. The problem of iron deficiency in mango trees and experiments to cure it in Israel. *J. Plant Nutr.* 7:282–90.

Karunanayake, K. O. L. C. 2007. Natural defence mechanisms in mango fruit and their potential in management of postharvest diseases. PhD thesis. Univ. Peridenya, Peridenya, Sri Lanka.

Kauser, A. G. 1959. Malformation of inflorescence in mango. *Punjab Fruit J.* 22:19–21.

Keith, L. M., Velasquez, M. E., and Zee, F. T. 2006. Identification and characterization of *Pestalotiopsis* spp. causing scab disease of guava, *Psidium guajava*, in Hawaii. *Plant Dis.* 90:16–23.

Kennelly, M. M., Cazorla, F. M., de Vicente, A., Ramos, C. and Sundin, G. W. 2007. *Pseudomonas syringae* diseases of fruit trees: Progress toward understanding and control. *Plant Dis.* 91:4–17.

Kile, G. A. 1993. Plant diseases caused by species of *Ceratocystis sensu stricto* and *Chalara*, pp. 173–83. In Wingfield, M. J., Seifert, K. A. and Webber, J. F. (Eds) *Ceratocystis and Ophiostoma: Taxonomy, Ecology and Pathogenicity*. APS Press: St. Paul, USA.

Kishun, R. 1994. Evaluation of phylloplane micro-organisms from mango against *Xanthomonas campestris* pv. *mangiferaeindicae*. Indian Phytopathol. 47:313.

Kishun, R. 1995. Detection and management of *Xanthomonas campestris* pv. *mangiferaeindicae*, pp. 173–82. In Varma, J. P., Verma, A. and Kumar, D. (Eds) *Detection of Plant Pathogens and their Management*. Angkar Publishers: New Dehli, India.

Kishun, R. and Sohi, H. S. 1983. Bacterial canker in mangoes. *Indian Farm. Dig.* 14:21–3.

Knodler, M., Reisenhauer, K., Schieber, A. and Carle, R. 2009. Quantitative determination of allergenic 5-alk(en)ylresorcinols in mango (*Mangifera indica* L.) peel, pulp, and fruit products by high-performance liquid chromatography. *J. Agric. Food Chem.* 57:3639–44.

Kobiler, I., Reved, R., Artez, L. and Prusky, D. 1998. Antifungal compounds regulating quiescent disease in mango, pp. 109–14. In Johnson, G. I., Higley, E. and Joyce, D. C. (Eds) *Australian Centre for International Agricultural Research (ACIAR) Proceedings 80*. ACIAR: Canberra, Australia.

Kueprakone, U., Saengkong, S., Pienpuck, K. and Choobumroong, W. 1986. *Phytophthora palmivora* (Butl.) Butl., the causal organism of black rot of mango seedlings. *Thai Agric. Res. J.* 4:67–73.

Kumar, J. and Beniwal, S. P. S. 1992. Mango malformation, pp. 357–93. In Kumar, J., Chaube, H. S., Singh, U. S. and Mukhopadhyay, A. N. (Eds), *Plant Diseases of International Importance. Volume III: Diseases of Fruit Crops*. Prentice Hall: Englewood Cliffs, USA.

Kvas, M., Steenkamp, E. T., Al Adawi, A. O., Deadman, M. L., Al Jahwari, A. A., Marasas, W. F. O., Wingfield, B. D., Ploetz, R. C. and Wingfield, M. J. 2008. *Fusarium mangiferae* associated with mango malformation in the Sultanate of Oman. *Eur. J. Plant Pathol.* 121:195–9.

Leslie, J. F. and Summerell, B. A. 2006. *The Fusarium Lab Manual*. Blackwell: Ames, USA.

Liew, E. C. Y., Laurence, M. H., Pearce, C., Shivas, R. G., Johnson, G. I., Tan, Y. P., Edwards, J., Perry, S., Cooke, A. W. and Summerell, B. A. 2016. Review of *Fusarium* species isolated in association with mango malformation in Australia. *Australas. Plant Pathol.* 45:547–59.

Lim, T. K. and Khoo, K. C. 1985. *Diseases and disorders of mango in Malaysia*. Tropical Press: Kuala Lumpur, Malaysia.
Lima, C. S., Costa, S. S., Campos, M. A. and Pfenning, L. H. 2006a. A new *Fusarium* species associated with mango malformation in Brazil. *Fitopatol. Bras.(S)* 31:191.
Lima, C. S., Pfenning, L. H., Costa, S. S., Campos, M. A. and Leslie, J. F. 2006b. A new biological species of the *Gibberella fujikuroi* complex associated with mango malformation in Brazil. *Fitopatol. Bras.(S)* 31:191.
Lima, C. S., Monteiro, J. H. A., Crespo, N. C., Costa, S. S., Leslie, J. F. and Pfenning, L. H. 2009. VCG and AFLP analyses identify the same groups in the causal agents of mango malformation in Brazil. *Eur. J. Plant Pathol.* 123:17–26.
Lima, C. S., Pfenning, L. H., Costa, S. S., Abreu, L. M. and Leslie, J. F. 2012. *Fusarium tupiense* sp. nov., a member of the *Gibberella fujikuroi* complex that causes mango malformation in Brazil. *Mycologia* 104:1408–19.
Liu, F., Wu, J.-B., Zhan, R.-L. and Ou, X.-C. 2016. Transcription profiling analysis of Mango – *Fusarium mangiferae* interaction. *Front. Microbiol.* 7:1443. doi: 10.3389/fmicb.2016.01443
Lonsdale, J. H. and Kotze, J. M. 1993. Chemical control of mango blossom diseases and the effect on fruit set and yield. *Plant Dis.* 77:558–62.
Lourd, M. and Keuli, S. D. 1975. Note sur un chancre a *Phytophthora* du manguier en Cote d'Ivoire. *Fruits* 30:541–4.
McGuire, R. G. 1991. Concomitant decay reductions when mangoes are treated with heat to control infestations of Caribbean fruit flies. *Plant Dis.* 75:946–9.
Malaguti, G. and de Reyes, C. 1964. A gall disease of cacao and mango in Venezuela caused by *Calonectria rigidiuscula*. Phytopathology 54:499.
Malik, M. T., Khan, S. M., Dasti, A. A. and Kazmi, M. R. 2005. First record of *Ceratocystis pimbriata* (sic) causal organism of mango sudden death in Pakistan. *Pakistan J. Phytopathol.* 17:187–91.
Manicom, B. Q. 1986. Factors affecting bacterial black spot of mangoes caused by *Xanthomonas campestris* pv. *mangiferaeindicae*. *Ann. Appl. Biol.* 109:129–35.
Manicom, B. Q. 1989. Blossom malformation of mango. *S. Afr. Mango Grow. Assoc. Yearb.* 10:11–12.
Manicom, B. Q. and Wallis, F. M. 1984. Further characterization of *Xantbomonas campestris* pv. *mangiferaeindicae*. *Intl. J. Sys. Bacteriol.* 34:77–9.
Manicom, B. Q. and Pruvost, O. P. 1994. Bacterial black spot, pp. 41–2. In Ploetz, R. C., Zentmyer, G. A., Nishijima, W. T., Rohrbach, K. G. and Ohr, H. D. (Eds), *Compendium of Tropical Fruit Diseases*. APS Press: St Paul, USA.
Marasas, W. F. O., Ploetz, R. C., Wingfield, M. J., Wingfield, B. D. and Steenkamp, E. T. 2006. Mango malformation disease and the associated *Fusarium* species. *Phytopathology* 96:667–72.
Marlatt, R. B., Knight Jr., R. J. and Goldweber, S. 1970. *Verticillium* wilt of mango (*Mangifera indica*) in Florida. *Plant Dis. Rptr.* 54:569–71.
Marques, M. W., Lima, N. B., de Morais Jr., M. A., Barbosa, M. A. G., Souza, B. O., Michereff, S. J., Phillips, A. J. L. and Câmara, M. P. S. 2013. Species of *Lasiodiplodia* associated with mango in Brazil. *Fungal Divers.* 61:181–93.
Martinez, J. M. P., Batzer, J., Ploetz, R. and Gleason, M. L. 2009. Avocado, banana, carambola and mango are hosts of members of the sooty blotch and flyspeck complex. *Phytopathology* 99:S102.
Matheron, M. E. and Matejka, J. C. 1988. Phytophthora crown and root rot on nursery- grown mango trees delivered to Arizona. *Phytopathology* 78:1572.
McSorley, R, Parrado, J. L. and Goldweber, S. 1980. Observations on a mango decline in South Florida. *Proc. Fla. State Hort. Soc.* 93:132–3.
Midha, S., and Patil, P. B. 2014. Genomic insights into the evolutionary origin of *Xanthomonas axonopodis* pv. *citri* and its ecological relatives. *Appl. Environ. Microbiol.* 80:6266–79.
Misra, A. K. and Prakash, O. 1992. Control of bacterial canker of mango by chemicals. *Pesticides* 18:32–3.
Mukherjee, S. K. 1997. Introduction: Botany and importance, pp. 1–19. In Litz, R. E. (Ed.) *The Mango: Botany, Production and Uses*. CAB International: Wallingford, UK.

Muller, E. M., Chelkowski, J. and Geiger, H. H. 1999. Species-specific PCR assays for the fungal pathogens *Fusarium moniliforme* and *Fusarium subglutinans* and their application to diagnose maize ear rot disease. *J. Phytopathol.* 147:497–508.

Muller, H. R. A. 1940. Overzicht van de belangrijkste Mangga-ziekten in Nederlandsch Indie (abstracted in Rev. Appl. Mycol. (1940) 19:355).

Narasimhan, M. J. 1954. Malformation of panicles in mango incited by spp. of *Eriophyes*. *Curr. Sci.* 23:297–8.

Narasimhan, M. J. 1959. Control of mango malformation disease. *Curr. Sci.* 28:254–5.

Nariani, T. K. and Seth, M. L. 1962. Role of eriophyid mites in causing malformation disease of mango. *Indian Phytopathol.* 15:231–4.

Nicholson, R. I. D. and van Staden, J. 1988. Cytokinins and mango flower malformation. I. Tentative identification of the complement in healthy and malformed inflorescences. *J. Plant Physiol.* 132:720–4.

Niehaus, E.-M., Munsterkotter, M., Proctor, R. H., Brown, D. W., Sharon, A., Idan, Y., Oren-Young, L., Sieber, C. M., Novak, O., Pencık, A., Tarkowska, D., Hromadova, K., Freeman, S., Maymon, M., Elazar, M., Youssef, S. A., El-Shabrawy, E. S. M., Shalaby, A. B. A., Houterman, P., Brock, N. L., Burkhardt, I., Tsavkelova, E. A., Dickschat, J. S., Galuszka, P., Guldener, U. and Tudzynski, B. 2016. Comparative 'Omics' of the *Fusarium fujikuroi* species complex highlights differences in genetic potential and metabolite synthesis. *Genome Biol. Evol.* 8:3574–99.

Nofal, M. A. and Haggag, W. M. 2006. Integrated management of powdery mildew of mango in Egypt. *Crop Prot.* 25:480–6.

O'Donnell, K., Cigelnik, E. and Nirenberg, H. I. 1998. Molecular systematics and phylogeography of the *Gibberella fujikuroi* species complex. *Mycologia* 90:465–93.

O'Donnell, K., Nirenberg, H. I., Aoki, T. and Cigelnik, E. 2000. A multigene phylogeny of the *Gibberella fujikuroi* species complex: Detection of additional phylogenetically distinct species. *Mycoscience* 41:61–78.

Oosthuyse, S. A. 1998. Cost reduction of powdery mildew control in mango with mono potassium phosphate.*S. Afr. Mango Grow. Assoc. Yearb.* 18:40–2.

Otero-Colina, G., Rodríguez-Alvarado, G., Fernández-Pavía, S., Maymon, M., Ploetz, R. C., Aoki, T., O'Donnell, K. and Freeman, S. 2010. Identification and characterization of a novel etiological agent of mango malformation disease in Mexico, *Fusarium mexicanum* sp. nov. *Phytopathology* 100:1176–84.

Palmateer, A. J., Ploetz, R. C. and van Santen, E. 2006. First occurrence of anthracnose caused by *Colletotrichum gloeosporioides* on Pitaya in South Florida. *Phytopathology* 96:S89.

Palo, M. A. 1933. Sclerotium seed rot and seedling rot of mango. *Philipp. J. Sci.* 52:237–61.

Palti, J., Pinkas, Y. and Chorin, M. 1974. Powdery mildew of mango. *Plant Dis. Reptr.* 58:45–9.

Pardo-De la Hoz, C. J., Calderon, C., Rincon, A. M., Cardenas, M., Danies, G., Lopez-Kleineb, L., Restrepo, S. and Jimenez, P. 2016. Species from the *Colletotrichum acutatum, Colletotrichum boninense* and *Colletotrichum gloeosporioides* species complexes associated with tree tomato and mango crops in Colombia. *Plant Pathol.* 65:227–37.

Patel, M. K., Kulkarni, Y. S. and Moniz, L. 1948. *Pseudomonas mangiferae-indicae*, pathogenic on mango. *Indian Phytopathol.* 1:147–52.

Peterson, R. A., Johnson, G. I., Schipke, L. G. and Cooke, A. W. 1990 Chemical control of stem end rot. *Acta Hort.* 291:304–7.

Phillips, A. J. L., Alves, A., Abdollahzadeh, J., Slippers, B., Wingfield, M. J., Groenewald, J. Z. and Crous, P. W. 2013. The Botryosphaeriaceae: Genera and species known from culture. *Stud. Mycol.* 76:51–167.

Piza, C. de T. (ed.) 1966. Anais do Simposio Sobre a Seca da Mangueira (abstracted in Rev. Appl. Mycol. (1967) 46:378).

Ploetz, R. C. 1994 Distribution and prevalence of *Fusarium subglutinans* in mango trees affected by malformation. *Can. J. Bot.* 72:7–9.

Ploetz, R. C. 2001. Malformation: A unique and important disease of mango, *Mangifera indica* L, pp. 233–47. In Summerell, B. A., Leslie, J. F., Backhouse, D. and Bryden, W. L. (Eds) *Fusarium: Paul E. Nelson Memorial Symposium*. APS Press: St Paul, USA.

Ploetz, R. C. 2003. Diseases of mango, pp. 327–63. In Ploetz, R. C. (Ed.) *Diseases of Tropical Fruit Crops*. CABI Publishing: Wallingford, UK.
Ploetz, R. C. 2017. The mango disease trilogy. *Acta Hort*. (In press).
Ploetz, R. C. and Freeman, S. 2009. Foliar, floral and soilborne diseases, pp. 231–302. In Litz, R. E. (Ed.) *The Mango: Botany, Production and Uses*. 2nd edition. CABI: Wallingford, UK.
Ploetz, R. C., Zentmyer, G. A., Nishijima, W., Rohrbach, K. and Ohr, H. D. (Eds). 1994. *Compendium of Tropical Fruit Diseases*. APS Press: St Paul, USA.
Ploetz, R. C., Benscher, D., Vazquez, A., Colls, A., Nagel, J. and Schaffer, B. 1996a. A re-evaluation of mango decline in Florida. *Plant Dis*. 80:664–8.
Ploetz, R. C, Vazquez, A. and Benscher, D. 1996b. First report of *Fusarium decemcellulare* as a pathogen of mango in the United States. *Plant Dis*. 80:1207.
Ploetz, R. C., Benscher, D., Dorey, A. J. and Vazquez, A. 2000. The epidemiology, control and cause of sooty blotch of carambola, *Averrhoa carambola* L., in South Florida. *Fruits* 55:241–52.
Ploetz, R., Zheng, Q., Vazquez, A. and Abdel Sattar, M. A. 2002. Current status and impact of mango malformation in Egypt. *Int. J. Pest Manage*. 48:279–85.
Ploetz, R. C., Hulcr, J., Wingfield, M. J. and de Beer, Z. W. 2013. Ambrosia and bark beetle-associated tree diseases: Black Swan events in tree pathology? *Plant Dis*. 95:856–72.
Pohronezney, K. and Marlatt, R. B. 1982. *Some Common Diseases of Mango in Florida*. Plant Pathology Fact Sheet PP-23. University of Florida: Gainesville.
Ponmurugan, P., Saravanan, D., Ramya, M., Srinivasan, T. R., Baby, U. I. and Ajay, D. 2009. Studies on *Cephaleuros parasiticus* Karst, a pathogenic alga causing red rust disease in tea plantations. *J. Plant. Crops* 37:70–3.
Prakash, O. 1988. Sooty mould disease of mango and its control. *Int. J. Trop. Plant Dis*. 9: 277–80.
Prakash, O. 1990. Report of the plant pathologist. Annual Report of CIHNP (CMRS), Lucknow for 1989–1990.
Prakash, O. 1991. Sooty mould disease of mango and its control. *Int. J. Trop. Plant Dis*. 9:277–80.
Prakash, O. and Singh, U. N. 1976. Basal rot of mango seedling caused by *Sclerotium rolfsii*. *Indian J. Mycol. Plant Pathol*. 6:75.
Prakash, O. and Singh, U. N. 1980. Root rot and damping off of mango seedlings caused by *Rhizoctonia solani*. *Indian J. Mycol. Plant Pathol*. 10:69.
Prakash, O. and Srivastava, K. C. 1987. *Mango Diseases and their Management – A World Review*. Tommorow's Printer: New Delhi, India.
Prakash, O. and Raoof, M. A. 1989. Die back disease of mango (*Mangifera indica*), its distribution, incidence, cause and management. *Fitopatol. Bras*. 14:207–15.
Prasad, A., Singh, H. and Shukla, T. N. 1965. Present status of mango malformation disease. *Indian J. Hort*. 22:254–65.
Prakash, O., Misra, A. K. and Raoof, M. A. 1994. Studies on mango bacterial canker disease. *Biol. Mem*. 20:95–107.
Prusky, D. 1994. Alternaria rot (black spot), pp. 34–5. In Ploetz, R. C., Zentmyer, G. A., Nishijima, W., Rohrbach, K. and Ohr, H. D. (Eds) *Compendium of Tropical Fruit Diseases*. APS Press: St Paul, USA.
Prusky, D., Fuch, Y. and Yanko, U. 1983. Assessment of latent infections as a basis for control of post harvest disease of mango. *Plant Dis*. 67:816–18.
Prusky, D. and Keen, N. T. 1993. Involvement of preformed antifungal compounds in the resistance of subtropical fruits to fungal decay. *Plant Dis*. 77:114–19.
Prusky, D., Shalom, Y., Kobiler, I., Akerman, M. and Fuchs, Y. 2002. The level of quiescent infection of *Alternaria alternata* in mango fruits at harvest determines the postharvest treatment applied for the control of rots during storage. *Postharvest Biol. Tech*. 25:339–47.
Prusky, D., Kobiler, I., Miyara, I. and Alkan, N. 2009. Fruit diseases, pp. 210–30. In Litz, R. E. (Ed.) *The Mango: Botany, Production and Uses*. 2nd edition. CABI: Wallingford, UK.
Pruvost, O., Boyer, C. and Grygiel, P. 2014. First report of *Xanthomonas citri* pv. *mangiferaeindicae* causing mango bacterial canker on *Mangifera indica* in Ivory Coast. *Plant Dis*. 98:1740.

Pruvost, O., Boyer, C., Vital, K., Verniere, C., Gagnevin, L., Austin, L. D., de Bruno, L. and Rey, J. Y. 2011a. First report in Ghana of *Xanthomonas citri* pv. *mangiferaeindicae* causing mango bacterial canker on *Mangifera indica*. *Plant Dis*. 95:774.

Pruvost, O., Boyer, C., Vital, K., Verniere, C., Gagnevin, L. and Somda, I. 2011b. First report in Burkina Faso of *Xanthomonas citri* pv. *mangiferaeindicae* causing bacterial canker on *Mangifera indica*. *Plant Dis*. 95:1312–13.

Pruvost, O., Boyer, C., Vital, K., Verniere, C., Gagnevin, L. and Traore, Y. N. 2012. First report in Mali of *Xanthomonas citri* pv. *mangiferaeindicae* causing mango bacterial canker on *Mangifera indica*. *Plant Dis*. 96:581.

Pruvost, O. and Luisetti, J. 1991a. Attempts to develop a biological control of bacterial black spot of mangoes. *Acta Hort*. 291:324–37.

Pruvost, O. and Luisetti, J. 1991b. Effect of time of inoculation with *Xantbomonas campestris* pv. *mangiferaeindicae* on mango fruits susceptibility. Epiphytic survival of *X. c.* pv. *mangiferaeindicae* on mango fruits in relation to disease development. *J. Phytopathol*. 133:139–51.

Pruvost, O., Couteau, A. and Luisetti, J. 1989. Efficacite de differentes formulations chemiques pour Jutter contre la maladie des taches noires de la mangue (*Xanthomonas campestris* pv. *mangiferaeindicae*). *Fruits* 44:343–50.

Pruvost, O., Couteau, A. and Luisetti, J. 1990. Development of bacterial black spot of mangoes and epiphytic populations of the pathogen (*Xanthomonas campestris* pv. *mangiferaeindicae*) under natural conditions in Reunion Island. *Fruits* 45:125–40.

Pruvost, O., Couteau, A. and Luisetti, J. 1992. Pepper tree (*Schinus terebenthifolius* Radii), a new host plant for *Xanthomonas campestris* pv. *mangiferaeindicae*. *J. Phytopathol*. 135:289–98.

Pruvost, O., Couteau, A., Vemiere C. and Luisetti, J. 1993. Epiphytic survival of *Xanthomonas campestris* pv. *mangiferaeindicae* on mango buds. *Acta Hort*. 341:337–44.

Pruvost, O., Roumagnac, P., Gaube, C., Chiroleu, F. and Gagnevin, L. 2005. New media for the semiselective isolation and enumeration of *Xanthomonas campestris* pv. *mangiferaeindicae*, the causal agent of mango bacterial black spot. *J. Appl. Microbiol*. 99:803–15.

Pruvost, O., Savelon, C., Boyer, C., Chiroleu, F., Gagnevin, L. and Jacques, M.-A. 2009. Populations of *Xanthomonas citri* pv. *mangiferaeindicae* from asymptomatic mango leaves are primarily endophytic. *Microb. Ecol*. 58:170–8.

Purseglove, J. 1968. *Tropical Crops: Dicotyledons*, Longman Publishers, Harlow, UK.

Ray, S. K. 2003. Efficacy of different fungicides for management of powdery mildew of mango in West Bengal. *J. Mycopathol. Res*. 41:153–5

Reckhaus, P. and Adamou, I. 1987. Hendersonula dieback of mango in Niger. Plant Dis. 71:1045.

Reuveni, M., Harpaz, M. and Reuveni, R. 1998. Integrated control of powdery mildew on field-grown mango trees by foliar sprays of mono-potassium phosphate fertilizer, sterol inhibitor fungicides and the strobilurin Kresoxym-methyl. *Eur. J. Plant Pathol*. 104:853–60.

Ribiero, I. J. A. 1980. Seca de mangueira. Agentes causais e estudio da molesta. Anais do I Simposio Brasiliero Sobre a Cultura de Mangueira, Sociedad Brasileira de Fruticultura, Jaboticobal, pp. 123–30.

Ribiero, I. J. A. 1997. Doenças da mangueira (*Mangifera indica* L.). In Kimati, H., Amorim, A., Bergamin-Filho, A., Camargo, L. E. A. and Rezende, J. A. M. (Eds) *Manual de Fitopatologia, Volume 2: Doenças das Plantas Cultivadas*. Agronômica Ceres: São Paulo, Brazil.

Ribiero, I. J. A., Rosetto, C. J., Donadio, L. C., Sabino, J. C. Martins, A. L. M. and Gallo, P. B. 1995. Mango wilt. XIV Selection of mango (*Mangifera indica* L.) rootstocks resistant to the mango wilt fungus *Ceratocystis fimbriata* Ell. & Halst. *Acta Hort*. 370:159–61.

Ridgeway, R. (Ed.) 1989. *Mango Pests and Disorders*. Department of Primary Industries: Brisbane, Australia.

Rivera-Vargas, L. Lugo-Noel, Y., McGovern, R. J., Sijo, T. and Davis, M. J. 2006. Occurrence and distribution of *Colletotrichum* spp. on mango (*Mangifera indica* L.) in Puerto Rico and Florida, USA. *Plant Pathol*. J. 5:191–8.

Rodríguez-Alvarado, G., Fernández-Pavía, S. P. and Ploetz, R. C. 2006. *Fusarium* sp. characterization causing mango malformation in Michoacan Mexico. *Phytopathology* 96:S99.

Rodríguez-Alvarado, G., Fernández-Pavía, S. P., Ploetz, R. C. and Valenzuela-Vázquez, M. 2008. A *Fusarium* sp. different from *Fusarium oxysporum* and *F. mangiferae* is associated with mango malformation in Michoacan, Mexico. *Plant Pathol.* 57:781

Rossetto, C. J., Ribeiro, I. J. A., Igue, T. and Gallo, P. B. 1996. Mango wilt: XV. Varietal resistance against two isolates of *Ceratocystis fimbriata*. *Bragantia* 55:117–21.

Rossman, A. Y., Samuels, G. J., Rogerson, C. T. and Lowen, R. 1998. Genera of Bionectriaceae, Hypocreaceae and Nectriaceae (Hypocreales, Ascomycetes). *Stud. Mycol.* 42:1–248.

Saeed, A. and Schlosser, E. 1972. Effect of some cultural practices on the incidence of mango malformation. *Z. Pflanzenkran Pflanzensch.* 79:349–51.

Saaiman, W. C. 1996. Biology and control of *Nattrassia mangiferae*. *Acta Hort.* 455:558–64.

Sanahuja, G., Palmateer, A. J., Crane, J. H. and Wasielewski, J. 2017. *Cephaleuros Species – Plant Parasitic Algae- on Tropical Fruit Trees in Florida*. EDIS publ. Univ. Florida: Gainesville (In press).

Sanahuja, G., Ploetz, R. C., Lopez, P., Konkol, J. L., Palmateer, A. J. and Pruvost, O. 2016. Bacterial canker of mango, *Mangifera indica*, caused by *Xanthomonas citri* pv. *mangiferaeindicae*, confirmed for the first time in the Americas. *Plant Dis.* 100:2520.

Sanders, G. M., Verschoor, J. A., van Wyngaard, S., Korsten, L. and Kotze, J. M. 1994. Production of monoclonal antibodies against *Xanthomonas campestris* pv. *mangiferaeindicae* and their use to investigate differences in virulence. *J. Appl. Bacteriol.* 77:509–18.

Santos, A. M. G., Lins, S. R. O., da Silva, J. M. and de Oliveira, S. M. A. 2015. Low doses of gamma radiation in the management of postharvest *Lasiodiplodia theobromae* in mangos. *Braz. J. Microbiol.* 46:841–7.

Schaffer, B. 1994. Decline, p. 43. In Ploetz, R. C., Zentmyer, G. A., Nishijima, W., Rohrbach, K. and Ohr, H. D. (Eds), *Compendium of Tropical Fruit Diseases*. APS Press: St Paul, USA.

Schaffer, B., Larson, K. D., Snyder, G. H. and Sanchez, C. A. 1988. Identification of mineral deficiencies associated with mango decline by DRIS. *HortScience* 23:617–19.

Schoeman, M. H., Manicom, B. Q. and Wingfield, M. J. 1995. Epidemiology of powdery mildew on mango blossoms. *Plant Dis.* 79:524–8.

Senghor, A. L., Sharma, K., Kumar, P. L. and Bandyopadhyay, R. 2012. First report of mango malformation disease caused by *Fusarium tupiense* in Senegal. *Plant Dis.* 96:1582.

Shivas, R. G., Tan, Y. P., Edwards, J., Dinh, Q., Maxwell, A., Andjic, V., Liberato, J. R., Anderson, C., Beasley, D. R., Bransgrove, K., Coates, L. M., Cowan, K., Daniel, R., Dean, J. R., Lomavatu, M. F., Mercado-Escueta, D., Mitchell, R. W., Thangavel, R., Tran-Nguyen, L. T. T. and Weir, B. S. 2016. *Colletotrichum* species in Australia. *Australas. Plant Pathol.* 45:447–64.

Silveira, S. F., Harrington, T. C., Mussi-Dias, V., Engelbrecht, C. J. B., Alfenas, A. C. and Silva, C. R. 2006. *Annona squamosa*, a new host of *Ceratocystis fimbriata*. *Fitopatol. Bras.* 31:394–7.

Singh, L. B. 1968. *The Mango: Botany, Cultivation, and Utilization*. Leonard Hill: London.

Singh, Z. and Dhillon, B. S. 1989. Hormonal changes associated with vegetative malformation of mango (*Mangifera indica* L.). *J. Phytopathol.* 125:193–7.

Singh, R. N., Majumder, P. K., Sharma, D. K., Sinha, G. C. and Bose, P. C. 1974. Effect of de-blossoming on the productivity of mango. *Sci. Hort.* 2:399–403.

Singh, Z., Singh, L., Arora, C. L. and Dhillon, B. S. 1994. Effect of cobalt, cadmium, and nickel as inhibitors of ethylene biosynthesis on floral malformation, yield, and fruit quality of mango. *J. Plant Nutr.* 17:1659–70.

Sinniah, G. D., Adikaram, N. K. B., Vithanage, I. S. K., Abayasekara, C. L., Maymon, M. and Freeman, S. 2013. First report of mango malformation disease caused by *Fusarium mangiferae* in Sri Lanka. *Plant Dis.* 97:427.

Slippers, B., Johnson, G. I., Crous, P. W., Coutinho, T. A., Wingfield, B. D. and Wingfield, M. J. 2005. Phylogenetic and morphological re-evaluation of the *Botryosphaeria* species causing diseases of *Mangifera indica*. *Mycologia* 97:99–110.

Slippers, B. and Wingfield, M. J. 2007. Botryosphaeriaceae as endophytes and latent pathogens of woody plants: Diversity, ecology and impact. *Fungal Biol. Rev.* 21:90–106.

Smith, P. F. and Scudder, G. K. 1951. Some studies of mineral deficiency symptoms in mango. *Proc. Fla. State Hort. Soc.* 64:243–8.

Snowdon, S. 1990. *Color Atlas of Post-harvest Diseases & Disorders of Fruits & Vegetables. Volume 1. General Introduction & Fruits.* CRC Press: Boca Raton, USA.

Sreenivasaprasad, S., Mills, P. R., Meehan, B. M. and Brown, A. E. 1996. Phylogeny and systematics of 18 *Colletotrichum* species based on ribosomal DNA spacer sequences. *Genome* 39:499–512.

Steenkamp, E. T., Wingfield, B. D., Coutinho, T. A., Wingfield, M. J. and Marasas, W. F. O. 1999. Differentiation of *Fusarium subglutinans* f. sp. *pini* by histone gene sequence data. *Appl. Environ. Microbiol.* 65:3401–6.

Steenkamp, E. T., Britz, H., Coutinho, T. A., Wingfield, B. D., Marasas, W. F. O., Wingfield, B. D., Marasas, W. F. O. and Wingfield, M. J. 2000. Molecular characterization of *Fusarium subglutinans* associated with mango malformation. Mol. *Plant Pathol.* 1:187–93.

Steyn, P. L., Viljoen, N. M. and Kotze, J. M. 1974. The causal organism of bacterial blackspot of mangoes. *Phytopathology* 64:1400–4.

Summanwar, A. S, Raychaudhuri, S. P. and Phatak, S. C. 1966. Association of the fungus *Fusarium moniliforme* Sheld. with the malformation in mango (*Mangifera indica* L.) *Indian Phytopathol.* 19:227–8.

Taba, S., Takaesu, K., Ooshiro, A., Moromizato, Z. and Takushi, T. 2004. Anthracnose of mango caused by *Colletotrichum acutatum. Jpn. J. Trop. Agric.* 48:57–61.

Tarnowski, T. L. 2009. Using Molecular Analysis to Investigate Phylogenetic Relationships in two tropical pathosystems: Witches' broom of cacao, caused by *Moniliophthora perniciosa*, and mango anthracnose, caused by *Colletotrichum spp.* Ph.D. dissertation. Univ. Florida, Gainesville, USA.

Tavares, S. C. C. de H., da Cruz, S. C., de Siqueira, J. G., Lima, M. L. C., Silva, P. C. G. C., de Neves, R. A. F. and Menezes, C. A. 2004. Alternative control to powdery mildew (*Erysiphe poligoni* d. c. Sin) on mango in submedio Sao Francisco river in Brazilian semi-arid region. *Acta Hort.* 645:471–4.

Terao, D., Campos, J. S. D., Benato, E. A. and Hashimoto, J. M. 2015. Alternative strategy on control of postharvest diseases of mango (*Mangifera indica* L.) by use of low dose of ultraviolet-C irradiation. *Food Eng. Rev.* 7:171–5.

Tiwari, R. K. S., Ashok, S., Rajput, M. L. and Bisen, R. K. 2006. Relative susceptibility of mango varieties to powdery mildew caused by *Oidium mangiferae. Adv. Plant Sci.* 19:181–3.

Trakunyingcharoen, T., Cheewangkoon, R., To-anun, C., Crous, P. W., van Niekerk, J. M. and Lombard, L. 2014. Botryosphaeriaceae associated with diseases of mango (*Mangifera indica*). *Australas. Plant Pathol.* 43:425–38.

Trakunyingcharoen, T., Lombard, L., Groenewald, J. Z., Cheewangkoon, R., To-Anun, C. and Crous, P. W. 2015. Caulicolous Botryosphaeriales from Thailand. *Persoonia* 34:87–99.

Tsao, P. H., Luzaran, P. B., de los Santos, A. B., Portales, L. A., Gochango, A. M. and Gruber, L. C. 1994. Phytophthora crown and foot rot of mango detected in Philippine nurseries. *Plant Dis.* 78:100.

Turner, P. D. 1960. Annual report of the West African Research Institute 1958–1959 (as cited in Chee, 1969).

Udayanga, D., Manamgoda, D. S., Liu, X., Chukeatirote, E. and Hyde, K. D. 2013.What are the common anthracnose pathogens of tropical fruits? *Fungal Divers.* 61:165–79.

Uppal, B. N., Patel, M. K. and Kamat, M. N. 1941. Powdery mildew of mango. *J. Univ. Bombay* 9:12–16.

van Staden, J. and Nicholson, R. J. D. 1989. Cytokinins and mango flower malformation. The cytokinin complement produced by *Fusarium moniliforme* and the ability of the fungus to incorporate (8–14C) adenine into cytokinins. *Physiol. Mol. Plant Pathol.* 35:423–31.

van Staden, J., Bayley, A. D. and Macrae, S. 1989. Cytokinins and mango flower malformation. III. The metabolism of (3H) iso-pentenyladenine and (8–14C) zeatin by *Fusarium moniliforme. Physiol. Mol. Plant Pathol.* 35:433–8.

van Wyk, M., Al Adawi, A. O., Wingfield, B. D., Al Subhi, A. M., Deadman, M. L. and Wingfield, M. J. 2005. DNA based characterization of *Ceratocystis fimbriata* isolates associated with mango decline in Oman. *Australas. Plant Pathol.* 34:587–90.

van Wyk, M., Al Adawi, A. O., Khan, I. A., Deadman, M. L., Wingfield, B. D., Ploetz, R. C. and Wingfield, M. J. 2007. *Ceratocystis manginecans sp. nov.*, causal agent of the destructive mango die-back disease in Oman and Pakistan. *Fungal Divers.* 27:213–30.

van Wyk, M., Wingfield, D. B., Al-Adawi, A. O., Rosseto, C. J., Ito, M. F. and Wingfield, M. J. 2011. Two new Ceratocystis species associated with mango disease in Brazil. *Mycotaxon* 117:381–404.

Varma, A., Lele, V. C., Raychauduri, S. P., Ram, A. and Sang, A. 1974. Mango malformation: A fungal disease. *Phytopathol. Z.* 70:254–7.

Verma, O. P. and Singh, R. D. 1970. Epidemiology of mango dieback caused by *Botyodiplodia theobromae* Pat. *Indian J. Agric. Sci.* 40:813–18.

Viegas, A. P. 1960. Mango blight. *Bragantia* 19:163–82 (abstracted in Rev. Appl. Mycol. (1963) 42:696).

Vieira, W. A. S., Michereff, S. J., de Morais, M. A. Jr., Hyde, K. D. and Câmara, M. P. S. 2015. Endophytic species of *Colletotrichum* associated with mango in northeastern Brazil. *Fungal Divers.* DOI: 10.1007/s13225-014-0293-6.

Viljoen, N. M. and Kotze, J. M. 1972. Bacterial black spot of mango. *Citrus Grow. Sub-Trop. Fruit J.* June, pp. 5–8.

Vivekananthana, R., Ravia, M., Saravanakumara, D., Kumar, N., Prakasama, V. and Samiyappana, R. 2004. Microbially induced defense related proteins against postharvest anthracnose infection in mango. *Crop Prot.* 23:1061–7.

Wagle, P. V. 1928. Studies in the shedding of mango flowers and fruits. Part I, Memoirs of the Department of Agriculture. *India Botanical Sci.* 8:219–49.

Waterhouse, G. M. 1970. *The Genus Phytophthora de Bary*, 2nd edn, Mycological Papers, no. 122. Commonwealth Mycological Institute: Kew, Surrey.

Webster, R. K. and Butler, E. E. 1967. A morphological and biological concept of the species *Ceratocystis fimbriata*. *Can. J. Bot.* 45:1457–68.

Weir, B. S., Johnston, P. R. and Damm, U. 2012. The *Colletotrichum gloeosporioides* species complex. *Stud. Mycol.* 73:115–80.

Weng, F. J. and Chuang, T. Y. 1995. Grouping of mango anthracnose fungus in Taiwan. *Plant Prot. Bull.* 3:295–309.

Williamson, S. M. and Sutton, T. B. 2000. Sooty blotch and flyspeck of apple: Etiology, biology, and control. *Plant Dis.* 84:714–24.

Wood, S. L. 1982. The bark and ambrosia beetles of North and Central America (Coleoptera: Scolytidae), a taxonomic monograph. *G. Basin Nat., Mem.* 6:1–1356.

Woudenberg, J. H. C., Seidl, M. F., Groenewald, J. Z., de Vries, M., Stielow, J. B., Thomma, B. P. H. J., and Crous, P. W. 2015. *Alternaria* section *Alternaria*: Species, formae speciales or pathotypes? *Stud. Mycol.* 82: 1–21.

Yasuhara-Bell, J., de Silva, A. S., Alvarez, A. M., Shimabuku, R. and Ko, M. 2013. First report in Hawai'i of *Xanthomonas citri* pv. *mangiferaeindicae* causing bacterial black spot on *Mangifera indica*. *Plant Dis.* 97:1244.

Youssef, S. A., Maymon, M., Zveibil, A., Klein-Gueta, D., Sztejnberg, A., Shalaby, A. A. and Freeman, S. 2007. Epidemiological aspects of mango malformation disease caused by *Fusarium mangiferae* and source of infection in seedlings cultivated in orchards in Egypt. *Plant Pathol.* 56:257–63.

Zainuri. 2006. Defence mechanisms and induced resistance in 'Kensington Prode' mango. PhD thesis, Univ. Queensland: Brisbane, Queensland, Australia.

Zainuri, J. D. C., Wearing, A. H., Coates, L. and Terry, L. 2001. Effects of phosphonate and salicyclic acid treatments on anthracnose disease, pp. 149–50. In *Australian Postharvest Horticultural Congress*. Brisbane, Queensland.

Zea-Bonilla, T., Martín-Sánchez, P. M., Hermoso, J. M., Carmona, M. P., Segundo, E. and Pérez-Jimenéz, R. M. 2007. First report of *Phytophthora citricola* on *Mangifera indica* in Spain. New Disease Reports. *Plant Pathol.* 56:356.

Zeng, K. F., Cao, J. K. and Jiang, W. 2006. Enhanced disease resistance in harvested mango (*Mangifera indica* L.) fruit by salicylic acid. *J. Sci. Food Agric.* 85:694–8.

Zeng, K. and Waibo, J. 2005. Induced resistance to anthracnose (*Colletotrichum gloeosoporiodes*) in mango by postharvest treatment with salicylic acid. *J. China Agric. Univ.* 10:36–40.

Zhan, R.-L., Yang, S.-J., Ho, H.-H., Liu, F., Zhao, Y.-L., Chang, J.-M. and He, Y.-B. 2010. Mango malformation disease in South China caused by *Fusarium proliferatum*. *J. Phytopathol.* 158:721–5.

Zhan, R.-L., Yang, S-.J., Liu, F., Zhao, Y.-L., Chang, J.-M. and He, Y.-B. 2012. First report of *Fusarium mangiferae* causing mango malformation in China. Plant Dis. 96:762.

Zheng, Q. and Ploetz, R. 2002. Genetic diversity in, and development of a PCR assay for identifying, the mango malformation pathogen. *Plant Pathol.* 51:208–16.

Zombre, C., Sankara, P., Ouedraogo, S. L., Wonni, I., Boyer, K., Boyer, C., Terville, M., Javegny, S., Allibert, A., Verniere, C. and Pruvost, O. 2016. Natural infection of cashew (*Anacardium occidentale*) by *Xanthomonas citri* pv. *mangiferaeindicae* in Burkina Faso. *Plant Dis.* 100:718–23.

Zombre, C., Sankara, P., Ouedraogo, S. L., Wonni, I., Pruvost, O., Boyer, C., Verniere, C., Adandonon, A., Vayssieres, J. F. and Ahohuendo, B. C. 2015. First report of *Xanthomonas citri* pv. *mangiferaeindicae* causing mango bacterial anker on *Mangifera indica* L. in Benin. *Plant Dis.* 99:1854.

Chapter 17

Integrated pest management and biological pest control strategies in mango cultivation

Stefano De Faveri, Department of Agriculture and Fisheries, Australia

1 Introduction

There have been many definitions of integrated pest management (IPM) over the years. Kogan (1998) reviewed the history behind the establishment of IPM. In the early twentieth century, pest management was based on the set of actions required to avoid, reduce or delay the impact of pests. However in the 1940s, with the advent of organosynthetic insecticides, the focus was on new chemistry to the detriment of biological control. In the 1960s, the term 'integrated control' was conceived where insecticides were used such that their usage did not interfere with natural pest control (Stern et al., 1959). The publication of Silent Spring (Carson, 1962) accelerated the acceptance of integrated control. In his review, Kogan (1998) suggested the following definition for IPM: 'IPM is a decision support system for the selection and use of pest control tactics, singly or harmoniously coordinated into a management system, based on cost/benefit analyses that take into account the interests of and impacts on producers, society and the environment'. This definition considers all relevant stakeholders; ultimately, decisions are based on economics/profit.

The environment may be impacted by the use of broad-spectrum insecticides that could impact waterways or wildlife. The consumer could also be affected by insecticide residues. A simplified definition could be as follows: 'A strategy where a variety of management options were considered and applied harmoniously to manage pests'. Whatever the definition, the ultimate aims are to (1) return a profit and (2) avoid using insecticides

http://dx.doi.org/10.19103/AS.2017.0026.21

wherever possible to benefit the environment, the applicator and the consumer. The aim is to produce good-quality fruits, in our case mangoes, at a minimal cost and with the least impact on the environment.

2 Problems with overreliance on insecticides

In the late nineteenth century and early twentieth century, crop protection was based on knowledge of pest biology and cultural practices. However with the advent of organosynthetic insecticides such as the organochlorides, dichlorodiphenyltrichloroethane and later the organophosphates, this shifted to routine applications of these broad-spectrum insecticides. The period between the 1940s and the 1960s is termed as the dark ages of pest control and this was highlighted by Carson (1962). Despite the concerns associated with the use of pesticides, there has been a dramatic increase in their use since the 1960s (Kogan, 1998). In 2012, 483 million kilograms of insecticides were used worldwide (Atwood and Paisley-Jones, 2017) compared to 434 million kilograms in 2006 (Grube et al., 2011), an increase of 49 million kilograms in six years.

Jeyaratnam (1990) estimated that over 25 million agricultural workers worldwide are unintentionally poisoned every year. The issues are more severe in developing countries where exposure control programmes are limited or non-existent (Alavanja, 2009). Pimentel (2005) outlines the health costs of pesticide applications. He outlines the neurological effects, respiratory and reproductive effects and development of cancer. Detectable levels of pesticides are often found in foods. Pimentel (2005) quotes figures of 73% and 90% in surveyed US food baskets containing fruits and vegetables and up to 5% of foods tested in 1997 contained residues above the FDA tolerance levels.

The environment has also suffered from indiscriminate and improper use of insecticides. Surface waters have been contaminated by insecticides from run off and leaching into groundwater. Run-off contributes to estuary decline and causes harm to non-target organisms. Poor insecticide application can also lead to non-target contamination. Aktar (2009) attributes USD 8 billion annual economic impact of pesticides on non-target species in developing countries. Hardy et al. (1996) chronicled the human and environmental health concerns from pesticide use. Pimentel (2005) estimates the environmental and economic costs of pesticide application in the United States. He concluded that indirect environmental and public health losses due to pesticide use cost USD 12 billion annually (based on 2005 calculations). Although he believed the figure would be double if other costs were included, he concluded that pesticide use would not be profitable if all costs were considered.

Insecticide resistance: The insecticide resistance action committee defines insecticide resistance as 'a heritable change in the sensitivity of a pest population that is reflected in the repeated failure of a product to achieve the expected level of control when used according to the label recommendation for that pest species'. Resistance occurs as a result of overuse or misuse of an insecticide against a pest, which results in the selection of resistant individuals and evolution of populations that are resistant to that insecticide. Cross resistance occurs when populations become resistant to different chemicals within the same group. This is the mode of action which is the same and targets the same site within the pest (http://www.irac-online.org/documents/moa-classification/?ext=pdf). Some examples of insecticide resistance within mango pests are the leafhoppers, *Idioscopus clypealis* (Lethierry) (Elbert et al., 2008) and Tea red spider mite, *Oligonychus coffeae* (Nietner) (Roy et al., 2009). The chilli thrips, *Scirtothrips dorsalis* Hood, have become a major issue for mangoes on the island of Mindanao in the Philippines resulting in reduced quality. Recent research by Medina (C. Medina, Entomologist, University Philippines, Los Banos, unpublished data, 2017) has shown major resistance to profenofos compared to other populations in the Philippines.

The cost of developing new pesticides has increased dramatically and has resulted in steady decline despite investments of nearly 12% of proceeds from agricultural sales (Hardy et al., 1996). The increased costs of development have resulted in very expensive products that may not be economically viable. This would be the case in developing countries and results in the use of older broad-spectrum insecticides. The costs associated with registering chemicals are also high, and chemical companies are reluctant to pursue label registrations for smaller industries as costs are usually greater than returns.

Bees are vital to the pollination of many crops. But most insecticides are harmful to bees. Pimentel (2005) estimated that pesticide-induced pollination losses contributed to 10% of all crop losses. The losses are attributed to death and lack of pollination when moving colonies to allow insecticide application as well as deaths related to moving. Lack of pollination results in poor yields and quality.

Predators and parasitoids play a major role in controlling arthropod pests. However, the use of broad-spectrum insecticides has a negative impact on beneficial insects. This then results in secondary/minor pests which are not affected by the pesticide being elevated to major pest status, which in turn results in more pesticides being applied. Pimentel (2005) estimated that the extra costs and losses related to pesticide-induced problems amounted to investing USD 257 million in extra pesticides, with an added USD 260 million in crop losses. In Queensland, broad-spectrum insecticide sprays directed at fruit fly will exacerbate mango scale and tea red spider mite. In fact, De Faveri (Entomologist, Department of Agriculture and Fisheries, Queensland, unpublished data) sprayed synthetic pyrethroids to a mango orchard to flare up mango scale in order to conduct insecticide trials.

3 Mango pests

Mangoes around the world have a range of pests to deal with. Pests affect leaves, flowers, fruit, twigs, branches and trunks. Not all pests cause economic losses and others are an issue when establishing seedlings. For example, the mango shoot caterpillar *Penicillaria jocosatrix* (Guenee) feeds on young leaves; however, the tree can tolerate moderate infestations without losing yields. But when establishing young trees, the pest will slow tree development. The mango leaf miner is also a pest of establishing trees with tolerable damage occurring in older trees (Anon., 2000; Brough et al., 1994; Cunningham, 1989).

Other pests cause economic losses; however, healthy populations of beneficial insects keep them in check and if carefully managed, the need for insecticide intervention is reduced. For example, biological control agents of the mango scale, *Aulacaspis tubercularis* (Newstead), include wasp parasitoids, ladybirds, thrips, predatory moths and so forth. Selective insecticides such as insect growth regulators (IGRs) and petroleum oils are registered for mango scale. De Faveri and Brown (1995) reported on research for alternative softer options for mango scale control. A case study on biological control agents associated with mango scale is described in Section 5.1.

There are also pests that have become prevalent due to reduction in use of certain broad-spectrum insecticides. One example is the banana spotting bug (BSB), *Amblypelta lutescens lutescens* Distant. This pest was controlled by the use of broad-spectrum insecticides against other pests such as mango scale. However with the use of selective insecticides against mango scales, it has become much more difficult to control, and the beneficial insects (egg parasitoids) associated with it are not effective. As a result, broad-spectrum insecticides are required and these cause flare-ups of secondary pests such as mites, mealybugs and scales. The pest is also difficult to monitor due to its cryptic behaviour and monitoring is often related to damage. However, the Department of Agriculture and Fisheries, Queensland, has developed a pheromone trap that is now available as a monitoring tool. Economic thresholds (ETs) have been recommended and this should result in better targeted control. Improved monitoring should result in locating hotspots which will reduce the need for cover sprays over the entire crop (I. Newton, Entomologist, Department of Agriculture and Fisheries, Queensland, pers. comm., 2017).

For some pests such as the mango seed weevil (MSW), *Sternochetus mangiferae* (Fabricius), and its relative the mango pulp weevil, *Sternochetus frigidus* (Fabricius), there are no known beneficial insects associated with their control (Waite, 2002). As a result, control is restricted to insecticides and cultural methods. The MSWs burrow directly through the flesh and into the seed, not leaving visible signs of entry (Anon., 2000). The mango pulp weevil, on the other hand, burrows and remains in the pulp and therefore results in unmarketable fruit (Dey and Pande, 1987).

Thrips have been recorded as pests of blossom that feed on nectaries and anthers and possibly reduce fruit set. Some of the species involved include *Frankliniella bispinosa* Morgan, *Frankliniella occidentalis* (Pergande), *Thrips imaginis* (Bagnall) and *Thrips palmi* (Karny). *Scirtothrips dorsalis* feed on flowers, fruit, young shoots and leaves (Waite, 2002).

Pests with trade implications include fruit fly. Being a tropical fruit, mangoes are susceptible to fruit flies that are present in most mango-growing regions in the world. Female fruit flies oviposit under the skin of mature fruit and larvae develop and tunnel through the pulp and initiate rots. If eggs are laid just prior to harvest, spoilage will occur in transit and at market (Bateman, 1972; Aluja, 1994; Christenson and Foote, 1960).

Many different fruit flies attack mangoes. In the Americas, the fruit fly pests of mango belong to the genus *Anastrepha* and include *Anastrepha chiclayae, Anastrepha distincta, Anastrepha fraterculus, Anastrepha ludens, Anastrepha obliqua, Anastrepha ocresia, Anastrepha serpentina, Anastrepha suspensa* and *Anastrepha striata* as the major pest fruit flies of mango (White and Elson-Harris, 1992). In Southeast Asia and the Pacific, the main pests of mango in the *Bactrocera* genus include *Bactrocera tryoni, Bactrocera neohumeralis, Bactrocera jarvisi, Bactrocera frauenfeldi, Bactrocera zonata, Bactrocera carambolae and Bactrocera dorsalis. Bactrocera tryoni, Bactrocera neohumeralis,* and *Bactrocera jarvisi* are the main pests in Australia (Hancock et al., 2000; Waite, 2002). The *Bactrocera dorsalis* group was split into a complex of fruit flies by Drew and Hancock in 1994 which included *Bactrocera dorsalis, Bactrocera philippinensis, Bactrocera papayae and Bactrocera carambolae* and in 2005, *Bactrocera invadens* was added (Drew et al., 2005). Drew and Romig (2013) synonymised *Bactrocera philippinensis* with *Bactrocera papayae* in 2013. In 2015, *Bactrocera dorsalis, Bactrocera papayae* and *Bactrocera invadens* were synonymised to the original species *Bactrocera dorsalis* (Schutze et al., 2014, 2015).

The authors concluded that *Bactrocera carambolae* remains a separate species. Since its discovery in Kenya in 2003, *Bactrocera dorsalis* has invaded most of sub-Saharan Africa.

The Mediterranean fruit fly *Ceratitis capitata*, which is found in Africa, South America, Mexico, Hawaii, Spain and other parts of the world including Western Australia, also affects mangoes. Other species in the genus that are pests of mango include *Ceratitis cosyra, Ceratitis rosa* and *Ceratitis catoirii, with Ceratitis rosa and Ceratitis catoirii* being the more damaging of the *Ceratitis* pests in mango in Africa (Waite, 2002).

A list of the major pests of mango is given below:

1 Mango scale, *Aulacaspis tubercularis*
2 Mango tipborer, *Chlumetia etuhtysticha*
3 Mango shoot caterpillar, *Penicillaria jocosatrix*
4 Mango cecid fly, multiple species
5 MSW, *Sternochetus mangiferae*
6 Mango pulp weevil, *Sternochetus frigidus*
7 Mango red-banded caterpillar, *Deanolis sublimbalis*
8 Mango citripestis, *Citripestis euthraphera*
9 Pink wax scale, *Ceroplastes rubens*
10 Mango mealybug, *Rastrococcus*
11 Mango planthopper, *Colgaroides acuminata*
12 Mango leafhoppers, multiple species
13 Tea red spider mite, *Olygonichus coffeae*
14 Tea mosquito bug, *Helopeltis* spp.
15 Mango bud mite, *Aceria mangiferae*
16 Red-banded thrips, *Selenothrips rubricinctus*
17 Flower caterpillars, multiple species
18 Fruit-spotting bug, *Amblypelta lutescens* and *Amblypelta nitida*
19 Coconut bug, *Pseudotheraptus wayi*
20 Fruit flies, *Bactrocera* spp., *Ceratitis* spp. and *Anastrepha* spp.

4 IPM options (biological, cultural and chemical)

IPM is based on three management techniques. They are biological control, cultural control and chemical control. A successful IPM system relies on a good pest and beneficial insect monitoring programme. Monitoring is the most important component of an IPM system. The aim is to detect and manage pest populations before they cause economic damage. Pest:biological control agent ratios are monitored throughout the season and decisions for pest management are based on these trends. Monitoring requires a certain level of experience which is provided by professional pest scouting services. However, professional scouts are not always required. Monitoring is based on random representative sampling of pests and beneficial insects and counting and recording them. Sampling plans are required so that monitoring is representative of the pest situation of the entire orchard. Monitoring will determine changes in pest levels, identify problem areas or 'hot spots' within the orchard, build up a history of pests and beneficial insects and ultimately provide accurate triggers for timing of interventions with economic benefits (Kogan, 1998; Smith et al., 1997; Anon., 2000).

To determine whether interventions based on monitoring results are necessary, they must trigger pest levels that cause economic losses. These levels are based on economic injury levels (EILs) and ET. The EIL is the level at which the cost of damage exceeds the cost of control. It is a mathematical formula that represents the magnitude of the pest-induced losses equal to the cost of the control tactic. The formula is EIL = C/VDK. C is the cost of control, V is the market value of the product, D is the loss in yield and K is the proportional reduction in the pest due to the control tactic (Pedigo et al., 1986; Norton et al., 1976). The EIL is fluid due to the volatility in market value and increases in control costs. The ET (sometimes called action threshold) is what determines the trigger for an intervention. It represents the pest population density whereby control measures must be initiated before the pest level reaches the stated EIL (Anon., 2000; Castle and Naranjo, 2009). The ETs should also take into account a variety of factors such as pest population trajectory, level of natural enemies that may impact the trajectory, climatic conditions and lead in time for intervention. The ET is lower than the EIL.

ET levels for certain pests have been established (Brough et al., 1994), with an updated version in 2000 (Anon., 2000) in Queensland and for the Northern Territory (Anon., 2010). Some pest consultants also derive their own thresholds based on experience and trial and error. The sampling method for the monitoring system should be simple and conducted regularly.

Traps are also important for pest monitoring. For example, fruit fly traps are an essential component of fruit fly area-wide management programmes (Anon., 2003). The Department of Agriculture and Fisheries in Queensland, Australia, has just released a pheromone trap to monitor BSB, *Amblypelta lutescens* trap (Organic Crop Protectants Pty Ltd). This pest is cryptic and monitoring had often been based on damage rather than presence of the insect. Gibb et al. (2007) developed a pheromone for the red-banded mango caterpillar, *Deanolis sublimbalis*. The pheromone was developed to detect and monitor the invasive pest in the quarantine area of Cape York Peninsula and the Torres Straits in Australia.

4.1 Biological control

Smith et al. (1997) defined biological control as the use of natural enemies (parasites, predators or pathogens) of pests. The aim is to establish and maintain populations of natural enemies which will keep pest populations below economically damaging levels. DeBach (1964) defined biological control as 'the action of parasites, predators or pathogens in maintaining organisms at a lower average than would occur in their absence'. Biological control can take the form of classical: The importation and establishment of exotic natural enemies for control of pests, inundative release: the release of mass-reared natural enemies to control a pest, primarily by the released individuals, not their progeny and inoculative release: the release of mass-reared natural enemies to control a pest by action of released individuals and progeny. For many pests, none of their natural enemies are bred by commercial insectaries either due to difficulties in breeding or economic reasons and therefore biological control relies on the maintenance of endemic natural enemies.

Biological control agents include predators that feed directly on their prey, parasitoids that lay their eggs on the host and larvae live inside their hosts and consume them (endoparasitoid) or lay eggs on the host but the larvae live outside the host and consume them (ectoparasitoid) and microbial pathogens (nematodes, fungi, bacteria, viruses and protozoa) which cause lethal infections.

Biological control agents can be host-specific or generalist. Generalist agents prey/parasitise insects from more than one family. Ladybirds and predatory bugs are examples of generalist predators. Parasitoids tend to be more specific. In general, ectoparasitoids are less host-specific than endoparasitoids (Gullan and Cranston, 2000).

Biological control of a range of mango pests is discussed below. While biological control of citrus pests has been very widely researched and there have been multiple releases of exotic parasitoids for pests in citrus and insectaries that specialise in beneficial insects for citrus pests that are in operation (Smith et al., 1997), biological control of mango pests has been far less researched with far less introductions for mango pests found in the literature.

In South Africa, the mango scale *Aulacaspis tubercularis* Newstead is a major pest with reports of losses up to 97% in neglected orchards being known (Daneel and Joubert, 2009). Only one parasitoid (*Encarsia citrina*) was recorded in South Africa and with some predators, biological control was ineffective and indiscriminate spraying of broad-spectrum insecticides occurred. A parasitoid, *Aphytis* sp., was then imported from Thailand, and a beetle, *Cybocephalus binotatus* Grouvelle, was also inadvertently imported in 1995 (Daneel and Joubert, 2009; Joubert et al., 2000). The parasitoid was then identified as *Aphytis chionaspis* Ren, by Neser and Prinsloo (Neser and Prinsloo, 2001). Labuschagne et al. (1996) had confirmed that both beneficial insects had been established in release areas. Subsequent surveys to assess the establishment of the parasitoids had found parasitism rates of up to 77% (Daneel and Joubert, 2009). The classical biological control of mango scale has helped progress IPM for mangoes in South Africa (Joubert et al., 2000).

The first written account of biological control was about the citrus ant *Oecophylla smaragdina* in 304 AD (Van Mele, 2008). Ants were sold in small rush mat bags to citrus growers (Huang and Yang, 1987). Weaver ants have been recorded as predators of pests in mango and other tropical fruits, in nuts and in forestry (Van Mele, 2008; Offenberg, 2015; Peng et al., 2012; Olotu et al., 2012; Peng and Christian, 2005). Weaver ants reduce pests by preying on them, by deterring them physically or by chemical deterrence (Offenberg, 2015). The use of weaver ants for biological control is ideal for resource-poor farmers due its low-cost and low-tech methodology (Offenberg, 2015). Olotu et al. (2012) claimed that *Oecophylla longinoda* effectively controlled the sucking pests, *Helopeltis* spp. and *Pseudotheraptus wayi* in cashews. These pests are also known to attack mangoes. Research trials by Peng and Christian (2005) demonstrated that flower and shoot damage by the leafhopper, *I. nitidulis*, was as low in the weaver ant, *Oecophylla smaragdina* treatment than the insecticide treatment and significantly lower than the untreated control. Peng et al. (2012) also demonstrated the effectiveness of *Oecophylla smaragdina* in controlling the BSB, *Amblypelta lutescens lutescens* (also a pest of mango) in African mahogany plantations in Australia. Abdulla et al. (2016) demonstrated that *Oecophylla longinoda* reduced infestation rates of *Sternochetus mangiferae* which were equal to insecticide-treated plots and much lower than the untreated control. Van Mele et al. (2007) claimed that weaver ants, *Oecophylla longinoda*, reduced infestation levels of fruit flies (*Ceratitis* spp. and *Bactrocera invadens* (now *dorsalis*)) in Benin. They claim that deterrence and ant disturbance were the most important causes for reducing fruit fly damage. The use of this generalist predator will play an important role in biological control of mango pests particularly for resource-poor farmers in developing countries.

Biological control practitioners must be cautious when introducing weaver ants as they also farm pests such as mealybugs and soft scales and also protect them from beneficial insects in return for honeydew that the pests secrete. There will also be detrimental consequences to the biodiversity of other biological control agents associated with mango pests (Bajhau et al., 2002; Smith et al., 1997; Astridge, 2004).

Fruit fly biological control has had limited success, with the most notable exceptions in Hawaii, Florida, Fiji and Southern Europe, but programmes in Australia, Mexico and Costa Rica were less successful (Purcell, 1998). In Australia, several species of parasitoids were released in the 1930s and the 1950s (Clarke et al., 2011). There has now been renewed interest in fruit fly biological control as part of area-wide management programmes. Surveys by Spinner et al. (2011) determined parasitism rates between 1.2% and 28.8% in inland New South Wales towns and they concluded that augmentative releases would be suitable in an IPM programme or systems approach. In Mexico, the Moscafrut facility released fifty million *Anastrepha ludens* pupae parasitised by *Diachasmimorpha longicaudata* weekly in 2005 which are released in several states in the country as part of the sterile insect release programme. The pupae are released in marginal areas such as backyards and results indicate a reduction in fruit flies entering commercial orchards (Montoya et al., 2007).

Other important generalist beneficial insects that are not listed include predatory mites, ladybirds and hover flies.

Microbial pathogens are also important in biological control of mango pests. Some entomopathogens known to be present against mango pests in north Queensland are listed in Table 1.

4.2 Cultural control

Cultural control refers to management practices which impact pests negatively. Cultural practices could either make conditions less favourable to pests, or help create a better environment for beneficial insects. Some examples of cultural management practices include pruning, dust control, crop health and providing shelter and food sources for beneficial insects. Some cultural practices may have opposing effects on different pests. For example, introducing predatory ants will help control *Lepidopterous* pests; however, they will also deter some beneficial insects and help farm certain *Homopteran* pests (Anon., 2000).

Different cultural control techniques and examples of their use for IPM are discussed in more detail below:

4.2.1 Canopy management

Pruning can be used to open canopies to:

1. Create a less favourable environment for pests by opening up the tree and let in more light, heat and wind movement.
2. Reduce populations of certain pests, for example, removing dead wood reduces borers.
3. Improve pesticide application, by aiding spray penetration.

Bondad (1989) recommended pruning to allow sunlight to prevent leafhoppers from sheltering in the tree. Mango planthoppers, *Colgaroides acuminata*, were less prevalent in well-pruned trees compared to dense foliage in a research orchard in Ayr north

Table 1 Entomopathogens found in north Queensland

Pathogen	Host
Hirsutella versicolor	*Colgaroides acuminata*
Nectria flammea	*Aulacaspis tubercularis*
Metarhizium anisopliae	Range of pests
Beauveria bassiana	Range of pests

Queensland (De Faveri, unpublished data). Waite et al. (2000) found that dense mango and carambola trees which allowed minimal light penetration sustained higher levels of BSB, *Amblypelta lutescens lutescens* damage than less dense trees. In a study evaluating IPM options in mangoes, Pinese et al. (2010) found that there was a significant interaction between pesticide applications and canopy management. Pruned and sprayed trees had less mango scale than unpruned, sprayed trees.

4.2.2 Resistant varieties

Certain varietal characteristics may be less favourable to pests than others. Nachiappan and Baksaran (1984) noted that certain Indian mango varieties were resistant to the leafhopper, *Amritodus atkinsoni*. The use of different rootstocks can also be valuable to control pests, for example, Vazquez-Luna et al. (2011) found that Manila mango grafted on Criollo rootstock was less susceptible to fruit flies *Anastrepha obliqua* infestation than Manila seedlings. This was attributed to the increased firmness, 3-Carene level and main flavonoids. Aluja et al. (2014) compared the susceptibility of 15 mango varieties to the fruit flies *Anastrepha ludens* and *Anastrepha obliqua*. The authors claim a possible relationship between fruit volatile profile and variable attractability to susceptible and resistant fruit.

Provision of undergrowth vegetation can enhance biological control by providing environmental requisites such as supplementary food (other pests), complementary food (honeydew, pollen, nectar) and shelter form climatic extremes (Pickett and Bugg, 1998). Conversely, some vegetation types could also provide shelter and alternate hosts for certain pests.

4.2.3 Physical barriers

Physical barriers can be used to protect fruit (bagging) and access to the tree (trunk bands). In the Philippines, growers wrap paper (telephone book pages) around fruit (Anon., 1994; Waite, 2002). There are however issues with fruit wrapping such as loss of blush in blush varieties due to reduced light (I. Bally, Horticulturalist, Department of Agriculture and Fisheries, Queensland, pers. comm., 2017) and increases in pests such as mealybugs due to protection from beneficial insects and a suitable microclimate (De Faveri, unpublished data, 2017). Waite (2002) claimed that expansion of fruit during development resulted in the end of the envelope opening of wrapped fruits and therefore provided access to fruit fly (Waite, 2002). Bondad (1985) demonstrated that only 46% of the fruit is protected from fruit fly. De Faveri (2016) observed that fruit flies had access to wrapped fruit approaching maturity due to open bags (unpublished data). New bags of non-woven material have been evaluated in the Philippines and appear to be an improvement on the paper bags with fewer insects and diseases (J. Eckman, Applied Horticulture Research, pers. comm., 2017).

Netting is also used to exclude pests. Exclusion netting must be fine enough to exclude the major pests and it should allow sufficient light penetration. Netting is not used much in mango due to high capital costs and insufficient returns.

Sanitation is an important cultural control for a range of pests, particularly resident pests with a limited host range. The MSW and pulp weevil are not known to travel distances and are restricted to mango as a host. Removal and destruction of infested fruit will reduce populations of the pests. The host range of the red-banded mango caterpillar is restricted to *Mangifera* spp., but the preference is for the cultivated mango, *Mangifera indica* (Waite, 2002). Although less limited by host range, the key to fruit fly management in IPM is to reduce populations; one of the components of a population reduction system is sanitation. Growers are encouraged to remove and destroy fallen fruit. Along with protein baiting and male annihilation, sanitation played a significant role in the Central Burnett in Queensland (Lloyd et al., 2010).

Along with sanitation, management and removal of alternate hosts are critical. Fruit flies in particular have a large host range, particularly fruit flies associated with mango. Wherever possible, alternate host plants should be removed. In the case of commercial crop hosts, these should be managed to reduce pressure on the mango crop.

Crop manipulation and harvesting practices will also reduce pest problems. For example, manipulation of the Carabao mango in the Philippines to flower early is recommended to reduce leafhopper populations (Anon., 1994). Mature, hard green fruit are less susceptible to mango infestation. Harvesting hard green, mature fruit and culling damaged fruit will reduce the risk of selling infested fruit (Hennessey and Schnell, 2001).

4.3 Chemical control

In an IPM programme, insecticides should only be used when all other options have been exhausted. Broad-spectrum insecticides should be avoided due to the deleterious effects on natural enemies. Selective insecticides will control the target pest and possibly some others, but importantly they will have a lesser impact on natural enemies. Decisions on what insecticide to be applied should be based on accurate identification of the pest and triggered by the ET. Only registered insecticides should be used and at the registered rate and correct withholding period. This will avoid exceeding the maximum residue limit (MRL) in the market and exposing consumers to health issues. Exporting growers should also consider the MRLs of the recipient country. Each country has different MRLs, with some MRLs being zero due to banning of that product in the recipient country. For example, dimethoate is registered for fruit fly control in Australia, but the MRL for exporting mangoes to the United States is zero. Weather conditions must also be taken into account as wind will cause drift onto non-target vegetation. Heat is also an issue that can result in phytotoxicity or rapid degradation of the insecticide, making it ineffective.

The applicator must also consider the consequences to beneficial insects and non-target pests of applying insecticides. For example, the use of most insecticides around flowering is harmful to pollinators. The application of synthetic pyrethroids is harmful to most beneficial insects and may cause non-target pests such as mites, scales and mealybugs to flare. Contact insecticides have a short-term effect; however, residual and systemic insecticides can affect beneficial insects long after application.

Even though considered selective or safe, selective insecticides may still impact beneficial insects. For example, the IGR registered for control of mango scale did not kill adult Coccinellids. However, Magagula and Samways (2000) reported that the IGRs affected egg hatch and immature stages of the Coccinellid, *Chilocorus nigritus* (Fabricius) in South Africa. Wang and Liu (2016) determined that the IGR's pyriproxyfen and fenoxycarb had a slight effect on adult survival of the parasitoid *Encarsia formosa*; however, the effect on immature stages was dramatic, and these effects should be weighed up against alternative options. The selective insecticide may be more disruptive than a contact insecticide, but less harmful than a residual insecticide.

Some pests have been constantly controlled by broad-spectrum insecticides which as in the case for BSB result in flare-ups of other pests. Fruit flies have been successfully controlled by dimethoate cover sprays applied close to harvest. The sprays are systemic and residual and result in annihilation of predators and parasitoids of mango scale, planthoppers and mites. Extra insecticides are then required to control pests that otherwise would be kept in check by biological control agents.

However, protein bait sprays applied to foliage have been successfully applied to manage fruit fly with relatively little impact on beneficial insects. Mango farmers in Queensland have not embraced protein bait spraying due to the blemish that protein baits cause to fruit. Department of Agriculture and Fisheries (DAF), Queensland, scientists have trialled applications of protein baits to tree trunks as a way of avoiding protein bait contact to fruit (De Faveri et al., 2015). A case study on fruit fly management is described in Section 5.3.

Issues with moving away from broad-spectrum insecticides are the resurgence of some pests that were controlled off-target by broad-spectrum insecticide applications for other pests. For example, in Queensland, the move away from the use of methidathion, a broad-spectrum insecticide used for scale control, resulted in increased pest activity of BSB. The reduction in broad-spectrum sprays also has an effect on management of the MSW, as for BSB, MSW is partially managed by broad-spectrum insecticides; however with selective insecticides against other pests, populations increase. Adult females lay eggs and create an incision where sap flows over the egg to protect it. The eggs are unsightly and the fruit quality is downgraded.

As discussed in Section 2, the overuse of insecticides can lead to insecticide resistance by the pest. Pests such as mites, thrips, whiteflies, aphids, moth larvae, etc., have developed resistance to insecticides. The IRAC website, http://www.irac-online.org/documents/moa-classification/?ext=pdf, has further information on major worldwide pests that have developed insecticide resistance. There are also far fewer insecticides being developed by chemical companies to replace current chemicals. Growers must use the available insecticides wisely to prolong their effective life.

5 Case studies

IPM in Queensland Australia

Mango pests in Queensland, Australia, are being managed by some level of IPM. Most growers employ pest consultants for advice on pest management. However, regular sprays are still applied to certain pests for differing reasons. The following case studies demonstrate the research to improve IPM for mango in north Queensland.

5.1 Biological control of mango scale (first reported in De Faveri et al. 2013)

The mango scale, *Aulacaspis tubercularis* Newstead (Fig. 1), is a major pest of Australian mangoes. Feeding damage by all stages of the scale on leaves and twigs results in necrosis and, in severe cases, premature leaf fall and twig and branch dieback. However, the greatest economic impact is due to scale feeding on maturing fruit resulting in superficial, permanent pinkish red discolouration referred by the industry as 'pink spot'. This reduces the price and marketability in domestic markets and renders the fruit unsuitable for export. All mango cultivars are susceptible.

In northern Australia, the scale breeds continuously throughout the year. Despite increased knowledge and improved pest management practices, residual scale-related blemish still remains the greatest cause of fruit rejection and downgrading estimated at between 5% and 10%, respectively. The residual damage levels vary significantly across the industry, primarily due to variable application efficiencies, tree size and variable canopy architecture, but overall fruit downgrading and rejection remain high.

Mango-scale control is expensive, with four to ten applications of primarily broad-spectrum insecticides per season being used commercially. These treatments also limit the potential benefits from endemic beneficial insects and are likely to limit the efficiency of any introduced exotic parasites. Effective and target-specific insecticides are available but costly and therefore have not been adopted widely.

Regular fortnightly monitoring was conducted on three IPM-practising properties in the northern Atherton Tablelands from March 2009 to December 2011 to determine the population dynamics of mango scale and the beneficial insects associated with it. In 2009, the ladybird *Rhyzobius lophanthae* was cultured and augmentatively released at two of

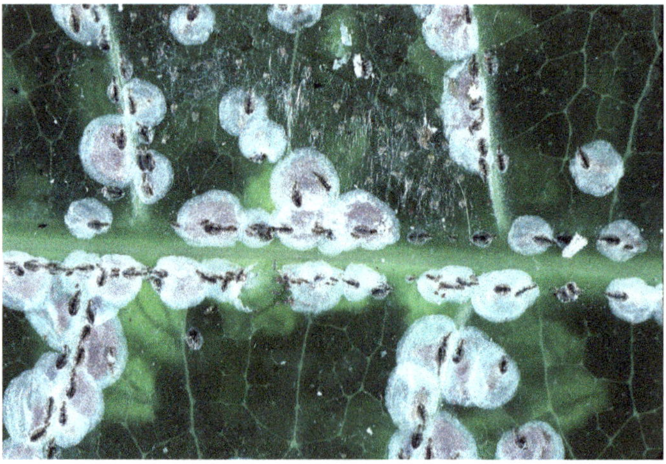

Figure 1 Mango scale, *Aulacaspis tubercularis*, on mango leaf.

the orchards. Assessments were conducted to determine whether augmentative releases of the ladybird would have an additive effect on biological control.

Random leaf samples were collected fortnightly. Mango scale and beneficial insects were counted under a low-power stereo microscope. Live beneficial insects and their parasitic and predatory effect on scale were recorded.

Parasitism rates were calculated for the endemic parasitoid wasps *Aphytis* sp. and *Encarsia citrina* over three sampling periods in October/November 2009.

Harvest assessments were conducted on fruit one week prior to harvest. Fruit were also selected at random from neighbouring IPM blocks in 2009. In 2010, no harvest data were collected due to the low mango yields from those crops. In 2011, harvest assessments were not compared to IPM blocks.

The results of female scale, parasitism and total biological control for all the sites over the three years of monitoring are presented in Fig. 2–4. All sites showed low scale counts early in the year up to April; however at Mutchilba in 2009, the scale counts remained low until May. In Dimbulah, scale counts were extremely low throughout 2010, but increased in late 2011.

Following the very low scale counts at the beginning of 2011 in Mutchilba, scale numbers increased rapidly, peaking at nine scales per leaf in August 2011. One insecticide application was made in June 2011 prior to flowering (Fig. 4) when the scale count was approximately 3.5 scales per leaf. The spray was applied to decrease scale populations prior to fruit set. The grower avoided spraying during flowering to protect pollinators. The insecticide used in this period had a minimal effect and the scale continued to increase. There was a negative effect on beneficial insects which decreased following the spray. Even though spray coverage was efficient, this registered product was not effective against the scale. Five applications of broad-spectrum insecticides were then conducted after flowering had finished in order to control the scale sufficiently prior to mango harvest. In all years, scale numbers increased prior to flowering; however, the peak occurred at different times each year (Fig. 4).

In all cases, there was a lag between the scale population increases and the biological control effect. Once biological control of the scale population becomes effective, the populations of both the pest and the beneficial insects decrease. This usually occurs late in the season (before crop maturity) and at times insecticide treatments are required.

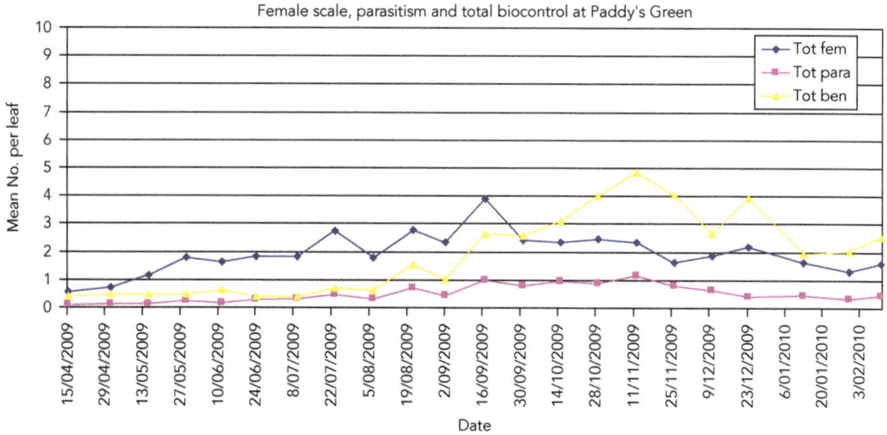

Figure 2 Number of female scales, parasitism and total biological control per leaf at Paddy's Green.

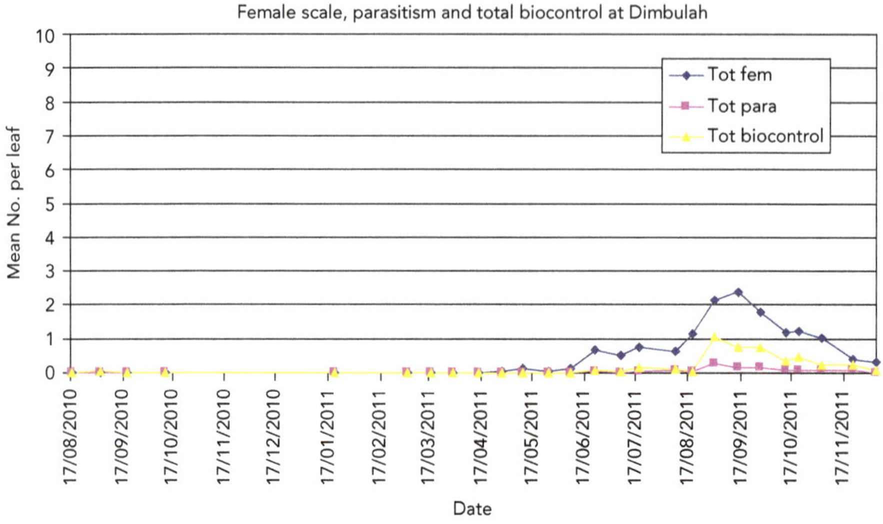

Figure 3 Number of female scales, parasitism and total biological control per leaf at Dimbulah.

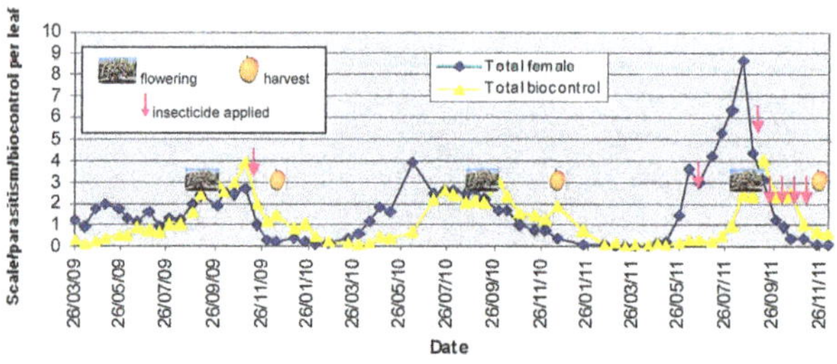

Figure 4 Number of female scales and total biological control at Mutchilba in 2009, 2010 and 2011.

At the Mutchilba site, one insecticide treatment was required in 2009, none in 2010 and six in 2011 (Fig. 4). No insecticides were required at Paddy's Green (Fig. 2) and only one application was required in 2011 at Dimbulah (Fig. 3).

Harvest assessments in 2009 showed minor blemishes at all sites, but this was similar to the adjacent IPM-managed blocks (Table 2). In 2011, there were no fruit with five or more scale/blemish that would result in fruit downgrade. However, six sprays at Mutchilba were required to achieve this result.

Parasitism rates during spring 2009 at Mutchilba and Paddy's Green are presented in Table 3. The parasitism rate was 43% *Encarsia citrina* and 0.43% *Aphytis* sp. at Mutchilba. At Mareeba, the rate was *Encarsia citrina* 21% and *Aphytis* sp. 7% at the same time of the year. However, the total effect of parasitism could not be confirmed as adult wasps also feed on their host as a protein source (Rosen and De Bach, 1979). Evidence of adult feeding is not distinguishable from natural mortality.

This study showed the importance of regular monitoring to target spray applications at critical times. Timing of spray applications will result in reduced insecticide applications,

Table 2 Harvest assessment data for three sites. There was insufficient fruit in 2010 to conduct a harvest assessment. There were no IPM blocks beside the biocontrol block in Dimbulah for comparison

Site	Year	Total no. of fruit	Total no. of downgraded fruit
Mutchilba biocontrol	2009	240	4
Mutchilba IPM	2009	240	0
Paddy's Green biocontrol	2009	240	13
Paddy's Green IPM	2009	240	10
Mutchilba biocontrol	2011	240	0
Dimbulah biocontrol	2011	240	0

Table 3 Mango-scale parasitism by *Aphytis* spp. and *Encarsia citrina* on the Atherton Tablelands in 2009

Locality	Number of scales	% parasitism *Aphytis* spp.	% parasitism *Encarsia citrina*
Mareeba (2009)	951	7.00	21
Mutchilba (2009)	1380	0.46	43

less harm to beneficial insects and better quality fruit. Pinese et al. (2010) stated that there were two population peaks tied in with tree phenology, particularly flowering and harvest, and not to local climate. However, the project data suggest that during periods of heavy rain, crawlers are washed off the leaves, resulting in low populations. Populations then begin to increase through the drier months. In 2011, which was a very wet year, the populations crashed and did not increase until May, compared to February in previous years. It was also noted that when the scale started to increase, it was at the stem terminals close to the leaf petioles (Fig. 5). It appears that the petiole protects crawlers from being washed off during periods of heavy rain.

Pinese et al. (2010) also suggested a prophylactic spray programme based on two population peaks linked to tree phenology. The programme consisted of a range of insecticides from soft to broad spectrum. This study showed the importance of monitoring, as population peaks do not always coincide at the same time or rate every year. There are times when biological control alone is effective at reducing mango-scale infestations. Insecticide intervention may be required; however if the timing is right and the beneficial insects are given a chance, this may be unnecessary. The selection of insecticide is critical, as demonstrated by the failure of an insecticide spray prior to flowering at Mutchilba in June 2011, which then required five follow-up sprays to control the resulting high-scale population numbers. The selection of softer chemistry, such as buprofezin and pyriproxyfen which are registered for mango scale, is also important in reducing the impact on beneficial insects.

There are quite a few beneficial insects associated with mango scale. The most numerous was the parasitoid *Encarsia citrina* (Fig. 6 and 7). In this study, the parasitoid achieved 43% parasitism, with *Encarsia citrina* having a greater impact than the endemic *Aphytis* species.

Figure 5 *Aulacaspis tubercularis* male clusters developing at leaf petioles and twigs following the cessation of heavy rain.

Figure 6 *Encarsia citrina* exit hole from female *Aulacaspis tubercularis*.

The ladybird *Rhyzobius lophanthae* was in low numbers throughout the study. However, the data showed high predation. It is difficult to establish which insect is responsible for the predation. Other predators such as predatory thrips (*Aleurodothrips fasciapennis*) and lacewings (*Mallada* sp.) were also recorded, but in numbers lower than that of the ladybird. It is assumed that *Rhyzobius lophanthae* was a more important predator in this study than the ladybird counts suggest.

A new predatory midge was discovered feeding on mango scale in this study (Fig. 8). The midge has been named *Dentifibula nigroapicalis* (Kolesik and De Faveri, 2014). The midge is associated with mango and mangrove scales. A further yet unnamed predatory midge of the same genus was also collected from mangoes. The midges were scarce, but they do play a part in overall biological control.

Figure 7 Mango-scale female mummified by presence of *Encarsia citrina*.

Figure 8 *Dentifibula nigroapicalis* larva in centre of image beside male *Aulacaspis tubercularis*.

The entomopathogenic fungus *Nectria flammea* (Fig. 9) was present at Paddy's Green during the wet season in 2009/10. The fungus plays a role in the overall biological control of mango scale, especially during the wet season: however, this was not quantified within this project.

The results have shown that mango-scale populations accelerate quicker than the endemic beneficial insects. The proposed introduction of *Aphytis chionaspis* would further enhance biological control of mango scale as the parasitoid has high fecundity and host specificity to *Aulacaspis tubercularis*. It is hopeful that insecticide sprays to control mango scale will not be necessary with the addition of an effective parasitoid, as shown in South Africa (Daneel and Dreyer, 1998). New parasitism rates as high as 80–90% could be likely. *Aphytis* spp. also kill as many scales through host feeding as through parasitism. These results are cryptic, as they are usually recorded as dead from unknown causes (Rosen and De Bach, 1979). Therefore, low levels of parasitisation actually equate to high mortality (e.g. 25% parasitisation could equate to 84% mortality).

Figure 9 Entomopathogenic fungus *Nectria flammea* on *Aulacaspis tubercularis*.

Insecticides were applied for other pests during this study. Endosulfan was used to control the BSB (*Amblypelta lutescens lutescens*), whilst it was registered. Dimethoate was applied for fruit fly control close to harvest for market access requirements. These insecticides have a negative impact on beneficial insects which can result in flare-ups of scale populations. Broad-spectrum insecticides have replaced Endosulfan; however, these are less effective and are applied at higher frequency. The increased use of these insecticides has reduced the effectiveness of mango-scale biological control. DAF, Queensland, has developed an aggregation pheromone which is now commercially available (Organic Crop Protectants) as a monitoring tool for BSB. The pheromone trap has the potential to be used in a lure and kill programme (I. Newton, Entomologist, Department of Agriculture and Fisheries, Queensland, pers. comm., 2017) and thus has less negative impact on natural enemies of mango scale.

5.2 Biological control of mango planthoppers *Colgaroides acuminata* (first reported in De Faveri et al. 2013)

Mango planthopper (*Colgaroides acuminata*) is a pest of increasing importance to mangoes. Immature and adult mango planthoppers congregate on twigs and leaf stalks and secrete large quantities of honeydew, which supports the growth of sooty mould (Smith et al., 1997). The honeydew flows down the stalk and onto the fruit. Sooty mould then develops on the fruit, which is difficult to clean, resulting in downgrading of affected fruit.

Mango planthopper has become an emerging problem due to the reduced use of broad-spectrum insecticides against major pests. The use of softer, more specific insecticides can result in positive effects on the populations of beneficial insects, but also can result in increased populations of non-target insect pests, including mango planthopper.

Increasing adoption of the use of biological control strategies against mango scale is also likely to result in mango planthopper becoming a more widespread issue for mango production. It is suggested that the use of biological control measures will be an important strategy for the control of mango planthopper in the future.

Beneficial insects found in mango orchards during this study included:

Dryinid wasps	*Dryinus australianus* (Fig. 10 and 12)
	Neodryinus koebelei (Fig. 11 and 12)
Syrphid fly	*Allobaccha siphanticida* (Fig. 13 and 14)
Egg parasitoid	*Achalcerinys* sp. (Fig. 15)

Other beneficial insects and entomopathogenic fungi recorded during this project:

Epipyropid moth	*Heteropsyche* sp.
Entomopathogenic fungus	*Hirsutella* sp.

Figure 10 *Dryinus australianus* adult.

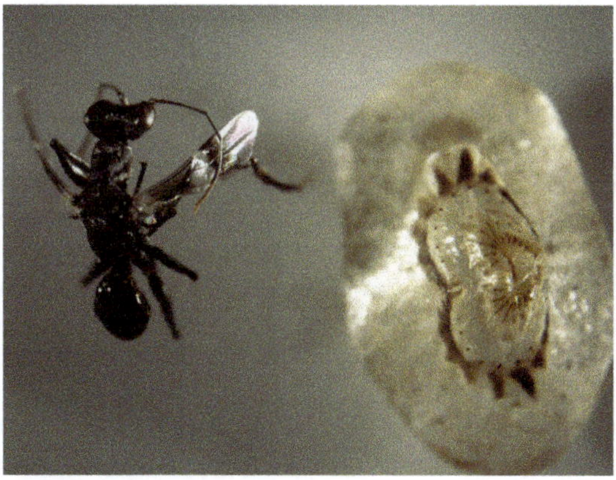

Figure 11 *Neodryinus koebelei* adult.

Figure 12 Dryinid larva on *Colgaroides acuminata* nymph.

Figure 13 *Allobaccha siphanticida* adult.

The egg parasitoid was not identified conclusively, but is likely to be *Achalcerinys* sp., which is responsible for up to 90% parasitism of mango planthopper egg rafts (Smith et al., 1997). Chambers (unpublished data, 2015) collected Encyrtid parasitoids from egg masses which were determined as *Cheiloneurus chlorodryini*, *Ooencyrtus minor* and sp. B near minor and the Platygastrid: *Aphaqnomerus pusillus*. Egg parasitoids have the greatest impact on mango planthopper with parasitism rates as high as 90% (Smith et al., 1997).

An Epipyropid moth (*Heteropsyche* sp.) was reared from samples collected at Mareeba. The entomopathogenic fungus *Hirsutella* sp. was also found on samples collected in Mareeba during the wet season. Smith et al. (1997) also state that spiders and assassin bugs commonly prey on planthoppers.

This preliminary study highlighted a number of beneficial insects that are involved in mango planthopper biological control. The suite of biological control agents has the potential to be effective against mango planthoppers if broad-spectrum insecticides are avoided.

Figure 14 *Allobaccha siphanticida* larva.

Figure 15 *Achalcerinys* sp. adults on *Colgaroides acuminata* egg mass.

5.3 Evaluating pasture species for shelter and food for beneficial insects (first reported in De Faveri et al. 2013)

Modern intensive agricultural systems lack biodiversity and in particular flowering plants that provide nutrients for natural enemies.

In north Queensland, mango orchards are usually well maintained with the inter-row mowed regularly and weeds in the intra-row controlled by herbicides. This environment is subsequently deficient in many of the nectar and/or pollen sources which are required by beneficial insects for their survival. Vegetative ground covers also provide shelter for beneficial insects. Heimpel and Rosenheim (1997) found that *Aphytis melinus* survived for three days without honey, compared to 30.5 days with honey, in caged trials. Nectar, like honey, is a good source of carbohydrates and flowering plants are a good source of nectar. In a study by Winkler et al. (2009), it was found that flowers that did not have accessible nectar had a negative impact on adult ladybird survival. Grasses are a particularly good source of pollen, which is also utilised by predatory mites and ladybirds. Improved understanding of the suitability of different plant species as nectar and pollen sources for beneficial insects is required to improve knowledge for selecting the most suitable plant species for these systems.

The aim of vegetation management within the orchard is to provide complementary and supplementary foods and shelter for beneficial insects. Complementary foods are carbohydrate-rich foods, such as pollen and nectar from flowering plants that sustain adult beneficial insects. Supplementary food sources provide food when the prey pest species is not available. Vegetation can also provide a modified climate for beneficial insects. Thick ground covers provide a cooler, more humid environment for beneficial insects than regularly mowed areas. Above- and below-ground biomass are also a source of insect biodiversity and microorganisms that would improve soil structure and health (Pickett and Bugg, 1998).

However, selection of unsuitable flowering plants could actually have a negative effect on beneficial insects. Previous flowering plant mixes have been based on arbitrary selections which can increase pest populations (Baggan and Gurr, 1998). Biodiversity of flowering plants alone does not necessarily result in suitable plants for natural enemies. Olsen and Wäckers (2007) found that cotton field edges established for quail conservation were not suitable for natural enemies. A generalist parasitoid, *Meteorus autographae*, was severely food limited and Tachinids and *Trichogramma* appeared to prefer the field margin to the cotton field. The authors suggested that adjusting the vegetation mix of the field margins with early flowering nectar-producing plants would result in improved conservation biological control.

This study evaluated several pasture species as shelters/nutrition sources for natural enemies and insect biodiversity. Potted plants were placed under mango trees and were vacuum-sampled one week after field placement. The mowed inter-row was used as the standard. The pasture grass, Rhodes grass, also known as *Chloris gayana*, had the most insects; however, these were mainly beetles and the number of parasitic wasps was quite low. The legume, Common Stylo, also known as *Stylosanthes hamata*, on the other hand, had greater numbers of wasps and insects other than beetles and ground dwellers, although this was not significant. The results show that grasses are preferred by beetles and therefore predatory ladybirds, whereas other insects that require carbohydrates prefer legume flowers (Table 4).

The results indicate that there was no significant difference between treatments for beneficial insects; however, the photoperiod-sensitive Common Stylo, which did not flower during the experiment, had more beneficial insects (Table 4). This plant is dense and perhaps provided greater shelter for beneficial insects.

The standard mowed treatment, consisting of a grass, *Brachiaria decumbens* and two legume species, *Stylosanthes hamata and Chamaecrista rotundifolia* which flower prolifically when mowed, supports a broad range of insects equal to or even greater than the pasture species tested alone (Table 4). If the inter-row was not mowed as regularly, there would potentially be greater beneficial insect abundance. An increased sward would provide shelter from the heat. The increased biomass would also benefit soil health by providing extra nutrients and support microorganisms. Staggering row mowing would provide increased shelter and flowers within the orchard. Mowing could commence on the first row and then skip two rows between each cut. The next mowing, about three weeks later, would include only the second row, again skipping two rows at a time. The

Table 4 Results of total insect numbers across all sampling periods for beneficial insects, beetles, wasps and total insects

Treatment	Total insects	Total insects *less* beetles and ground dwellers	Total beneficials	Wasps	Beetles
Butterfly pea	74.3 c	50.3	39.7	8.0	21.0 b
Rhodes grass	156.0 ab	52.7	30.3	5.7	98.3 a
Common stylo	121.3 abc	76.0	53.3	10.0	37.7 b
Verano stylo	93.3 bc	65.3	45.3	6.3	10.7 b
Wynn cassia	121.0 abc	74.0	44.3	8.0	9.0 b
Standard mowed	185.0 a	92.3	46.3	18.7	9.7 b
p-value	0.041	ns	ns	ns	0.03
LSD 95%	67.47	32.68	23.56	9.33	28.98

ns = not significant ($p > 0.05$).
Means followed by the same letter are not significantly different.

third mowing would commence on the third row, again skipping two rows in between. This system would provide flowers and shelter at all times within the orchard. A similar mowing regime is promoted in citrus IPM to support predatory mites (Dan Papacek, pers. comm., 2013).

5.4 Farm-wide fruit fly management in Bowen, North Queensland, Australia (first reported in De Faveri et al. 2013)

Fruit fly management in the field in Queensland has previously relied on relatively cheap effective cover sprays to control Australian pest fruit fly species. These chemicals have now been banned or restricted. Consequently, growers require alternative control options for fruit fly. Protein baits have been trialled in mango in the past, but cause a phytotoxic reaction to the sensitive mango skins.

This project aimed to test a farm-wide management systems approach to control fruit fly attacks in mangoes based on protein baiting, male annihilation and packhouse grading.

The system included: (a) male annihilation technique (MAT) using male lures and toxicant insecticide (www.bugsforbugs.com.au) distributed twice annually, coinciding with fruit set and effective for six months (August and November) for cue lure (for *Bactrocera tryoni*/*Bactrocera neohumeralis*) at the rate of 10/ha, and once annually for zingerone (for *Bactrocera jarvisi*) at the rate of 5/ha; (b) female fruit fly control using thickened protein bait applied to tree trunks to avoid contact with fruit; (c) monitoring male populations at each trial site with cue lure and zingerone traps at the treatment and control blocks and (d) packhouse grading.

Planned weekly protein trunk treatments were applied by the cooperating grower at Bowen 4–6 weeks prior to harvest.

In 2013/14, the number of pest fruit flies in traps was lower in the treated 'Honey Gold' mango cultivar compared with the untreated R2E2 area (Fig. 16 and 17), demonstrating fruit fly population reduction from lure and protein treatments.

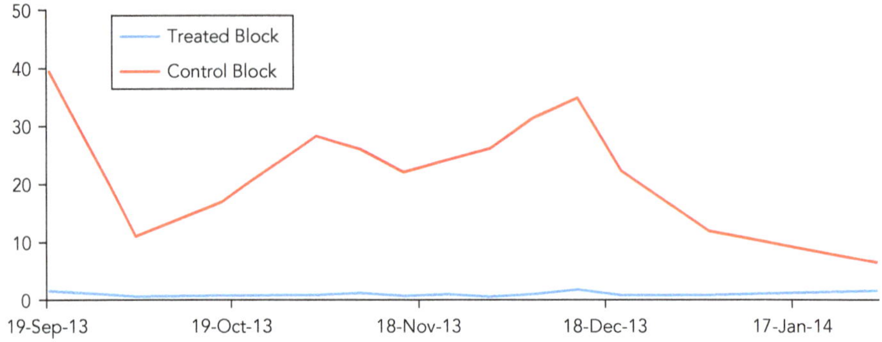

Figure 16 Mean number of male *Bactrocera tryoni/Bactrocera neohumeralis* per trap per day (FTD) collected from traps in mango orchard areas at Bowen untreated and treated with MAT lures and protein baits 2013/14.

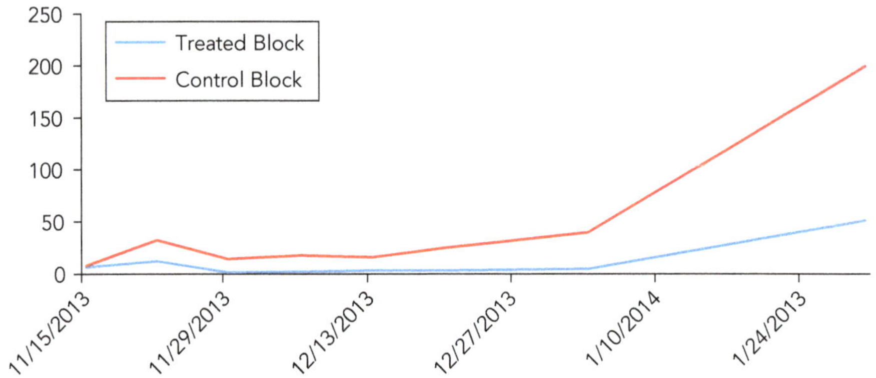

Figure 17 Mean number of male *Bactrocera jarvisi* per trap per day (FTD) collected from traps in mango orchard areas at Bowen untreated and treated with MAT lures and protein baits 2013/14.

In 2014/15, the previously untreated R2E2 mango cultivar area was also treated. On almost all trap clearance occasions, the number of pest fruit flies in traps was lower in the treated 'Honey Gold' and 'R2E2' areas compared with the untreated area (Fig. 18 and 19), demonstrating fruit fly population reduction from lure and protein treatments. *Bactrocera tryoni* and *Bactrocera neohumeralis* trap numbers of the'R2E2' area were higher than those of the 'Honey Gold' and untreated areas during the first two weeks of monitoring, but declined to lower numbers than the untreated area in subsequent weeks. *Bactrocera jarvisi* numbers at the treatment areas were lower than the control area throughout the trial. Trap numbers of both treatment areas increased after 6th January coinciding with completion of protein baiting and the removal of MAT cups in preparation for tree pruning.

In 2013/14, 2230 fruit from the treated block were assessed for fruit fly eggs and larval presence. The total sample, which included 1974 graded fruit and 256 reject fruit, was

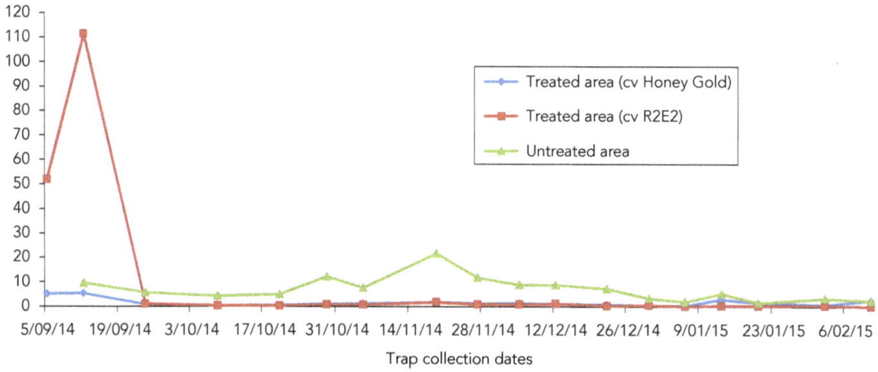

Figure 18 Mean number of male *Bactrocera tryoni/Bactrocera neohumeralis* per trap per day (FTD) collected from traps in mango orchard areas at Bowen untreated and treated with MAT lures and protein baits 2014/15.

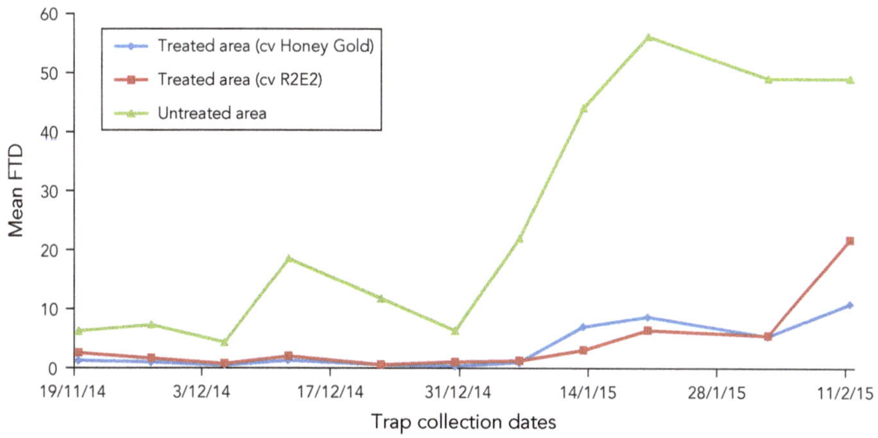

Figure 19 Mean number of male *Bactrocera jarvisi* per trap per day (FTD) collected from traps in mango orchard areas at Bowen untreated and treated with MAT lures and protein baits.

inspected for eggs and larvae, and a 300-fruit subsample (bulk grade; 3rd grade green mature, slight skin blemish) was incubated. There were no eggs or larvae found in any fruit grade assessed visually or by incubation. A total of 929 fruit from the untreated area, 875 first-grade and 54 rejected fruit, were inspected for fruit fly oviposition site punctures (stings). Six of the reject untreated fruit had apparent fruit fly stings.

In 2014/15, 1968 'R2E2' and 1997 'Honey Gold' fruit were assessed for fruit fly eggs and larval presence (Table 5). The proportion of R2E2 fruit with damaged skin was higher than that of 'Honey Gold' (42% *cf* 7%). Larvae were only found in four R2E2 fruit (0.2% of the total fruit sampled); all four had damaged skin, possibly the result of bird feeding. Three of the infested fruit were 30–50% yellow, and one was mature green.

Table 5 Number of fruit of two maturity grades sampled from two farm-wide management treatment areas at Bowen and the number of fruit infested by fruit fly in 2014/15

	Intact skin		Damaged skin		
Fruit maturity category	Total fruit	Number infested	Total fruit	Number infested	Number pupae
Treated area cv. R2E2					
Mature green	758	0	427	1	3
30–50% yellow	389	0	394	3	7
Total fruit assessed	1147		821		
Treated area cv. Honey Gold					
Mature green	1011	0	62	0	0
30–50% yellow	837	0	87	0	0
Total fruit assessed	1848		149		

The farm-wide management system was successfully implemented at Bowen. Fruit fly populations were reduced and damage was limited to four fruits that had previous skin damage. The Bowen grower considered the system feasible and successful. The grower plans to continue the systems approach following the success of the project.

5.5 Concluding remarks

The four case studies demonstrate the research that has been carried out to improve the sustainability of IPM in mangoes in Queensland by reducing reliance on insecticides.

The case study for mango scale (5.1) demonstrates the importance of regular monitoring and shows that at times, beneficial insects maintain scale populations below economic levels, negating the need for insecticide intervention. It also illustrates the repercussions of failed spray applications, particularly around flowering time when follow-up sprays are detrimental to pollinators. There is a suite of beneficial insects associated with mango-scale control; however, well-timed insecticide interventions are still required. The introduction of the parasitoid, *Aphytis chionaspis* from South Africa, would improve mango-scale control and possibly negate the need for insecticide sprays.

In case study 5.2, the beneficial insects associated with the mango planthopper have been documented. Further research would determine their effectiveness.

In case study 5.3, candidate flowering plants were evaluated as shelter and food sources for beneficial insects. Regularly mowed orchards usually provide little shelter and food for beneficial insects, which results in lack of or early mortality of beneficial insects. Allowing grasses and other plants to flower will provide shelter and food resources that will help sustain beneficial insects.

In case study 5.4, the implementation of the farm-wide management systems approach reduces the need for broad-spectrum fruit fly sprays. The project demonstrated that the farm-wide systems approach to fruit fly management can be successfully implemented and is a sustainable alternative to broad-spectrum insecticide application.

6 Future trends and conclusion

IPM in mangoes is not very well developed and the literature on the topic is scant. However, work in the area is evolving. The work by the South African researchers to introduce *Aphytis chionaspis* and *Cybocephalus binotatus* to control mango scale has resulted in the reduction in insecticide use and progress in IPM implementation (Daneel and Joubert, 2000; Joubert et al., 2009; Labuschagne et al., 1996; Neser and Prinsloo, 2001). As indicated before, there is also some very important work in mango-producing countries to utilise weaver ants in the management of pests. The challenges in using weaver ants include farm worker's well-being as the ants are protective and attack and sting workers in hordes, and ensuring that pests such as soft scales and mealybugs do not become more prevalent due to the mutual relationship between the pest which produces honeydew for the ant and the ant protects the pest from natural enemies.

Benign or soft solutions must be found for some of the intractable pests such as the mango seed and mango pulp weevils. Current options are based on heavy sprays of broad-spectrum insecticides or systemic neonicotinoid drenches. Broad-spectrum insecticides have a detrimental effect on beneficial insects and cause flare-ups of secondary pests such as mites and *Homopteran* bugs. The full effects of neonicotinoid drenches are unknown. Drenches could affect pollinators and other beneficial insects. It is uncertain whether annual use could result in increased residues in subsequent years.

IPM must also be implemented to reduce insecticide resistance. Some of the pests of concern for insecticide resistance include the mango leafhoppers, thrips and mites. Overuse of insecticides in the southern Philippines has resulted in insecticide resistance by thrips (C. Medina, Entomologist, University Philippines, Los Banos, unpublished data, 2017). Mite resistance is known worldwide and although not documented, mango leafhoppers are proving difficult to control. Overuse of broad-spectrum cocktails of insecticides in developing countries will result in increased insecticide resistance.

Increased adoption of IPM strategies will result in benefits to the environment, human safety and a reduction in insecticide resistance. The environmental effects of insecticide usage have been well documented, with contaminated waterways and non-target effects. Human health has been affected by insecticide poisoning through usage and food contamination. The costs of developing new chemistry have increased dramatically, resulting in fewer, expensive new insecticides. Despite all these issues, the use of insecticides has increased dramatically.

Future IPM systems should be developed with a holistic approach to pest management. The health and survival of beneficial insects should be considered. The most obvious ways to improve beneficial insect health is to stop using broad-spectrum insecticides and only use selective insecticides as a last resort. However, landscape management should play a greater role in the conservation and health of beneficial insects. Flowering plants provide shelter and food for beneficial insects increasing their survival and enticing them to stay within the orchard. More emphasis needs to be placed on growing flowering plants that are allowed to flower within the orchard.

7 Where to look for further information

Bugs for Bugs https://bugsforbugs.com.au/
Insecticide Resistance Action Committee http://www.irac-online.org/
International Organisation for Biological Control http://www.iobc-global.org/
Australian Mango Industry Association http://www.industry.mangoes.net.au/

8 References

Abdulla, N., Rwegasiri, G., Jensen, K., Mwatalala, M. and Offenberg, J. (2016). Control of mango seed weevil (Sternochetus mangiferae) using the African weaver ant (Oecophylla longinoda Latreille) (Hymenoptera: Formicidae). *Journal of Applied Entomology*, 140, 500–6.

Aktar, M., Sengupta, D. and Chowdury, A. (2009). Impact of pesticide use in agriculture: Their benefits and hazards. *Interdisciplinary Toxicology*, 2, 1–12.

Alavanja, M. (2009). Introduction: Pesticides use and exposure extensive worldwide. *Review of Environmental Health*, 24, 303–9.

Aluja, M. (1994). Bionomics and management of Anastrepha. *Annual Review of Entomology*, 39, 155–78.

Aluja, M., Arredondo, J., Diaz-Fleisher, F., Birke, A., Rull, J., Niogret, J. and Epsky, N. (2014). Susceptibility of 15 mango (Sapindales: Anacardiaceae) cultivars to the attack by Anastrepha ludens and Anastrepha obliqua (Diptera: Tephritidae) and the role of underdeveloped fruit as pest reservoirs: Management implications. *Journal of Economic Entomology*, 107, 375–88.

Anon. (1994). Major insect pests and diseases of mango and their control. In: *The Mango Technical Committee. The Philippines Recommends for Mango*. PCARRD and PAFRI, Los Banos, Laguna, p. 124.

Anon. (2000). Mango pest management workshop manual. Horticulture series, Mango Care.

Anon. (2003). *Trapping Guidelines for Area-wide Fruit Fly Programmes*. International Atomic Energy Agency, Vienna, p. 125.

Anon. (2010). *Field Guide to Pests, Beneficials, Diseases and Disorders of Mangoes*. Department of Resources, Darwin, NT.

Astridge, D. (2004). Rambutan identification and monitoring the major insects and mites: Their biology and ecology. *Agdex*, 217/10, 98.

Atwood, D. and Paisley-Jones, C. (2017). *Pesticides Industry Sales and Usage – 2008–2012 Market Estimates*. US EPA, Washington, DC.

Baggen, L. R. and Gurr, G. M. (1998). The influence of food on *Copidosoma koehleri* (Hymenoptera: Encyrtidae), and the use of flowering plants as a habitat management tool to enhance biological control of potato moth, *Phthorimaea opercullela* (Lepidoptera: Gelechiidae). *Biological Control*, 11, 9–17.

Bajhau, H., Baker. I. and Smith, S. (2002). *Pests, Diseases and Disorders of Mangoes in the Northern Territory: An Illustrated Guide*. Northern Territory Government, Department of Business, Industry and Resource Development, p. 99.

Bondad, N. (1989). *The Mango, Especially as Observed in the Philippines*. Rex printing company, Quezon City, Philippines, p. 402.

Brough, E., Elder, R. and Beavis, C. (1994). Managing insects and mites in horticultural crops. Information series Q194010.

Carson, R. (1962). *Silent Spring*. Houghton Mifflin, Boston, MA, p. 368.

Castle, S. and Naranjo, S. (2009). Sampling plans, selective insecticides and sustainability: The case for IPM as informed pest management. *Pest Management Science*, 65, 1321–8

Clarke, A., Powell, K., Weldon. C. and Taylor, P. (2011). The ecology of Bactrocera tryoni (Diptera: Tephritidae): What do we know to assist pest management? *Annals of Applied Biology*, 158, 26–54.

Cunningham, I. (1989). Pest. In: Bagshaw, J. (Ed.), *Mango Pests and Disorders*. Department of Primary Industries Information Series, Q189007. Department of Primary Industries, Brisbane, pp. 10–21.

Daneel, M. and Joubert, P. (2009). Biological control of the mango scale Aulacaspis tubercularis Newstead (Coccidae: Diaspididae) by a parasitoid Aphytis chionaspis Ren (Hymentoptera: Aphelineadae). *Acta Horticulturae*, 820, 567–74.

De Faveri, S. and Brown, J. (1995). Effect of buprofezin, fluvalinate, Lo-Vis and imidachloprid against the mango scale, Aulacaspis tubercularis (Newstead) mango. In: Holmes, R. (Ed.), *Mango 2000 Proceedings of Marketing and Production Workshop*. Department of Primary Industries, Brisbane, Townsville, Queensland Australia, pp. 153–9.

De Faveri, S., Missenden, B., Cook, L. and Wright, C. (2013). Biological control options for mango scale. HAL project MG08003, p. 43.

De Faveri, S., O'Farrell, P., Senior, L., Lowe, G., Subramanium, S., Chambers, D., Subramanium, V. and Robertson, J. (2015). Farm-wide fruit fly management systems for the east coast of Australia, HIA project MT12050, p. 73.

DeBach, P. (1964). *Biological Control of Insect Pests and Weeds*. Chapman and Hall, London, p. 844.

Dey, K. and Pand, Y. (1987). Evaluation of certain non-insecticidal methods of reducing infestation of the mango nut weevil, Sternochetaus gravis (F.) in India. *Tropical Pest Management*, 33, 27–8.

Drew, R. and Romig., M. (2013) *Tropical Fruit Flies of South East Asia*. CABI, Wallingford, p. 653.

Drew, R., Tsuruta, K. and White, I. (2005). A new species of pest fruit fly (Diptera: Tephritidae: Dacinae) from Sri Lanka and Africa. *African Entomology*, 13, 149–54.

Edwards, C. (2000). Ecologically based use of inesecticides, pp. 103–30. In: Rechcigl, J. and Rechcigl, N. (Eds), *Insect Pest Management: Techniques for Environmental Protection*. CRC Press LLC, Boca Raton, FL.

Elbert, A., Haas, M., Springer, B., Thielert, W. and Nauen, R. (2008). Applied aspects of neonicotinoid uses in crop protection'. *Pest Management Science*, 64, 1099–105.

Gibb, A., Pinese, B., Tenakanai, D., Kawi, A., Bunn, B., Ramankutty, P. and Suckling, M. (2007). (Z)-11-hexadecenal and (3Z,6Z,9Z)-tricosatriene: Sex pheromone components of the red banded mango caterpillar Deanolis sublimbalis. *Journal of Chemical Ecology*, 33, 579–89.

Grube, A., Donaldson, D., Kiely, T. and Wu, L. (2011). *Pesticides Industry Sales and Usage*. US EPA, Washington DC.

Hancock, D., Hamacek, E., Lloyd, A. and Elson-Harris, M. (2000). The distribution and host plants of fruit flies (Dipetera: Tephritidae) in Australia. Department of Primary Industries Information Series,Q199067, p. 75.

Hardy, R., Beachy, R., Browning, H., Caulder, J., Charudattan, R., Faulkner, P., Gould, F., Hinkle, M., Jaffee, B., Knudson, M., Lewis, W., Loper, J., Mahr, D. and Van Alfen, N. (1996). *Ecologically Based Pest Management: New Solutions for a New Century*. National Academy Press, Washington, DC, p. 146.

Heimpel, G. and Rosenheim, J. (1997). Adult feeding and lifetime reproductive success in the parasitoid Aphytis melinus. *Entomologia Experimentalis et Applicata*, 83, 305–15.

Hennessey, M. and Schnell, R. (1995) Screening Florida mango fruits from germplasm for resistance to immature stages of Caribbean fruit fly. *Insect Science Application*, 16, 119–21.

Hennessey, M. and Schnell, R. (2001) Resistance of immature mango fruits to Caribbean fruit fly (Diptera: Tephritidae). *Florida Entomologist*, 84, 318–19.

IRAC international MoA working group (2016). IRAC MoA (Insecticide Resistance Action Committee) IRAC mode of action classification scheme, Version 8.2, March 2017 [online]. http://www.irac-online.org/documents/moa-classification/?ext=pdf [6 June 2017].

Jeyaratnam, J. (1990). Acute pesticide poisoning: A major global health problem. *World Health Statistics Quarterly*, 43, 139–44.

Joubert, P., Daneel, M. and Grove, T. (2000). Progress towards integrated pest management (IPM) on mangoes in South Africa. *Acta Horticulturae*, 509, 811–17.

Kogan, M. (1998). Integrated pest management: Historical perspectives and contemporary developments. *Annual Review of Entomology*, 43, 243–70.

Kolesik, P. and De Faveri, S. (2014). A new gall midge of the genus *Dentifibula* (Diptera: Cecidomyiidae), a predator of the scale insect *Aulacaspis australis* feeding on mangrove *Bruguiera gymnorrhiza*. *Austral Entomology*, 53, 99–103.

Labuschagne, T., Daneel, M. and De Beer, M. (1996). Establishment of Aphytis sp. (Hymenoptera: Aphelinidae) and Cybocephalus, binotatus Grouvelle (Coleoptera: Nitidulidae) in mango orchards in South Africa for control of mango scale, Aulacaspis tubercularis Newstead (Homoptera Diaspididae). *South African Mango Growers Yearbook*, 16, 21–2.

Lloyd, A., Hamacek, E., Kopitke, R., Peek, T., Wyatt, P., Neale, C., Eelkema, M. and Gu, H. (2010). Area-wide management of fruit flies (Diptera: Tephritidae) in the Central Burnett district of Queensland, Australia. *Crop Protection*, 29, 462–9.

Montoya, P., Cancino, J., Zenil, M., Santiago, G. and Gutierrez, J. (2007). The augmentative biological control component in the Mexican national campaign against Anastrepha spp. fruit flies, pp. 661–70. In: Vreysen, M. J. B., Robinson, A. S. and Hendrichs, J. (Eds), *Area-Wide Control of Insect Pests*. Springer, Dordrecht, The Netherlands.

Nachiappan, R. and Baksaran, P. (1984). Quantum of feeding and survival of mango leafhopper adults on inflorescence of certain varieties of mango. *Indian Journal of Agricultural Science*, 54, 312–14.

Neser, O. and Prinsloo, G. (2001). Aphytis chionaspis (Hymenoptera: Aphelinidae), a parasitoid introduced to South Africa from Thailand for the control of mango scale, Aulacaspis tubercularis (Hemiptera: Diaspididae). *African Entomology*, 9, 199–201

Norton, G. (1976). Analysis of decision making in crop protection. *Agro-Ecosystems*, 3, 27–44.

Offenberg, J. (2015). Ants as tools in sustainable agriculture. *Journal of Applied Ecology*, 52, 1197–205.

Olsen, D. and Wäckers, F. L. (2007). Management of field margins to maximize multiple ecological services. *Journal of Applied Ecology*, 44, 13–21.

Olutu, M., du Plessis, H., Seguni, Z. and Maniania, K. (2012). Efficacy of the African weaver ant Oecophylla longinoda (Hymenoptera: Formicidae) in the control of Helopeltis spp. (Hemiptera: Miridae) and Pseudotheraptis wayi (Hemiptera: Coreidae) in cashew crops in Tanzania. *Pest Management Science*, 69, 911–18.

Pedigo, L., Hutchins, S. and Higley L. (1986). Economic injury levels in theory and practice. *Annual Review of Entomology*, 31, 341–68.

Peng, R. and Christian, K. (2005). The control efficacy of the weaver ant, Oecophylla smaragdina (Hymenoptera: Formicidae) on the mango leafhopper, Idioscopus nitidulus (Hemiptera: Cicadellidae) in mango orchards in the Northern Territory. *International Journal of Pest Management*, 51, 297–304

Peng, R., Christian, K. and Reilly, D. (2012). Biological control of the fruit-spotting bug Amblypelta lutescens using weaver ants Oecophylla smaragdina on African mahoganies in Australia. *Agricultural and Forest Entomology*, 14, 428–33.

Pickett, C. and Bugg, R. (1998). *Enhancing Biological Control: Habitat Management to Promote Natural Enemies of Agricultural Pests*. University of California Press, Berkley, CA, p. 421.

Pimentel, D. (2005). Environmental and economic costs of the application of pesticides primarily in the United States. *Environment, Development and Sustainability*, 7, 229–52.

Pinese, B., De Faveri, S., Smith, D., Neale, C., Missenden, B., Freebairn, C. and Bauer, R. (2010). Mango Plant Protection Phase 1. Final Report HAL project-FR02050.

Rosen, D. and DeBach, P. (1979). *Species of Aphytis of the World (Hymenoptera: Aphelinidae) Series*. Entomologica 17. Dr W Junk BV, The Hague, The Netherlands.

Roy, S., Mukhopadhyay, A. and Gurusubramanian, G. (2009). Monitoring the resistance of red spider mite (Oligonychus coffeae Nietner) of tea to commonly used acaricides from the Darjeeling foothills and plains of North Bengal, India. *Resistant Pest Management Newsletter (A Biannual Newsletter of the Center for Integrated Plant Systems (CIPS) in Cooperation with the Insecticide Resistance Action Committee (IRAC) and the Western Regional Coordinating Committee (WRCC-60)*, 18, 10–14

Schutze, M., Mahmood, K., Pavasovic, A., Bo, W., Newman, J., Clarke, A., Krosch, M. and Cameron, S. (2014). One and the same: Integrative taxonomic evidence that Bactrocera invadens (Diptera: Tephritidae) is the same species as the Oriental fruit fly Bactrocera dorsalis. *Systematic Entomology*, 40, 472–86.

Schutze, M., Aketarawong, N., Amornsak, W., Armstrong, K., Augustinos, A., Barr, N., Bo, W., Bourtzis, K., Boykin, L., Cáceres, C., Cameron, S., Chapman, T. Chinvinijkul, S., Chomič, A., De Meyer, M., Drosopoulou, E., Englezou, E., Ekesi, S., Gariou-papalexiou, A., Geib, S., Hailstones, D., Hasanuzzaman, M., Haymer, D. W., Hee, A., Hendrichs, J., Jessup, A., Ji, Q., Khamis, F., Krosch, M., Leblanc, L., Mahmood, K., Malacrida, A., Mavragani-tsipidou, P., Mwatawala, M., Nishida, R., Ono, H., Reyes, J., Rubinoff, D., San Jose, M., Shelly, T., Srikachar, S., Tan, K., Thanaphum, S., Haq, I., Vijaysegaran, S. Wee, S., Yesmin, F., Zacharopoulou, A. and Clarke, A. (2015). Synonymization of key pest species within the *Bactrocera dorsalis* species complex (Diptera: Tephritidae): Taxonomic changes based on a review of 20 years of integrative morphological, molecular, cytogenetic, behavioural and chemoecological data. *Systematic Entomology*, 40, 456–71.

Smith, D., Beattie, G. and Broadley, R. (1997). Citrus pests and their natural enemies. Information Series Q197030.

Snowball, G., Wilson, F., Campbell, T. and Lukins, R. (1962). The utilization of parasites of Oriental fruit fly, (Dacus dorsalis) against Queensland fruit fly (Strumeta tryoni). *Australian Journal of Agricultural Research*, 13, 443–60.

Spinner, J., Cowling, A., Gurr, G., Jessup, A. and Reynolds, O. (2011). Parasitoid fauna of Queensland fruit fly, Bactrocera tryoni (Froggatt) in inland New South Wales, Australia and their potential for use in augmentative biological control. *Australian Journal of Entomology*, 50, 445–52.

Stern, V., Smith, R., van den Bosch, R. and Hagen, K. (1959). The integrated control concept. *Hilgardia*, 29, 81–101.

Van Mele, P. (2008). A historical review of research on the weaver ant Oecophylla in biological control. *Agricultural and Forestry Entomology*, 10, 13–22.

Van Mele, P., Vayssieres, J., Van Tellingen, E. and Vroliks, J. (2007). Effects of an African weaver ant, Oecophylla longinoda, in controlling mango fruit flies (Diptera: Tephritidae) in Benin. *Biological and Microbial Control*, 100, 695–701.

Vazquez-Luna, A., Rivera-Cabrera, F., Perez-Florez, A. and Diaz-Sobac, R. (2011). Effect of rootstock on mango susceptibility to infestation by Anastrepha Obliqua. *Horticultural Entomology*, 104, 1991–8.

Vreysen, M., Robinson, A. and Hendricks, J. (Eds) (2007). *Area-wide Control of Insect Pests: From Research to Field Implementation*. Springer, Dodrecht, The Netherlands, p. 789.

Waite, G. (2002). Pest and pollinators of mango, pp. 103–29. In: Pena, J., Sharp, J. and Wysoki, M. (Eds), *Tropical Fruit Pests and Pollinators*. CAB International, Wallingford, UK.

Waite, G., Fay, H., Hood, S., Campbell, G, Parker, R., De Faveri, S., Maddox, C., Adkins, M. and Wade, M. (2000). *Final Report HG97010 Ecology and Behaviour of Fruitspotting Bugs*. Horticulture Research and Development Corporation, Sydney, p. 148.

Wang, Q. and Liu, T. (2016). Effects of three insect growth regulators on Encarsia Formosa (Hymenoptera: Aphelinidae), an ectoparasitoid of Bemisia tabaci (Hemiptera: Aleyrodoidae). *Journal of Economic Entomology*, 109, 2290–7.

White, I. and Elson-Harris, M. (1992). *Fruit Flies of Economic Significance: Their Identification and Bionomics*. CAB International, Wallingford, UK, p. 601.

Winkler, K., Wackers, F. L., Kaufman, L. V., Larraz, V. and Van Lenteren, J. C. (2009). Nectar exploitation by herbivores and their host parasitoids is a function of flower species and relative humidity. *Biological Control*, 50, 299–306.

Index

www.ingramcontent.com/pod-product-compliance
Ingram Content Group UK Ltd.
Pitfield, Milton Keynes, MK11 3LW, UK
UKHW021008120926
8556IPUK00028B/141

9 781786 761323